Multi-Dimensional Liquid Chromatography

Two-dimensional liquid chromatography (2D-LC) is finding increasingly wide application principally due to the analysis of mixtures of moderate to high complexity. Many industries are developing increasingly complex products that are challenging the separation capabilities of state-of-the-art 1D-LC and need new analytical methodologies with substantially higher resolving power, and 2D-LC meets that need.

This text, organized by two leaders in the field, establishes a sound fundamental basis for the principles of the technique, followed by a discussion of important practical considerations. The book begins with an introduction to multi-dimensional separations and a discussion of the history and development of the technique over the past 40 years, followed by several chapters that provide a theoretical basis for development of 2D-LC methods, including foundational concepts regarding separation complementarity, undersampling, and dynamics of liquid chromatography separations. Instrumentation for 2D-LC is discussed extensively, including practical aspects such as interface selection and setup. Building on this foundation, two separate chapters are focused on method development for non-comprehensive and comprehensive separations, followed by a chapter dedicated to data analysis. Finally, applications of 2D-LC in several fields ranging from pharmaceutical analysis to polymer science are summarized.

The book is an important resource for both students and practitioners who are already using 2D-LC or are interested in getting started in the field.

Key Features:

- Demonstrates the conditions under which a 2D-LC method should be considered as an alternative to a 1D-LC method.
- Establishes a sound fundamental basis of the principles of the technique, followed by guidelines for method optimization.
- Provides a single source for technical knowledge advances and practical guidance described in recent literature.
- Assists with the initial decision to develop a 2D-LC method.
- Guides the reader in developing a high-quality method that meets the needs of their application.

Chromatographic Science Series

A Series of Textbooks and Reference Books
Editor: Nelu Grinberg
Founding Editor: Jack Cazes

The **Chromatographic Science Series** offers an in-depth treatment of the latest developments and applications in the field of separation sciences. The series enjoys a broad international readership in part due to the accomplished list of authors and editors that have contributed their expertise to series books covering both classic and cutting-edge topics.

Handbook of HPLC: Second Edition
Edited by Danilo Corradini and consulting editor Terry M. Phillips

High Performance Liquid Chromatography in Phytochemical Analysis
Edited by Monika Waksmundzka-Hajnos and Joseph Sherma

Hydrophilic Interaction Liquid Chromatography (HILIC) and Advanced Applications
Edited by Perry G. Wang and Weixuan He

Hyphenated and Alternative Methods of Detection in Chromatography
Edited by R. Andrew Shalliker

LC-NMR: Expanding the Limits of Structure Elucidation
Nina C. Gonnella

Thin Layer Chromatography in Drug Analysis
Edited by Łukasz Komsta, Monika Waksmundzka-Hajnos, and Joseph Sherma

Pharmaceutical Industry Practices on Genotoxic Impurities
Edited by Heewon Lee

Advanced Separations by Specialized Sorbents
Edited by Ecaterina Stela Dragan

High Performance Liquid Chromatography in Pesticide Residue Analysis
Edited by Tomasz Tuzimski and Joseph Sherma

Planar Chromatography-Mass Spectrometry
Edited by Teresa Kowalska, Mieczysław Sajewicz, and Joseph Sherma

Chemometrics in Chromatography
Edited by Łukasz Komsta, Yvan Vander Heyden, and Joseph Sherma

Chromatographic Techniques in the Forensic Analysis of Designer Drugs
Edited by Teresa Kowalska, Mieczyslaw Sajewicz, and Joseph Sherma

Determination of Target Xenobiotics and Unknown Compounds Residue in Food, Environmental, and Biological Samples
Edited by Tomasz Tuzimski and Joseph Sherma

LC-NMR: Expanding the Limits of Structure Elucidation, Second Edition
Nina C. Gonnella

Multi-Dimensional Liquid Chromatography: Principles, Practice, and Applications
Edited by Dwight R. Stoll and Peter W. Carr

For more information about this series, please visit: www.crcpress.com/Chromatographic-Science-Series/book-series/CRCCHROMASCI

Multi-Dimensional Liquid Chromatography

Principles, Practice, and Applications

Edited by
Dwight R. Stoll and Peter W. Carr

CRC Press
Taylor & Francis Group
Boca Raton London New York

CRC Press is an imprint of the
Taylor & Francis Group, an **informa** business

First edition published 2023
by CRC Press
6000 Broken Sound Parkway NW, Suite 300, Boca Raton, FL 33487-2742

and by CRC Press
4 Park Square, Milton Park, Abingdon, Oxon, OX14 4RN

CRC Press is an imprint of Taylor & Francis Group, LLC

ISBN: 9780367547660 (hbk)
ISBN: 9780367547745 (pbk)
ISBN: 9781003090557 (ebk)

DOI: 10.1201/9781003090557

Typeset in Times
by Newgen Publishing UK

Contents

Preface

In thinking about this monograph on multi-dimensional liquid chromatography as a rapidly developing area of chemical analysis, we decided to address a broad group of readers. In general, we believe that our reader should already be well versed in the most basic concepts of chromatography including some general concepts of resolution, plate count, retention, and selectivity, which are foundational for all forms of separation science. Other background knowledge is helpful as well, including topics closely related to liquid chromatography such as the principles involved in choosing an appropriate mobile phase solvent type, strength, and composition. We assumed that the reader is familiar with the most commonly used form of LC, namely reversed-phase chromatography (RPC) with chemically bonded stationary phases. The book is intended for use in advanced academic courses, and/or those who are interested in applying multi-dimensional LC in research or applications. Readers with significant experience with conventional one-dimensional LC (1D-LC) will be able to jump right into the content, whereas those with little experience with 1D-LC are advised to spend more time with Chapter 2 which goes into considerable detail on topics related to optimization of 1D-LC. The book could be used as a reference text in conjunction with an intensive short course on multi-dimensional liquid chromatography. Additionally, it should be useful to scientists already familiar with the principles of, and instrumentation for multi-dimensional LC, but not as familiar with its use in specific areas of application, including pharmaceutical analysis (including biopharmaceutical analysis), chemical analysis (e.g., polymers and surfactants), food and beverage analysis, and enantio-selective separations where it has been fruitfully employed. A novel aspect of the book is that it treats the theory and application of both the *comprehensive* form of 2D-LC, as exemplified by its use in proteomics, as well as the various forms of *non-comprehensive* 2D-LC (i.e., multiple heartcut and selective comprehensive) that have become very popular in pharmaceutical analysis in recent years.

A few words about the organization of the book and how the reader might most efficiently approach its content are in order. The first four chapters are primarily concerned with an introduction to multi-dimensional chromatography, fundamental principles that must be understood for successful method development, and the instrumentation needed to implement the various methodologies and ameliorate some of the intrinsic difficulties of 2D-LC that are either not present in 1D-LC, or not as difficult to remediate as they are in 2D-LC. Chapter 1 is aimed at the big picture and introductory material, the history of multi-dimensional LC, and the language of multi-dimensional separations.

Chapter 2 focuses on speed and performance (resolving power) in 1D-LC, and how they are optimized when using either isocratic and mobile phase gradient elution conditions. The main text of this chapter assumes that the reader is rather familiar with the dynamic aspects of peak broadening, however we have addressed many fundamental issues of column dynamics in detail in an appendix to this chapter. We advise those who are not at all familiar with or need to brush up on the details of the physico-chemical processes that cause peaks to broaden to read or perhaps study the appendix before diving directly into Chapter 2. The appendix includes a good deal of relatively recent fundamental research on column dynamics, especially that of the Desmet group.

Chapter 3 is aimed at the core principles of two-dimensional liquid chromatography. This includes the all-important topic of the peak capacity of 2D-LC, including the so-called product rule as established by Giddings. Effective 2D peak capacities depend strongly on the undersampling of the first dimension separation due to the challenges associated with very fast second dimension separations. Additionally, effective peak capacities are affected by incomplete usage of the 2D separation space, which is related to the complementarity of the two individual separations that make up a 2D separation. Finally, various approaches to the optimization of the 2D peak capacity and the impact on detection sensitivity are discussed.

Chapter 4 is focused on the instrumentation used for 2D-LC, and the various practical aspects one must consider when developing 2D-LC methods (e.g., the acceptable extent of sampling loop filling). Recent increases in the commercial offerings of hardware and software for 2D-LC, and a deepening of understanding about how 2D-LC instrumentation works, have been some of the most important developments in the field in the last ten years. Readers will enjoy learning not only about different instrument setups that have been used over the years as the field has developed, but also recent research results on fundamental aspects of practical importance such as the effect of interface design on column lifetime and the use of shifting gradients in the second dimension of 2D separations.

Chapter 5 is dedicated to the topic of selecting separation modes (e.g., normal phase, reversed-phase, HILIC, etc.) and selectivities (e.g., phenyl, C18) for use in multi-dimensional LC separations. This is an important topic that has many facets due to the vast array of conditions that can be used in LC, ranging from salty aqueous mobile phases for ion-exchange separations to non-aqueous mobile phases used for normal phase separations. The chapter not only reviews the strengths and weaknesses of each separation mode in the context of 2D-LC, but also very importantly addresses which combinations of modes are most likely to yield successful 2D separations.

Chapters 6 and 7 are focused on approaches to method development for non-comprehensive (Chapter 6) and comprehensive (Chapter 7) 2D-LC separations. Although method development is arguably one of the least developed areas of multi-dimensional LC, it is also currently of tremendous importance as more users get into the field and look to apply the technique to an increasing array of application targets for which methods may not exist.

Chapter 8 is dedicated to data analysis for multi-dimensional liquid chromatography. Since many data analysis approaches for 2D-LC are derived from, or depend on approaches developed for 1D-LC, much of the chapter is focused on these principles, and recent advances in this area. The chapter then goes on to discuss challenges that are unique to data analysis for 2D-LC, and the approaches various groups have developed to address them.

Chapters 9 through 13 are application oriented, covering applications of multi-dimensional LC in analysis of synthetic pharmaceuticals (i.e., traditionally "small molecules"; Chapter 9) and biologics (i.e., traditionally "biopharmaceuticals", Chapter 10), general chemical analysis (Chapter 11), food, beverage, and natural product analysis (Chapter 12), and separations of chiral and structurally similar compounds (Chapter 13). Each of these chapters focuses mainly on the state of the art of application of 2D-LC in the particular area, with an emphasis on the most recent five years of research.

Finally, we would like to acknowledge our indebtedness to our colleagues who have contributed to the realization of this book through their support – both directly and indirectly – and collaborations with us over the years. The following were major players in our early published research work in 2D-LC, especially in the decade from 2005 to 2015, at the University of Minnesota and Gustavus Adolphus College: Professors Joe Davis and Sarah Rutan, and Drs. Xiaoli Wang, Adam Schellinger, and Marcelo Filgueira. We must also express our thanks to Drs. William E. Barber, Stephan Buckenmaier, Monika Dittmann, and Konstantin Shoykhet, Tom van der Goor, and Thomas Doerr, and Klaus Witt of Agilent Technologies. Each has been supportive of our endeavors in multi-dimensional separations over the years, and generously shared their expertises without which our work could not have been done. Lastly, we are very grateful to Dr. Daniel Meston and Tina Dahlseid for their careful review of the manuscript.

Dwight R. Stoll
Peter W. Carr
June 14, 2022

Contributors

Stephan M.C. Buckenmaier
Agilent Technologies
Waldbronn, Germany

Peter W. Carr
University of Minnesota
Minneapolis, MN, USA

Carlos Calderón
Escuela de Química, Universidad
  de Costa Rica
San José, Mercedes, Costa Rica

André de Villiers
Stellenbosch University
Stellenbosch, South Africa

Zachary Dunn
Merck, Dohme, and Sharpe
Kenilworth, NJ, USA

Alexandre Goyon
Genentech
San Francisco, CA, USA

Michael Lämmerhofer
University of Tübingen
Tübingen, Germany

Sascha Lege
Agilent Technologies
Waldbronn, Germany

Gabriel Mazzi Leme
Gustavus Adolphus College
Saint Peter, MN, USA

Magriet Muller
Stellenbosch University
Stellenbosch, South Africa

Bob W.J. Pirok
University of Amsterdam
Amsterdam, Netherlands

Matthias Pursch
Dow, Inc
Wiesbaden, Germany

Douglas D. Richardson
Merck, Dohme, and Sharpe
Kenilworth, NJ, USA

Sarah C. Rutan
Virginia Commonwealth University
Richmond, VA, USA

Gregory O. Staples
Agilent Technologies; Current – Thermo Fisher
  Scientific
Santa Clara, CA, USA

Dwight R. Stoll
Gustavus Adolphus College
Saint Peter, MN, USA

Peilin Yang
Dow, Inc
Lake Jackson, TX, USA

Kelly Zhang
Genentech
San Francisco, CA, USA

1 Introduction to Two-Dimensional Liquid Chromatography

Dwight R. Stoll and Peter W. Carr

CONTENTS

DOI: 10.1201/9781003090557-1

1.1 THE BASIC CONCEPT OF MULTI-DIMENSIONAL SEPARATION

A useful way of thinking about the value of a multi-dimensional separation, whether it involves two, three, or even more dimensions, is that the added dimensions of separation enable us to improve the resolution of the separation started by the first dimension. This idea is illustrated in Figure 1.1 where we consider the separation of a hypothetical mixture of four compounds, some of which have different molecular sizes, and some of which carry different charges in solution. Since some of the molecules vary by size, one way to approach the analysis would be to separate the mixture using a conventional one-dimensional (1D) size-based separation such as size-exclusion chromatography (SEC). The resulting separation might look like that shown in Figure 1.1A, where two of the four molecules are nicely separated, but the other two overlap because they have similar sizes/molecular weights, and this separation is not selective toward differences in charge. An alternative approach, still based on a conventional 1D separation, would be to use a charge-based separation such as ion-exchange chromatography (IEX). The resulting separation might look like that shown in Figure 1.1B, where now the beige and light blue molecules are separated, however now we have three molecules coeluting early in the separation. This is because this separation has no selectivity for variations in the sizes of the analytes.

The next logical step then, is to combine these two separation mechanisms together in a single analysis such that we capitalize on both selectivities instead of just one of them. The outcome of doing so might look like that illustrated in Figure 1.1C, where all four components of the sample are now separated. Moreover, we can achieve complete resolution of all the components of the sample without significantly adding to the analysis time.

As with many things in science, having an idea about how things should work, and then actually realizing those results in practice, can be two very different things. Nevertheless, nowadays it is fairly straightforward to execute a separation of the type illustrated in Figure 1.1 using modern instrumentation. There are in fact several different ways of achieving this type of separation, and a major portion of this book is devoted to explaining these different approaches. In the interest of simplifying this introductory section we show one simple approach here. First, in Figure 1.2 we show

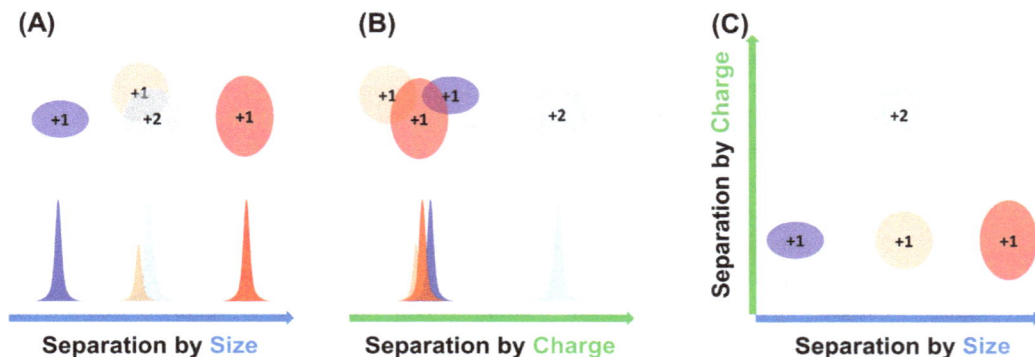

FIGURE 1.1   Illustration of the basic concept of a two-dimensional separation.

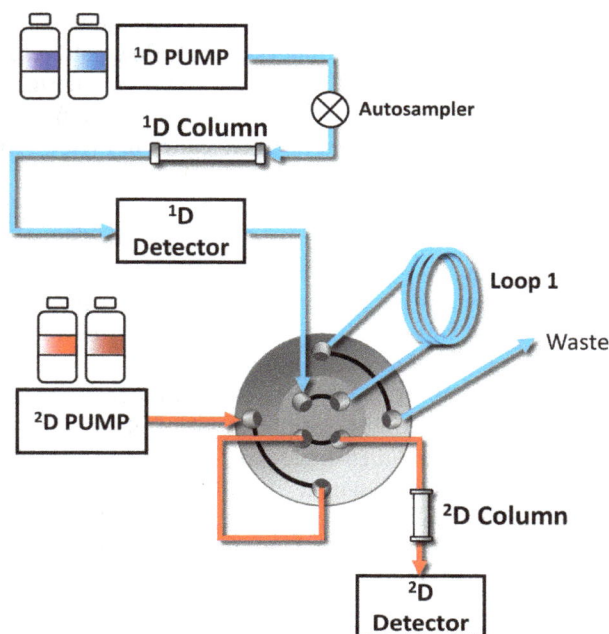

FIGURE 1.2   Basic block diagram for instrumentation used for online 2D-LC.

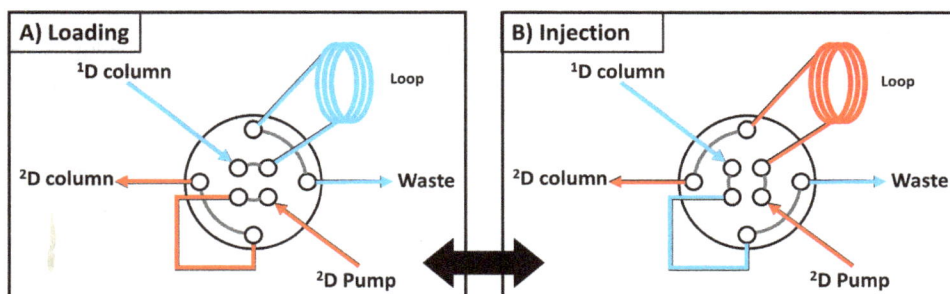

FIGURE 1.3   Illustration of flow paths through the two dimensions of a 2D-LC system corresponding to the two positions of an 8-port / 2-position that acts as an interface between the two dimensions.

a basic block diagram of the type of instrument used for contemporary online 2D-LC separations. This diagram shows the flow paths through the two dimensions of the system, and that the two dimensions are connected by an interface – in this case, an 8-port / 2-position valve.

Now, to understand how a separation like that shown in Figure 1.1C can be realized, one needs to appreciate how the flow paths in the system change as the interface valve rotates. Figure 1.3 shows the flow paths through the first and second dimensions of a 2D-LC system corresponding to the two positions of an 8-port / 2-position valve that acts as an interface between the two dimensions. In Position A the effluent from the first dimension (^{1}D) column flows through the sample loop fixed to the valve, and the outlet of this flow path is connected to waste. Upon reaching the time in the separation where the analyte of interest eluting from the ^{1}D column has entered the entrance-side of the loop, but not yet started exiting the waste-side, the valve is rotated to establish the flow paths illustrated in Figure 1.3 as Position B. In the example shown in Figure 1.1A, this would occur after the mixture of beige and light blue molecules eluting from the ^{1}D column has entered the loop. This action connects the sample loop to the second dimension (^{2}D) flow path resulting in displacement of the collected fraction of ^{1}D effluent into the ^{2}D column, which we refer to as the injection step in the second dimension. The analytes contained in that fraction then elute from the ^{2}D column having been separated, resulting in an overall separation like that shown in Figure 1.1C.

1.2   SCOPE OF THIS BOOK

The focus of this book is on online, two-dimensional separations involving LC separations in both dimensions, primarily because this is the type of multi-dimensional separation currently of greatest interest to most practitioners, and this will likely be the case for the foreseeable future. Three-dimensional separations have been demonstrated experimentally [1–3], discussed from a theoretical point of view [3–5], and are currently a major research area for the Schoenmakers group [6, 7]. However, this is a relatively niche area. For completeness we also briefly discuss offline multi-dimensional separations, as well as two-dimensional separations involving something other than LC in one of the dimensions (e.g., gas chromatography, capillary electrophoresis, or supercritical fluid chromatography; see Section 1.11). We assume that the reader is familiar with fundamental concepts in LC as discussed, for example, in Snyder's *Introduction to Modern Liquid Chromatography* [8, 9].

1.3   TERMINOLOGY

In this book we have adopted the terminology and symbol set for multi-dimensional separations advocated by Marriott and Schoenmakers [10, 11]. The reader is referred to their exhaustive lists of terms in case of questions. In the interest of clarity, we review a few of the more important tenets of their approach here.

First, an Arabic numeral superscripted on the left side of a chromatographic symbol such as t_r or N is used to communicate which dimension of separation is being referred to. For example, the symbol ^1t_r indicates we are referring to the retention time of an analyte in the first dimension of our system, whereas the symbol ^2N would imply we are referring to the plate number of the column used in the second dimension of our system.

Second, the symbols 1D, 2D, ^{1}D, and ^{2}D are reserved for very specific uses, mostly for the purpose of making descriptions less cumbersome, as follows:

- The symbol **1D** is used to replace the modifier "one-dimensional". For example, we replace "one-dimensional separation of peptides" with "1D separation of peptides". In cases where we describe **one-dimensional liquid chromatography**, we use **1D-LC**.
- The symbol **2D** is used to replace the modifier "two-dimensional". For example, we replace "two-dimensional separation of proteins" with "2D separation of proteins". In cases where we describe **two-dimensional liquid chromatography**, we use **2D-LC**.
- The symbols **^{1}D** and **^{2}D** are used to replace the modifiers "first dimension" and "second dimension", respectively. For example, when referring to the "**first dimension** column", we can instead write "**^{1}D** column". Similarly, when referring to the "**second dimension** pump", we can instead write "**^{2}D** pump".

We make every attempt to avoid switching the uses of 1D and ^{1}D, and 2D and ^{2}D, as this would introduce a great deal of confusion.

In Section 1.5 we discuss different modes of online 2D-LC separation in detail. Here, we simply introduce the symbols used for each of the modes, since they are used frequently throughout the text.

LC-LC: This symbol is used to denote **heartcutting** 2D-LC, where a single fraction of ^{1}D effluent is transferred to the ^{2}D column for further separation.

mLC-LC: This symbol is used to denote **multiple heartcutting** 2D-LC, where a single fraction of ^{1}D effluent derived from a given region of the ^{1}D separation is transferred to the ^{2}D column, but this process is repeated multiple times over the course of the 2D separation. In the literature, mLC-LC is also sometimes referred to as MHC.

LC×LC: This symbol is used to denote a **comprehensive** 2D separation where the effluent from the ^{1}D separation is fractionated and transferred at regular intervals over the course of the entire 2D separation. In this way, all of the ^{1}D effluent (or a representative fraction of it) is transferred to the ^{2}D column at some point in the 2D separation.

sLC×LC: This symbol is used to denote a **selective comprehensive** 2D separation. sLC×LC is like LC×LC in that *multiple fractions* of ^{1}D effluent are collected from a particular region of the ^{1}D separation, but also like LC-LC in that it is focused on *particular regions* of the ^{1}D separation. This mode of operation is also sometimes referred to as "high-resolution sampling" in the literature.

A complete glossary of terms and symbols used in this book is provided at the end of the book.

1.4   WHY DO WE NEED MORE THAN ONE DIMENSION OF SEPARATION?

One-dimensional LC separations are very powerful, and the application areas where they are used, which range from quantitation of amino acids in bioreactors to identification of drug metabolites in urine, is a testament to the remarkable versatility and effectiveness of 1D-LC. However, like all other analytical methods, it cannot solve every problem. And so in short, the answer to the question – why do we need more than one dimension of separation? – is that sometimes a 1D separation is unable to meet our objectives for a particular analytical problem. The 1D separation may be too slow, not sensitive enough, or not able to resolve two or more analytes to the extent we would like. In these situations a serious consideration of 2D separation is warranted, and increasingly 2D separations provide solutions to real problems that are not practically viable with a 1D separation.

Several theoretical frameworks have been introduced to help understand the value of multi-dimensional separations in the context of contemporary separation science. The most successful of these approaches has been the Statistical Overlap Theory (SOT), which was initially developed by Davis and Giddings for 1D chromatography [12], and has since been applied to multi-dimensional separations [13]. SOT is a completely stochastic model, and we can find examples of specific cases that appear to contradict predictions of the model. Nevertheless, SOT provides a useful means for comparing the potential effectiveness of 1D and multi-dimensional separations for different analytical problems *in general*. One of the stunning predictions of SOT is that the ratio of peak capacity of a 1D separation ($n_{c,1D}$) to the number of constituents of a sample requiring separation (m) that is required to reach a 90% probability of completely resolving the mixture is 200. For example, suppose we have a sample containing ten randomly retained detectable sample constituents. SOT tells us that to have a 90% chance of resolving them without any method optimization, the peak capacity of our separation must be 2,000 (see Eqs. 9 and 23 of ref. [12]). Now, this particular prediction would not be so bad if we could realize 1D separations with peak capacities in the millions within reasonable analysis times; however, we cannot do this with existing technology. In fact, we cannot even come close. In a recent contribution we derived and discussed an expression for the estimation of maximal 1D-LC peak capacities under fully optimized conditions (including optimization of particle size, column length, eluent velocity, and the ratio of gradient time to column dead time) for a given molecule type at a particular column temperature and pressure drop across the column [14]. In other words, this expression gives us a sense for the most peak capacity we can hope to generate in a given analysis time by 1D-LC. Using this expression (see Eq. 1.38 of ref [14]) for a typical small molecule (ca. 200 Da) separated under RP conditions at 40°C and a maximum pressure drop of 800 bar we can expect a peak capacity of about 300 in an analysis time of one hour. Since this peak capacity is much less than the 2,000 needed to have a 90% chance of fully resolving the 10-component mixture, our chances of doing so will only be about 50%.

To make matters worse, in 1D chromatography simply trying harder does not help much, mostly because the peak capacity depends on analysis time very weakly ($n_{c,1D} \, \alpha \, t^{1/4}$) [14]. For example, under the conditions discussed above, *doubling the 1D peak capacity from 300 to 600 would require a 16-fold increase in analysis time to 16 hours*; this is prohibitively long for all practical work, and still does not reach our peak capacity goal of 2,000 in this example. In 2D-LC, on the other hand, peak capacities on the order of 2,000 are now routinely achieved in one hour or less [15]. Finally, the primary message here is not that 1D-LC is not useful, but that 2D-LC is more temporally efficient than 1D-LC for generating the resolving power needed to tackle samples of moderate to high complexity in reasonable analysis times.

This way of thinking about the comparison of 1D- and 2D-LC is grounded in a completely stochastic view of LC separations where resolving power (peak capacity) is the most convenient measure of separation performance, and is most well suited to LCxLC separations and analyses of complex samples. In Section 1.6 we will discuss in detail other types of analytical problems where the more relevant measure of performance is the resolution of specific sets of peaks.

1.5   TYPES AND MODES OF 2D-LC SEPARATION

1.5.1   TYPES OF 2D SEPARATION

1.5.1.1   Serial Coupling of Columns is Not Two-Dimensional Separation

Before getting into the details of the types and modes of actual 2D-LC separations, it is worth briefly discussing the fact that coupling two columns with different chemistry – serially, but in a 1D-LC system – is very different from 2D-LC as we will discuss later in this book, and the two must not be confused [16]. In serially coupled 1D-LC, the idea is that two or more columns of different stationary phase chemistry are connected in series with the goal of adjusting the overall selectivity of the separation to meet a particular analytical goal. This coupled-column arrangement is used in a way similar to the way a single column would be used in a 1D-LC instrument – preceded by a pump and sample injector, and followed by a detector of some kind. Theoretical frameworks to guide method development using coupled columns have been developed and described extensively in the literature [17–19]. Although coupling columns with different selectivities can be used to tune the overall selectivity of the resulting 1D-LC separation, it is still a 1D-LC separation with all the kinetic limitations characteristic of 1D-LC systems as described above in Section 1.4. In other words, serially coupling two columns of different selectivity can dramatically change the overall selectivity of the system, but it cannot dramatically change the resolving power of the system as measured by peak capacity. For example, suppose we consider two situations: A) serially coupling two columns of the same chemistry (A + A = C); and 2) serially coupling two columns of different chemistries (A + B = D). If columns A and B both have 5,000 plates, then both of the serially coupled columns C and D will have 10,000 plates. This means that the limitations we encounter in conventional 1D-LC – namely that increasing analysis time or column length only weakly improves peak capacity – also apply here. And so for complex mixtures, serially coupling columns of different selectivity changes analyte band spacing, but does not increase peak capacity.

Real 2D-LC separations as described in Sections 1.1 and 1.3 are different. In the case of serially coupled columns, all of the analytes that elute from the first column also travel through the following columns. In other words, the order in which analytes elute from the last column is dependent on the order in which they elute from the preceding columns. In 2D-LC this is not the case. Although in some cases we can only minimize this dependence rather than eliminate it (see discussion of undersampling in Section 3.4), in most 2D-LC applications the separation of analytes in a particular fraction of ^{1}D effluent by the ^{2}D separation is independent of what happened in the ^{1}D separation both prior to, and after, the fraction of ^{1}D effluent was transferred to the second dimension. Consider for example a sample that has 20 compounds in it, 18 of which are easy to separate, but two of which are hard to separate because they are isomers. In a 1D-LC separation this is a difficult problem to solve because of the limited peak capacity of 1D-LC separations. Coupling the initial separation with one of different selectivity may enable the resolution of the isomers, but in most cases introduces one or more new areas of peak overlap that did not exist with the single column because all of the analytes in the sample are subject to the different selectivity, not just the two isomers. In 2D-LC, however, this problem is easily solved. We first rely on a ^{1}D separation to resolve 18 of the 20 compounds. Then, we transfer ^{1}D effluent containing the overlapping isomers to a ^{2}D column that is selective for the separation of isomers. This ^{2}D separation is in most cases a trivial problem, because the ^{1}D effluent fraction that is transferred only contains two compounds.

1.5.1.2 Offline vs. Online Separation

The two broadest categories of 2D-LC separation are *offline* and *online* separations. In the case of offline separation, fractions of ^{1}D effluent are collected in containers such as vials or wellplates, and subjected to a second separation at a later time. This is physically different from the online approach illustrated in Figure 1.2, where the ^{1}D effluent fraction of interest is collected in a sample loop that is physically connected to the 2D-LC system such that the fraction never leaves this flow path before being injected into the ^{2}D column. In the offline separation, the collected fractions are subjected to further separation by injecting them into a second separation system at some later time. In principle this time can be relatively short (seconds) or very long (minutes, days, or perhaps even years); indeed, this is one of the main advantages of offline separations – that we need not know *a priori* which fractions we intend to separate further, nor by which means. If we collect fractions today, and have a look at the data tomorrow, we can further separate selected fractions next month at a convenient time. Moreover, if after looking at the data from the first set of offline ^{2}D separations we decide that a different separation mechanism is needed, we can simply go back to the stored set of ^{1}D fractions and separate them again using a different set of ^{2}D conditions. There are two other major advantages of the offline approach over online approaches. First, the separations used in the two dimensions can be optimized nearly completely independently, particularly with respect to the kinetics of the separations. This allows each separation to be carried out using a set of conditions (i.e., particle size, column length, and eluent velocity) that is optimal for the length of that separation [20]. This in turn enables the offline 2D-LC separation to be truly optimized for maximal absolute peak capacity, whereas in the case of online 2D-LC separations there are always compromises that must be made that lead to lower peak capacities. These details for the comparison of offline and online 2D-LC separations have been discussed in detail by Guiochon and coworkers [21]. Second, the instrumentation required for offline 2D-LC separations tends to be simpler than that required for online separations. Whereas online 2D-LC separations require at a minimum an additional pump and column, as well as some device for interfacing the two dimensions (e.g., see Figure 1.3), with offline separations the same instrument (i.e., sampler, pump, column, and detector) can be used for the ^{1}D separation as for the ^{2}D separation. Of course, some device is required to collect the fractions of ^{1}D effluent from the ^{1}D column; for example, this could be as simple as the analyst holding a sample vial to collect the effluent, or as sophisticated as a liquid handling robot that can organize the fractions into wellplates.

In current practice we note that the vast majority of all 2D-LC separations that are done globally are executed in the online manner. We believe there are two main reasons for this. First, as a matter of convenience it is simpler with current instrumentation to inject the sample once into an instrument designed for online 2D-LC and know that all of the data produced by the ^{2}D detector belongs to that sample. In an offline 2D-LC experiment one must keep track of where the fractions go, where they are stored, and so on. This may be straightforward if they are analyzed immediately after the ^{1}D separation has finished, but will be much more complicated if they are stored for days or months at time. Second, and perhaps more importantly, as soon as the ^{1}D effluent leaves the system and is collected into storage containers, it is prone to contamination from the container and/or the atmosphere, and the collected material may degrade or transform over time [22]. Indeed, the concern over the introduction of artifacts during collection and concentration of the ^{1}D effluent was raised over four decades ago when Erni and Frei proposed that 2D-LC separations be executed online using instrumentation like that shown in Figure 1.2 [23].

1.5.1.3   2D-LC Separations with Discontinuous ^{1}D Elution

One of the major challenges in developing online 2D-LC separations is that the desired timescales for the ^{1}D and ^{2}D separations are usually quite different. As is discussed in Section 3.4, the ^{2}D separation should be fast enough to enable fractionation of a ^{1}D peak to avoid loss of ^{1}D resolution due to undersampling the ^{1}D effluent. This is not hard in the case of LC-LC separations, but it is typically extremely challenging in the case of LC×LC separations. The concepts of Stop-Flow [21, 24–30] and Pulsed-Elution [31] for 2D-LC have been developed in response to this challenge. In the case of Stop-Flow 2D separation, the flow through the ^{1}D separation is stopped after a fraction of ^{1}D effluent has been collected to allow separation of that fraction using the ^{2}D column before resuming the ^{1}D flow and collecting more fractions. This process of turning the flow off and on can be repeated multiple times over the course of the 2D separation. In principle this provides the analyst with unlimited time for ^{2}D separations. This is similar to the benefit of offline 2D separations, but in an online system.

Whereas in Stop-Flow 2D separations the ^{1}D mobile phase composition is not changed during the period when the ^{1}D flow is stopped, in Pulsed-Elution 2D separation the ^{1}D eluent composition is changed without changing the ^{1}D flow rate. Specifically, after collection of a certain portion of ^{1}D effluent containing analytes of interest (e.g., 50 µL) into a sampling loop, the ^{1}D eluent is switched to a much weaker elution strength (e.g., more water for a RP separation) such that no additional analyte elutes until the eluting mobile phase composition is restored. The period over which the ^{1}D eluent is held at the weak strength is determined by the analyst, and this time can be used for ^{2}D separation of the collected ^{1}D fraction over a period of minutes. At end of the ^{2}D separation, the composition of the ^{1}D eluent is restored to elution strength, the next fraction of ^{1}D effluent is collected, and so on. Diagrams showing further details related to Stop-Flow and Pulsed-Elution 2D separations are given in Chapter 4.

1.5.2   Modes of Online 2D-LC Separation

In Section 1.3 we introduced terms for different modes of online 2D-LC separation. Here we describe these modes in conceptual terms and a bit more detail, as they will be referred to frequently from this point forward. The instrumentation required to execute these modes in practice is discussed in greater detail in Chapter 4. The four most commonly used modes of 2D-LC in practice today are LC-LC, mLC-LC, sLC×LC, and LC×LC. The similarities and differences between them are illustrated graphically in Figure 1.4.

1.5.2.1   Single Heartcut 2D-LC (LC-LC)

As shown in Figure 1.2, in LC-LC a fraction of ^{1}D effluent containing the particular analytes of interest is collected and injected into the ^{2}D column for further separation. The ^{2}D separation may be needed to resolve two or more analytes that coelute in the ^{1}D separation, or to separate one or more analytes from other constituents of the ^{1}D effluent. The latter is sometimes referred to broadly as "desalting" or "solvent exchange" and has become quite popular in pharmaceutical analysis (see Section 9.4). This is the simplest mode of online 2D-LC separation, both in terms of execution and interpretation of the results. It has been practiced for more than four decades [32] to great effect in a variety of application areas ranging from reaction monitoring [33] to biopharmaceutical analysis [34].

1.5.2.2 Multiple Heartcut 2D-LC (mLC-LC)

If analytes of particular interest elute in very different regions of the ^{1}D separation, then a simple LC-LC setup will not provide the capabilities needed to resolve the sample as desired. If the analytes elute in parts of the ^{1}D separation that are very far apart in time, then simply repeating the heartcut operation shown in Figure 1.4A several times throughout the ^{1}D separation will suffice. However, this becomes more difficult as the regions of interest become closer in time, such that there is not sufficient time to complete the ^{2}D separation of the first ^{1}D fraction of interest before the second ^{1}D fraction must be collected. This problem motivated the advent of a version of what is now commonly referred to as multiple heartcutting (mLC-LC) 2D-LC by Zhang and coworkers [35]. In this type of separation illustrated in Figure 1.4B, the idea is that a second fraction of ^{1}D effluent can be stored temporarily in a sampling loop or sorbent cartridge until the ^{2}D separation of the prior ^{1}D effluent fraction is finished. As with Stop-Flow and Pulsed-Elution 2D-LC, this type of mLC-LC affords the user a great deal of flexibility in method development, but without the need to change the ^{1}D eluent flow rate or composition.

1.5.2.3 Selective Comprehensive 2D-LC (sLC×LC)

The mLC-LC approach is very powerful, and is employed in a variety of practical applications including chemical [36] and pharmaceutical analysis [35, 37]. However, one major limitation arises

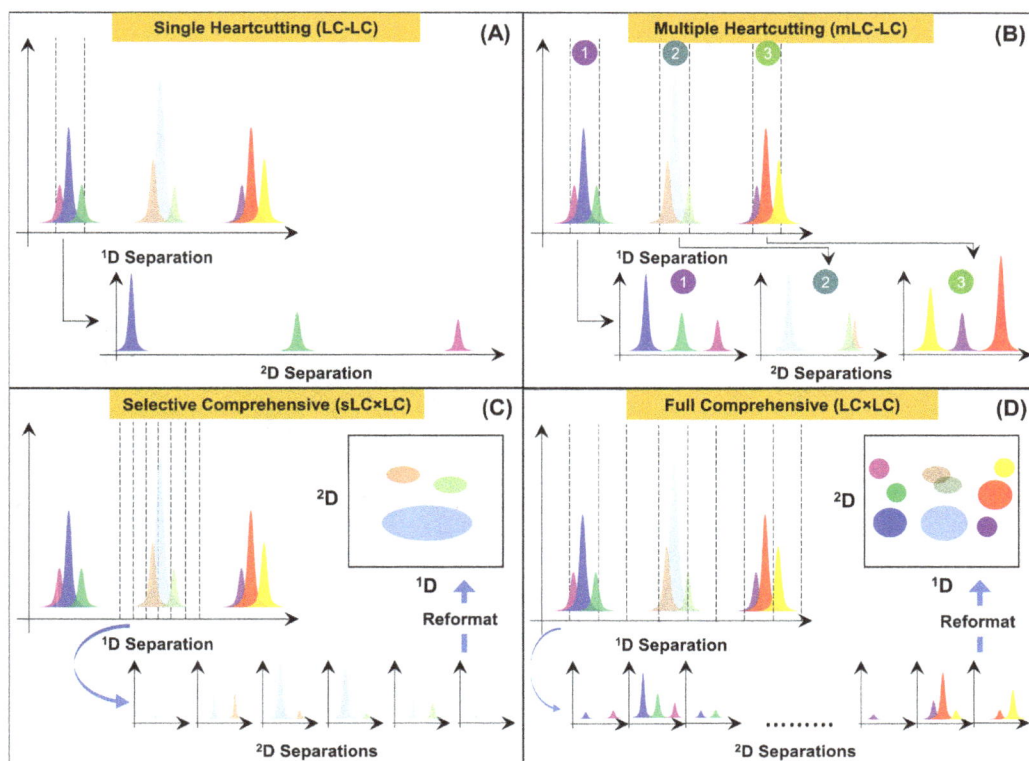

FIGURE 1.4 Illustration of the conceptual similarities and differences between the most commonly used modes of 2D-LC separation. Adapted with permission from [43]. Copyright 2019 from Recent Advances in Two-Dimensional Liquid Chromatography for the Characterization of Monoclonal Antibodies and Other Therapeutic Proteins by D. Stoll, K. Zhang, G. Staples, A. Beck. Reproduced by permission of Taylor and Francis Group, LLC, a division of Informa plc.

from the fact that this approach relies in principle on the collection of analytes from the ^{1}D separation in a single fraction of ^{1}D effluent. A problem arises in this case when analytes that were separated by the ^{1}D column are collected in a single fraction of effluent and mixed back together upon transfer to the ^{2}D separation. This is especially problematic if the ^{2}D separation does not have sufficient selectivity to separate them again. This type of situation is illustrated graphically in Figure 1.4B. The process of remixing analytes in the 2D-LC interface is a consequence of what we refer to as "undersampling" of the ^{1}D separation; it is important that it is minimized, and is discussed in much more detail in Section 3.4. Moreover, this problem was a significant source of motivation for our development of the sLC×LC approach several years ago [38–40]. The mLC-LC and sLC×LC approaches are similar in the sense that they both enable *targeted* use of the ^{2}D separation power to further resolve analytes eluting from the ^{1}D separation in particular regions. The primary difference between them is that the sLC×LC approach enables the user to at least minimize, if not practically eliminate, the undersampling problem by fractionating the ^{1}D effluent containing analytes of interest into several small fractions rather than a single larger one. This is illustrated in Figure 1.4C, where the smaller fraction volumes retain the separation of the beige and green components provided by the ^{1}D separation. This means that they remain separated in the 2D chromatogram, despite the fact that the ^{2}D separation could not resolve them on its own. This approach is also being applied in diverse areas including chemical [36] and pharmaceutical analysis [41].

1.5.2.1 Fully Comprehensive 2D-LC (LC×LC)

The fourth mode, illustrated in Figure 1.4D, is referred to as fully comprehensive 2D-LC (LC×LC). As the name suggests, the primary goal of LC×LC separations is to obtain a comprehensive profile of all the constituents of a sample, usually in an untargeted way. In this case the ^{1}D effluent is fractionated at regular intervals across most of the ^{1}D analysis time. One of the major benefits of this approach is that the analyst need not know *a priori* which regions of the ^{1}D separation to focus on.

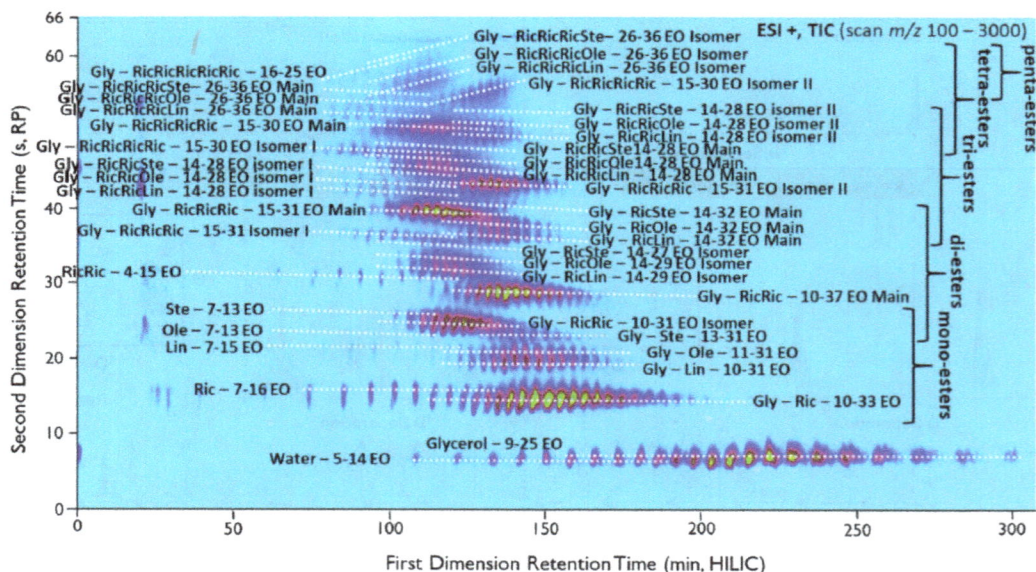

FIGURE 1.5 LC×LC-HRMS separation of castor oil ethoxylate from the work of Groeneveld *et al.* Reprinted from *Journal of Chromatography, A*, 1569, G. Groeneveld, M. Dunkle, M. Rinken, A. Gargano, A. de Niet, Characterization of complex polyether polyols using comprehensive two-dimensional liquid chromatography hyphenated to high-resolution mass spectrometry, 128–138, Copyright 2018, with permission from Elsevier.

In this way the sampling of the first separation is unbiased and can be implemented with minimal information about the ^{1}D peak pattern. There are also significant challenges associated with this approach, however, because the ^{2}D separation time for each fraction of ^{1}D effluent is directly linked to the rate at which the ^{1}D is sampled. This naturally results in a number of strong interdependencies between the conditions used for the ^{1}D and ^{2}D separations, and presents several serious challenges during method development and optimization. These topics are discussed in much more detail in Chapter 5. In spite of these challenges, however, LC×LC has provided spectacular separations of highly complex materials that are impossible to achieve using 1D-LC in analysis times that are practically relevant. One such separation is shown in Figure 1.5, which shows the separation of hundreds of different polyol species present in a castor oil ethoxylate sample [42].

1.6   SCOPE OF APPLICATION FOR 2D-LC IN THE CONTEXT OF ANALYTICAL CHEMISTRY

We argue that up until the late 2000s the general perception of the chromatography community was that 2D-LC was most well suited to the analysis of very complex materials. These 2D separations could be carried out in a highly targeted (LC-LC) or untargeted (LC×LC) manner, but the sample matrices that were typically considered were very complex. In our own research in the early 2000s this perspective strongly influenced our research aims to focus on improving the speed of LC×LC separations. In doing so we felt that improving the speed of LC×LC could make it competitive with gradient elution 1D-LC separations of moderately complex materials. However, we also felt that it was unlikely that 2D-LC would be deployed very often for analyses of less than 20 minutes or so [44]. However, with the introduction and development of mLC-LC and sLC×LC in the early 2010s, this perception has changed rapidly. Now, 2D-LC is looked upon more frequently as a solution for the analysis of materials that are moderately complex, or even samples that contain as few as 10 to 20 compounds, but happen to have certain groups (pairs, triads, etc.) that are inherently difficult to separate because of their particular chemistries. Indeed, some recently developed LC-LC methods are intended for high throughput applications with analysis times of five minutes, or even less [45]. The following list of types of 2D-LC separations is an attempt to group and categorize the highly diverse applications of 2D-LC that have emerged recently in addition to the ones that have been considered more often historically.

1.6.1   2-for-1 Methods

The idea of a 2-for-1 method is that with a single 2D-LC method composed of ^{1}D and ^{2}D separations A and B we can obtain the same information, or perhaps even more, than we could with two separate 1D-LC methods based on separations A and B. Implementing such a scheme comes with significant benefits. Implementing two 1D-LC methods to obtain this information requires two separate instruments, operated by two analysts, maintained by two service technicians. Two different samples must be injected, and ultimately these analyses result in two different datafiles and reports. However, if we have a single 2D-LC method to replace them, we only need one of each of these things – one analyst, one instrument, one datafile, and so on.

The best example of this strategy at this time is the development of LC-LC methods for characterization of monoclonal antibodies (mAbs) that involve a ^{1}D affinity separation based on Protein A modified stationary phases, and a ^{2}D sized-based separation such as size-exclusion chromatography (SEC) [45–48]. In this particular case the ^{1}D Protein A separation is used to very selectively retain the mAb protein and separate it from other constituents of a complex sample, typically cell culture supernatant. Because this separation is highly selective and captures all of the mAb protein as a single peak, it can be coupled with a relatively non-specific detector, such as UV absorbance, to get reliable information about the concentration of protein in the sample. The fraction of ^{1}D effluent is then transferred to the ^{2}D SEC separation where different forms of the mAb protein are separated according to their sizes. Most of the protein is usually present as a monomer with a nominal mass of about 150 kDa, but measurable fractions of the protein will be present as high molecular weight species (e.g., dimers, trimers, etc.) or low molecular weight species (e.g., a protein missing one light chain). This particular application is discussed in more detail in Section 10.2.1.1. It is a compelling example of how a single 2D-LC method can replace two separate 1D-LC methods without loss of analytical information, and without significantly increasing analysis time.

1.6.2 METHODS FOR ENHANCING THE SELECTIVITY OF 1D-LC

One way of viewing the value of adding a second dimension of separation is that it can enhance the selectivity of an existing 1D separation. At this time we believe the most compelling example of this is the addition of a ^{2}D chiral separation to an existing 1D achiral separation. This idea has been demonstrated several times using LC-LC and mLC-LC concepts (see Sections 9.5 and 13.2), but recently has also been demonstrated in a LC×LC format as a result of the recent improvements in the speed of chiral separations [41]. One such example from the work of Regalado and coworkers is shown in Figure 1.6. In this particular case chiral separations are used in both dimensions, but this example still illustrates the point quite nicely. Chiral separations are generally highly selective for resolving compounds that have very similar structures. Indeed, we see from the ^{1}D chromatogram here that the ^{1}D column separates most of the molecules in this mixture well. However, for the pair of molecules that coelute in the first dimension at about 73 min, the addition of the ^{2}D chiral separation having complementary selectivity nicely and efficiently extends the selectivity of the separation to fully resolve the analyte mixture.

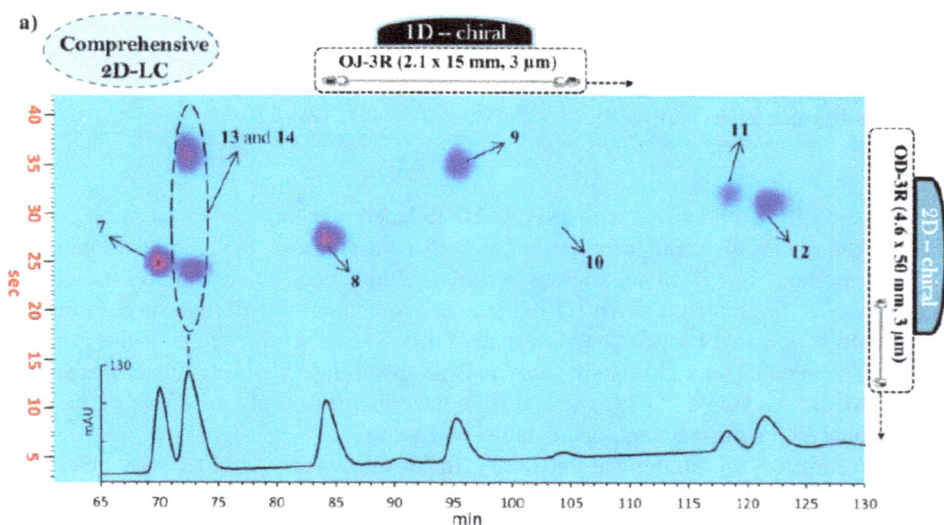

FIGURE 1.6 LC×LC separation of the isomers of an organic synthetic intermediate, using a chiral separation in the second dimension. Reprinted with permission from C. Barhate, E. Regalado, N. Contrella, J. Lee, J. Jo, A. Makarov, D. Armstrong, C. Welch, Ultrafast chiral chromatography as the second dimension in two-dimensional liquid chromatography experiments, *Analytical Chemistry* 89 (2017) 3545–3553. https://doi.org/10.1021/acs.analchem.6b04834. Copyright 2017 American Chemical Society.

1.6.3 Methods for Profiling Complex Materials

Historically, the area where the most effort has been expended to advance the performance of 2D-LC has been methods aimed at profiling very complex materials. This includes, for example, those methods used in proteomics, metabolomics, and polymer characterization. Taking proteomics as an example, the goal of LCxLC methods deployed in this area is to separate and identify as many peptides as possible from mixtures that contain tens or even hundreds of thousands of different peptides. As discussed above in Section 1.4 we cannot hope to chromatographically resolve them all because the mixture is just too complex. Rather, the goal is to obtain as much separation as possible in an analysis time that is deemed reasonable. Numerous examples of this type of 2D-LC separation are discussed in subsequent chapters focused on specific application areas. It is likely that separations of this kind will continue to be a rich research area for the foreseeable future simply because there are no practical alternatives for liquid phase separation of highly complex mixtures of non-volatile compounds.

1.6.4 Methods for Extraction of Chemical Information from Patterns in 2D Chromatograms

One of the features of 2D-GC that has facilitated its development and adoption in the chemical industry is that GCxGC chromatograms often contain peak patterns that carry chemical information [49, 50]. For example, in the GCxGC chromatogram shown in Figure 1.7 obtained from the separation of cod liver oil FAMEs (fatty acid methyl esters) we see that peaks appear in a highly

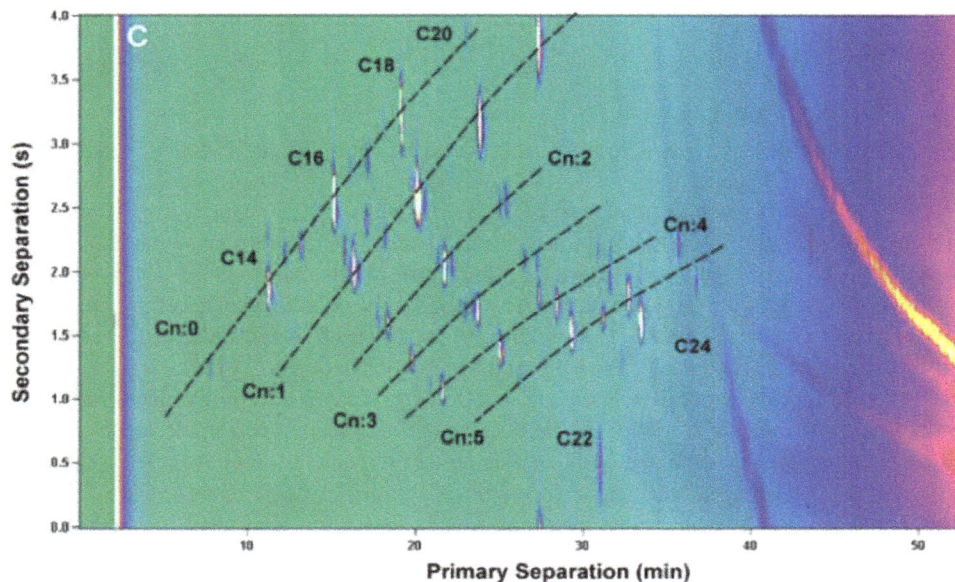

FIGURE 1.7 Chromatogram from the GCxGC separation of cod liver oil FAMEs (fatty acid methyl esters). Peaks appear in a highly structured pattern such that the number of carbons and double bonds in each FAME can be inferred from the retention coordinates of the peak. Reprinted by permission from Springer Nature: *Analytical and Bioanalytical Chemistry* (Projection of multi-dimensional GC data into alternative dimensions – exploiting sample dimensionality and structured retention patterns, J. Harynuk, B. Vlaeminck, P. Zaher, P. Marriott), Copyright (2006).

structured array such that the number of carbons and double bonds in each FAME can be inferred from its location in the array. Although such patterns certainly exist in 1D chromatograms, they are harder to interpret because multiple dimensions of information are projected onto a single retention axis. In 2D chromatograms these patterns become more evident and are more readily useful. Whereas such retention patterns are quite common in 2D-GC, they are much less common in 2D-LC. This is probably due to a variety of factors that are different between GC and LC, including the influence of the mobile phase on selectivity, complexity of retention mechanisms, and the nature of analyte mixtures that tend to be separated by LC. Nevertheless, some examples are beginning to emerge in the 2D-LC literature that such patterning is possible, and very useful, in 2D-LC as well. For example, Elsner et al. showed HILIC×RP separations of surfactants that were highly structured with retention depending on the type of polar end group and the length of alkyl chains [51]. As shown in Figure 1.6, Groenveld et al. have demonstrated HILIC×RP separations of complex polyether polyols that yield beautifully structured chromatograms where retention patterns carry information about the level of ethoxylation and hydrophobicity of the analytes [42]. Finally, in our own work on the separation of glycoprotein subunits by HILIC×RP we have shown that the observed retention patterns carry information about the size of glycans attached to the protein and the type of protein subunit that is detected [52]. Continued development of 2D-LC separations with these kinds of patterns will add more analytical power to the continually expanding 2D-LC toolbox.

1.6.5 METHODS FOR INHERENTLY DIFFICULT ANALYSES

The final category of applications of 2D-LC that we'd like to discuss here is focused on inherently difficult analyses. These are applications where there is currently no single 1D-LC method that provides the required information about the sample within a reasonable amount of time, and with reasonable effort. Though this is not a comprehensive list, we provide three examples of this type of application of 2D-LC.

1.6.5.1 Desalting and/or Solvent Exchange

A relatively straightforward, but incredibly useful application of online 2D-LC is to use the ²D separation as a means of "cleaning up" analytes as they elute from the ¹D column prior to their detection by some means that would otherwise be incompatible with the properties of the ¹D effluent. For example, a very common problem in the pharmaceutical industry involves the identification of unknown impurity peaks eluting from a RP column in a buffer that contains a non-volatile, MS-incompatible salt (e.g., sodium phosphate). An increasingly popular way to efficiently solve this problem is to add a MS-compatible ²D separation to the existing 1D method [53]. Analytes eluting from the ¹D column are separated from the non-volatile salts of the ¹D effluent such that the ²D effluent can be directly admitted into a mass spectrometer. In this case a diversion valve is added to the flow path between the outlet of the ²D column and the mass spectrometer such that undesirable effluent constituents can be diverted to waste around the dead time of each ²D separation. This then enables near real-time identification of unknown impurities [37]. The same kind of principle could be applied to other applications involving non-MS detectors that have special requirements for effluent properties. Concrete examples of this type of application are discussed in more detail in Section 9.4.

1.6.5.2 Separations of Analytes with a Large Range in Physicochemical Properties

Some samples contain mixtures of analytes with such a wide range of physicochemical properties that they are very difficult to separate using a single 1D-LC method. For example, antibody-drug conjugate (ADC) materials contain antibody proteins (~150 kDa), protein conjugated to a small molecule drug, and free small molecule drug (~300 Da). Whereas the small molecule drug is most easily quantified by RP separation, the stationary phases that are the most fit for this purpose may be irreversibly harmed by direct injection of ADC samples due to irreversible adsorption of the large antibody protein. Li *et al.* showed that this challenge can be effectively addressed using online 2D-LC by first separating the components of the ADC sample by size using SEC, and then quantifying the free small molecule drug in a 2D separation using a RP column that is particularly well suited to the small molecule separation [54].

In another more recent example, Bäuer *et al.* demonstrated that a sLC×LC method using HILIC and RP separations in the first and second dimensions can be used to efficiently quantify both water- and fat-soluble vitamins in a single analysis. Whereas using 1D-LC separations with either HILIC or RP columns results in poor retention and resolution of either the fat- or water-soluble vitamins, respectively, combining the two complementary separations in a 2D format enables full resolution and quantification of the full mixture in a single analysis [55].

1.6.5.3 Separation of All the D-/L- Enantiomers of a Large Number of Amino Acids

As a final example in this category of 2D-LC separation, consider the separation of all the D- and L- enantiomers of a mixture of amino acids. Achiral separations of 20 or more amino acids by

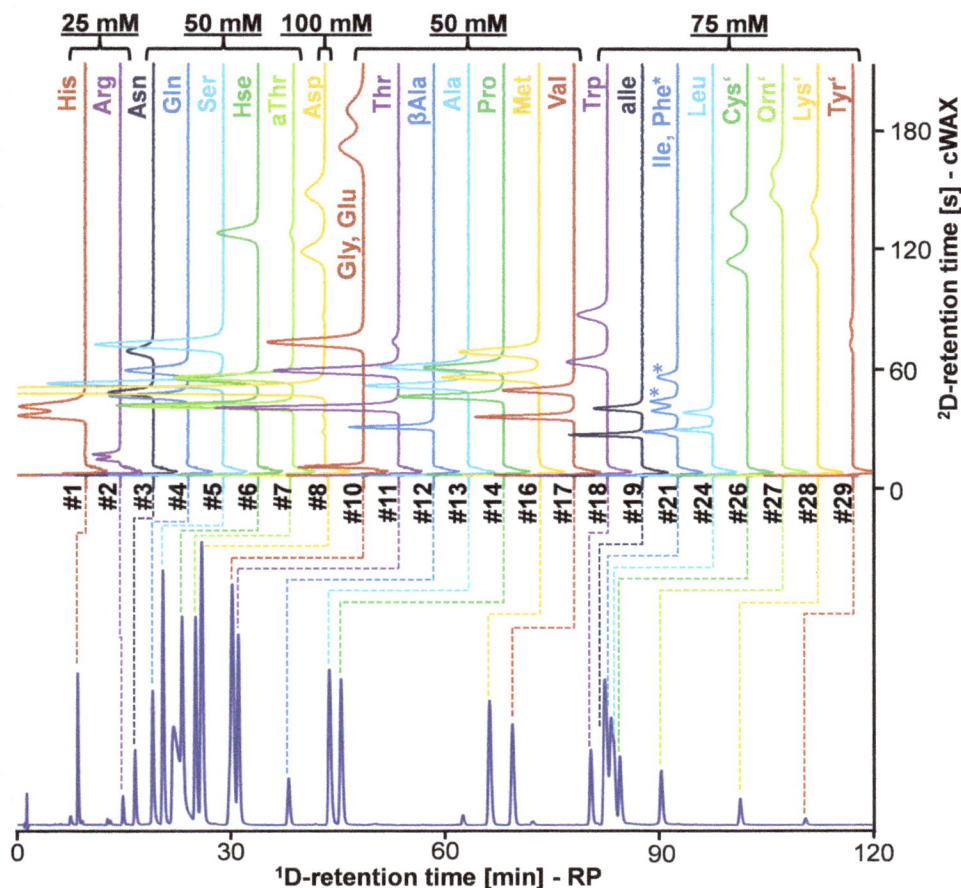

FIGURE 1.8 Online 2D-LC separation of the 2,4-dinitrobenzene (DNP) derivatives of the D- and L-enantiomers of all 20 proteinogenic amino acids (plus allo-threonine, allo-isoleucine, homoserine, ornithine, and β-alanine). The ¹D separation is carried out in the achiral RP mode and the ²D separation in the chiral mode. Source: Figure based on ref. [56] and kindly provided by M. Lämmerhofer.

1D-LC are difficult in the first place because of the peak capacity limitations of 1D-LC as discussed above in Section 1.4. If we then add to this challenge the desire to also resolve both the D- and L- enantiomers of each amino acid, this becomes an intractable problem for 1D-LC, at least within a reasonable analysis time. However, this challenge can be effectively addressed in a practically useful analysis time using online 2D-LC. The chromatogram shown in Figure 1.8, from the work of Woiwode *et al.*, shows the complete resolution of all the D- and L- enantiomers of 25 amino acids in an analysis time of 2 hours. This type of capability is particularly exciting and should motivate new lines of biochemical research that previously may have seemed impractical.

1.7   HISTORY OF 2D-LC DEVELOPMENT

This history of the development of 2D-LC is rich and interesting, with many of the influential researchers working in very different fields and organizations. Indeed, it is too rich to capture in a comprehensive way here; instead, the following sections briefly describe some of the highlights and most influential events over the past four decades.

1.7.1   1970s–1980s: INTRODUCTION OF CONCEPTUAL FRAMEWORKS FOR MULTI-DIMENSIONAL SEPARATIONS, AND INSTRUMENTATION FOR ONLINE LC-LC AND LC×LC

As early as the 1940s the extension of paper chromatography from one- to two-dimensional separations was shown to be quite useful for the separation of complex mixtures [57]. Over the next two decades this idea was transferred to 2D separations involving column chromatography. At this stage the separations were carried out in an offline manner (see Section 1.5.1.2). The most influential experimental papers during this period were by Johnson, Majors, and coworkers who described online LC-LC separations [32], and Erni and Frei who described a first implementation of online LC×LC [23]. In parallel with these experimental efforts to push technology for 2D separations forward, a number of groups were contributing to a conceptual framework for thinking about how and when 2D separations could be useful to analytical chemistry. In 1970 Snyder and coworkers compared 2D separations to 1D separation modes that were cutting-edge approaches at that time, including solvent gradient elution and temperature programming [58]. Already at this point they recognized that: "The real potential of coupled-column operation is in the use of columns of different composition such that solute k values vary among the different columns in a coupled-column set" (note: the use of the term "coupled-column" in this chapter actually implies what we would call 2D separation today, which is very different from the notion of serial coupling of columns as discussed in Section 1.5.1). This way of thinking remains highly relevant to 2D-LC nearly 50 years later. A few years later Karger, Snyder, and Horvath discussed the implication of multi-dimensional separation for the peak capacities that could be achieved in LC [59]. Giddings then followed with detailed explanations of key concepts for 2D separations, laying the groundwork that contemporary discussion in the field stands on today [60, 61]. Around the same time, Guiochon and coworkers discussed the potential for 2D liquid phase separations in a planar format at moderate pressure, and the conditions that would be needed for successful implementation of such devices [62, 63]. It is interesting that these have not yet been realized in contemporary practice, although experimental work on this continues, with some groups even pushing the concept to a third dimension of separation [6].

1.7.2   1990s: AUTOMATED LC×LC, AND THE RISE OF LC×LC IN PROTEOMICS

One of the most frequently cited papers on 2D-LC (317 citations as of December 2021) in this era described the work of Bushey and Jorgenson in 1990 to develop a fully automated instrument for LC×LC with the control infrastructure needed to coordinate the activities of the components of the

^{1}D and ^{2}D separations with the interface valve needed to transfer ^{1}D effluent to the ^{2}D column [64]. The work was focused on the separation of proteins using IEX in the first dimension, followed by SEC in the second dimension. In the following decade the Jorgenson group elaborated on this work in several highly innovative ways, including three-dimensional separations of proteins using capillary electrophoresis as the terminal separation mode [65], and the use of two ^{2}D columns operated in parallel [66]. This work captured the imagination of many groups in the proteomics community, and the decade ended with the description of the so-called Multi-dimensional Protein Identification Technology (MudPIT) by the Yates group [67]. This paper has been cited a remarkable 1,531 times (as of December 2021) and has had a tremendous influence on proteomics and related fields.

1.7.3 2000s: Development of a Solid Theoretical Foundation

In our view the development of 2D-LC in the decade 2000–2010 was dominated by advances in theory. This began just prior to 2000 with the publication of two seminal papers by Schure – one by himself [68], and one with Murphy and Foley [69]. These two papers had a dramatic influence on the trajectory of research during this time. The first paper, published in 1998, described what has come to be known as the undersampling problem in multi-dimensional separations [69]. This was not only important as a theoretical construct, but it made clear to the research community that we were going to have to figure out how to do very fast separations in the second dimension of 2D-LC separations if the technique was going to be useful in more than a handful of niche applications. This topic is discussed in detail in Section 3.4. The second paper highlighted the fact that in 2D separations one of the major problems with 1D chromatography – namely that under most conditions the separation process dilutes the analyte – is compounded, because the dilution process happens twice [68]. This means that unless we take steps to counteract this effect, the detection sensitivity of a 2D method will be worse than that for a 1D method. The good news is that we are learning how to cope with the dilution problem, and contemporary 2D separations do not suffer from this limitation as much as they did in the past. This topic is discussed in more detail in Section 3.6.

The two other major areas of theoretical development during this period were focused on methods for assessment of the usage of 2D separation spaces, and optimization of 2D separation methods. The former is discussed in detail in Section 3.3. Several methods for gauging the usage of 2D separation spaces have been developed by different groups [70, 71], and today practitioners can easily leverage this prior research in their work to develop effective applications of 2D-LC. Optimization of 2D-LC methods gained significant attention in the 2000s, and this work continues to be a very important topic of research today. As theories for undersampling and usage of the 2D separation space became established, several groups worked to bring these concepts together under a single framework [72] that would enable holistic optimization of 2D-LC methods according to performance metrics including analysis time, peak capacity, and detection sensitivity (see also Section 1.9). This has enabled realistic comparisons of the performance of 1D and 2D methods [73–77], and systematic, algorithm-driven approaches to the complex problem of optimizing 2D separations [78–80]. This topic is addressed in detail in Section 3.5 and 7.5.

1.7.4 2010s: Reduction to Practice

In the current decade we are witnessing a significant expansion of the group of scientists using 2D-LC. Whereas prior to 2010 2D-LC was mainly used by *bona fide* experts, predominantly in academic research laboratories, recent advances in commercially available instrument hardware and software for 2D-LC have made the technique more accessible to use by scientists with a wider range of interests. During the same period, advances in technology and methodology such as the development of mLC-LC, sLC×LC, and new modulation strategies have also dramatically improved the flexibility and performance of 2D-LC [81]. The net result of all this progress is that we now see

2D-LC methods being deployed in diverse application areas ranging from environmental analysis, to food and biopharmaceutical analysis [81]. The next big step in the maturation of 2D-LC may be a move toward use of 2D-LC in quality control laboratories, in regulated environments [82, 83]. It will be interesting to see how this develops in the near future.

1.8   OPPORTUNITIES AND CHALLENGES

1.8.1   CHALLENGES

We have learned from the past 15 years of development of 2D-LC that there are unique challenges we encounter with the technique – mostly having to do with 2D separations – that simply are not relevant in 1D-LC, and thus solutions to these challenges have to be developed from scratch. A short list of some of the topics that need to be addressed most urgently follows.

1) ^{2}D column lifetime – In our own experience we have observed that the lifetimes of columns used in the second dimension of 2D-LC systems are much shorter than what we expect based on experience with similar columns in 1D-LC. Many people we talk with that are more than casually involved in research with 2D-LC report the same type of problem. Although this is rarely discussed in formal settings [84], this challenge has not been fully addressed. We are optimistic that it is a solvable problem, but much more research is needed to fully understand and solve the problem.
2) Low dispersion ^{2}D detectors – Current instrument and column technologies enable ^{2}D separations with peak widths approaching those observed in 2D-GC [85]. This means they are very narrow in time units (e.g., 100 ms at half-height), and sometimes also in volume units (e.g., peak variances on the order of 2 μL^2). Such narrow peaks present a challenge if we are to avoid serious post-column peak broadening, particularly in the detector.
3) Sophisticated software to support 2D-LC method development – The development of software to support development of 1D-LC methods is relatively mature. Sometimes this software supports instrument control for the purpose of obtaining and processing results from scouting methods. In other cases instrument-independent software can support discovery of optimal separation conditions in a more general way (e.g., DryLab). Currently there is no easy-to-use software that supports method development for 2D-LC – commercial or otherwise – although there is significant research activity ongoing now that will likely provide the foundational knowledge for such software in the future [86].
4) Data analysis – Today we can develop 2D-LC methods that are capable of separating hundreds – perhaps even thousands – of compounds, however we do not have methods that enable reliable and efficient analysis of the resulting data, especially for LC×LC coupled with MS detection. This is a major barrier to wider implementation of LC×LC. In other words, we do not have the tools needed to translate rich LC×LC datafiles (> 10 GB per separation in many cases where high-resolution MS is used) into chemical information (i.e., concentrations of identifiable chemical species) about our samples. Specific details related to the nature of this challenge are discussed below in Section 1.10, and in Chapter 8.

1.8.2   OPPORTUNITIES

In spite of the challenges highlighted in the preceding section, there are also tremendous opportunities lying ahead for the further development of 2D-LC and its use to address challenging problems in science. Following is a short list of some of the more exciting possibilities.

1) "Universal" methods – There is tremendous interest in so-called "universal" or "platform" methods, where the idea is that a 2D-LC method with very high total selectivity and/or

resolving power could be broadly applied to a variety of samples without the need for intensive method development for the problem at hand [87]. Clearly the success of such methods will require significant development to increase the likelihood that the method will indeed be sufficient for all the samples that will be encountered in a particular application space. However, the potential upside, which would allow for a significant reduction in method development efforts for new samples, is very attractive. The importance of this concept for application of 2D-LC in the pharmaceutical industry is discussed in more detail in Chapter 9.

2) LC–chemistry–LC – A very interesting concept that has tremendous potential involves changing the chemistry of analytes between the ^{1}D and ^{2}D separations. Recent examples of this idea include the work of Gstöttner *et al.* who demonstrated the utility of digesting proteins with trypsin between the ^{1}D and ^{2}D separations of intact proteins and the resulting peptides. This was done in a completely automated, online system [88]. In a completely different application, Pirok *et al.* have described a 2D-LC application for the characterization of nanoparticles [89]. This method first separates the particles according to size using hydrodynamic chromatography, then dissolves the particle into its constituent polymers in the interface between the separation dimensions using tetrahydrofuran, and finally separates the constituent polymers by size in the second dimension. These innovative approaches surely will inspire the development of similar approaches for other applications in the future.

3) Simplification of instrumentation and operation – Although there have been tremendous advances in commercially available instrumentation and software for 2D-LC in the last decade, many prospective users of 2D-LC still find these tools complex, and even overwhelming. An important opportunity lies ahead for the simplification of instrumentation and its operation, and streamlining of the supporting software. As with other powerful analytical technologies (e.g., mass spectrometry), improving ease-of-use is a natural step in the development and maturation of the technology, and one we should expect going forward.

1.9   METHOD DEVELOPMENT GOALS FOR 2D-LC

As with most analytical methods, method development in chromatography involves tradeoffs between various performance metrics, and ultimately the analyst must make compromises. Perhaps the best known of these compromises is that between analysis time (speed) and resolution. When working at the pressure limit of an instrument with a fixed particle size we can increase chromatographic resolution by using a longer column, but this always comes at the cost of increased analysis time [20]. Such tradeoffs are relevant in 2D-LC as well, but there are more of them, and typically they are more complex because of the larger number of variables involved (e.g., two flow rates, two column lengths, etc.). Although the theory underlying these tradeoffs has largely been known for some time [78, 90–92], Sarrut *et al.* recently provided a visual way of understanding the relationships between variables in LC×LC experiments using the diagram shown in Figure 1.9 [76]. For example, one of the most practically significant relationships between operating parameters in a typical LC×LC experiment is the linkage between the ^{1}D flow rate (^1F), the volume of ^{1}D effluent injected into the ^{2}D column (^2V_i), the sampling time (t_s), and the ^{2}D gradient time (^2t_g).

Numerous interrelationships shown in Figure 1.9 and the related compromises are discussed in detail in Chapters 3, 6, and 7 on method development for 2D-LC. For now, it is most important to understand the major tradeoffs that are made in 2D-LC, illustrated in Figure 1.10, as this provides a useful framework within which many of the details of the remainder of this book can be understood and appreciated. For many years the primary metric of performance that was discussed in 2D-LC research circles was peak capacity. Recognizing that the speed of 2D-LC had to be improved in order for it to be adopted for routine use, researchers began discussing more the tradeoff between peak capacity and analysis time. The third factor to enter the discussion was detection sensitivity (often discussed in terms of dilution factors). And now, as we see more and

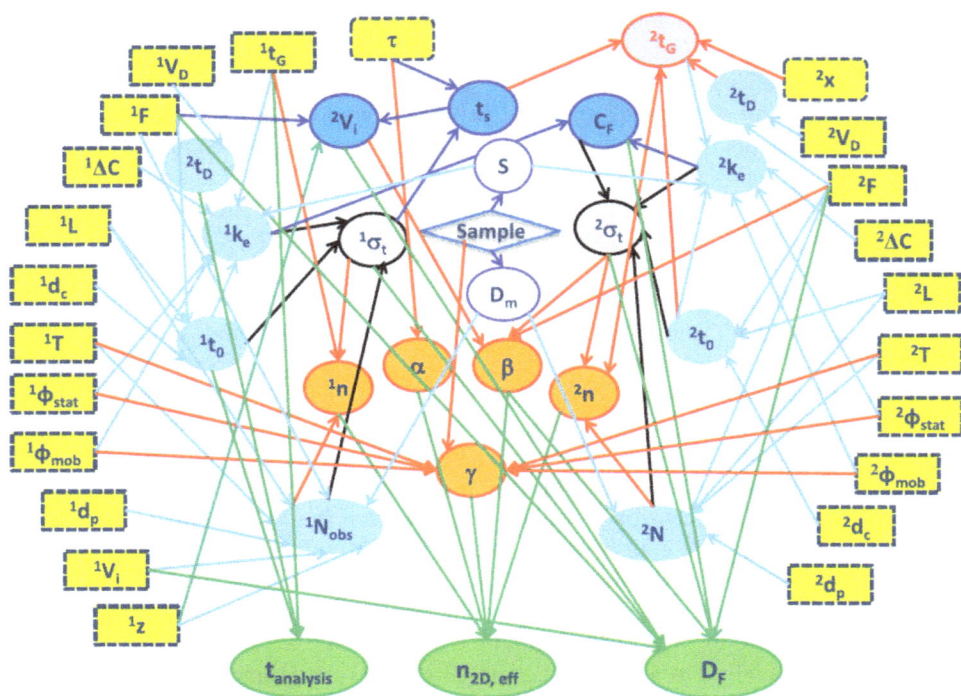

FIGURE 1.9 Illustration of the complex set of relationships between operating parameters in LC×LC separations. Reprinted from *Journal of Chromatography, A*, 1421, M. Sarrut, A. D'Attoma, S. Heinisch, Optimization of conditions in on-line comprehensive two-dimensional reversed phase liquid chromatography. Experimental comparison with one-dimensional reversed phase liquid chromatography for the separation of peptides, 48–59, Copyright (2015), with permission from Elsevier.

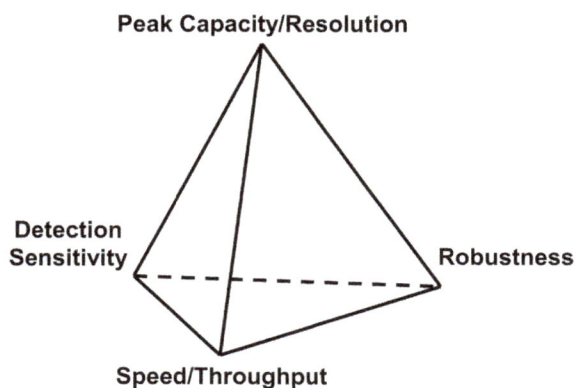

FIGURE 1.10 Illustration of the compromises between different performance metrics for 2D-LC.

more a transfer of some 2D-LC methods from expert researchers in the technology to less-expert practitioners, there are more frequent discussions of additional methodological metrics such as reliability and robustness [82].

When considering developing a new 2D-LC method for a particular application, the very first consideration should be what mode of 2D-LC separation will be used. As discussed in Section 1.5.2, today we have good options to effectively address all types of applications using 2D-LC. However, not all modes are useful for all applications. For example, a mLC-LC method might be the best choice for quantitation of peptide biomarkers in serum, but LC×LC would be a far better choice for an untargeted experiment aimed at discovering a new biomarker. Once this initial, important choice of 2D-LC separation mode is made, then the analyst must decide which of the performance metrics shown in Figure 1.10 is most important, and think about how subsequent decisions in method development will affect this particular metric, keeping in mind that compromises in 2D-LC are real and must be made, whether the analyst is aware of them or not.

1.10 DATA ANALYSIS

One of the major differences between 1D- and 2D-LC is the overall data analysis process. While there is much in common among the two methods, including establishing the baseline under peaks, the start and stop points of peaks, and filtering the signals to enhance the signal-to-noise ratio, there are some major differences.

Perhaps the most obvious yet extremely important difference, at least from the perspective of quantitation, is that in 1D-LC the determination of the peak size (as either height or area) of a peak containing only a single chemical component requires the determination of only one peak area or height. In stark contrast when the first dimension of the a 2D-LC separation is sampled multiple times (e.g., following the guideline of Murphy, Foley, and Schure [69]), the ^{2}D sampling time must be considerably less than the ^{1}D peak width. This then means that a single chemical component peak from the first dimension will be broken up into several ^{2}D peaks, each of which must be properly quantified. We need to bear in mind that one objective in doing 2D-LC is that each fraction of ^{1}D effluent be further separated by the second dimension into additional components that had coeluted in the ^{1}D separation. Consequently, there may well be many more peaks to be quantified in 2D-LC, each requiring definition of a local baseline, start and stop points, and filtering. It should also be pointed out that the fast gradients generally employed for the ^{2}D separation generate baseline signals which are steeper and more complicated than in a single slow 1D gradient separation. The peak quantitation process is considerably more complicated than appears at first glance, especially when there are significant run-to-run variations in either the ^{1}D or ^{2}D retention times (see Section 8.3). Indeed, we are only just now coming to grips with making the reproducible measurement of peak size in 2D-LC as reliable as that in 1D-LC. In the case of non-comprehensive 2D-LC separations (i.e., LC-LC, mLC-LC, and sLC×LC) considerable progress has been made in recent years, and now %RSDs for quantitation under 1% can be expected [93]. However, with little overstatement, highly automated quantitation with high precision is still problematic for LC×LC, and is arguably the "Achilles's heel" of LC×LC at this point in time. While a good deal of progress in understanding the issue has been made [94], considerable work remains.

Despite the challenges described above, we must not overlook the profound importance of the multivariate structure of data obtained from 2D-LC experiments, which allows the use of a number of very powerful data analysis methods including a wide variety of curve resolution techniques (see Section 8.5). As is well known for 1D chromatography, the minimum resolution needed to observe two maxima for two equal height chemical components with no help from advanced data analysis techniques is 0.5. However, since a 2D chromatographic separation inherently has a multivariate data structure the average 2D resolution for randomly spaced components to see two maxima without use of a multivariate detector drops to 0.26. When additional help is available from a multivariate detector even less resolution is needed [95]. Considering that the cost of increasing chromatographic resolution is usually a significant increase in analysis time, the ability to effectively increase resolution by chemometric means is highly valuable indeed. A great deal of additional research is needed in the area of data analysis to enable better the utilization of the information inherent in data produced by 2D-LC separations coupled with multivariate detectors.

1.11 COUPLING LC WITH OTHER SEPARATION MODES IN A 2D FORMAT

A detailed discussion of the advantages and disadvantages of coupling LC to other separation modes in a 2D format is beyond the scope of this book. However, we do point out here several examples of work in these areas for reference.

1.11.1 COUPLING WITH CAPILLARY ELECTROPHORESIS

In the same year (1990) that Bushey and Jorgenson published their seminal paper describing automated instrumentation for online LC×LC [64], they also described instrumentation for coupling LC and capillary zone electrophoresis in an automated online system (LC×CZE) [96]. In the following years there has been a significant amount of research activity involving different electrophoretic separation modes, as well as different approaches for interfacing LC and CE separations. This work has been thoroughly reviewed recently by Ranjbar *et al.* [97]. One particularly interesting development in this area has been the use of micro free-flow electrophoresis (μ-FFE) as a second dimension for a LC×CE separation [98, 99]. Because of the continuous nature of the μ-FFE separation, in principle this approach provides a path to significant mitigation of the undersampling problem that is so challenging in 2D-LC separation.

1.11.2 COUPLING WITH SUPERCRITICAL FLUID CHROMATOGRAPHY

The idea of coupling LC with SFC in a 2D format is attractive because of the potential for high speed ^{2}D separations, the unique selectivity of SFC for some applications (e.g., enantiomer separations), and the low organic solvent consumption of SFC in general. An online coupling of normal phase LC (NPLC) and SFC has been demonstrated for separation of fungus extracts [100]. Online coupling of RPLC and SFC has also been demonstrated for the separation of hydrocarbons [101], and more recently pharmaceutical molecules [102–104], and depolymerized lignin [105]. As with coupling LC and CE, an active area of research for coupling LC and SFC is concerned with the interfacing of these two separations that work with very different eluents.

1.11.3 COUPLING WITH GAS CHROMATOGRAPHY

The concept of coupling LC and GC separations in an online 2D format was pursued starting in the 1980s with the work of Cram *et al.* [106]. As with coupling LC to CE and SFC, here too the major challenge with coupling LC and GC lies in the interface between the two separations. In this case, obviously the large amount of liquid mobile phase must be removed from the LC effluent prior to ^{2}D GC separation. Nevertheless, LC-GC separations were put to good use in the following decades, where the ^{1}D LC separation has been used mainly as a high performance sample cleanup device that allows isolation of analytes of interest into specific fractions of the sample prior to injection into the 2D GC separation. Interested readers are referred to the recent review by Purcaro *et al.* [107].

REFERENCES

[1]   A. Furusho, R. Koga, T. Akita, M. Mita, T. Kimura, K. Hamase, Three-dimensional high-performance liquid chromatographic determination of asn, ser, ala and pro enantiomers in the plasma of patients with chronic kidney disease, Anal. Chem. (2019). https://doi.org/10.1021/acs.analchem.9b01615.

[2]   A.W. Moore, J.W. Jorgenson, Comprehensive three-dimensional separation of peptides using size exclusion chromatography/reversed phase liquid chromatography/optically gated capillary zone electrophoresis, Anal. Chem. 67 (1995) 3456–3463. https://doi.org/10.1021/ac00115a014.

[3]   S.W. Simpkins, J.W. Bedard, S.R. Groskreutz, M.M. Swenson, T.E. Liskutin, D.R. Stoll, Targeted three-dimensional liquid chromatography: A versatile tool for quantitative trace analysis in complex matrices, J. Chromatogr., A. 1217 (2010) 7648–7660. https://doi.org/10.1016/j.chroma.2010.09.023.

[4] M.R. Schure, J.M. Davis, Orthogonality measurements for multidimensional chromatography in three and higher dimensional separations, J. Chromatogr. A. 1523 (2017) 148–161. https://doi.org/10.1016/j.chroma.2017.06.036.

[5] B. Wouters, E. Davydova, S. Wouters, G. Vivo-Truyols, P.J. Schoenmakers, S. Eeltink, Towards ultra-high peak capacities and peak-production rates using spatial three-dimensional liquid chromatography, Lab Chip. 15 (2015) 4415–4422. https://doi.org/10.1039/C5LC01169H.

[6] E. Davydova, P.J. Schoenmakers, G. Vivó-Truyols, Study on the performance of different types of three-dimensional chromatographic systems, J. Chromatogr. A. 1271 (2013) 137–143. https://doi.org/10.1016/j.chroma.2012.11.043.

[7] N. Abdulhussain, S. Nawada, P. Schoenmakers, Latest trends on the future of three-dimensional separations in chromatography, Chem. Rev. 121 (2021). https://doi.org/10.1021/acs.chemrev.0c01244.

[8] L.R. Snyder, J.J. Kirkland, J.W. Dolan, Introduction to Modern Liquid Chromatography, 3rd ed, Wiley, Hoboken, NJ 2010.

[9] L.R. Snyder, J.W. Dolan, High-performance Gradient Elution: The Practical Application of the Linear-solvent-strength Model, Wiley, Hoboken, NJ, 2007.

[10] P.J. Marriott, P.J. Schoenmakers, Z. Wu, Nomenclature and conventions in comprehensive multidimensional chromatography – An update, LC-GC Eur. 25 (2012) 266, 268, 270, 272–275.

[11] P.J. Schoenmakers, P.J. Marriott, J. Beens, Nomenclature and conventions in comprehensive multidimensional chromatography, LC-GC Eur. 16 (2003) 335–336, 338–339.

[12] J.M. Davis, J. Calvin. Giddings, Statistical theory of component overlap in multicomponent chromatograms, Anal. Chem. 55 (1983) 418–424. https://doi.org/10.1021/ac00254a003.

[13] J.M. Davis, Statistical theory of spot overlap in two-dimensional separations, Anal. Chem. 63 (1991) 2141–2152. https://doi.org/10.1021/ac00019a014.

[14] P.W. Carr, D.R. Stoll, A study of peak capacity optimization in one-dimensional gradient elution reversed-phase chromatography: a memorial to Eli Grushka, in: N. Grinberg (Ed.), Advances in Chromatography, CRC Press, Boca Raton, FL, 2018: pp. 1–21.

[15] D. Stoll, J. Cohen, P. Carr, Fast, comprehensive online two-dimensional high performance liquid chromatography through the use of high temperature ultra-fast gradient elution reversed-phase liquid chromatography, J. Chromatogr. A. 1122 (2006) 123–137. https://doi.org/10.1016/j.chroma.2006.04.058.

[16] D.R. Stoll, P.W. Carr, Two-dimensional liquid chromatography: A state of the art tutorial, Anal. Chem. 89 (2017) 519–531. https://doi.org/10.1021/acs.analchem.6b03506.

[17] L.N. Jeong, S.C. Rutan, Simulation of elution profiles in liquid chromatography – III. Stationary phase gradients, J. Chromatogr. A. 1564 (2018) 128–136. https://doi.org/10.1016/j.chroma.2018.06.007.

[18] Y. Mao, P.W. Carr, Adjusting selectivity in liquid chromatography by use of the thermally tuned tandem column concept, Anal. Chem. 72 (2000) 110–118. https://doi.org/10.1021/ac990638x.

[19] S.L. Weatherbee, M.M. Collinson, Stationary phase gradients in liquid chromatography, in: Advances in Chromatography, CRC Press, Boca Raton, FL , 2022: pp. 75–120.

[20] P.W. Carr, X. Wang, D.R. Stoll, Effect of pressure, particle size, and time on optimizing performance in liquid chromatography, Anal. Chem. 81 (2009) 5342–5353. https://doi.org/10.1021/ac9001244.

[21] G. Guiochon, N. Marchetti, K. Mriziq, R. Shalliker, Implementations of two-dimensional liquid chromatography, J. Chromatogr. A. 1189 (2008) 109–168. https://doi.org/10.1016/j.chroma.2008.01.086.

[22] J. Camperi, L. Dai, D. Guillarme, C. Stella, Development of a 3D-LC/MS workflow for fast, automated, and effective characterization of glycosylation patterns of biotherapeutic products, Anal. Chem. 92 (2020) 4357–4363. https://doi.org/10.1021/acs.analchem.9b05193.

[23] F. Erni, R. Frei, Two-dimensional column liquid chromatographic technique for resolution of complex mixtures, J. Chromatogr. A. 149 (1978) 561–569. https://doi.org/10.1016/S0021-9673(00)81011-0.

[24] J. Xu, L. Zheng, L. Lin, B. Sun, G. Su, M. Zhao, Stop-flow reversed phase liquid chromatography×size-exclusion chromatography for separation of peptides, Anal. Chim. Acta. 1018 (2018) 119–126. https://doi.org/10.1016/j.aca.2018.02.025.

[25] J. Xu, D. Sun-Waterhouse, C. Qiu, M. Zhao, B. Sun, L. Lin, G. Su, Additional band broadening of peptides in the first size-exclusion chromatographic dimension of an automated stop-flow two-dimensional high performance liquid chromatography, J. Chromatogr. A. 1521 (2017) 80–89. https://doi.org/10.1016/j.chroma.2017.09.025.

[26] X. Hou, J. Ma, X. He, L. Chen, S. Wang, L. He, A stop-flow two-dimensional liquid chromatography method for determination of food additives in yogurt, Anal. Methods. 7 (2015) 2141–2148. https://doi.org/10.1039/C4AY02855D.

[27] K.M. Kalili, A. de Villiers, Systematic optimisation and evaluation of on-line, off-line and stop-flow comprehensive hydrophilic interaction chromatography×reversed phase liquid chromatographic analysis of procyanidins. Part II: Application to cocoa procyanidins, J. Chromatogr. A. 1289 (2013) 69–79. https://doi.org/10.1016/j.chroma.2013.03.009.

[28] K.M. Kalili, A. de Villiers, Systematic optimisation and evaluation of on-line, off-line and stop-flow comprehensive hydrophilic interaction chromatography×reversed phase liquid chromatographic analysis of procyanidins, Part I: Theoretical considerations, J. Chromatogr. A. 1289 (2013) 58–68. https://doi.org/10.1016/j.chroma.2013.03.008.

[29] P. Dugo, N. Fawzy, F. Cichello, F. Cacciola, P. Donato, L. Mondello, Stop-flow comprehensive two-dimensional liquid chromatography combined with mass spectrometric detection for phospholipid analysis, J. Chromatogr. A. 1278 (2013) 46–53. https://doi.org/10.1016/j.chroma.2012.12.042.

[30] F. Bedani, W.Th. Kok, H.-G. Janssen, A theoretical basis for parameter selection and instrument design in comprehensive size-exclusion chromatography×liquid chromatography, J. Chromatogr. A. 1133 (2006) 126–134. https://doi.org/10.1016/j.chroma.2006.08.048.

[31] S.S. Jakobsen, S. Verdier, C.R. Mallet, J.H. Christensen, N.J. Nielsen, Increasing flexibility in two-dimensional liquid chromatography by pulsed elution of the first dimension (pulsed-elution 2D-LC): A proof of concept, Anal. Chem. 89 (2017) 8723–8730. https://doi.org/10.1021/acs.analchem.7b00758.

[32] E.L. Johnson, R. Gloor, R.E. Majors, Coupled column chromatography employing exclusion and a reversed phase. A potential general approach to sequential analysis., J. Chromatogr. 149 (1978) 571–585. https://doi.org/10.1016/S0021-9673(00)81012-2.

[33] S. Ma, N. Grinberg, N. Haddad, S. Rodriguez, C.A. Busacca, K. Fandrick, H. Lee, J.J. Song, N. Yee, D. Krishnamurthy, C.H. Senanayake, J. Wang, J. Trenck, S. Mendonsa, P.R. Claise, R.J. Gilman, T.H. Evers, Heart-cutting two-dimensional ultrahigh-pressure liquid chromatography for process development: Asymmetric reaction monitoring, Org. Process Res. Dev. 17 (2013) 806–810. https://doi.org/10.1021/op300266j.

[34] R.E. Birdsall, S.M. McCarthy, M.C. Janin-Bussat, M. Perez, J.-F. Haeuw, W. Chen, A. Beck, A sensitive multidimensional method for the detection, characterization, and quantification of trace free drug species in antibody-drug conjugate samples using mass spectral detection, MAbs. 8 (2016) 306–317. https://doi.org/10.1080/19420862.2015.1116659.

[35] K. Zhang, Y. Li, M. Tsang, N.P. Chetwyn, Analysis of pharmaceutical impurities using multi-heartcutting 2D LC coupled with UV-charged aerosol MS detection: Liquid Chromatography, J. Sep. Sci. 36 (2013) 2986–2992. https://doi.org/10.1002/jssc.201300493.

[36] M. Pursch, P. Lewer, S. Buckenmaier, Resolving co-elution problems of components in complex mixtures by multiple heart-cutting 2D-LC, Chromatographia. 80 (2017) 31–38. https://doi.org/10.1007/s10337-016-3214-x.

[37] P. Petersson, K. Haselmann, S. Buckenmaier, Multiple heart-cutting two dimensional liquid chromatography mass spectrometry: Towards real time determination of related impurities of biopharmaceuticals in salt based separation methods, J. Chromatogr. A. 1468 (2016) 95–101. https://doi.org/10.1016/j.chroma.2016.09.023.

[38] S.R. Groskreutz, M.M. Swenson, L.B. Secor, D.R. Stoll, Selective comprehensive multi-dimensional separation for resolution enhancement in high performance liquid chromatography, Part II–Applications, J. Chromatogr. A. 1228 (2012) 41–50. https://doi.org/10.1016/j.chroma.2011.06.038.

[39] S.R. Groskreutz, M.M. Swenson, L.B. Secor, D.R. Stoll, Selective comprehensive multi-dimensional separation for resolution enhancement in high performance liquid chromatography, Part I–Principles and instrumentation, J. Chromatogr. A. 1228 (2012) 31–40. https://doi.org/10.1016/j.chroma.2011.06.035.

[40] E.D. Larson, S.R. Groskreutz, D.C. Harmes, I. Gibbs-Hall, S.P. Trudo, R.C. Allen, S.C. Rutan, D.R. Stoll, Development of selective comprehensive two-dimensional liquid chromatography with parallel first-dimension sampling and second-dimension separation – application to the quantitative analysis of furanocoumarins in apiaceious vegetables., Anal. Bioanal. Chem. 405 (2013) 4639–4653. https://doi.org/10.1007/s00216-013-6758-8.

[41] C.L. Barhate, E.L. Regalado, N.D. Contrella, J. Lee, J. Jo, A.A. Makarov, D.W. Armstrong, C.J. Welch, Ultrafast chiral chromatography as the second dimension in two-dimensional liquid chromatography experiments, Anal. Chem. 89 (2017) 3545–3553. https://doi.org/10.1021/acs.analchem.6b04834.

[42] G. Groeneveld, M.N. Dunkle, M. Rinken, A.F.G. Gargano, A. de Niet, M. Pursch, E.P.C. Mes, P.J. Schoenmakers, Characterization of complex polyether polyols using comprehensive two-dimensional liquid chromatography hyphenated to high-resolution mass spectrometry, J. Chromatogr. A. 1569 (2018) 128–138. https://doi.org/10.1016/j.chroma.2018.07.054.

[43] D.R. Stoll, K. Zhang, G.O. Staples, A. Beck, Recent advances in two-dimensional liquid chromatography for the characterization of monoclonal antibodies and other therapeutic proteins, in: Advances in Chromatography, 58 (2018) 29–63.

[44] D.R. Stoll, Fast, comprehensive two-dimensional liquid chromatography, Ph.D. Dissertation, University of Minnesota, 2007.

[45] Z. Dunn, J. Desai, G.M. Leme, D. Stoll, D. Richardson, Rapid two-dimensional Protein-A size exclusion chromatography for determination of titer and aggregation for monoclonal antibodies in harvested cell culture fluid samples, MAbs. 12 (2020). https://doi.org/10.1080/19420862.2019.1702263.

[46] L. Wang, H.K. Trang, J. Desai, Z.D. Dunn, D.D. Richardson, R.K. Marcus, Fiber-based HIC capture loop for coupling of protein A and size exclusion chromatography in a two-dimensional separation of monoclonal antibodies, Anal. Chim. Acta. 1098 (2020) 190–200. https://doi.org/10.1016/j.aca.2019.11.023.

[47] A. Williams, E.K. Read, C.D. Agarabi, S. Lute, K.A. Brorson, Automated 2D-HPLC method for characterization of protein aggregation with in-line fraction collection device, J. Chromatogr. B. 1046 (2017) 122–130. https://doi.org/10.1016/j.jchromb.2017.01.021.

[48] K. Sandra, M. Steenbeke, I. Vandenheede, G. Vanhoenacker, P. Sandra, The versatility of heart-cutting and comprehensive two-dimensional liquid chromatography in monoclonal antibody clone selection, J. Chromatogr. A. 1523 (2017) 283–292. https://doi.org/10.1016/j.chroma.2017.06.052.

[49] P. Quinto Tranchida, R. Costa, P. Donato, D. Sciarrone, C. Ragonese, P. Dugo, G. Dugo, L. Mondello, Acquisition of deeper knowledge on the human plasma fatty acid profile exploiting comprehensive 2-D GC, J. Sep. Sci. 31 (2008) 3347–3351. https://doi.org/10.1002/jssc.200800289.

[50] M. Adahchour, J. Beens, R. Vreuls, A. Batenburg, U. Brinkman, Comprehensive two-dimensional gas chromatography of complex samples by using a "reversed-type" column combination: application to food analysis, J. Chromatogr. A. 1054 (2004) 47–55. https://doi.org/10.1016/S0021-9673(04)01288-9.

[51] V. Elsner, S. Laun, D. Melchior, M. Köhler, O.J. Schmitz, Analysis of fatty alcohol derivatives with comprehensive two-dimensional liquid chromatography coupled with mass spectrometry, J. Chromatogr. A. 1268 (2012) 22–28. https://doi.org/10.1016/j.chroma.2012.09.072.

[52] D.R. Stoll, D.C. Harmes, G.O. Staples, O.G. Potter, C.T. Dammann, D. Guillarme, A. Beck, Development of comprehensive online two-dimensional liquid chromatography-mass spectrometry using hydrophilic interaction and reversed-phase separations for rapid and deep profiling of therapeutic antibodies, Anal. Chem. 90 (2018) 5923–5929. https://doi.org/10.1021/acs.analchem.8b00776.

[53] H. Luo, W. Zhong, J. Yang, P. Zhuang, F. Meng, J. Caldwell, B. Mao, C.J. Welch, 2D-LC as an on-line desalting tool allowing peptide identification directly from MS unfriendly HPLC methods, J. Pharm. Biomed. Anal. 137 (2017) 139–145. https://doi.org/10.1016/j.jpba.2016.11.012.

[54] Y. Li, C. Gu, J. Gruenhagen, K. Zhang, P. Yehl, N.P. Chetwyn, C.D. Medley, A size exclusion-reversed phase two dimensional-liquid chromatography methodology for stability and small molecule related species in antibody drug conjugates, J. Chromatogr. A. 1393 (2015) 81–88. https://doi.org/10.1016/j.chroma.2015.03.027.

[55] S. Bäurer, W. Guo, S. Polnick, M. Lämmerhofer, Simultaneous separation of water- and fat-soluble vitamins by selective comprehensive HILIC×RPLC (high-resolution sampling) and active solvent modulation, Chromatographia. 82 (2019) 167–180. https://doi.org/10.1007/s10337-018-3615-0.

[56] U. Woiwode, S. Neubauer, W. Lindner, S. Buckenmaier, M. Lämmerhofer, Enantioselective multiple heartcut two-dimensional ultra-high-performance liquid chromatography method with a Coreshell chiral stationary phase in the second dimension for analysis of all proteinogenic amino acids in a single run, J. Chromatogr. A. 1562 (2018) 69–77. https://doi.org/10.1016/j.chroma.2018.05.062.

[57] C.E. Dent, A study of the behaviour of some sixty amino-acids and other ninhydrin-reacting substances on phenol-"collidine" filter-paper chromatograms, with notes as to the occurrence of some of them in biological fluids, Biochem. J. 43 (1948) 169–180. https://doi.org/10.1042/bj0430169.

[58] L.R. Snyder, Comparisons of normal elution, coupled-columns, and solvent, flow or temperature programming in liquid chromatography, J. Chromatogr. Sci. 8 (1970) 692–706. https://doi.org/10.1093/chromsci/8.12.692.

[59]   B. Karger, L. Snyder, C. Horvath, Multistep separation schemes for complex samples, in: An Introduction to Separation Science, 1st ed., John Wiley & Sons, Inc., New York, 1973: pp. 558–562.

[60]   J.C. Giddings, Concepts and comparisons in multidimensional separation, HRC CC, J. High Resolut. Chromatogr. Commun. 10 (1987) 319–323. https://doi.org/10.1002/jhrc.1240100517.

[61]   J.C. Giddings, Two-dimensional separations: Concept and promise, Anal. Chem. 56 (1984) 1258A–1270A. https://doi.org/10.1021/ac00276a003.

[62]   G. Guiochon, M.F. Gonnord, M. Zakaria, L.A. Beaver, A.M. Siouffi, Chromatography with a two-dimensional column, Chromatographia. 17 (1983) 121–124. https://doi.org/10.1007/BF02271033.

[63]   G. Guiochon, L.A. Beaver, M.F. Gonnord, A.M. Siouffi, M. Zakaria, Theoretical investigation of the potentialities of the use of a multidimensional column in chromatography, J. Chromatogr. A. 255 (1983) 415–437. https://doi.org/10.1016/S0021-9673(01)88298-4.

[64]   M.M. Bushey, J.W. Jorgenson, Automated instrumentation for comprehensive two-dimensional high-performance liquid chromatography of proteins, Anal. Chem. 62 (1990) 161–167. https://doi.org/10.1021/ac00201a015.

[65]   A.W. Moore, J.W. Jorgenson, Comprehensive three-dimensional separation of peptides using size exclusion chromatography/reversed phase liquid chromatography/optically gated capillary zone electrophoresis, Anal. Chem. 67 (1995) 3456–3463. https://doi.org/10.1021/ac00115a014.

[66]   G.J. Opiteck, S.M. Ramirez, J.W. Jorgenson, M.A. Moseley, Comprehensive two-dimensional high-performance liquid chromatography for the isolation of overexpressed proteins and proteome mapping, Anal. Biochem. 258 (1998) 349–361. https://doi.org/10.1006/abio.1998.2588.

[67]   D.A. Wolters, M.P. Washburn, J.R. Yates, An automated multidimensional protein identification technology for shotgun proteomics, Anal. Chem. 73 (2001) 5683–5690. https://doi.org/10.1021/ac010617e.

[68]   M.R. Schure, Limit of detection, dilution factors, and technique compatibility in multidimensional separations utilizing chromatography, capillary electrophoresis, and field-flow fractionation, Anal. Chem. 71 (1999) 1645–1657. https://doi.org/10.1021/ac981128q.

[69]   R.E. Murphy, M.R. Schure, J.P. Foley, Effect of sampling rate on resolution in comprehensive two-dimensional liquid chromatography, Anal. Chem. 70 (1998) 1585–1594. https://doi.org/10.1021/ac971184b.

[70]   M.R. Schure, J.M. Davis, Orthogonal separations: Comparison of orthogonality metrics by statistical analysis, J. Chromatogr. A. 1414 (2015) 60–76. https://doi.org/10.1016/j.chroma.2015.08.029.

[71]   M. Gilar, J. Fridrich, M.R. Schure, A. Jaworski, Comparison of orthogonality estimation methods for the two-dimensional separations of peptides, Anal. Chem. 84 (2012) 8722–8732. https://doi.org/10.1021/ac3020214.

[72]   F. Bedani, P.J. Schoenmakers, H.-G. Janssen, Theories to support method development in comprehensive two-dimensional liquid chromatography – A review, J. Sep. Sci. 35 (2012) 1697–1711. https://doi.org/10.1002/jssc.201200070.

[73]   D.R. Stoll, X. Wang, P.W. Carr, Comparison of the practical resolving power of one- and two-dimensional high-performance liquid chromatography analysis of metabolomic samples, Anal. Chem. 80 (2008) 268–278. https://doi.org/10.1021/ac701676b.

[74]   Y. Huang, H. Gu, M. Filgueira, P.W. Carr, An experimental study of sampling time effects on the resolving power of on-line two-dimensional high performance liquid chromatography, J. Chromatogr., A. 1218 (2011) 2984–2994. https://doi.org/10.1016/j.chroma.2011.03.032.

[75]   L.W. Potts, P.W. Carr, Analysis of the temporal performance of one versus on-line comprehensive two-dimensional liquid chromatography, J. Chromatogr., A. 1310 (2013) 37–44. https://doi.org/10.1016/j.chroma.2013.07.102.

[76]   M. Sarrut, A. D'Attoma, S. Heinisch, Optimization of conditions in on-line comprehensive two-dimensional reversed phase liquid chromatography. Experimental comparison with one-dimensional reversed phase liquid chromatography for the separation of peptides, J. Chromatogr. A. 1421 (2015) 48–59. https://doi.org/10.1016/j.chroma.2015.08.052.

[77]   M. Sarrut, F. Rouvière, S. Heinisch, Theoretical and experimental comparison of one dimensional versus on-line comprehensive two dimensional liquid chromatography for optimized sub-hour separations of complex peptide samples, J. Chromatogr. A. 1498 (2017) 183–195. https://doi.org/10.1016/j.chroma.2017.01.054.

[78] G. Vivó-Truyols, Sj. van der Wal, P.J. Schoenmakers, Comprehensive study on the optimization of online two-dimensional liquid chromatographic systems considering losses in theoretical peak capacity in first and second dimensions: A Pareto-optimality approach, Anal. Chem. 82 (2010) 8525–8536. https://doi.org/10.1021/ac101420f.

[79] B.W.J. Pirok, A.F.G. Gargano, P.J. Schoenmakers, Optimizing separations in on-line comprehensive two-dimensional liquid chromatography, J. Sep. Sci. 41 (2017) 68–98. https://doi.org/10.1002/jssc.201700863.

[80] M. Muller, A.G.J. Tredoux, A. de Villiers, Predictive kinetic optimisation of hydrophilic interaction chromatography×reversed phase liquid chromatography separations: Experimental verification and application to phenolic analysis, J. Chromatogr. A. 1571 (2018) 107–120. https://doi.org/10.1016/j.chroma.2018.08.004.

[81] B.W.J. Pirok, D. Stoll R., P.J. Schoenmakers, Recent developments in two-dimensional liquid chromatography – Fundamental improvements for practical applications, Anal. Chem. 91 (2019) 240–263. https://doi.org/10.1021/acs.analchem.8b04841.

[82] S.H. Yang, J. Wang, K. Zhang, Validation of a two-dimensional liquid chromatography method for quality control testing of pharmaceutical materials, J. Chromatogr. A. 1492 (2017) 89–97. https://doi.org/10.1016/j.chroma.2017.02.074.

[83] E. Largy, A. Catrain, G. Van Vynckt, A. Delobel, 2D-LC–MS for the analysis of monoclonal antibodies and antibody–drug conjugates in a regulated environment, Current Trends in Mass Spectrometry 14 (2016) 29–35.

[84] E.S. Talus, K.E. Witt, D.R. Stoll, Effect of pressure pulses at the interface valve on the stability of second dimension columns in online comprehensive two-dimensional liquid chromatography, J. Chromatogr. A. 1378 (2015) 50–57. https://doi.org/10.1016/j.chroma.2014.12.019.

[85] D.R. Stoll, H.R. Lhotka, D.C. Harmes, B. Madigan, J.J. Hsiao, G.O. Staples, High resolution two-dimensional liquid chromatography coupled with mass spectrometry for robust and sensitive characterization of therapeutic antibodies at the peptide level, J. Chromatogr. B. 1134–1135 (2019). https://doi.org/10.1016/j.jchromb.2019.121832.

[86] B.W.J. Pirok, S. Pous-Torres, C. Ortiz-Bolsico, G. Vivó-Truyols, P.J. Schoenmakers, Program for the interpretive optimization of two-dimensional resolution, J. Chromatogr. A. 1450 (2016) 29–37. https://doi.org/10.1016/j.chroma.2016.04.061.

[87] D. Stoll, Introduction to two-dimensional liquid chromatography – Theory and practice, in: M. Holcapek, Wm.C. Byrdwell (Eds.), Handbook of Advanced Chromatography /Mass Spectrometry Techniques, Elsevier, London, 2017: pp. 227–286.

[88] C.J. Gstöttner, D. Klemm, M. Haberger, A. Bathke, H. Wegele, C.H. Bell, R. Kopf, Fast and automated characterization of antibody variants with 4D-HPLC/MS, Anal. Chem. 90 (2017) 2119–2125. https://doi.org/10.1021/acs.analchem.7b04372.

[89] B.W.J. Pirok, N. Abdulhussain, T. Aalbers, B. Wouters, R.A.H. Peters, P.J. Schoenmakers, Nanoparticle analysis by online comprehensive two-dimensional liquid chromatography combining hydrodynamic chromatography and size-exclusion chromatography with intermediate sample transformation, Anal. Chem. 89 (2017) 9167–9174. https://doi.org/10.1021/acs.analchem.7b01906.

[90] P.J. Schoenmakers, G. Vivó-Truyols, W.M.C. Decrop, A protocol for designing comprehensive two-dimensional liquid chromatography separation systems, J. Chromatogr. A. 1120 (2006) 282–290. https://doi.org/10.1016/j.chroma.2005.11.039.

[91] D.R. Stoll, X. Li, X. Wang, P.W. Carr, S.E.G. Porter, S.C. Rutan, Fast, comprehensive two-dimensional liquid chromatography, J. Chromatogr., A. 1168 (2007) 3–43. https://doi.org/10.1016/j.chroma.2007.08.054.

[92] R.E. Murphy, M.R. Schure, Method development in multidimensional liquid chromatography, in: M. Schure, S.A. Cohen (Eds.), Multidimensional Liquid Chromatography: Theory and Applications in Industrial Chemistry and the Life Sciences, Wiley-Interscience, Hoboken, NJ, 2008: pp. 127–145.

[93] M. Pursch, S. Buckenmaier, Loop-based multiple heart-cutting two-dimensional liquid chromatography for target analysis in complex matrices, Anal. Chem. 87 (2015) 5310–5317. https://doi.org/10.1021/acs.analchem.5b00492.

[94] D.F. Thekkudan, S.C. Rutan, P.W. Carr, A study of the precision and accuracy of peak quantification in comprehensive two-dimensional liquid chromatography in time, J. Chromatogr., A. 1217 (2010) 4313–4327. https://doi.org/10.1016/j.chroma.2010.04.039.

[95]   J.M. Davis, S.C. Rutan, P.W. Carr, Relationship between selectivity and average resolution in comprehensive two-dimensional separations with spectroscopic detection, J. Chromatogr. A. 1218 (2011) 5819–5828. https://doi.org/10.1016/j.chroma.2011.06.086.

[96]   M.M. Bushey, J.W. Jorgenson, Automated instrumentation for comprehensive two-dimensional high-performance liquid chromatography/capillary zone electrophoresis, Anal. Chem. 62 (1990) 978–984. https://doi.org/10.1021/ac00209a002.

[97]   L. Ranjbar, J.P. Foley, M.C. Breadmore, Multidimensional liquid-phase separations combining both chromatography and electrophoresis – A review, Anal. Chim. Acta. 950 (2017) 7–31. https://doi.org/10.1016/j.aca.2016.10.025.

[98]   A.C. Johnson, M.T. Bowser, High-speed, comprehensive, two dimensional separations of peptides and small molecule biological amines using capillary electrophoresis coupled with micro free flow electrophoresis, Anal. Chem. 89 (2017) 1665–1673. https://doi.org/10.1021/acs.analchem.6b03768.

[99]   M. Geiger, N.W. Frost, M.T. Bowser, Comprehensive multidimensional separations of peptides using nano-liquid chromatography coupled with micro free flow electrophoresis, Anal. Chem. 86 (2014) 5136–5142. https://doi.org/10.1021/ac500939q.

[100] L. Gao, J. Zhang, W. Zhang, Y. Shan, Z. Liang, L. Zhang, Y. Huo, Y. Zhang, Integration of normal phase liquid chromatography with supercritical fluid chromatography for analysis of fruiting bodies of Ganoderma lucidum, J. Sep. Science. 33 (2010) 3817–3821. https://doi.org/10.1002/jssc.201000453.

[101] H.J. Cortes, R.M. Campbell, R.P. Himes, C.D. Pfeiffer, On-line coupled liquid chromatography and capillary supercritical fluid chromatography: Large-volume injection system for capillary SFC, J. Microcolumn Sep. 4 (1992) 239–244. https://doi.org/10.1002/mcs.1220040310.

[102] M. Iguiniz, E. Corbel, N. Roques, S. Heinisch, On-line coupling of achiral reversed phase liquid chromatography and chiral supercritical fluid chromatography for the analysis of pharmaceutical compounds, J. Pharm. Biomed. Anal. 159 (2018) 237–244. https://doi.org/10.1016/j.jpba.2018.06.058.

[103] M. Goel, E. Larson, C.J. Venkatramani, M.A. Al-Sayah, Optimization of a two-dimensional liquid chromatography-supercritical fluid chromatography-mass spectrometry (2D-LC-SFC-MS) system to assess "in-vivo" inter-conversion of chiral drug molecules, J. Chromatogr. B. 1084 (2018) 89–95. https://doi.org/10.1016/j.jchromb.2018.03.029.

[104] C.J. Venkatramani, M. Al-Sayah, G. Li, M. Goel, J. Girotti, L. Zang, L. Wigman, P. Yehl, N. Chetwyn, Simultaneous achiral-chiral analysis of pharmaceutical compounds using two-dimensional reversed phase liquid chromatography-supercritical fluid chromatography, Talanta. 148 (2016) 548–555. https://doi.org/10.1016/j.talanta.2015.10.054.

[105] M. Sun, M. Sandahl, C. Turner, Comprehensive on-line two-dimensional liquid chromatography×supercritical fluid chromatography with trapping column-assisted modulation for depolymerised lignin analysis, J. Chromatogr. A. 1541 (2018) 21–30. https://doi.org/10.1016/j.chroma.2018.02.008.

[106] R.E. Majors, Multidimensional high performance liquid chromatography., J. Chromatogr. Sci. 18 (1980) 571–579.

[107] G. Purcaro, S. Moret, L. Conte, Sample pre-fractionation of environmental and food samples using LC-GC multidimensional techniques, Trends Analyt. Chem. 43 (2013) 146–160. https://doi.org/10.1016/j.trac.2012.10.007.

2 Speed and Performance in Liquid Chromatography

Peter W. Carr and Dwight R. Stoll

CONTENTS

DOI: 10.1201/9781003090557-2

The fastest possible analysis with any column design will be achieved with a column operated at the maximum possible pressure drop and having a length as to give the plate number necessary to perform the desired separation.

G. Guiochon [1]

2.1 INTRODUCTION

In this chapter, we will deal with the factors that control the speed and resolving power of liquid chromatography, including important aspects of gradient elution chromatography. The focus of this chapter is 1D separations; this knowledge can be leveraged in the two dimensions of 2D-LC separations independently. However, many of the factors that dictate the performance of 2D separations are interdependent. These interactions are discussed separately, primarily in Chapters 3, 6, and 7. Considerable attention must be paid to the factors controlling the speed of 1D-LC because it is clear that the overall speed and resolving power of 2D-LC, especially that of comprehensive 2D-LC, are critically dependent on the speed of the second dimension system, as is discussed in Section 3.5. Speed in LC is often gauged as the ratio of column dead time (t_0) to the plate count (N). Clearly, it is important that we understand the factors that control both analysis time and resolving power. As Guiochon has reminded us, the only two ways to pay for resolving power are by increasing analysis time or increasing operating pressure [1].

The above quote from Guiochon is truly the "Rosetta Stone" for designing fast, isocratic 1D chromatographic separations and therefore 2D-LC as well. Consequently, much of this chapter is aimed at putting meat on the bones of the skeleton of his statement. Filling out the bare bones requires adding a bit more detail to the quotation. Specifically, *column design* means a column with a given material in the column that is being run at the maximum possible system pressure where the mobile phase velocity and column length have been adjusted to give the desired plate count and resolution within the shortest possible analysis time. The process of optimizing speed and resolution is complicated because there are many strong interactions between the operating variables that impact speed and resolution. These variables include the mobile phase velocity, column length, particle size (and type), as well as the mobile phase composition and temperature, which control retention and viscosity, and the solute's diffusivity in the mobile and stationary phases. These parameters influence both the plate number and the pressure drop across the column. An often-overlooked consideration that influences speed in 1D-LC is the selectivity (α) of the separation. It should be evident that the higher is the chromatographic selectivity of the "critical pair" of peaks, that is, the pair of peaks having the poorest resolution [2], the less time the separation will require.

Before proceeding, we want to remind the reader as stated in the preface to the book that Chapter 2 has been written for those who are familiar with the basic principles of the dynamic aspects of peak broadening. Thus, we assume knowledge of the van Deemter and related equations that describe the various diffusive and convective processes that are responsible for peak broadening. We also assume knowledge of the relationships between flow and pressure (e.g., the Kozeny-Carman equation), and the factors such as mobile phase viscosity, packing particle size, temperature, and so on, that determine the pressure drop across the column. Similarly, because peak broadening depends so strongly on the analyte's molecular diffusion coefficient (D_m), we assume knowledge of the Wilke-Chang equation for estimating this property. Much of the assumed background is presented and discussed in the appendix that immediately follows this chapter (Appendix A). Where necessary we will refer to equations in the appendix as Eq. A.1, A.2, and so on. For those not conversant with these issues we suggest that before diving into Chapter 2 they become familiar with the material in the appendix.

2.2   OVERVIEW OF ASPECTS OF PERFORMANCE OPTIMIZATION

When one sets out to optimize an LC separation, they must decide which column technologies they will leverage (e.g., small particles and high pressure, high temperatures), and what optimization procedure (e.g., one-, two-, or three-parameter optimization, kinetic plots) they will use to identify optimal conditions for their application. In this section we provide an overview of these aspects before discussing them in detail in Sections 2.3 and 2.4. An overview of the various approaches to improve performance (speed, resolution) in LC and the system parameters that can be adjusted is given in Table 2.1. This table is discussed in some detail below.

To understand the basic principles that determine speed in LC there are three fundamental issues to keep in mind. First, is the relationship between the height equivalent to a theoretical plate (*HETP* or *H* for short, see Glossary) and the *interstitial* mobile phase velocity (u_e). This controls the number of theoretical plates on a column of given length and particle size. Second, we need to know the relationship between the required pressure to operate a column of length (*L*) packed with particles of a given size (d_p). Last, we need to account for the relationship between column dead time, column length and mobile phase velocity.

There are fundamentally three different frameworks we can use to optimize the separation process:

a. **One-parameter or van Deemter optimization** – For a column of arbitrarily fixed length and particle size, dismissing any concern for both the required pressure drop and desired analysis time, we want to know the mobile phase velocity (*u*) that optimizes the plate number. This is the velocity at which *H* is a minimum [3]. This is not a practical approach because the required system pressure, volumetric flow rate, and desired analysis time may not be achievable with the column and system hardware at hand.

b. **Two-parameter or Poppe optimization** – For some arbitrary but pre-selected particle size, maximum available pressure drop (P_{max}), and fixed column dead time one simultaneously varies both the column length and the eluent velocity to maximize *N*. This is the approach espoused by H. Poppe [4]. The Poppe type optimization is the optimization method that leads to the most practical recommendations. One chooses a commercially available particle size and/or type (e.g., totally porous particles (TPPs) vs. superficially porous particles (SPPs)). However, the two-parameter approach does make the assumption that one can obtain columns in exactly the optimum length. Fortunately, using a column within about 1 cm of the precisely optimum length makes little difference except for when very short columns are needed (see below). As we will see *the best mobile phase velocity under these conditions is usually not the same as the van Deemter optimum velocity that minimizes H.*

c. **Three-parameter or Knox-Saleem-Halasz (K-S-H) optimization** – In this approach, one co-optimizes the particle size, column length, and mobile phase velocity. All three constraints (minimized analysis time, maximized plate count, and using all the available pressure drop) are satisfied simultaneously. Clearly, this is a three-parameter optimization; it is the approach most clearly described by Knox and Saleem [5], mathematically proven by Guiochon using the method of LaGrange multipliers [1], later described in more detail by Halasz and Gorlitz

TABLE 2.1
Various Ways to Increase Speed and Resolution in One-dimensional Liquid Chromatography

Characteristics	Higher Pressure	Higher Temperature	Superficially Porous Particles	Non-Porous Particles	Monoliths	Chemometric Curve Deconvolution
Primary Effect on Speed	Drives fluid faster. Used in conjunction with smaller particles and narrower columns (2.1 mm or less).	Reduces viscosity and thus pressure drop; drives fluid faster at same pressure. Improves interphase mass transfer. Can be combined with smaller particles.	Lower reduced plate height through improved eddy dispersion, improved interphase mass transfer. Narrower particle size distribution.	Greatly improved stationary phase mass transfer.	Minimal. Generally low flow resistance; best for producing very high plate numbers through use of long columns.	Decreases resolution needed for analytical purposes and thus can speed up analysis considerably.
Requirements	When used with smaller particles and narrower columns requires re-engineering of system to minimize extra-column broadening.	Generally requires narrow columns (2.1 mm or less). No PEEK in columns.	Improved extra-column dispersion and high-speed detectors.	Performance is greatly enhanced by use of very small ($\ll 2$ μm) particles and high pressures.	Conventional LC but at high flow rate.	Requires extreme reproducibility in peak shape and retention time.
Equipment	Requires high pressure instrument, fast detectors.	Requires appropriate mobile phase pre-heater and column temperature control.	Low extra-column dispersion is almost as important as with sub-2 μm TPPs.	Generally requires high pressure, low-dispersion equipment.	No special needs.	No special needs.
Equipment Cost	High	Low	Low	High when very small particles are used.	Low	Low
Effect on relative retention	Can be significant for large molecules.	Considerable	Minimal	Minimal	None	None
Mass Loadability	Good	Good	Somewhat less	Much less	Good	N/A

Available Phases	Numerous	Numerous	Very limited		Limited	N/A
Major Advantages	Can improve speed significantly. Due to use of narrower and shorter columns, solvent consumption is decreased.	Can be done with conventional HPLC equipment.			Can easily serially couple columns to increase column length.	No real changes needed in instrumentation.
Major Disadvantages	Cost of equipment. Generally, requires narrow column (2.1 mm or less).	Slightly decreased mass loadability and retention factors.	Very low mass loadability and much less retention. Very small particles are hard to pack well.	Requires stable mobile and stationary phases, and analytes.	Generally low flow resistance; best for producing very high plate numbers. Not many choices in stationary phases.	Mathematically complex. Must have very precise retention to fully leverage some algorithms used.

[6], and then extended by Meyer [7] to gradient elution chromatography. Most interestingly, *the best mobile phase velocity to use in this case turns out to be the van Deemter optimum velocity – i.e., the velocity that gives the van Deemter minimum H.* In principle, this approach gives the *absolute maximum possible plate count* (performance, resolution) *in the shortest possible time*, or conversely the *shortest possible analysis time needed to achieve a fixed plate count* while using the maximum pressure drop supported by the instrument. Obviously, as required by this type of optimization, we cannot do so by using only commercial particle sizes or column lengths as only a few particle sizes are commercially available. However, recent work by Matula [8] has shown that the difference between the two- and three-parameter optimization approaches is at most only a few percent (5–10% in plate count at a given dead time) *provided that when using the two-parameter approach we change from the smaller to the larger particle diameter at the right system t_0.*

In the early days of chromatography, many others contributed significantly to our understanding of performance and speed in both GC and LC, most especially Giddings [9], and Guiochon [1]. More recently, we [10, 11] and particularly Desmet and his group [12–16] through their development of kinetic plots have added to the understanding and limitations of the various optimization processes. Major stumbling blocks to understanding how one should optimize performance are the very different results obtained when performance is optimized with freely varying particle size (three-parameter optimization) and when only the column length and velocity are varied while using an arbitrary pre-set particle size (i.e., two-parameter optimization). These differences will be discussed in detail below. Their consequences cannot be overstated.

The past 15 years or so have seen a series of major developments aimed at improving both the efficiency (N, plate number) and the speed (t_0/N, time to generate a unit of efficiency) of LC. In many ways, both the basic instrumentation and column technology of LC have changed significantly after a nearly two-decades-long plateau wherein only a series of incremental improvements took place. Numerous approaches to increasing the speed of LC have evolved during this period including the development of sub-3 μm particles (both TPP and SPP), and various types of monolithic columns comprised of either a single piece of organic polymer [17, 18] or a porous silica-based rod [19, 20]. Both significantly higher pressure [21, 22] and higher temperature [23–25] operation of chemically and thermally stable columns [26, 27] have also appeared during this period.

Many of these new column technologies have mandated extensive instrument modifications to enable full utilization of the potential of the smaller particles, including significantly increased pressure capabilities, dramatic reductions in extra-column dispersion, improved (lower dead-volume) fittings, and lower diameter connection tubing, as well as faster and lower volume detectors. Indeed, many of these developments were anticipated in Giddings's seminal paper [9] in which he studied the limits of separation speed and resolution in GC and LC. He showed that increasing the system pressure up to some critical pressure could either increase the plate count, at a specified analysis time, or decrease analysis time at a given plate count. Quite importantly, Giddings made clear that "the comparative speed of separation (that is of LC vs. GC) depends, to a large extent, on the relative viscosity and diffusivity of liquids and gases" [9]. The positive impact of using elevated temperatures has been obvious for quite some time and may well have led Giddings to look at the possibilities inherent in supercritical fluid chromatography. The recognition of the drastic dependence of the optimized analysis time on particle diameter and system pressure dates to the seminal work of Knox [28–30], Halasz [31], and Purnell [32].

In the following sections we discuss in detail several different approaches to improving speed and efficiency in LC.

2.3 IMPACT OF COLUMN AND INSTRUMENT TECHNOLOGIES ON SEPARATION PERFORMANCE

2.3.1 SMALLER PARTICLES AND HIGHER PRESSURES

Many believe that using smaller particles, all else being equal, *must invariably improve perform-ance in LC*. It is generally true that the particle size required to yield a specific number of plates in a given analysis time decreases as the target analysis time is decreased. However, one must also take into account that while working at a fixed pressure equal to the maximum available system pressure, the mobile phase velocity and column length must both be adjusted (i.e., re-optimized) to keep the pressure constant as one switches to a smaller particle size. To fully capitalize on the potential gain in speed provided by using smaller packing materials requires a concomitant increase in the avail-able pressure drop provided by the pump. When the available pressure drop is increased, it is the combination of higher pressure drops and smaller particles that leads to significant gains in speed, not merely the use of smaller particles *per se*. This has been the chief driving force for the dramatic increases in maximum pressure capabilities of commercial systems over the last two decades (i.e., 400 to 1,500 bar).

The pioneering work of Jorgenson [33, 34] and his coworkers on ultra-high pressure LC was followed by Waters Corporation's introduction in 2004 of equipment capable of exceeding the long-standing 400-bar pressure limit of commercial LC hardware. Indeed, currently several manufacturers now offer "ultra-high pressure" equipment (i.e., UHPLC systems) along with a suite of stationary phase chemistries on sub-2 μm particles. It is now clear that the advantages of smaller TPPs for faster separations simply cannot be realized without also improving the hardware by greatly decreasing its extra-column contribution to band broadening relative to that which has been common for the past 20 years. The need to use different LC instrumentation is the foremost barrier to the wider, more routine use of sub-2 μm particles. Another challenge to the use of even smaller particles under UHPLC conditions is the intra-column band dispersion due to the viscous (frictional) heating of the mobile phase. Halasz [35] appears to have been the first to quantify the effect of frictional heating and describe its impact on peak broadening. He indicated that the *ultimate limit of speed in LC is in large measure due to the viscous heating problem* and that particles smaller than 1 μm would probably not be useful except in very narrow capillaries (1 mm in diameter, or less). Clearly, we are rapidly approaching this limit.

The Wirth group has leaped over the 1 μm barrier to the use of highly monodisperse 300 and 500 nm colloidal particles packed in ordered (crystal-like) arrays in capillaries on the order of 100 μm i.d. Under slip-flow conditions, which has been demonstrated for these columns under some conditions [36–38], column permeability is enhanced and concomitantly the pressure required to drive the mobile phase through the column is greatly reduced.

Under perfect adiabatic column conditions both the axial and radial temperature gradients increase in proportion to the pressure drop; however, it is the radial gradient that has the greatest influence on peak broadening. For a mobile phase with a heat capacity and density between that of water and methanol, the axial temperature change under *adiabatic* conditions is between 0.02 and 0.05 °C/bar. Thus, operating at a pressure of 400 bar can easily generate a temperature differential upwards of 10 °C along the length of a typical analytical LC column. This axial temperature differ-ential does not lead to serious peak broadening *per se*, though it does lead to systematic variations in measured retention volume as the flow rate is varied. The real source of peak broadening is the radial temperature gradient that causes radial gradients in the local axial analyte zone velocity. These are due in part to the thermally induced radial gradient in mobile phase viscosity [39–41], but more so to radial variation in analyte retention [24, 25]. Generally, more strongly retained analytes have larger enthalpies of retention and thus are more sensitive to these issues [42].

The best way to deal with the frictional heating induced dispersion issue is to use narrower diameter columns. This is precisely what is done in high resolution electrophoresis by using capillary tubes [43]. Consequently, columns packed with small (< 2 μm) particles are almost always used in columns that are 2.1 mm in diameter and are frequently shorter in length than columns with bigger particles (e.g., > 3 μm); as a result, peaks produced by these columns are exceedingly narrow in time units, and low in volume (see Table 2.2 in Section 2.4.1 below). Recalling that the extra-column broadening must be less than about 25% of the peak width in order that N not be diminished by more than 10%, extra-column effects must be very small. This is the principal reason for the need to re-design all LC hardware including injectors, tubing, fittings, and detectors when one wants to use very small particles in short, narrow columns to do high speed LC. Indeed, although use of SPPs partially ameliorates the pressure and self-heating problems because larger particles can be used to obtain plate heights close to those obtained with smaller TPPs, they still require use of an LC system with rather low extra-column volumes when 2.1 mm (or smaller) diameter columns are used.

2.3.2 HIGHER TEMPERATURES

Giddings' seminal paper on the relative speed and performance of LC and GC [9] made clear that the key reason why GC is inherently faster than LC is because analyte diffusion is faster in gases than in liquids. The main reason that elevated temperatures can be used to improve the speed of LC is that increasing the mobile phase temperature decreases its viscosity, thereby both decreasing the pressure drop across the column, and increasing analyte diffusivity in the mobile and stationary phases. The increase in diffusion speed is beneficial when velocities are higher than the van Deemter optimum velocity. Obviously, elevated temperature LC is, in terms of speed, more like GC, and less like low temperature LC. Increased analyte diffusivity decreases the slope of the C-branch (i.e., the high velocity region) of the van Deemter curve, allowing one to work at higher eluent velocities without the serious loss of plate count (N) generally associated with use of mobile phase velocities higher than the van Deemter optimum velocity. We refer the interested reader to the Appendix and especially plots of the van Deemter equation (see Eqs. A.1–A.3 and Figure A.1) for discussion of the C-term. Additionally, the lower mobile phase viscosity allows the use of both longer columns and/or smaller particles at the same pressure, or to use the same column length and particle diameter operated at a higher velocity, as compared to the use of near-ambient temperatures.

Since the pioneering work of Horvath's group [23] on elevated temperatures in LC, a variety of thermally stable stationary phases (e.g., improved stability silica bonded phases, and other metal oxide phases, typically coated with polymers) have become commercially available [44]. However, in contradistinction to using higher pressures and smaller particles, elevated temperature LC can be done with conventional LC hardware augmented with a heating system that minimizes the effect of mobile phase/column thermal mismatch broadening. The thermal mismatch problem was studied quite some time ago by Poppe and others [24, 25]. It is vital that the eluent entering the column be within about 5 K of the desired column temperature. The book by Teutenberg reviews the essential aspects of using higher temperatures in LC [44]. Some of the key issues are related to the stationary and mobile phases, and analyte stability [26, 45]. Many users needlessly suffer from severe "hyper-thermophobia", but as was shown by Horvath [23] analyte instability is not really a big problem in the context of fast separations because the analytes are only exposed to higher temperatures for quite short times. Indeed, in 2D-LC work it is the second dimension column that benefits the most by being operated at higher temperatures, perhaps even greater than 100 °C, but under optimal conditions the analyte residence time in the hot zone is often only on the order of 20 to 60 s [46].

2.3.3 SUPERFICIALLY POROUS PARTICLES (SPPs)

The idea of using SPPs, which were initially called pellicular particles, was motivated by the desire to improve the speed of LC by using larger particles than sub-2 μm TPPs without increasing peak broadening due to slow stationary phase mass inside the particle, and concomitantly avoid the need for ultra-high pressures. These objectives can be satisfied simultaneously by using a particle with a solid core surrounded by a porous layer, or shell (also called a "pellicle"). The synthesis and use of such particles was pioneered by C. Horvath, and J.J. Kirkland in the early days of HPLC [47, 48]. Mass transfer inside the particle, which is putatively strictly diffusive (not even slightly convective, except when the pores in the particles are quite large relative to the particle diameter), only takes place in the thin, porous shell, which is typically ¼ to ½ of the overall particle diameter, and thus should be much faster than in a TPP of the same size, thereby decreasing the stationary phase mass transfer resistance contribution to the plate height (see discussion of C_s term in Appendix section A.4). Of course, retention and analyte loadability are smaller for SPPs compared to TPPs of the same outer diameter, but with reasonable shell/core ratios adequate retention and acceptable loadability are obtained for analytical, if not preparative purposes. Interest in these particles has grown tremendously over the past 15 years following the introduction of sub-3 μm SPPs by Advanced Materials Technologies. Using these small SPPs, reduced plate heights of 1.5 (or less) in 4.6 mm i.d. columns have been reported; this was a significant advance because well-packed columns with conventional TPPs seldom yield reduced plate heights below 2.0 [49]. Consequently, sub-3 μm SPPs often perform as well as sub-2 μm TPPs but at lower operating pressures [50–54]. Additional SPP particles have since been introduced by many mainstream manufacturers.

Many factors are thought to contribute to the improved performance of SPPs:

1. The particle size distribution of SPPs is narrower, sometimes only 5% based in the width of the distribution relative to the mean, compared to that of high quality TPPs, typically about 19% [53]. This better distribution might well improve radial and axial packed bed homogeneity and thereby decrease the van Deemter A-term; however, this remains a controversial issue [54].

2. The decreased internal particle porosity and the intrinsic geometry (i.e., pore connectivity) of SPPs inhibits longitudinal diffusion of the analyte in the stationary phase and thus can give smaller van Deemter B-coefficients [50, 55]. This characteristic might well be a highly advantageous feature because it enables the use of longer, possibly coupled, columns to obtain higher plate counts at low velocities to improve resolution for highly complex samples and might be competitive with monolithic columns for this purpose [50].

   As discussed above, the principal motivation for development of SPPs was to restrict the analyte diffusional equilibration time *to the porous shell region* to decrease it relative to that in a TTP. In principle, this should speed up the stationary phase mass transfer process. However, Gritti *et al.* showed that the van Deemter C-coefficients were, at least for low molecular weight solutes, similar for some SPP and TPP architectures [54]. One putative explanation for this observation is that mass transfer outside the particle (i.e., that accounted for by the mobile phase mass transfer coefficient (C_m)), and *not in the stationary phase* (C_s), is the dominant contributor to the measured or experimentally determined C-term (see below). Indeed, Knox has argued very convincingly that the internal C_s-term is often very small if not totally negligible, and typically less than 0.02 for small molecules regardless of particle architecture [57]. Alternatively, it is possible that the similarity of the total C-terms of SPPs and TPPs arises from a velocity dependence of the C-term that is inflated by using the velocity independent A-term when experimental data are fitted to the van Deemter equation that assumes a velocity independent A-term. This is discussed in some detail in the Appendix (see sections

A.2 and A.4 for effect of different velocity dependencies of eddy dispersion on the estimate of C_m obtained by fitting data to equations that model the overall dependence of plate height on velocity).

3. The thermal conductivity of SPPs is greater than that of TPPs thereby lessening residual radial temperature gradients induced by viscous heating at high mobile phase velocities [58].

4. Molecules with high molecular weights and thus low diffusion coefficients (e.g., larger peptides, proteins and nucleic acids) will benefit from the improved internal diffusional equilibration times [59, 60]. This has been shown using SPPs with wider pores to facilitate diffusion of high molecular weight analytes [61].

It certainly must be understood that the higher plate numbers of SPPs along with their lower porosities and thus lower dead volumes when compared to the same diameter TPPs place greater demands on instrument performance. However, at least until fairly recently, SPPs as small as the smallest TPPs were not available and consequently due to the lower h_{min} of SPPs one could use them in wider diameter columns than one could with TPPs without suffering from the frictional heating induced dispersion that afflicts TTPs unless packed in narrow (e.g., 2.1 mm) columns. Such narrow columns with extremely low volume peaks require instruments with extraordinarily low extra-column dispersion. The advent of sub-3 μm SPPs also requires their use in such narrow columns. In any case the high cost of the instruments with minute extra-column dispersion is offset in part by the much lower cost of the mobile phases needed to operate narrow columns.

2.3.4 Monolithic Columns

Another significant development in stationary phase materials for LC in the early 2000s was the introduction of monolithic type columns [62–67] comprised of a single rod of silica or organic polymer encapsulated in a capillary or wrapped in a tubular structure. One of the prevailing ideas at that time was that monoliths would be ideal for very fast high-resolution analysis because of their quite high permeability (see Appendix section A.5) and thus low pressure drops compared to columns packed with sub-3 μm particles in packed columns. Additionally, peak broadening due to the C_s-term should be decreased because the microstructures that comprise the monolithic rod are only 1-2μm in diameter. According to Guiochon's [62] magisterial review of chromatographic monoliths, the origins of such materials date back to the earliest days of chromatography, and include paper and thin layer plates, as well as uses in gas chromatography. This area gained more attention in the mid-1990s with the introduction of silica and polymer monoliths to HPLC [66, 67] (also see reviews by Svec [63] and Tanaka *et al.* [64, 65]).

However, the high permeability provided by the open macropore structure of the porous rods results in a deterioration in the overall mass transfer properties (i.e., an increased C_m-term), giving less than expected efficiencies on a plates per meter basis. Considering the above discussion of the overall C-term of SPPs vs TPPs for low molecular weight analytes, this result for monoliths is to be expected, at least to some extent. Theoretical studies of this compromise by Desmet *et al.* [68] have led them to the conclusion that monoliths are generally better suited to high efficiency (i.e., very large N) separations involving long or coupled columns and high plate counts, rather than fast separations where plate count is sacrificed for speed. Specifically, they state that the plate count threshold above which monoliths compare more favorably to packed particle beds is about 30,000 [68]. This conclusion was arrived at before the meteoric rise of small SPPs and given their superior column dynamics, the performance threshold for the superiority of monoliths is probably now shifted to even higher plate counts and longer analysis times [62].

Colon *et al.* compared the performance of monoliths to that of sub-3 μm SPPs and sub-2 μm TPPs [69]. Monoliths gave optimum H values of about 8 μm, which is about equal to that of typical 3 μm TPPs. At high mobile phase velocities, considerably better efficiencies were obtained on both the TPPs and SPPs as compared to monoliths. However, further improvements can be expected in monolith technology as shown by the superior performance of second-generation silica monoliths versus the first-generation materials [62, 69, 70].

2.3.5 NON-POROUS PARTICLES

Compared to SPPs and TPPs, with fully non-porous particles (FNPs) the possibility of intraparticle (i.e., stationary phase) diffusion is eliminated, which means that the C_s contribution to the plate height is negligible, leaving the possibility for lower plate heights for FNPs compared to TPPs, especially at high mobile phase velocities and particularly for high molecular weight analytes such as proteins. Monodisperse non-porous silica particles can be made easily in the sub-2 μm range using the well-known Stöber process, and can provide very high plate counts under UHPLC conditions [71–73]. However, their chief drawback in practice is their much lower surface areas (typically about 1/20th of the surface area of TPPs), which in turn result in much lower mass loadability and retention compared to their TPP and SPP counterparts.

Wu *et al.* studied the mass transfer properties of FNPs and TPPs of similar particle sizes with both low and high molecular weight molecules. Similar efficiencies for small molecules were observed on both types of particles but considerably higher plate counts were seen for proteins on the FNPs at high velocities [72, 73]. This result suggested that FNPs are most attractive for separations of strongly retained – at least in RPLC – high molecular weight compounds provided that the low loading capacity of FNPs is not so low that detection of the possibly minute amount of sample becomes limiting.

2.4 APPROACHES TO OPTIMIZING SPEED AND PERFORMANCE

As pointed out above it is quite true that the maximum plate count for a *specific* LC column is reached at the minimum in a plot of H versus u. However, it should be evident that the above statement is seriously limited by several assumptions. The statement implies that a column of given length and inner diameter packed with particles of a specific size is being optimized *without regard to the characteristics of the entire chromatographic system* including both the pump's pressure and volumetric flow limits. When the system (i.e., column and pump) is being optimized then we must consider that the column length, particle size and mobile phase velocity are simultaneously adjusted within the limitations of the maximum available pressure and flow.

In their fundamental contributions to the theory of optimization in HPLC Knox [5] and Guiochon [1] showed that the production of any desired plate count in the least time occurs at the linear interstitial velocity corresponding to the van Deemter optimum. However, this happens *only if the particle diameter and column length are co-optimize*d to work at the pressure drop corresponding to the maximum allowed by the pump and that the flow rate required by a column of the diameter employed at the optimum velocity can be provided.

As is implicit in Guiochon's review, and evidently at the heart of Poppe's analysis [4] of speed in LC, we must conclude that neither the fastest separations, nor the highest plate counts, are obtained by working at the van Deemter optimum if the particle size is arbitrarily set before both the column length and eluent velocity are simultaneously optimized. We will call the velocity and column length that produce the highest plate count at a given analysis time and system pressure the "Poppe optimum conditions". The Poppe optimum velocity and the van Deemter optimum velocity will only be the same when one uses columns of the optimal length packed with particles of the optimal size as given by Guiochon's analysis (see details below). Something that must be understood and borne in mind is the important difference between these two approaches to optimizing column performance. They certainly are not the same, and the failure to understand completely the differences between them and their consequences is a source of significant confusion in the literature on optimizing plate count and speed in LC.

Another very widely held, yet erroneous concept, is that a given plate count can *always* be achieved in less time by decreasing the particle size, or alternatively that more plates can be obtained in the same time by using smaller particles. That is, it is an error to think that speed is *always improved by using the smallest available particles*. The basic question is: under practically useful conditions, how should one optimize plate count and speed to solve real problems? Most separation tasks can be classified as follows:

1. Given some desired resolution (and thus a specific N for the resolution of a critical pair of target analytes), how does one generate the requisite N in the least possible time?
2. Given a specific desired analysis time (e.g., a high-throughput application, or separation of a very complex sample) how can one maximize the number of plates generated in that time, and thus optimize the productivity (speed) of the separation (plates/time)?

In their early studies of optimization of separations Giddings [9], Knox [5, 28], and Guiochon [1] developed approaches that responded well to these questions. In the pre-UHPLC era, fewer choices than we now have had to be made to optimize separations. The maximum pressure drops provided by pumps were about 400 bar, most columns and packings could only be operated at ambient temperatures, most packing materials were TPPs (although SPPs also have a long history [47, 48]), but fewer particle sizes were available. Today we are blessed (or cursed) with many more commercially available options, including: a greater range of instrument capabilities in terms of both pressures and temperature ranges, quite a few particle sizes from about 1.3 to 10 μm, multiple pore sizes, multiple particle types (i.e., FNP, SPP, and TPP), and monolithic columns.

Optimization is related to the compromise between efficiency (as measured by N) and column dead time. Poppe popularized the idea of plotting log t_0/N against log N, which nicely captures the speed vs. efficiency trades-off in a way that is easy to visualize [4]. Several groups have used Poppe type plots and modifications of it known as kinetic plots [74–77]. Some of these incorporate Knox's separation impedance concept [78], which is discussed in Section 2.5, and various derivatives of it using other metrics on the two axes to show different practical issues with respect to optimization practice.

A study by Desmet [75] gives a succinct overview of the theory of optimization, which spans some 40 years, beginning with the work of Giddings, Knox, and Guiochon. The practical utility of such plots (including both "Poppe" and "kinetic" plots) has been shown by Desmet and coworkers [75] and several additional groups [76, 77, 79]. Their work indicates good agreement between theoretical predictions (e.g., from a set of kinetic parameters) and experimental data obtained using real columns used under practical conditions. Moreover, the Desmet group has improved on Poppe's original plot by their development of a methodology in which H vs. u data from experimental measurements are transformed by use of simple equations and scaling parameters to produce a suite of "kinetic plots". Desmet *et al.* have also extended Poppe's work by including the impact of extra-column dispersion in the calculated "kinetic" curves. Extra-column effects obviously become increasingly important as the particle and column technology are improved, thereby producing temporally narrower peaks of lower volume [80, 81].

Our objective at this point in this chapter is to give a set of easily understood, yet accurate algebraic equations that clearly show the interplay between particle size, column length, mobile phase velocity, and pressure, in optimizing LC. The three distinct optimization procedures outlined above in Section 2.2 are considered in order of increasing generality. First, only the mobile phase velocity is allowed to vary assuming both a fixed particle size and column length; this is basically the simple, pure "van Deemter" or one-parameter optimization. Second, both the velocity and column length are optimized for a fixed, arbitrarily chosen particle size. We will call this the "Poppe" or two-parameter optimization. It is, of course, the only approach that yields immediately actionable results, because one can start with the choice of a particle size that is commercially available. Third, mobile phase velocity, column length, and particle size are all concurrently varied. This approach leads to the *theoretically limiting maximum possible speed*. That is, t_0/N is absolutely minimized at a specified t_0. We will call this the "Knox-Saleem-Halasz (K-S-H)" [5,6] or three-parameter optimization because these pioneers developed the basic equations that solved the problem. The controlling equations (subject to the desired analysis time and maximum available system pressure) are derived below; they give the *optimum* velocity, length, and particle size. These equations complement Desmet *et al.* in that they facilitate the fast, yet accurate prediction of the influence of experimental parameters such as the maximum operating pressure, mobile phase viscosity, column permeability, and the various dynamic characteristics of the column (i.e., the van Deemter A, B, and C parameters, see Appendix A.1) on speed and performance (resolution).

Specifically, we will show that, at best, one can only improve the analysis time by the inverse first power of the maximum operating pressure. That is, *under fully optimized conditions* when one doubles the maximum system pressure the best one can do is to decrease the analysis time by a factor of two. Claims of greater performance gains by only increasing pressure are invalid. Perhaps most startling to many, we will show that once the particle size is set, increasing the maximum available pressure does not improve the *limiting* speed of analysis. Last, we will use the three-parameter equations to determine the impact of increasing the operating pressure on the optimum particle size in the limit of very fast (1 s < t_0 < 10 s) separations.

Here we point out once more that readers who are not well acquainted with the theory of column dynamics and the basic factors that control column pressure drop, column frictional heating by mobile phase flow through small particles, and so on, are advised to thoroughly review the material in the Appendix before proceeding. If it is simply a question of recalling the meaning of common chromatographic symbols, then the Glossary at the end of the book should be consulted. We have made every effort to make the glossary comprehensive.

2.4.1 ONE-PARAMETER OPTIMIZATION

The simplest possible optimization procedure is one where we have a column of given length and particle size and merely wish to determine the mobile phase velocity that will maximize the plate count without regard to any other limitations. The van Deemter equation (see Eqs. A.1 and A.2) does a good job of describing the dependence of plate height on the interstitial mobile phase velocity provided we work at velocities that are within a factor of two to three on either side of the minimum plate height. Using the van Deemter equation in its reduced form (see Eq. A.7), it is easily shown that the reduced interstitial mobile phase velocity ($v_{e,opt}$) and reduced plate height (h_{opt}) at the minimum in the curve are given by:

$$v_{e,opt} = \sqrt{\frac{B}{C}} \tag{2.1}$$

$$h_{min} = A + 2\sqrt{B \cdot C} \tag{2.2}$$

where A, B, and C are the fitting coefficients of the van Deemter equation in reduced coordinates. We will subsequently refer to $v_{e,opt}$ as the *van Deemter* optimum velocity. For columns packed with different particle sizes – but the same packing quality – the values of the reduced van Deemter coefficients (A, B and C) will be the same. This works reasonably well for real columns as shown by the similarity of the plots of h versus v in Figure A.1B for three different particle sizes in the range of 1.8 to 5.0 μm. Thus, in contrast plots of variables with dimensions, such as H and u_e, the minimum *reduced plate height* and optimum *reduced velocity* are almost independent of particle size. This is an approximation, and several factors can lead to serious deviations:

1. It assumes that particles of all sizes can be packed equally well. This assumption becomes more tenuous as particles get smaller, especially below 2 μm. The major problem is the increased importance of surface energy causing particle agglomeration during packing. The packing of columns is far from an exact science.
2. For various reasons it easier to pack capillary columns (< 200 μm tubes) with quite low *h* values (some as low as 1.05) than it is to pack narrow tubes (2.1 mm columns) [82, 83]. Additionally, low values of h_{min} are easier to obtain with 4.6 mm i.d. columns than they are with 2.1 mm i.d. columns because extra-column dispersion effects – in both time and volume – have less impact on plate height measurements for large volume columns than with narrower columns.
3. The heat generated in columns packed with small particles is better dissipated in narrow columns (2.1 mm i.d.) and capillaries than it is in conventional bore (4.6 mm i.d.) tubes. Thus, peak broadening due to frictional heating of the mobile phase in the column and thermal mismatch between the mobile phase entering the column and the column's temperature are more serious issues in conventional tube diameters and may not be negligible [24, 25].

The net result of these factors leads to an apparent increase in the eddy dispersion coefficient (the *A* term). However, these factors also lead to a stronger dependence of *h* on *v*. When the van Deemter equation is used this results in an increase in the apparent *C* term. If the Knox form were used (see Eq. A.12) only the *A* term would increase. We point out that it is common to see van Deemter fit *C* terms as small as 0.02 to 0.04 in 4.6 mm i.d. columns packed with 5 μm particles for analytes that are sufficiently well retained that extra-column dispersion is minor or negligible. It is, in our opinion, quite rare to see such small *C* terms for sub-2 μm particles. Part of this is likely due to the exceedingly high demands placed on the instrument by the very small peak volumes and very narrow peak widths in time units required by use of very small particles. However, with the use of 2.1 mm columns it is difficult to say that frictional heating is not a significant cause of peak broadening.

Given all of the preceding caveats, we will assume that the fitting coefficients are independent of particle size for the rest of this chapter. It is an easy matter to arrive at the following results:

$$u_{e,opt} = \frac{D_m}{d_p} \cdot \sqrt{\frac{B}{C}} \tag{2.3}$$

$$H_{min} = (A + 2\sqrt{BC}) \cdot d_p \tag{2.4}$$

$$N_{opt} = \frac{L}{(A + 2\sqrt{BC}) \cdot d_p} \tag{2.5}$$

Inspection of Eq. 2.3 tells us that the optimum velocity will increase with the analyte's diffusion coefficient. Note *B* and *C* above are based on the relationship between the reduced plate height and reduced interstitial velocity, and can be computed from Eqs. A.17–A.19. The C_m and C_s terms so defined depend only weakly on diffusion coefficients, and the *B* term depends on the ratio of two diffusion coefficients. Working at constant retention factor, we see that an increase in temperature will increase D_m and thus the optimum van Deemter velocity ($u_{e,opt}$). Measurements made under very carefully controlled conditions (see Figure 2.1) so as to utterly negate any thermal mismatch broadening clearly show that the optimum velocity increases rather significantly as the temperature

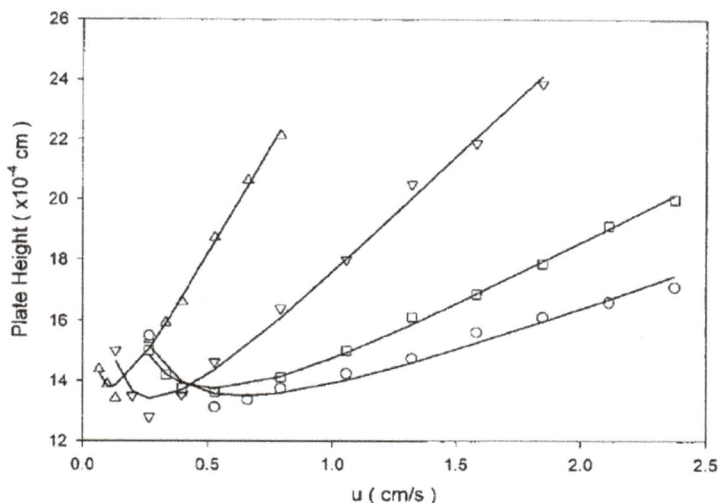

FIGURE 2.1 Plate height vs. mobile phase linear velocity at different temperatures for moderately retained solutes, experimental conditions are given in Table 2.2 of [84]. Reprinted with permission from B. Yan, J. Zhao, J. Brown, J. Blackwell, P. Carr, High-temperature ultrafast liquid chromatography, *Analytical Chemistry* 72 (2000) 1253–1262.

is raised from 25 °C to 150 °C. The values of *A*, *B*, and *C* given in Table 2.2 are taken from fits of *H vs. u* (see Eq. A.2) not *h vs. v*. It is evident that *A* is essentially independent of temperature, *B* increases and *C* decreases with increasing temperature as they should in view of Eq. A.17–A.19 in the Appendix. *Statements that temperature does not increase the optimum velocity are incorrect.* We will return to the details of the effect of temperature on speed and plate count later.

2.4.2 Two-Parameter (Poppe) Optimization

We now assume that we deliberately pick particles of a specific size (e.g., 1.8, 3.0 μm, etc.) and a system with a specific maximum operating pressure. The pragmatically important question is: for a given set of operating parameters (particle diameter, temperature, solvent composition) what mobile phase velocity and column length maximize the plate count? This is a rather different question than that posed in the previous section.

Guiochon has comprehensively reviewed the various conditions under which one might want to optimize various kinds of column "performance" including optimization of analytical sensitivity, amount of solvent consumed, and so on [1]. For present purposes, let us restrict the consideration to choosing the optimum column length and velocity for a column packed with a pre-determined (fixed) particle size. Again, we set as our goal *the highest plate number that can be achieved at a desired column dead time, and pre-determined system pressure, and a pre-set particle diameter.*

2.4.2.1 Introduction to Poppe Plots

The above is precisely the problem solved by Poppe in his classic paper [4]. He described an iterative computational procedure for determining the specific column length and velocity that *maximize* the plate count for a given maximum pressure drop and analysis time for columns filled with the particle size defined at the outset. However, as shown in Section 2.4.2.2, Poppe's iterative approach is unnecessary because a closed-form algebraic solution that gives exactly the same

TABLE 2.2
Effect of Temperature on Column Dynamics

	Experimental Conditions				van Deemter Equation Coefficients[a]			
T (°C)	Mobile Phase (% ACN (v/v))	Analyte	D_m x 10⁴ (cm²/s)[b]	k	A x 10³ (cm)	B x 10⁴ (cm²/s)	C x 10⁴ (cm²/s)	u_{opt} (cm/s)
25	40	acetophenone	0.10	0.50	1.0 ± 0.03	0.28 ± 0.03	1.2 ± 0.06	0.2
90	40	acetophenone	0.25	0.26	1.1 ± 0.06	0.58 ± 0.11	0.46 ± 0.05	0.4
120	30	acetophenone	0.41	0.15	1.0 ± 0.08	1.4 ± 0.22	0.37 ± 0.04	0.7
150	25	acetophenone	0.59	0.10	1.1 ± 0.02	1.5 ± 0.05	0.27 ± 0.01	0.8
25	40	octanophenone	0.08	3.87	1.1 ± 0.04	0.18 ± 0.03	1.4 ± 0.06	0.1
80	40	decanophenone	0.15	3.15	0.90 ± 0.05	0.6 ± 0.09	0.80 ± 0.03	0.3
120	30	decanophenone	0.25	5.70	0.91 ± 0.03	1.2 ± 0.08	0.44 ± 0.01	0.6
150	25	decanophenone	0.36	1.65	1.0 ± 0.05	1.3 ± 0.08	0.31 ± 0.03	0.7
25	40	decanophenone	0.06	12.2	1.0 ± 0.06	0.23 ± 0.05	1.9 ± 0.09	0.1
80	40	dodecanophenone	0.14	7.39	0.93 ± 0.04	0.63 ± 0.07	0.77 ± 0.03	0.3
120	30	tetradecanophenone	0.22	12.3	1.0 ± 0.05	1.0 ± 0.12	0.38 ± 0.02	0.6
150	25	tetradecanophenone	0.31	7.00	1.1 ± 0.02	1.3 ± 0.07	0.20 ± 0.08	0.8

Source: Reprinted with permission from B. Yan, J. Zhao, J. Brown, J. Blackwell, P. Carr, High-temperature ultrafast liquid chromatography, *Analytical Chemistry*. 72 (2000) 1253–1262.

Note:
a) Parameter ± one standard deviation. a) Estimated analyte diffusion coefficient in the mobile phase at the conditions shown.

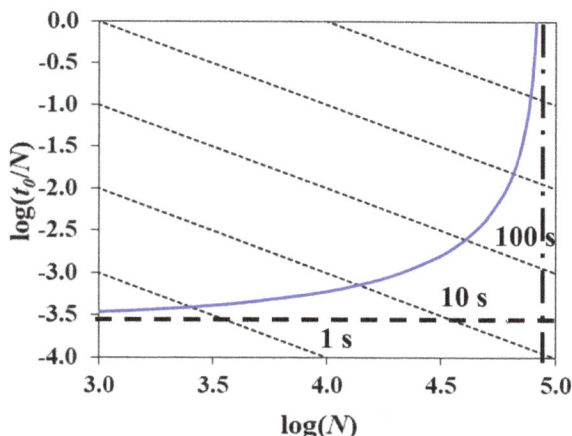

FIGURE 2.2   This is a typical Poppe plot, with the curve calculated as described below with Eqs. 2.8–2.10 using the following parameters: P_{max}, 400 bar; T, 40 °C; η, 0.69 cP; %ACN, 25; D_m, $7.0×10^{-6}$ cm^2/s; Φ, 500; ε_e, 0.38; ε_t, 0.3; λ, 0.67; A, 0.95; B, 7.0 ($k = 5$); C, 0.04. Most importantly $d_p = 1.8$ μm. The diffusion coefficient was chosen to be that of a small drug molecule (MW = 300–350).

results is possible as discussed below and pictured in Figure 2.2 [11]. In this section we give only a qualitative description of the procedure to differentiate Poppe's approach from subsequent work that involves direct calculation of the curves using analytical expressions without resorting to the iterative calculation approach.

Figure 2.2 is a plot of log t_0/N vs. log N. The logarithmic scale allows the visualization of the system behavior over a very wide range of performance. Each point on the curve is computed by first choosing and fixing the parameters P_{max}, d_p, T, ε_e, mobile phase composition, and D_m. Choosing these parameters determines the mobile phase viscosity. The column characteristics are defined by setting A, B, and C as well as ε_e. Now one chooses a system dead time (t_0). At this point, the column length and velocity (u_e) are varied until N is maximized. The curve is generated by systematically varying the dead time and repeating the computation. For convenience a series of "isochrones" are drawn. The system dead time is fixed at the indicated values along each isochrones (see the diagonal lines indicating specific t_0 values).

It is evident that starting at very short time (the left most side of the figure) the value of t_0/N is smallest. This means that we are getting the fastest performance. Obviously if t_0/N is smallest then N/t_0 is maximized. Indeed, inspection shows that the plot is approaching a limiting value of t_0/N. No matter how small t_0 is made t_0/N *does not get any* smaller. However, we see that in this region N is small. As the dead time is increased the speed decreases – that is, t_0/N gets bigger. Thus, there is a very definite trade-off between speed and resolving power (plate count). *You can have either high speed or high plate counts; you cannot have both simultaneously.*

As we move to the right the Poppe curve swings upward and approaches a vertical asymptote. That is, no matter how much time one is willing to spend one cannot get more than some maximum number of plates. We will discuss the limiting values of N and t_0/N in more detail shortly. It must absolutely be understood that *every point on the Poppe curve corresponds to a different pair of values of L and u_e*. To make this clear consult Table 2.3.

The region above and to the left of the Poppe curve is completely accessible as it corresponds to running the system at a pressure less than the maximum possible pressure deliverable by the pump. Conversely the region below and to the right of the Poppe curve is inaccessible as it corresponds to performance that could only be achieved by operating at a pressure greater than the maximum available system pressure, given that all other system parameters are fixed.

TABLE 2.3

Typical Poppe Optimization Results for a Fixed Particle Size of 1.8 μm[a]

t_0 (s)	N^* (plates)	t_0/N^* (s/plate)	L^* (cm)	u_e^* (cm/s)	F^b (mL/min)	F^c (mL/min)
1	2566	3.9×10^{-4}	1.7	2.49	9.45	1.97
5	8880	5.6×10^{-4}	3.7	1.12	4.23	0.88
10	13960	7.2×10^{-4}	5.3	0.79	2.99	0.62
50	31550	1.6×10^{-3}	11.8	0.35	1.34	0.28
100	40470	2.5×10^{-3}	16.7	0.25	0.94	0.20
500	59320	8.4×10^{-3}	37.4	0.11	0.42	0.09
1000	65590	1.5×10^{-2}	52.9	0.08	0.30	0.06

Notes:
a) All conditions are per Figure 2.2; values were calculated using Eqs. 2.8–2.10.
b) Flow rate assumes a 4.6 mm i.d. column.
c) Flow rate assumes a 2.1 mm i.d. column.

The Poppe plot in Figure 2.2 has a horizontal asymptote that illustrates the limiting speed (see heavy dashed line) at short time. Additionally, the plot also has a vertical asymptote that illustrates the limiting plate count (see dashed dot line) at long time. The numerical values and the factors controlling these limits will become clear shortly.

2.4.2.2 An Alternative Way to Compute Poppe Plots

The relationship between dead time, column length and mobile phase velocity of an unretained species (u_0) is:

$$L = u_0 \cdot t_0 = \frac{\varepsilon_e}{\varepsilon_T} u_e \cdot t_0 = \lambda \cdot u_e \cdot t_0 \tag{2.6}$$

Here ε_T and ε_e are the total and interstitial porosities of the column, respectively, and the parameter λ ($= \varepsilon_e/\varepsilon_T$) is defined for mathematical convenience. Clearly, if we desire that the "performance" meets some desired time constraint, an additional relationship is established between the mobile phase velocity and column length that just did not exist when optimization is done by simply choosing the van Deemter optimum velocity. Equation 2.6 is a very important relationship that must be imposed beforehand, *not afterwards*, when trying to achieve the "optimum" plate count for a real system.

In order to assure that the choice of L and u_e are practically achievable we now must also require that the system pressure not exceed some maximum. The relationship between pressure drop, column length, mobile phase velocity and viscosity, and particle size is discussed in Appendix A.5. Combining Eqs. A.29 and A.30 yields a more compact relationship:

$$\Delta P = \Phi \cdot \eta \frac{u_e \cdot L}{d_p^2} \tag{2.7}$$

It is found that Φ, a dimensionless semi-empirical factor, is typically in the range of about 500–1000 [85] for reasonably well-packed random beds. We will use a value of 500 here, which is consistent with recent experimental data [55]. This parameter will vary with how the column is packed, the particle shape, and the degree of polydispersity of the particles and especially the interstitial porosity. Section A.5 provides additional discussion of the conditions when Eq. 2.7 is valid. Equation 2.7, like Eq. 2.6, also establishes a relationship between mobile phase velocity and column length that was not considered when Eqs. 2.3–2.5 were developed. Evidently, if we preselect a value of d_p (particle size), they are no longer free to choose any mobile phase velocity and column length. Rather the velocity and length must be chosen to simultaneously satisfy Eqs. 2.6 and 2.7; that is, once d_p is assigned the number of plates can be optimized only by varying the column length and mobile phase velocity such that they simultaneously satisfy Eqs. 2.6 and 2.7. Indeed, as soon as we decide to satisfy these equations with some desired time and pressure, *the column length and velocity are fully determined*. It must be so because we only have two equations and two unknowns (L and u_e). This lack of freedom means that the plate count is also established as soon as these two conditions are imposed.

Because the equations relating velocity and length to pressure and time are linear, the required values of column length and velocity are easily obtained by simple algebra; *an iterative procedure is not needed*. By solving Eqs. 2.6 and 2.7 we obtain explicit equations that dictate the values of L and u_e that optimize N and satisfy the time and pressure constraints. We denote these special values of L and u_e with an asterisk: L^* and u_e^*. These are the Poppe optimum column length and optimum mobile phase velocity. The optimum velocity can be used to compute H and given the optimum L one can compute the Poppe maximum plate count (see Eq. 2.10).

$$u_e^* = \sqrt{\frac{P_{max}}{\Phi \cdot \eta \cdot \lambda \cdot t_o}} d_p = \psi \cdot d_p \cdot t_o^{-1/2} \text{ with } \psi = \sqrt{\frac{P_{max}}{\Phi \cdot \eta \cdot \lambda}} \tag{2.8}$$

$$L^* = \sqrt{\frac{P_{max} \cdot \lambda \cdot t_o}{\Phi \cdot \eta}} d_p = \psi \cdot \lambda \cdot t_o^{1/2} \cdot d_p \tag{2.9}$$

$$N^* = \frac{L^*}{A \cdot d_p + \dfrac{B \cdot D_m}{u_e^*} + \dfrac{C \cdot d_p^2 \cdot u_e^*}{D_m}} = \frac{\psi \cdot \lambda \cdot t_0^{1/2} \cdot d_p}{A \cdot d_p + \dfrac{B \cdot D_m}{\psi \cdot d_p \cdot t_0^{-1/2}} + \dfrac{C \cdot d_p^3 \cdot \psi \cdot t_0^{-1/2}}{D_m}}$$

$$= \frac{\psi \cdot \lambda \cdot d_p}{\dfrac{A \cdot d_p}{t_0^{1/2}} + \dfrac{B \cdot D_m}{\psi \cdot d_p} + \dfrac{C \cdot d_p^3 \cdot \psi}{D_m \cdot t_0}} \tag{2.10}$$

Note that we use u_e^* and N^* to indicate that these values are not hypothetical optima. They are real and achievable; that is, they are conditioned on the actual operational parameters including particle size, maximum pressure drop, analysis time, and implicitly on the mobile phase composition and temperature through their dependence on the mobile phase viscosity and analyte diffusion coefficient.

In contrast to the van Deemter optimum velocity that gives N_{opt}, the mobile phase velocity that maximizes N^* at the desired time and pressure does not depend on the column's dynamic parameters (A, B, and C) as is the case in Eq. 2.5. Rather, the Poppe optimum velocity has nothing whatever to do with Eq. 2.1. However, note that the resulting plate count (N^*) *does depend* on the column's dynamic parameters. As stated above, this problem is precisely that addressed by Poppe. We point out that Eqs. 2.8–2.10 allow the exact generation of the Poppe curve in Figure 2.2 without recourse to any iterative scheme. This should not have been surprising as all the equations involved are linear. That is, once one decides on d_p, t_o, and P_{max} *there are no degrees of freedom left*, and thus both L^* and u_e^* are fixed. Clearly, once the mobile phase velocity and column length are fixed then so is N^*; one gets exactly the same values of N^* using either Eqs. 2.8–2.10 or Poppe's iterative method.

2.4.2.3   Long-Time and Short-Time Limits of LC as Revealed by the Poppe Plot

It is evident in Figure 2.2 that as t_0 becomes very large a limiting value of N^* is reached (see the vertical asymptote). Inspection of Eq. 2.10 at very large values of t_0 shows that the equation reduces to:

$$N^*_{\text{lim}} = \frac{\psi^2 \cdot \lambda \cdot d_p^2}{B \cdot D_m} = \frac{P_{\text{max}} \cdot d_p^2}{\Phi \cdot \eta \cdot B \cdot D_m} \tag{2.11}$$

This is the maximum possible plate count that can be generated at P_{max} regardless of the amount of time invested [11]. Clearly, N^*_{lim} depends only on the van Deemter B parameter indicating that in the limit of maximum plate count the only important broadening process is longitudinal diffusion. Equation 2.11 clearly indicates that the larger are the particles and the smaller is the analyte diffusion coefficient the larger is the number of plates that can be generated. There is clearly a first power dependence on P_{max}. Thus, in this limit doubling the pressure will double the plate count but that is all it can possibly do. We point out that the limiting plate count is reached only at very long times. Indeed, it is reached at vastly impractically long times especially given that the time metric is the dead time. Recalling that the Poppe plot only holds for isocratic chromatography we anticipate that in practice the last peak might elute not later than about 10–20 times the dead time.

Now consider the other end of the time scale. We note (see Figure 2.2) that as t_0 becomes very small the Poppe curve becomes very flat. This is understood by rearranging Eq. 2.10 to the form:

$$N^* = \frac{\psi \cdot \lambda \cdot d_p \cdot t_0}{A \cdot d_p \cdot t_0^{1/2} + \dfrac{B \cdot D_m \cdot t_0}{\psi \cdot d_p} + \dfrac{C \cdot d_p^3 \cdot \psi}{D_m}} \tag{2.12}$$

The short analysis time limit for the plate count (denoted N^{o*}_{lim}) can easily be derived from Eq. 2.12. As t_0 becomes very small it is obvious that the last term in the denominator becomes dominant resulting in the following limiting value:

$$N^{o*}_{\text{lim}} = \frac{\lambda \cdot t_0 \cdot D_m}{C \cdot d_p^2} \tag{2.13}$$

Eq. 2.13 tells us that *in the limit of high speed small particles and fast diffusion give better plate counts*. This is in stark contradistinction to Eq. 2.11 for the limiting plate count at long time where just the opposite conditions apply. However, in the context of recent developments and overreaching claims for ultra-high pressure chromatography [86,87] we see that the *high speed limiting plate count is completely independent of the maximum pressure available* although the plate count definitely depends inversely on the square of the particle size. We also note that in the high speed limit the plate count is directly proportional to the dead time. Thus, the more time one allows for the analysis, the better is the plate count. Clearly, the smaller is the C term, the better is the plate count at the limiting high speed. Finally, we note that the high speed limit may only be reached under conditions that are currently difficult to reach in practice given that columns less than 5 mm in length are not commercially available.

The limiting speed (taken here as $\dfrac{t_0}{N^{o*}_{\text{lim}}}$) is easily found by rearranging Eq. 2.13 [11]:

$$\frac{t_0}{N^{o*}_{\text{lim}}} = \frac{C \cdot d_p^2}{\lambda \cdot D_m} \tag{2.14}$$

These two equations make it quite clear that *in the limit of very fast analysis neither the plate count nor the speed depend on system pressure.* At the very least, the implication is that elevating the available system pressure does not universally provide a meaningful improvement in the speed or plate count. Given that at very long times the plate count is proportional to P_{max} and at very short time the plate count is independent of P_{max} we conclude that *at intermediate times the plate count will depend on less than the first power of P_{max}.* It is also interesting to note that Eq. 2.14 shows no dependence of the limiting speed either on the dimensionless flow resistance parameter (Φ) or on viscosity (η). However, there is an inverse dependence on the analyte diffusion coefficient thus we can achieve higher speeds with smaller analytes at higher temperatures, and in light of the Wilke-Chang equation, in mobile phases of lower viscosity (see Section A.7). For this reason, acetonitrile-water mixtures are preferred to methanol-water mixtures for high-speed chromatography.

2.4.3 IMPACT OF SYSTEM PARAMETERS ON POPPE PLOTS, THE SPEED AND LIMITING PLATE COUNT IN LC

We turn now to an exploration by means of Poppe plots of the impact of different system parameters on the speed and limiting plate count. The most recent developments have been the introduction of smaller particles and higher pressure.

2.4.3.1 Impact of High Pressure on Speed of LC

We look first at the impact of a higher system pressure on the performance achievable with fully porous 1.8 μm particles. The Poppe plots at 400 and 1200 bar are shown in Figure 2.3A. First, we see that in the limit of very high speeds all three curves approach the same horizontal asymptote, that is, limiting value of t_0/N. This is exactly as predicted by Eq. 2.14. Thus, the ultimate speed of LC is not affected by the available pressure. As per Table 2.3 as we move to the right the optimum column length increases but the optimum velocity decreases in inverse proportion to the length. This is also easily seen by combining Eqs. 2.9 and 2.10.

$$u_e^* = \frac{(\psi \cdot d_p)^2}{L^*} = \frac{P_{max} \cdot d_p^2}{\Phi \cdot \eta \cdot \lambda \cdot L^*} \tag{2.15}$$

This is an important equation as it relates to what can be a limitation of modern LC and that is the flow rate limitations of modern equipment. The flow rate corresponding to u_e^* is:

$$F^* = \frac{\pi \cdot R_{col}^2 \cdot \varepsilon_T \cdot P_{max} \cdot d_p^2}{\Phi \cdot \eta \cdot L^*} \tag{2.16}$$

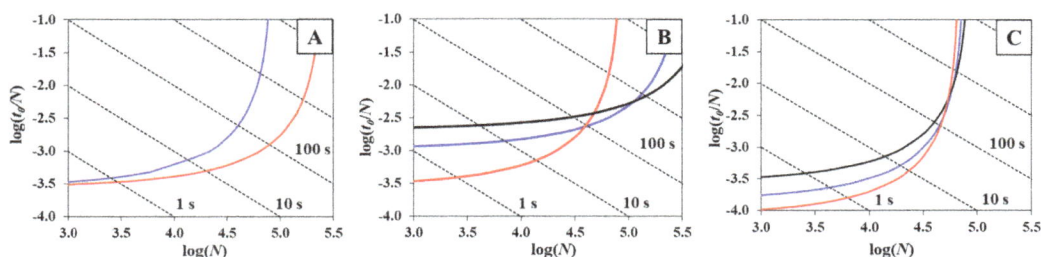

FIGURE 2.3 Effects of operating conditions on Poppe curves. All parameters are as in Figure 2.2, except for changes indicated here. A) Effect of pressure: 400 (blue) and 1200 (red) bar. B) Effect of particle size: 1.8 (red), 3.5 (blue), and 5.0 (black) μm. C) Effect of temperature: 40 (black), 80 (blue), and 120 (red) °C.

Values of F^* for 2.1 and 4.6 mm i.d. columns are given in Table 2.3. It is evident that flow rate limits of 5 mL/min are easily exceeded. However, it should be borne in mind that the deleterious band broadening effects of self-heating are such that almost no very high speed chromatography is done in tubes wider than 2.1 mm.

Second, as the pressure is increased the maximum possible plate count (N^*_{lim}) increases in agreement with Eq. 2.11. As the dead time is increased the speed (t_d/N) becomes a stronger function of pressure and it is in this region where one sees the great benefit of very high pressures. Inspection of the three Poppe plots shows that the use of high pressure extends the limiting speed into regions of shorter dead time. This is where the higher pressure really makes itself useful.

2.4.3.2   Impact of Particle Size on the Speed of LC

This is a much more complicated variable than the system pressure. A set of three Poppe plots, holding everything but particle size constant, are shown in Figure 2.3B. All plots were computed at P_{max} = 400 bar. The shape and general features of the three curves are independent of the operating conditions. That is, the curves are always convex and the curves always show the "crossover" phenomena. This is perhaps their most important feature. What the crossovers clearly indicate is that *the smallest particles are not necessarily the best particles (highest speed, highest plate count) to use.* This may seem like it contravenes conventional wisdom, but it is nonetheless true. It is, however, true that the *smallest particles are always the best particles to use when one is trying to achieve the fastest possible analysis*.

Why do we see this crossover phenomenon? There are two factors:

1. At the shortest possible times, the speed (t_d/N) will always be best for the smaller of two particle sizes. Thus, the horizontal asymptote of the Poppe curve for the smaller particle will lie below that of the larger particles. This is evident in Eq. 2.14.
2. At very long time, the larger of two particle sizes will produce more plates than the smaller of two particles. This is made clear by Eq. 2.11.

Cleary the two curves must crossover. This brings up an important issue. Assuming that everything possible has been done to optimize the chromatographic selectivity it is necessary to generate a minimum number of plates to achieve some minimum acceptable resolution. There are four possible scenarios:

1. The number of plates is to the left of the crossover point. In this case, the smaller particles *should be used*, as the separation time will be shorter.
2. The number of plates needed cannot be generated as it is to the right of the vertical asymptote for the smaller particles but to the left of the vertical asymptote for the larger particle. Clearly, the larger particles *must be used*.
3. The number of plates needed is to the right of the crossover but to the left of the vertical asymptote for the small particles. Although the number of plates needed is achievable with the smaller particles the larger particles *should be used*, as the separation time will be shorter.
4. The number of plates needed is to the right of the larger particles then still larger particles *must be used*.

The importance of choosing the right particle size is underscored in Table 2.4.

Here we can clearly see that the smallest particle size does not always give the fastest analysis. If one needs more than about 50,000 plates the 3.5 μm particles will do the job faster than will 1.8 μm particles. The actual crossover time (see Figure 2.3B) is about 89 s. In fact, 1.8 μm particles cannot generate 100,000 plates given an infinite amount of time as this plate count under these conditions is above the limiting value for this size particle regardless of the column length and velocity used.

TABLE 2.4
Effect of Particle Size (TPP) on Analysis Time Needed to Reach a Specified Plate Number[a]

Plates Required	t_0 Required (s)		
	1.8 μm	3.5 μm	5.0 μm
100,000	infinite	490	528
50,000	215	131	192
20,000	18.8	35.5	60.5
10,000	5.95	15.0	27.0
5,000	2.29	6.70	12.5
2,000	0.74	2.45	4.7
1,000	0.34	1.15	2.3

Note:
a) All conditions are per Figure 2.2; values were calculated using Eqs. 2.8–2.10.

On the other hand, 5 μm particles are never superior to 3.5 μm particles as long as one needs fewer than 100,000 plates. Since such high plate counts are seldom required, we have not extended the calculation. The crossover time is 685 s. The crossover times and plate counts for different size particles depend on all of the operating conditions (P_{max}, D_m, d_{p1} and d_{p2}) as well as the van Deemter parameters B and C, so one should not fixate on the data in Table 2.4 as applicable under all conditions. In general, the lower the requisite plate count, the more likely very small particles should be used and the faster the analysis can be done.

2.4.3.3 Impact of Column Temperature on Speed of LC

Temperature is definitely an important system variable in terms of separation speed. Its effect on the Poppe curve is shown in Figure 2.3C. This curve should be contrasted with the effect of pressure shown in Figure 2.3A. There are some radical differences. First, we see that increasing the temperature has its biggest effect in the realm of ultrafast LC. That is, times scales less than 100 s down to tenths of seconds. Even at a time scale of 1 s there is almost one-half of an order of magnitude improvement as temperature is increased from 40 to 120 °C. Clearly, the speed of LC (t_0/N) is significantly enhanced by increasing the temperature. However, a price is paid for this improvement. As temperature is increased the long-time limiting plate count decreases. It is not quite correct to say that this results from the increase in the diffusion coefficient as the temperature is raised. Examination of Eq. 2.11 shows that the long-time limiting plate count depends on the product of the viscosity and the diffusion coefficient (the so-called Walden product). However, use of the Wilke-Chang equation (see Appendix A.7, Eq. A.38) shows that:

$$\frac{D_m(T_2) \cdot \eta(T_2)}{D_m(T_1) \cdot \eta(T_1)} = \frac{T_2}{T_1} \tag{2.17}$$

Thus, the Walden product increases with the absolute temperature. That is, the diffusion coefficient increases with temperature just a bit faster than the viscosity decreases.

The limiting speed of LC ($\frac{t_0}{N_{lim}^{o*}}$, see Eq. 2.14) improves as the diffusion coefficient of the analyte increases with temperature. The dominant effect is obviously through the effect of temperature on viscosity. Thus, the extent to which an increase in temperature can speed up LC depends on mobile phase composition and in RPLC on the nature of the organic modifier. Inspection of Eq. 2.14 suggests that simultaneously increasing temperature and decreasing particle size should be a very effective way to speed up HPLC. It is sometimes necessary in high speed two-dimensional liquid chromatography to do gradient elution with column dead times as short as 1 s. Comparison of Figures 2.3A and 2.3C show that in the limit of such fast LC increased temperature has a much greater impact than increased system pressure. This is evident in Table 2.5. Note that if we assume the retention factor of the last peak is 5, then a dead time of 10 s corresponds to an elution time of nearly 1 min for the last peak. However, at slightly longer dead times the effect of increasing system pressure is definitely felt and pressure becomes a better way to increase speed than temperature.

2.4.3.4   Effect of Column Parameters on Speed of LC

It is obvious that the van Deemter C term will have the greatest impact on the speed of LC (see Eq. 2.14). This is clearly shown in Figure 2.4A. At a dead time of 1s the speed (t_0/N) improves 2.4-fold as the C term is decreased three-fold from 0.12 to 0.04. As anticipated the C term has no effect on the limiting plate count at long time.

Exactly the reverse happens as the B term varies (see Figure 2.4B). That is, B has no effect at all on the limiting high speed (see Eq. 2.14) but, of course, a decrease in B definitely improves the long-time limiting plate count. This figure should be examined closely. At dead times as short as 10 s a decrease in the B-term definitely improves the speed. Thus, in analyses even as short as 1 min ($k = 5$)

TABLE 2.5
Effect of Temperature and Pressure on the Speed of LC at Very Short Analysis Time[a]

| Pressure (bar) | Temperature (°C) | Plate Number | | |
		$t_0 = 1$ s	$t_0 = 10$ s	$t_0 = 30$ s
400	40	2570	13,960	25,270
400	80	4330	19,310	31,190
400	120	6330	23,530	34,740
1200	40	2950	19,940	41,890

Note:
a) All conditions are per Figure 2.2 except for the variations in temperature and pressure; values were calculated using Eqs. 2.8–2.10.

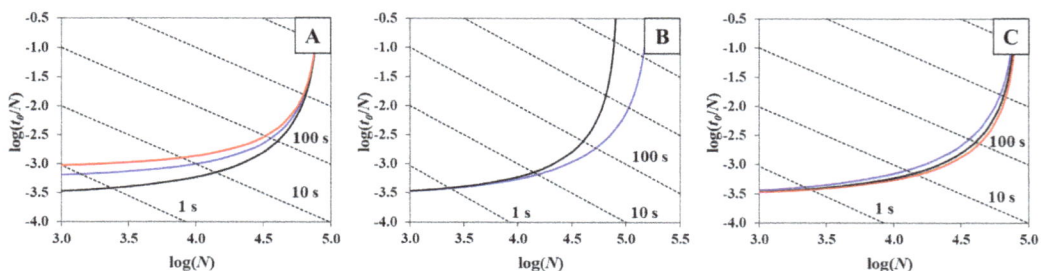

FIGURE 2.4   Effect of van Deemter terms on Poppe curves. All conditions as in Figure 2.2 except as indicated. A) Effect of C-term: 0.04 (black), 0.08 (blue), and 0.12 (red). B) Effect of B-term: 7.0 (black), or 3.5 (blue). C) Effect of A-term: 0.95 (black), 1.2 (blue), and 0.80 (red).

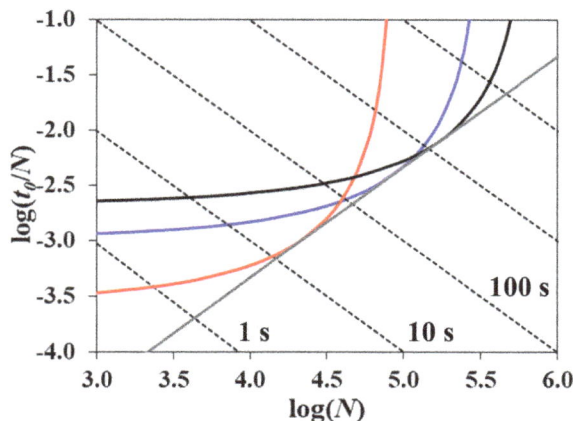

FIGURE 2.7 Knox-Saleem-Halasz line (grey) superimposed on Figure 2.4B.

Any point on the Poppe curve to the left of the point of tangency indicates that the speed or plate count could be improved by using a smaller particle. Conversely, any point on a Poppe curve to the right of the tangent point tells us that performance can be improved by using a larger particle. At a given set of conditions it is not possible to move below the K-S-H line. It is the best that can possibly be done.

2.4.4.1 Implications of the Three-Parameter (K-S-H) Optimization

Equation 2.22 is particularly valuable and provides very significant insights into the best possible performance. Holding N, P_{max}, or t_0 constant yields a triad of simple yet instructive relationships (Eqs. 2.23–2.25). First, if we vary P_{max}, and allow t_0 to change with a given K-S-H optimized column (thus according to Eqs. 2.1 and 2.2 v_{opt} and h_{min} will be fixed) it follows from Eq. 2.21 that:

$$P_{max,2} \cdot t_{0,2} = P_{max,1} \cdot t_{0,1} \tag{2.23}$$

Thus, it is evident that if nothing but the system pressure is changed *the best one can hope for is a proportional improvement in analysis time*. Let us be clear. If the maximum available pressure is doubled, *the best one can hope* for at constant plate count under the conditions of K-S-H optimization is a two-fold decrease in analysis time. Obviously, the gain will be less if one is working under conditions far removed from the optimum particle size or column length.

On the other hand, if we assume a fixed P_{max} we find that:

$$N_{max,2} = N_{max,1} \cdot \sqrt{\frac{t_{0,2}}{t_{0,1}}} \tag{2.24}$$

which shows that under fixed conditions (temperature, composition, pressure) plate count increases with the square root of the analysis time. *If we want to double the plate number we must increase the analysis time four-fold*. In isocratic chromatography resolution varies with the square root of the plate count, thus it is evident that *huge increases in analysis time are needed to have any significant effect on resolution*. This is one of the major driving forces for the interest in 2D-LC, especially LC×LC for the analysis of complex samples.

Finally, at constant t_0 we find:

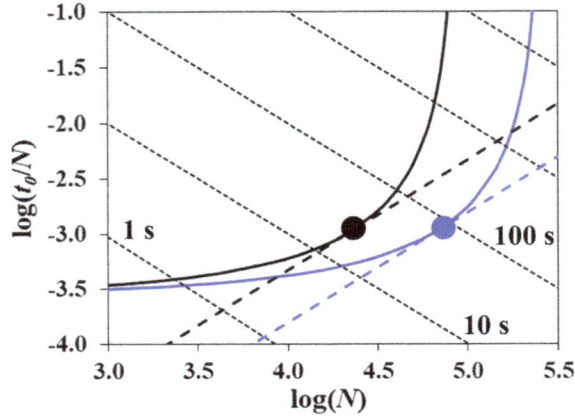

FIGURE 2.8 Poppe (solid lines) and Knox-Saleem-Halasz (dashed lines) plots showing the impact of pressure. All conditions are as in Figure 2.2. The black and blue curves are for pressures of 400 and 1200 bar, respectively. The filled symbols indicate the points where $u_e^* = u_{e,opt}$, in which case each "Poppe curve" is exactly tangent its respective "Knox-Saleem-Halasz line".

$$N_{max,2} = N_{max,1} \cdot \sqrt{\frac{P_{max,2}}{P_{max,1}}} \tag{2.25}$$

Thus, Eq. 2.25 makes it clear that under fully optimized conditions (K-S-H optimization), the maximum plate count achievable in a given analysis time depends on the *square root of the maximum available pressure, P_{max}*. Thus, *if we want to double the plate number the pressure must be increased four-fold.*

The pressure dependence is evident in Figure 2.8. Two Poppe plots and two "K-S-H lines" at two different values of P_{max} are shown. The K-S-H lines are parallel and thus the distance between them is independent of time. However, once a particular particle size is chosen (the two-parameter case), the dependence of N^* on P_{max} depends on the analysis time; for the practitioner, this is an inconvenient truth. We see that at very long times the relationship between N^* and P_{max} tends to a first-order dependence, which is consistent with Eq. 2.10. It is critical to recognize that although this pressure dependence is stronger than the square root dependence that results from the K-S-H optimization (Eq. 2.25), this is only because a particular particle size has been chosen. When the particle size is optimized and thus allowed to vary (i.e., increase at long times) the dependence will relax back toward the square root relationship.

The second area of interest is the region of analysis times where the Poppe curves are almost tangent to the K-S-H lines. In this region the pre-selected 1.8 μm particle size used to calculate the Poppe curves is close to the d_p^* values that result from Eq. 2.19 for the same value of P_{max} and thus the relationship between N^* and P_{max} is very close to the square root dependence discussed in the preceding paragraph. This means that *when one is using the most appropriate particle size the plate count will only increase by 1.7-fold when the pressure drop is tripled from 400 bar to 1200 bar.*

A third area of interest in the Poppe plot is that of very fast analyses ($t_0 < 5$ s). In this region we see that both "Poppe curves" in Figure 2.8 approach the same asymptotic limiting speed, despite the significant difference in P_{max} (400 vs. 1200 bar). This behavior is in exact accord with Eq. 2.14, and emphasizes the point that once one decides to use a particular particles size, increasing P_{max} has very little impact on the limiting speed (t_0/N) at very short time.

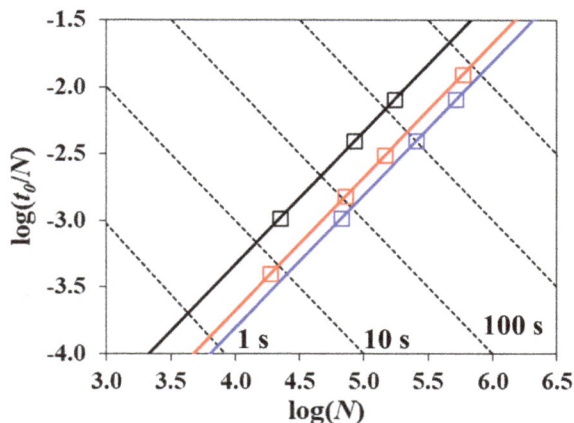

FIGURE 2.9 Knox-Saleem-Halasz plots showing the effect of pressure and temperature on speed. The points on each line indicate the conditions where 1.8, 3.0, 3.5, 5.0, and 10 μm particles are optimal. Conditions are as in Figure 2.2, except for temperature and pressure, as follows: 400 bar / 40°C (black), 1200 bar / 40°C (blue), and 400 bar / 100°C (red).

2.4.4.2 Comparison of Pressure and Temperature Effects on Speed Based on Knox-Saleem-Halasz Optimization

The K-S-H line affords a very clean way to show the relationship between speed, plate count and particle size. In Figure 2.9 we show the three K-S-H lines corresponding to the pressure-temperature pairs of 400 bar-40 °C, 1200 bar-40 °C, and 400 bar-100 °C. Further, the points on the line indicate from left to right (bottom to top) 1.8, 3.0, 3.5, 5.0, and 10 μm particles respectively. We note that as either the pressure or temperature are increased, the K-S-H line drops showing an overall increase in speed.

The 60 °C increase in temperature is not quite as effective for increasing speed as the increase in pressure to 800 bar. The position of the optimal size particles relative to the isochrones is very important. If we look at the data for the 1.8 μm particles at 400 bar and 40 °C we see that below a dead time of 20 s or so we need particles smaller than 1.8 μm. Because these are not widely available commercially, we cannot presently achieve results as good as allowed by the K-S-H limit. Unfortunately increasing the pressure increases the shortest dead time at which 1.8 μm particles are optimum and thus one will fall rather short of the K-S-H limit at very high speeds. In contradistinction, an increase in temperature shifts the 1.8 μm point to shorter dead time, increasing our access to nearly optimal commercially available particles. At the other extreme we see that 5 μm particles only become optimal at very long dead times and 10 μm particles are of virtually no practical use as the separations would require impractically long times.

The optimum column length and velocities are given in Table 2.7 for the three sets of conditions. We see that increasing the pressure requires longer columns at the optimum particle sizes whereas increasing temperature decreases the length of the optimum size column. These results clearly point out the need for a systematic strategy for choosing the most practical particle sizes and column lengths for accomplishing a specific analytical goal with a specific instrument and a given set of its limitations and assay boundary conditions.

2.5 THE KINETIC PLOT – A VARIANT OF THE POPPE PLOT

Sometime after Poppe introduced the above approach to visualize the effect of various system parameters on speed and resolving power, Desmet [12, 50, 60, 75, 81] introduced a set of alternative

TABLE 2.7
Knox-Saleem-Halasz Optimized Results for Selected Commercially Available Particle Sizes [a,b]

	d_p (μm)	t_0 (s)	u_e (cm/s)	L (cm)	N
400 bar, 40 °C	1.8	23.5	0.51	8.1	22,400
	3.0	182	0.31	37.7	62,500
	3.5	335	0.26	59.5	84,800
	5.0	1,400	0.19	174.0	173,000
	10	22,400	0.09	1392	693,000
1200 bar, 40 °C	1.8	70	0.52	24.2	67,100
	3.0	545	0.31	112.9	187,000
	3.5	1,010	0.26	179.3	255,000
	5.0	4,200	0.19	522.1	520,000
	10	67,200	0.09	4177	2,080,000
400 bar, 120 °C	1.8	5.5	1.75	6.4	17,900
	3.0	42.5	1.05	29.9	50,000
	3.5	79	0.90	47.6	67,700
	5.0	328	0.63	138.5	138,000
	10	5,250	0.31	1109	552,000

Notes:
a) All conditions are per Figure 2.2; values were calculated using Eqs. 2.18–2.21.
b) The optimum reduced velocity and minimum reduced plate height are 13.2 and 2.0, respectively, under all conditions.

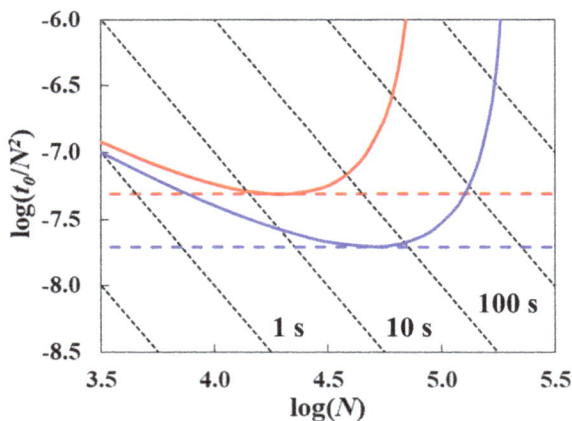

FIGURE 2.10 Basic kinetic plot of Desmet using the same data as in Figure 2.8. The red lines are the kinetic plot (solid – two-parameter optimization) and K-S-H line (dashed – three-parameter optimization) for P_{max} = 400 bar. The blue lines are the corresponding plots for P_{max} = 1200 bar. The dashed black lines are isochrones.

ways to plot the same results. Figure 2.10 is the "kinetic plot" for the same column under the same conditions as used in Figure 2.3A.

Each type of plot (Poppe and kinetic) has advantages for understanding the optimization process. The coordinates of t_0/N vs. N used in Figure 2.8 are the same as those used by Poppe [4] whereas in Figure 2.10 the coordinates of t_0/N^2 vs. N are the same as those used by Desmet [12]. The reader will recognize that if we multiply t_0/N^2 by $\Delta P/\eta$ one gets Knox's separation impedance, usually denoted E [78].

$$E = \left(\frac{\Delta P}{\eta} \right) \left[\frac{t_0}{N^2} \right] \qquad (2.26)$$

Thus at constant pressure and viscosity a plot of t_0/N^2 vs. N is really the same as a plot of E vs. N. If we compare two chromatographic devices (say a packed bed and a monolith), the device with the lower curve over any particular range in dead time or plate count has the lower separation impedance and is therefore the superior device. An advantage of the Poppe-style plot is that the limiting speed (the horizontal asymptote, see Figure 2.2) and the limiting plate count (the vertical asymptote) are quite evident. An advantage of the Desmet-style plot is that the curves are similar in shape to van Deemter H vs. u curves (albeit with the high velocity on the left in this case), and thus kinetic curves clearly have a minimum and two ascending branches that correspond to the B term (right) or C term (left) dominating performance. The filled points in Figure 2.8 correspond to the special points where the K-S-H lines are tangent to the "Poppe curves". Another important attribute of kinetic plots is that their minimum coincides with the minimum in the plot of h vs. v, consequently the K-S-H line must be tangent to the kinetic plot at the minimum in the kinetic plot. On the other hand, in Figure 2.8 it is not evident where the K-S-H limit is reached on each "Poppe curve" until the K-S-H line is actually calculated and plotted to see where it touches the Poppe curves. If we extrapolate the two kinetic curves at the different pressures to very short time, it is evident that they converge to give the same dependence of t_0/N^2 on N, that is, they have the same speed as required by Eq. 2.14.

2.6 ANALYSIS TIME AND RESOLUTION

One must never forget that the fundamental objective of a separation is not merely to obtain as many plates as possible in a given time, or to minimize the time needed to get some desired number of plates, but rather to achieve the desired degree of resolution. We can easily connect resolution to the analysis time by first recalling the general resolution equation [89]:

$$R_s = \frac{\sqrt{N}}{4} \frac{\alpha - 1}{\alpha} \frac{k}{1+k} \qquad (2.27)$$

Now we can take N as being equal to the K-S-H maximum plate number (N_{max}, see Eq. 2.21). A slight rearrangement quickly gives:

$$t_0 = 256 \frac{\Phi \cdot \eta \cdot h_{min}^2}{P_{max} \cdot \lambda} \left(R_s \frac{\alpha}{\alpha - 1} \frac{1+k}{k} \right)^4 \qquad (2.28)$$

We see that the analysis time is a *very strong function of the desired resolution*. In other words, improving the resolution by increasing the plate count via the analysis time is a very costly proposition. On the other hand, Eq. 2.28 also tells us of the extreme benefits to be gained by optimizing the separation selectivity (α). For example, suppose the α of the critical pair is 1.05 and by manipulating those factors that control it (stationary phase, eluent composition, pH, and temperature), α becomes 1.10, then at constant resolution the analysis time can be decreased by about 16-fold. This calculation assumes that the retention time of the most retained peak varies with the column dead time.

2.6.1 OPTIMUM RETENTION FACTOR AND SPEED

If we assume that the last peak in the chromatogram is the *critical peak*, that is, a member of the pair of *least well separated peaks*, then we can crudely estimate its retention time by multiplying the dead time obtained from Eq. 2.28 by $(1 + k)$ thus:

$$t_{R,last} = 256 \frac{\Phi \cdot \eta \cdot h_{min}^2}{P_{max} \cdot \lambda} \left(R_s \frac{\alpha}{\alpha - 1} \right)^4 \frac{(1+k)^5}{k^4} \tag{2.29}$$

If we now assume that h_{min} does not vary with retention factor – an assumption which is quite crude (see Section A.2) – then $t_{R,last}$ will vary as $(1+k)^5/k^4$. The minimum value of the last retention time occurs when the retention factor is exactly 4.0. Thus, the best results will be obtained with the conditions chosen so that the desired resolution and selectivity position the critical pair to have a retention factor of about 4.0.

2.7 COMPARISON OF SPEED AND PERFORMANCE PREDICTED BY TWO-AND THREE-PARAMETER OPTIMIZATION METHODS

It is clear that the three-parameter approach always predicts better speed and performance than does two-parameter optimization (see Figure 2.10, and [11])) provided that particles of the optimum size exist for all analysis times of interest. Clearly, only a limited number of particle sizes are commercially available. It is therefore important to understand how much performance is lost at each analysis time when the best particle size is not used. Matula and Carr [8] addressed this problem. As shown in Figure 2.11, which provides a closer look at the regions where the curves cross in Figure 2.7, the largest loss in performance occurs at the crossing point of adjacent Poppe curves drawn for the closest pair of available particle sizes. That is, starting at the crossover time as one moves towards the Poppe curve pertaining to the smaller particles, the columns packed with the smaller particles perform better at shorter analysis time than do the larger particles. Conversely, again assuming one is at the crossover time, if one moves towards the Poppe curve for the larger particles, the performance of the column packed with the larger particles improves relative to the column packed with the smaller particles.

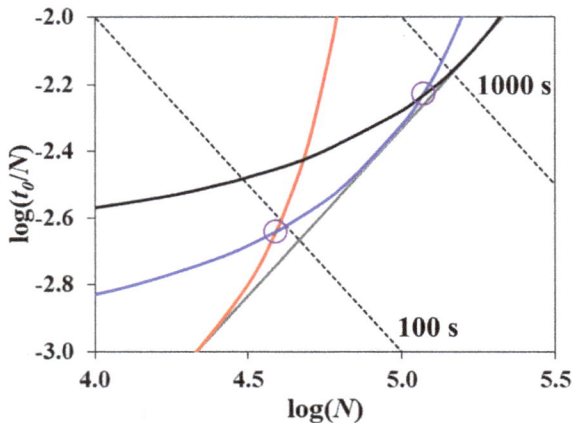

FIGURE 2.11 All conditions are the same as Figure 2.7 with the axes adjusted to focus on the crossover points and demonstrate that these intersections occur at the point of maximum deviations from the Knox-Saleem-Halasz line.

Matula showed that the ratio of the plate count of two-parameter optimization (N^*_{2P}) to three-parameter optimization (N_{3p}) at the crossing time of two Poppe curves is given by:

$$\frac{N^*_{2P}}{N_{3P}} = \frac{\dfrac{A}{\sqrt{BC}}+2}{\dfrac{A}{\sqrt{BC}}+\dfrac{d_{p,2}}{d_{p,1}}+\dfrac{d_{p,1}}{d_{p,2}}} \tag{2.30}$$

Clearly, the loss in plate count depends strongly on the ratio of the two particle sizes used. The greatest ratio for commercial particles occurs for pairing 5 µm with 10 µm particles. This two-fold ratio causes a loss of only 11%. The corresponding loss in resolution for isocratic and gradient chromatography depends on the square root of N and thus the maximum loss in resolution is quite small provided that we change particle sizes at the crossing time of the two Poppe curves. As will be shown below, use of the much simpler three-parameter (Eq. 2.21) as compared to the two-parameter equation (Eq. 2.12) for the dependence of N on analysis time is quite helpful and revealing when used to estimate gradient elution peak capacity (see Section 2.10 and Eq. 2.60).

2.8   INTRODUCTION TO PEAK CAPACITY

Peak capacity (denoted n_c) is of central importance, especially for 2D separations. The severely limited peak capacity of 1D chromatography is the principle reason why the high peak capacities, and especially the high speed of producing peaks enabled by 2D chromatography has become so important in separation science [90–94]. Peak capacity is the chief metric of separating power when dealing with complex samples – that is, those containing hundreds or more analytes. Its significance for polymer analysis, proteomics, and metabolomics can scarcely be exaggerated, as discussed in subsequent chapters. The biggest problem with chromatograms crowded with too many peaks is that any attempt to fix *band-spacing* in one region of a chromatogram results in a diminution in the room available for separation elsewhere. The only reliable solution is to increase n_c. Simply adjusting the chromatographic selectivity, that is, the band-spacing, will not help. We must then increase the available separation space relative to average space per peak; this quantity is directly related to n_c.

Peak capacity is defined as the *maximum* number of *well-resolved* peaks that can fit in a separation space defined as the earliest time (t_0) at which a peak can elute and the time at which the nth (last) peak elutes (t_n). Evidently, all peaks ought to have the same resolution (R_s) otherwise some will be too well separated, thereby wasting space. The requisite resolution is usually taken as 1.0. Here, B and A denote the later and earlier of a pair of adjacent peaks, t_R and σ are the retention times and standard deviations of these peaks:

$$R_s = \frac{t_{R,B}-t_{R,A}}{2\left(\sigma_A+\sigma_B\right)} \tag{2.31}$$

When R_s is set to 1.0, and the two peaks have about the same width, their retention times will differ by 4σ. In any given separation space, the total number of 4σ intervals is finite and controls the maximum number of peaks that can fit in the space. Giddings concluded that under conditions of time-invariant elution the 4σ width is:

$$4\sigma_i = 4\frac{t_{R,i}}{\sqrt{N}} \tag{2.32}$$

The $i^{th} + 1$ peak must elute at a time relative to the ith peak equal to the average of the two peak widths:

$$t_{R,i+1} - t_{R,i} = \frac{2\left(t_{R,i+1} + t_{R,i}\right)}{\sqrt{N}}$$ (2.33)

This result assumes that all analytes have the same N. We know that this is not exactly true because several broadening processes vary with analyte characteristics, especially their retention factors (k). Some authors suggest using N for an unretained species while others use the average of representative peaks. When we use a resolution other than 1.0, then Eq. 2.33 must be rewritten as:

$$t_{R,i+1} - t_{R,i} = \frac{2R_s\left(t_{R,i+1} + t_{R,i}\right)}{\sqrt{N}}$$ (2.34)

This readily becomes:

$$\frac{t_{R,i+1}}{t_{R,i}} = \frac{1 + \dfrac{2R_s}{\sqrt{N}}}{1 - \dfrac{2R_s}{\sqrt{N}}}$$ (2.35)

When the first peak elutes at $t_{R,1}$ equal to t_0, then the elution of the last peak relative to the first peak is the $n_c - 1$ products of the ratio on the right-hand side of Eq. 2.35:

$$\frac{t_{R,n}}{t_0} = \left[\frac{1 + \dfrac{2R_s}{\sqrt{N}}}{1 - \dfrac{2R_s}{\sqrt{N}}}\right]^{(n_c - 1)}$$ (2.36)

Solving Eq. 2.36 for n_c by taking the logarithm of both sides and rearranging gives

$$n_c = 1 + \frac{\ln\left(\dfrac{t_{R,n}}{t_0}\right)}{\ln\left[\dfrac{\left(1 + \dfrac{2R_s}{\sqrt{N}}\right)}{\left(1 - \dfrac{2R_s}{\sqrt{N}}\right)}\right]}$$ (2.37)

where n_c is the number of peaks that fit between $t_{R,n}$ and t_0 with all resolutions equal to R_s. When R_s is about 1 and N is large (>100), the argument of the logarithm in the denominator can be approximated as $1 + \dfrac{4R_s}{\sqrt{N}}$ with high accuracy (ca. 1%), thus:

$$n_c \approx 1 + \frac{\ln\left(\dfrac{t_{R,n}}{t_0}\right)}{\ln\left(1 + \dfrac{4R_s}{\sqrt{N}}\right)} \tag{2.38}$$

Under almost all conditions of importance $\dfrac{4R_s}{\sqrt{N}} \ll 1$; thus $\ln\left(1 + \dfrac{4R_s}{\sqrt{N}}\right) \approx \dfrac{4R_s}{\sqrt{N}}$, consequently:

$$n_c \approx 1 + \frac{\sqrt{N}}{4R_s} \cdot \ln\left(\frac{t_{R,n}}{t_0}\right) \tag{2.39}$$

In Grushka's seminal paper [90] he developed an alternative approach to estimating gradient elution peak capacity based on his observation that the number of peaks in a small increment of time is given by:

$$dn_c = \frac{dt}{4R_s\sigma} \tag{2.40}$$

On integrating Eq. 2.40 and substituting 4σ from Eq. 2.32 he got:

$$\int_1^n dn_c = \int_{t_0}^{t_{R,n}} \frac{dt}{4R_s\sigma} = \frac{\sqrt{N}}{4R_s} \int_{t_0}^{t_{R,n}} \frac{dt}{t} \tag{2.41}$$

Consequently, for time-invariant isocratic chromatography:

$$n_c = 1 + \frac{\sqrt{N}}{4R_s} \cdot \ln\left(\frac{t_{R,n}}{t_0}\right) \tag{2.42}$$

Horvath pointed out [93] that by using the appropriate temporally programmed elution conditions the widths of all peaks will be about equal, thus doing the integral in Eq. 2.41 assuming a constant σ equal to that of a peak that elutes at t_0 gives:

$$dn_c = \frac{\sqrt{N}}{4R_s t_0} dt \tag{2.43}$$

Finally, integration as done with the same limits as in Eq. 2.42 gives:

$$n_c = 1 + \frac{\sqrt{N}}{4R_s t_0}\left(t_{R,n} - t_0\right) \tag{2.44}$$

Inspection of the above equations clearly indicates the higher peak capacity for gradient elution LC. *This strongly suggests that in 2D-LC we should generally operate both dimensions of 2D-LC with*

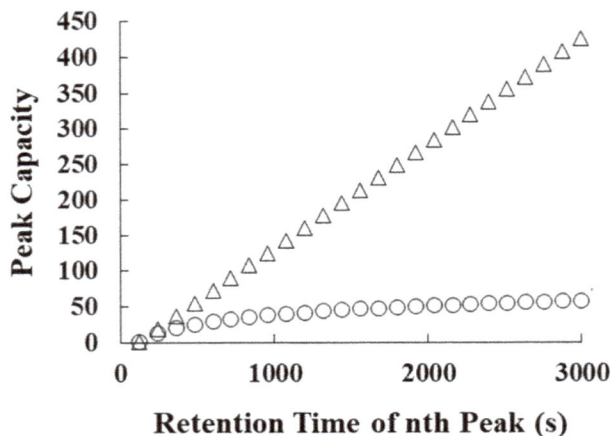

FIGURE 2.12 Comparison of peak capacity of programmed (Δ) and time-invariant ($\circ$) elution methods. $N = 5000$ and $t_0 = 120$ s for both types of elution. Reprinted with permission from [95]. Copyright 2018 from A study of peak capacity optimization in one-dimensional gradient elution reversed-phase chromatography: a memorial to Eli Grushka by P. Carr, D. Stoll. Reproduced by permission of Taylor and Francis Group, LLC, a division of Informa plc.

gradient elution conditions whenever warranted by the sample characteristics. Figure 2.12 shows that programmed elution produces more peak capacity per unit time than does time-invariant elution.

2.8.1 LIMITATIONS OF THE PEAK CAPACITY CONCEPT

In quantifying the concept of peak capacity several simplifying assumptions are made:

1. All analytes have the same N value
2. Under time-varying elution all peaks have the same peak width (σ)

It is well known that this second assumption is really only true for species that are strongly retained under the initial elution conditions (see the discussion concerning Eqs. 2.52 and 2.53 below). However, neither of these limitations is particularly serious relative to the assumption that the analytes are spaced by an interval equal to $2R_s(\sigma_i + \sigma_{i+1})$. We will never see peaks spread so uniformly. Clearly, the peak capacity concept is fundamentally hypothetical. It only sets the *upper limit* to the number of peaks that can be observed [90]. As Giddings [91] said: "One has no real hope of resolving 100 components on a column with n = 100". Questions about the number of peak maxima (singlets and multiplets) that one will observe *on average* with a sample containing *m randomly spaced* analytes were first answered in a truly classic paper by Davis and Giddings [96] in their seminal work on statistical overlap theory (SOT).

The most important result of SOT is that the expected value of the number of singlet (pure component) peaks (s) that will be seen on average is much less than n_c. The total number of peaks (p) – that is, singlets, doublets, and so on – is a larger number but still much less than n_c. Under the best circumstances, *on average*, one will only see *singlet* peaks numbering less than 18% of n_c and a *total* number of peaks that is, on average, only 37% of n_c. The maximum number of peaks occurs when m is equal to n_c and the maximum number of singlets is seen when $m = n_c/2$. Evidently, as a chromatogram becomes more crowded, the number of maxima must diminish because more peaks become fused as seen in Figure 2.13. The results of SOT prevent accurate quantitative work *with*

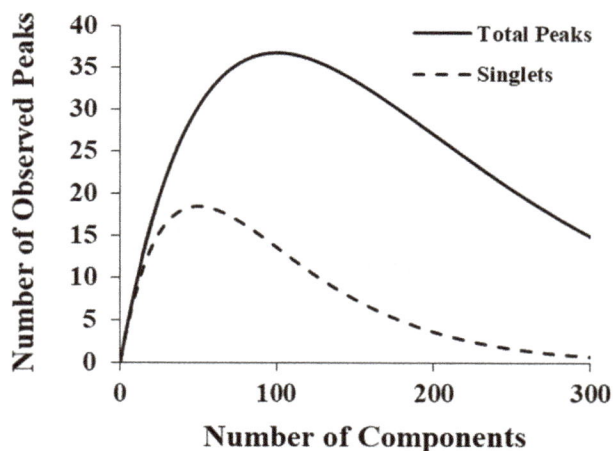

FIGURE 2.13 Davis–Giddings SOT prediction of the average number of observed peaks (p) and singlets (s) for a separation with a peak capacity of 100. Reprinted with permission from [95]. Copyright 2018 from A study of peak capacity optimization in one-dimensional gradient elution reversed-phase chromatography: a memorial to Eli Grushka by P. Carr, D. Stoll. Reproduced by permission of Taylor and Francis Group, LLC, a division of Informa plc.

minimal method development by adjusting only the relative band selectivity. Conventional method development really is a fight to decrease the separation entropy by bringing order to the relative band spacings. As mentioned above at some point a chromatogram becomes so crowded that adjusting the selectivity becomes a losing battle when the *m/n* ratio becomes less than 1.0.

The predictions of SOT have motivated the use of 2D separation methods in many areas of analytical chemistry. As explained by Giddings [94], when one has hundreds or even thousands of components there is no option but to increase the peak capacity. We will go into how this is physically accomplished in Chapters 3 and 4. We hasten to point out that 2D separations as described by Giddings are fundamentally different than separations using tandem (serially) coupled columns.

However, as will become quite apparent in Chapter 3, a major interest in 2D separations is how one can maximize the speed of peak capacity production of both the first and second dimension separations. In view of Figure 2.12 we will be talking about gradient elution and focus our attention on reversed-phase chromatography (RPLC) as it is one of the most common modes of all LC.

2.9   GRADIENT ELUTION REVERSED-PHASE LIQUID CHROMATOGRAPHY

Many chromatographers have contributed to the theory of gradient chromatography [7, 95, 97, 98]; however, most relevant here is the work of Uwe Neue [99–101]. At the outset it is widely agreed that the logarithm of the retention factor (k) is a *quasi-linear* function of the volume fraction of the organic modifier in the eluent (ϕ).

$$\ln(k) = \ln(k_w) - S \cdot \phi \tag{2.45}$$

Here, $\ln(k_w)$ is a hypothetical, strongly analyte-dependent, retention factor in pure water and S is an analyte-dependent sensitivity coefficient that controls how k varies with ϕ. In linear solvent strength theory (LSST), ϕ is a linear function of time:

$$\phi(t) = \phi_0 + (\phi_f - \phi_0)\frac{t}{t_g} = \phi_0 + \Delta\phi\frac{t}{t_g} \tag{2.46}$$

The terms ϕ_0, ϕ_f and t_g represent the initial and final mobile phase compositions, and the gradient time. Further, we assume that the gradient propagates at the same rate as the mobile phase. This means that the strong component of the mobile phase does not sorb into the stationary phase. This is an important assumption and has been challenged for fast gradients [98, 102]. Assuming that the gradient enters the column at the same time as the sample – that is, there is zero gradient delay – the analyte retention time will be:

$$t_r = t_0(1 + \frac{1}{b}\ln(bk_0 + 1)) \tag{2.47}$$

where k_0 is the retention factor at time equal to zero, and b is the *dimensionless gradient slope*.

$$b = \frac{S\Delta\phi t_0}{t_g} \tag{2.48}$$

Defining a *gradient retention factor* (k_g):

$$k_g \equiv \frac{t_r - t_0}{t_0} = \frac{1}{b}\ln\left(bk_0 + 1\right) \tag{2.49}$$

and then substituting b from Eq. 2.48 we get

$$k_g \equiv \frac{t_r - t_0}{t_0} = \frac{t_g}{S\Delta\phi t_0}\ln\left(\frac{S\Delta\phi t_0}{t_g} \cdot k_0 + 1\right) \tag{2.50}$$

Next, we consider the peak width. The relations are somewhat less certain than those for the retention time. The following equation is deemed theoretically correct by most authors [101, 103]:

$$\sigma = \frac{t_0}{\sqrt{N}}(1 + k_e)G(p) \tag{2.51}$$

The term $\frac{t_0}{\sqrt{N}}$ corresponds to σ for a peak eluting at the column dead time, and k_e is the *local* analyte retention factor at the column exit. Under LSST conditions, it can be shown that:

$$k_e = \frac{k_0}{bk_0 + 1} \tag{2.52}$$

A very important case occurs when k_0 is large:

$$k_e \approx \frac{1}{b} \tag{2.53}$$

Because analyte-to-analyte differences in k_0 are generally much larger than those in S, variations in k_e can be small provided k_0 is large, so for mathematical simplicity S is held constant when the integration in Eq. 2.41 is done in virtually all theoretical work. This is a significant weakness of the theory because S does vary between analytes and frequently tends to increase with retention.

Taken together, the first two terms in Eq. 2.51 correspond to what happens in isothermal/isocratic chromatography. The last term $G(p)$ is called the *gradient (peak) compression factor*. $G(p)$ accounts for the fact that the front of a peak is always in a slightly weaker eluent than the rear of a peak and thus the peak will be somewhat compressed. Poppe [103] showed that this term is given by Eq. 2.54:

$$G(p) = \sqrt{\frac{1 + p + \dfrac{p^2}{3}}{\left(1 + p\right)^2}} \qquad (2.54)$$

where p is defined as

$$p = \frac{k_0 \cdot b}{1 + k_0}; \text{ note when } k_0 \gg 1, \; p \approx b \qquad (2.55)$$

Obviously, if $b = 0$ then $p = 0$ which corresponds to isocratic chromatography. $G(p)$ must then be equal to 1.0 (no zone compression). At the other extreme, as p becomes exceedingly large (corresponding to very fast gradients), $G(p)$ approaches $\sqrt{1/3}$ – that is, about 0.58. Consequently, the factor $G(p)$ can only reduce the peak width by at most a factor of almost 2.

In his theoretical work [100, 101] on gradient peak capacity, Neue started with Grushka's integral (Eq. 2.41), but he used the value of σ for gradient elution obtained from Eq. 2.51 with $G(p)$ set equal to unity. Also both N and S were assumed to be the same for all analytes and the retention time of the last peak was taken as $t_0 + t_g$. This results in Eq. 2.56:

$$n_c \approx 1 + \frac{\sqrt{N}}{4} \frac{1}{1+b} \ln\left(\frac{1+b}{b} \exp\left(bk_g\right) - \frac{1}{b}\right) = 1 + \frac{\sqrt{N}}{4} \frac{1}{1+b} \ln\left(\frac{1+b}{b} \exp\left(S\Delta\phi\right) - \frac{1}{b}\right) \qquad (2.56)$$

We will call this equation *Neue's exact equation*. In deriving it Neue used Eq. 2.52; however, a much simpler equation is obtained when one assumes that all the solutes are well retained in the initial eluent ($bk_0 \gg 1$), and thus k_e is taken as that given by Eq. 2.53.

In this case, the integral simplifies:

$$n_c = 1 + \frac{\sqrt{N}}{4R} \frac{t_{r,n} - t_0}{t_0} \frac{1}{1 + k_e} \approx 1 + \frac{\sqrt{N}}{4R} \frac{t_{r,n} - t_0}{t_0} \frac{b}{1 + b} \qquad (2.57)$$

We will return later to see what happens when Eq. 2.54 is used for $G(p)$ in the recent work by both Gritti *et al.* [104], and Blumberg and Desmet [105–108]. As done previously, Eq. 2.57 is put in final form by taking the retention time of the last peak equal to be $t_g + t_0$ (Figure 2.14).

$$n_c \approx 1 + \frac{\sqrt{N}}{4R_s} \frac{t_g}{t_0} \frac{b}{1 + b} \approx 1 + \frac{\sqrt{N}}{4R_s} \frac{S \cdot \Delta\phi \cdot t_g}{t_g + S \cdot \Delta\phi \cdot t_0} \qquad (2.58)$$

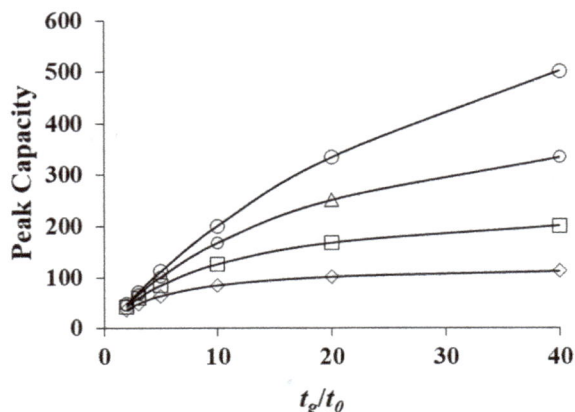

FIGURE 2.14   Peak capacity of reversed-phase gradient elution liquid chromatography according to Neue's approximate equation. In all cases, $\Delta\phi = 1$, with $S = 5$ ($\lozenge$), $S = 10$ ($\square$), $S = 20$ ($\triangle$), and $S = 40$ ($\circ$). The last peak has a retention time of $t_g + t_0$. Reprinted with permission from [95]. Copyright 2018 from A study of peak capacity optimization in one-dimensional gradient elution reversed-phase chromatography: a memorial to Eli Grushka by P. Carr, D. Stoll. Reproduced by permission of Taylor and Francis Group, LLC, a division of Informa plc.

TABLE 2.8
Errors Resulting (% Difference[a]) from the Use of the Neue's Approximate versus Exact Equations

t_0/t_g	$S\Delta\phi$			
	5	10	20	40
0.5	6.1	1.7	0.5	0.1
0.2	12.0	3.9	1.1	0.3
0.1	17.8	6.4	2.0	0.6
0.05	24.2	9.8	3.3	1.0
0.025	30.3	13.8	5.2	1.7

Source: Reprinted with permission froeprinted with permission from [95]. Copyright 2018 from A study of peak capacity optimization in one-dimensional gradient elution reversed-phase chromatography: a memorial to Eli Grushka by P. Carr, D. Stoll. Reproduced by permission of Taylor and Francis Group, LLC, a division of Informa plc.

a) %Difference = 100·(exact–approximate)/exact.

We will refer to both forms in Eq. 2.58 as *Neue's approximate equation*. Clearly, the peak capacity increases monotonically with gradient time and is higher for analytes, such as peptides, proteins, and low polarity macromolecules, that have larger values of S and mixtures that require a wide range in mobile phase composition. The question arises as to how much error is made when Eq. 2.53 is used or k_e instead of Eq. 2.52. As shown in Table 2.8 the error is rather small for fast gradients but becomes more substantial for slow gradients. Fortunately, as $S\Delta\phi$ increases the error decreases. A value of $S = 5$ is actually pretty small and would be typical of a molecule of the size of benzene ($MW = 78$); however, S generally gets bigger as the analyte MW increases. Clearly, Eq. 2.58 gives a reasonable estimate that is always conservative, as Neue's exact Eq. 2.57 gives higher peak capacities (see Table 2.8).

2.10 SPEED IN LIQUID CHROMATOGRAPHY AND OPTIMIZATION OF PEAK CAPACITY

In order to compute the peak capacity we use the maximum possible value of the plate count obtained using the K-S-H optimization method (see Eq. 2.21, which can be written most compactly using ψ as defined by Eq. 2.8):

$$N_{max} = \frac{L^*}{h_{min} \cdot d_p^*} = \sqrt{\frac{P_{max} \cdot \lambda \cdot t_0}{\Phi \cdot \eta}} \frac{1}{h_{min}} = \frac{\psi \cdot \lambda}{h_{min}} \cdot \sqrt{t_0} \tag{2.59}$$

On the other hand, if one merely optimizes the eluent velocity and column length at some *arbitrary but available particle size* one will sacrifice some efficiency. As discussed above, Matula [8] showed that when one changes particle size from one available particle size to the next larger available size *at the appropriate time*, the lost performance only amounts to about 11%. Thus, following Meyer [7] using Eq. 2.59 for N in the equation for gradient peak capacity causes little error.

We now substitute Eq. 2.59 in Eq. 2.58 to get:

$$n_c \approx 1 + \frac{1}{4R_s} \sqrt{\frac{\psi \cdot \lambda}{h_{min}}} t_0^{1/4} \frac{S \cdot \Delta\phi \cdot t_g}{t_g + S \cdot \Delta\phi \cdot t_0} = 1 + \frac{1}{4R_s} \sqrt{\frac{\psi \cdot \lambda}{h_{min}}} \frac{S \cdot \Delta\phi \cdot t_0^{1/4}}{1 + b} \tag{2.60}$$

Note that b depends on t_0, thus the dead time is in both the numerator and denominator of Eq. 2.60, which suggests that there is an optimum value for t_0. Differentiation of Eq. 2.60 leads to the result that there is a maximum peak capacity (see Figure 2.15) when:

$$t_g = 3 \cdot S\Delta\phi \cdot t_0 = t_{g,opt} \tag{2.61}$$

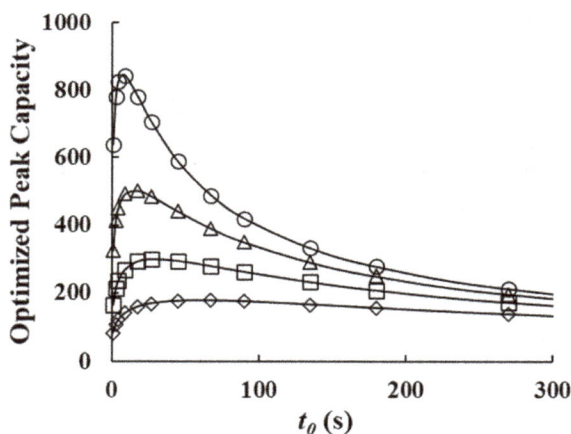

FIGURE 2.15 Dependence of optimized peak capacity versus dead time at fixed gradient time. Conditions: $t_g = 900$ s, curves: $S\Delta\phi = 5$ (◊), $S\Delta\phi = 10$ (□), $S\Delta\phi = 20$ (△), and $S\Delta\phi = 40$ (○). Conditions: P_{max}, 400 bar; T, 40°C; η, 0.69 with acetonitrile at a volume fraction of 0.25 in water; D_m, 9.0×10⁻⁶ cm²/s; Φ, 500; λ, 0.67. The reduced A, B, and C terms in the van Deemter equation are 0.95, 7.0, and 0.040, respectively. Reprinted with permission from [95]. Copyright 2018 from A study of peak capacity optimization in one-dimensional gradient elution reversed-phase chromatography: a memorial to Eli Grushka by P. Carr, D. Stoll. Reproduced by permission of Taylor and Francis Group, LLC, a division of Informa plc.

Of course, Eq. 2.61 also tells us that an optimum value of the dimensionless gradient slope parameter (*b*) must exist:

$$b_{opt} = \frac{S \cdot \Delta\phi \cdot t_0}{t_{g,opt}} = \frac{1}{3} \qquad (2.62)$$

Substitution of b_{opt} in Eq. 2.60 gives the amazingly simple result for the maximum peak capacity:

$$n_{c,max} \approx 1 + \frac{3}{16R_s}\sqrt{\frac{\psi \cdot \lambda}{h_{min}}} S \cdot \Delta\phi \cdot t_0^{1/4} \approx 1 + \frac{0.1875}{R_s}\sqrt{\frac{\psi \cdot \lambda}{h_{min}}} S \cdot \Delta\phi \cdot t_0^{1/4} \qquad (2.63)$$

Alternatively, in terms of t_g we get:

$$n_{c,max} \approx 1 + \frac{1}{16R_s}\sqrt{\frac{\psi \cdot \lambda}{h_{min}}}(3S \cdot \Delta\phi)^{3/4} t_g^{1/4} \approx 1 + 0.1424 \frac{1}{R_s}\sqrt{\frac{\psi \cdot \lambda}{h_{min}}}(S \cdot \Delta\phi)^{3/4} t_g^{1/4} \qquad (2.64)$$

There seems to be some disagreement on the value of the maximum possible peak capacity, and the existence and value of the optimum dimensionless gradient slope [20, 49]. The studies of Desmet and Blumberg [24] also arrive at the conclusion that there *must be an optimum gradient rate*. However, because they include the gradient compression factor $G(p)$ in their calculation of the peak width, the optimum *b* is a function of k_0.

The most important thing that Eqs. 2.63 and 2.64 teach is that *the peak capacity under optimum conditions only increases with the one-fourth power of time whether it is gauged by either the column dead time or the gradient time*. It should now be clear that Eq. 2.58 is somewhat misleading in two regards. First, the initial rate of increase in the optimum peak capacity with t_g is not linear with t_g as Eq. 2.58 suggests, and second, the optimized peak capacity does not approach a horizontal asymptote but rather it increases indefinitely with the amount of time invested. These issues result because Eq. 2.58 assumes that the column length, velocity, and particle size are all held fixed as t_g is increased, whereas it is well known that as t_g is increased, one needs to increase the column length to maximize the peak capacity [109]. These results, as a function of the dead time, are shown in Figure 2.16. The slope of these log-log plots is essentially 0.25, as unity is much smaller than the second time-dependent term in Eqs. 2.63 and 2.64. Plots of the maximum possible peak capacity versus the gradient time are more complex in that the optimum value of t_g corresponding to a given t_0 depends on $S\Delta\phi$. Instead, we give maximum peak capacity results versus t_g in tabular form (Tables 2.9 and 2.10).

The aforementioned results tell us that the optimized peak capacity increases almost exactly in proportion to the one-fourth power of both the column dead time and the gradient time. Further, we see that peak capacity increases with increases in pressure or temperature (see Table 2.10). In addition, increasing both temperature and pressure simultaneously has a cumulative effect. As Neue's approximate equation gives lower peak capacities than his more exact equation, the numerical values given here are somewhat conservative; however, the maximum difference between the two equations is only about 11% (Table 2.8) when $S\Delta\phi = 10$ and gets smaller as $S\Delta\phi$ increases. Finally, neither Eq. 2.62, nor 2.63 include the gradient compression term G(p) that was included in the work of Desmet and Blumberg [23, 24]. Since compression reduces the peak width for some peaks it follows that even higher results can be expected (see Table 2.11).

It is often stated that the peak capacity under gradient elution varies with the square root of the column length. This is not true under optimized conditions. Jorgenson [50] rightly points out that this assumes that all the conditions are held constant in the developing Eq. 2.58. Keep in mind that K-S-H optimization of N requires that the column length, particle diameter, and mobile phase

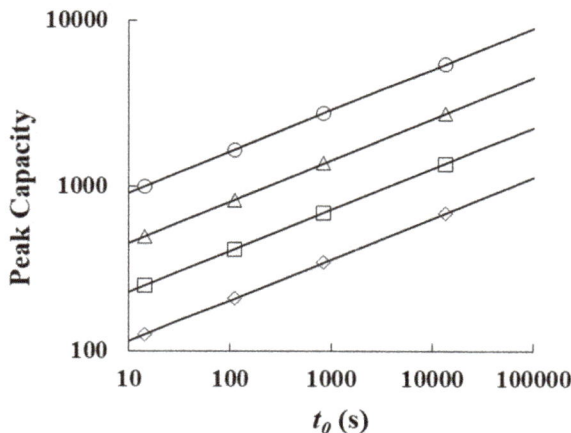

FIGURE 2.16 Optimum peak capacity according to Eq. 2.69. The points marked on the curve correspond to 1.8, 3.0, 5.0, and 10 μm particle diameters from left to right. All conditions as in Figure 2.15. $S\Delta\phi = 5$ (◊), $S\Delta\phi = 10$ (□), $S\Delta\phi = 20$ (△), and $S\Delta\phi = 40$ (○). Reprinted with permission from [95]. Copyright 2018 from A Study of Peak Capacity Optimization in One-dimensional Gradient Elution Reversed-phase Chromatography: A Memorial to Eli Grushka by P. Carr, D. Stoll. Reproduced by permission of Taylor and Francis Group, LLC, a division of Informa plc.

TABLE 2.9
Maximum Possible Peak Capacity[a] as a Function of Gradient Time

t_g (min)	$S\Delta\phi$			
	5	10	20	40
5	140	230	385	640
15	180	300	500	840
30	210	360	600	1000
60	250	420	710	1200
120	300	500	850	1400

Source: Reprinted with permission from [95]. Copyright 2018 from A study of peak capacity optimization in one-dimensional gradient elution reversed-phase chromatography: a memorial to Eli Grushka by P. Carr, D. Stoll. Reproduced by permission of Taylor and Francis Group, LLC, a division of Informa plc.

a) All conditions are per Figure 2.2; values were calculated using Eq. 2.64. Requires $t_g = 3 \cdot S\Delta\phi \cdot t_0$.

velocity must be continuously and simultaneously varied as the analysis timescale is increased (see Eqs. 2.18–2.20). Assuming K-S-H optimization (Section 2.4.4), the equation relating N_{max} to L^* is easily derived:

$$N_{max} = \frac{1}{h_{min}} \sqrt[3]{\frac{P_{max}}{\Phi \cdot \eta \cdot D_m}} \left[\frac{C}{B}\right]^{1/6} (L^*)^{2/3} \qquad (2.65)$$

Since the peak capacity depends on $\sqrt{N_{max}}$ (see Eq. 2.58), it follows that the optimized peak capacity varies with $(L^*)^{1/3}$; thus, to just double the peak capacity the column length needs to be increased

TABLE 2.10

Effect of Temperature and Pressure on Maximum Possible Peak Capacity[a]

Temperature (°C)	40		120	
Pressure (bar)	400	1200	400	1200
t_g (min)	Peak Capacity			
5	230	300	290	385
15	300	390	385	505
30	360	470	460	600
60	420	560	540	715
120	500	660	645	850

Source: Reprinted with permission from [95]. Copyright 2018 from A study of peak capacity optimization in one-dimensional gradient elution reversed-phase chromatography: a memorial to Eli Grushka by P. Carr, D. Stoll. Reproduced by permission of Taylor and Francis Group, LLC, a division of Informa plc.

Note:
a) All conditions are as in Figure 2.2, except that the viscosity and analyte diffusion coefficient are adjusted as needed to account for the elevated temperatures; $S\Delta\phi = 10$.

TABLE 2.11

Comparison of Peak Capacities With and Without the Gradient Compression Factor[a]

$S\Delta\phi$	t_g (min)	n, with $G(p) = 1$	n, with $G(p) =$ Eq. 2.54	Ratio
5	15	121.6	136.5	1.122
10	30	216.5	243.4	1.124
20	60	403.9	456.7	1.131
30	90	591.3	669.9	1.133
40	120	778.7	883.6	1.135

Source: Reprinted with permission from [95]. Copyright 2018 from A study of peak capacity optimization in one-dimensional gradient elution reversed-phase chromatography: a memorial to Eli Grushka by P. Carr, D. Stoll. Reproduced by permission of Taylor and Francis Group, LLC, a division of Informa plc.

eight-fold, whereas the analysis time must be increased 16-fold. Interestingly, this agrees with the findings of Blumberg and Desmet [108].

It is also interesting to examine the dependence of N_{max} on the optimum particle size $\left(d_p^*\right)$ and optimum interstitial mobile phase velocity $\left(u_e^*\right)$:

$$N_{max} = \frac{1}{h_{min}} \left[\frac{P_{max}}{\Phi \cdot \eta \cdot D_m} \right] \left[\frac{C}{B} \right]^{1/2} \left(d_p^* \right)^2 \tag{2.66}$$

$$N_{max} = \frac{1}{h_{min}} \left[\frac{P_{max} \cdot D_m}{\Phi \cdot \eta} \right] \left[\frac{B}{C} \right]^{1/2} \frac{1}{\left(u_e^* \right)^2} \tag{2.67}$$

These two equations tell us that as t_0 is increased we must simultaneously increase both the column length and particle size, while decreasing the eluent velocity to achieve the maximum possible plate count. As the peak capacity increases with the square root of N_{max} it follows that, to double the peak capacity the plate count must be quadrupled, consequently the particle size must be doubled. Similarly, to double the peak capacity the eluent velocity must be decreased two-fold.

It is clearly very *expensive* to buy peak capacity by increasing the analysis time in 1D gradient elution. This underscores the importance of 2D methods when high peak capacities are needed. Another important point is that we assume all solutes have the same diffusion coefficient. A somewhat surprising result of K-S-H optimization is that while all the three optimization variables (mobile phase velocity, column length, and particle size – see Eqs. 2.18–2.20) vary with the analyte's diffusion coefficient, N_{max} (see Eq. 2.59) does not.

2.11   EFFECT OF THE GRADIENT COMPRESSION FACTOR ON THE PEAK CAPACITY

As mentioned earlier, Blumberg and Desmet included $G(p)$ in their approach to computing the peak capacity. Again, based on Grushka's integral formulation (Eq. 2.40) the peak capacity will now be given by

$$\int_1^n dn_c = \int_{t_0}^{t_{R,n}} \frac{dt}{4R_s \cdot \sigma} = \frac{\sqrt{N}}{4R_s \cdot t_0} \int_{t_0}^{t_{R,n}} \frac{dt}{(1+k_e)G(p)} \tag{2.68}$$

As both k_e and $G(p)$ depend on k_0, and we expect them to vary substantially as time progresses across the gradient, the actual integral is rather complex. Table 2.12 shows the influence of the initial retention factor and the dimensionless gradient slope on the extent of gradient compression.

At the optimum value of b ($=1/3$) as given by Eq. 2.62 we see that $G(p)$ only varies by about 7.7% as k_0 varies from 1.0 to an essentially infinitely large value. The limiting value of $G(p)$ as k_0 becomes very large is (see Eq. 2.54):

$$G(p) \to \sqrt{\frac{1+b+\dfrac{b^2}{3}}{(1+b)^2}} \tag{2.69}$$

TABLE 2.12
Dependence of the Gradient Compression Factor ($G(p)$) on k_o and b

k_0	0.01	0.05	0.1	0.333	0.5	1
1	0.9975	0.988	0.976	0.930	0.902	0.839
2	0.9967	0.984	0.969	0.911	0.878	0.808
5	0.9959	0.980	0.962	0.894	0.857	0.784
10	0.9955	0.978	0.959	0.886	0.849	0.774
30	0.9952	0.977	0.956	0.881	0.842	0.767
100	0.9951	0.977	0.955	0.879	0.840	0.765
10,000	0.9951	0.976	0.955	0.878	0.839	0.764

Source: Reprinted with permission from [95]. Copyright 2018 from A study of peak capacity optimization in one-dimensional gradient elution reversed-phase chromatography: a memorial to Eli Grushka by P. Carr, D. Stoll. Reproduced by permission of Taylor and Francis Group, LLC, a division of Informa plc.

If we also use the limiting value of k_e as is done to get Neue's approximate peak capacity (Eq. 2.53) then the denominator in the integral of Eq. 2.68 becomes independent of k_0 and consequently of time, so we can easily do the integral to obtain:

$$n_c \approx 1 + \frac{\sqrt{N}}{4R_s} \frac{t_g}{t_0} \frac{b}{\sqrt{1+b+\dfrac{b^2}{3}}} \approx 1 + \frac{\sqrt{N}}{4R_s} \frac{S \cdot \Delta\phi}{\sqrt{1+b+\dfrac{b^2}{3}}} \tag{2.70}$$

Now, when we substitute Eq. 2.59 for N_{max} we get

$$n_c \approx 1 + \frac{1}{4R_s} \sqrt{\frac{\psi \cdot \lambda}{h_{min}}} \frac{S \cdot \Delta\phi \cdot t_0^{1/4}}{\sqrt{1+b+\dfrac{b^2}{3}}} \tag{2.71}$$

Once again there is an optimum value of b that maximizes n_c; in this case it is:

$$b_{opt} = 0.618 \tag{2.72}$$

And thus:

$$n_{c,max} \approx 1 + \frac{0.1894}{R_s} \cdot \sqrt{\frac{\psi \cdot \lambda}{h_{min}}} \cdot S \cdot \Delta\phi \cdot t_0^{1/4} \tag{2.73}$$

Equation 2.73 should be compared with Eq. 2.63. The difference in the two approaches (based on Neue [25]) in comparison to Desmet–Blumberg [108] is barely greater than 1%. It should be recalled that according to Table 2.8 the compression effect is at a maximum when k_0 is very large. Thus, we do not anticipate any big difference between Neue's more approximate method and the exact treatment of Desmet and Blumberg when k_0 is large. Combining Eq. 2.71 with Eq. 2.72 gives the analog to Eq. 2.64:

$$n_c \approx 1 + \frac{0.1679}{R_s} \sqrt{\frac{\psi \cdot \lambda}{h_{min}}} (S \cdot \Delta\phi)^{3/4} t_g^{1/4} \tag{2.74}$$

In terms of t_g the Desmet–Blumberg treatment at the limiting value of k_0 gives an increased peak capacity of about 17% when Eq. 2.74 is compared to Eq. 2.64.

The aforementioned approach assumes that k_0 is always very large, so that $G(p)$ is given by Eq. 2.69. The exact closed-form integral is quite complex but Eq. 2.68 is easily integrated numerically. One can then compare Neue's exact equation with $G(p) = 1$ to the exact numerical integral with k_e and p given by Eqs 2.52 and 2.55, respectively. The comparison is most significant at the optimum value of b given by using Neue's approximate equation (i.e., $b_{opt} = 1/3$).

We see that there is only a 10%–13% error between the Desmet–Blumberg result and Neue's exact result. Obviously, when gradient compression is included, the peaks are narrower and higher peak capacities will be obtained. Nonetheless, we see that what we have called Neue's exact equation (Eq. 2.56) and the exact treatment of Desmet and Blumberg (Eq. 2.54) are numerically quite similar.

It must be understood that Neue's integrations, which lead to Eqs 2.56 and 2.57, and that of Desmet and Blumberg, which gives Eq. 2.66, all assume that N and b (and thus S) are numerically the same for all analytes in a given sample. Furthermore, Desmet and Blumberg [107] assume a value of

S of about 10 for a *typical* analyte of MW 100–500. Generally, the n-alkylbenzenes form a homologous series which for the series from benzene to pentadecyl benzene (MW = 78 to 289) both S, and thus b, vary almost three-fold. The analyte's polarity also has a significant effect. Consider solutes ranging from *n*-benzylformamide (S = 2.83, MW = 144) to benzophenone (S = 8.45, MW = 183). Clearly, the errors caused by the assumption that N and b are constant under the integration are very likely considerably bigger than when we ignore gradient compression; consequently, we feel that k_e and p can reasonably be approximated as $1/b$ and b, respectively (see Eqs 2.53 and 2.57).

2.12   SUMMARY

The most important points in this chapter relevant to optimizing the speed and resolving power of liquid chromatography are as follows. These points are valuable guiding principles that can be used to make decisions about parameters (e.g., column dimensions, particle sizes, flow rates) during development of 2D-LC methods as discussed in Chapters 3, 6, and 7. Since the first and second dimension separations are typically carried out on quite different timescales (i.e., tens of minutes and tens of seconds, respectively), it is important to understand that the optimal parameters for the first dimension can be quite different from the optimal parameters for the second dimension.

There are two main approaches to optimization that are practically relevant:

1. Two-parameter (also referred to here as Poppe) optimization – one chooses a particle size, operating conditions (temperature, maximum available pressure, eluent composition) and varies the mobile phase velocity and column length to maximize the plate count for a given analysis time (proportional to the column dead time).
2. Three-parameter (also referred to here as Knox-Saleem-Halasz or K-S-H) optimization – the particle size, column length and mobile phase velocity are simultaneously adjusted to maximize the plate count for a given analysis time.

In both cases, the best separation speed or plate count is not always obtained with the smallest available particles. Generally larger particles yield better separations at long analysis times (with long columns and low mobile phase velocities), and smaller particles yield the best separations at short times (with short columns and high mobile phase velocities).

The absolute best results in terms of speed and resolution are obtained using three-parameter (K-S-H) optimization. Under these conditions, the following relationships hold:

* doubling the plate count requires a four-fold increase in the pressure drop across the column;
* when the operating pressure is held constant, doubling the plate count requires a four-fold increase in the analysis time;
* when the plate count is held constant, decreasing the analysis time by a factor of two requires a four-fold increase in the pressure; and
* the maximum achievable plate count increases with the square of the particle diameter.

When optimizing using the two-parameter (Poppe) approach and the particle size is fixed, separation speed as measured by the dead time/plate count ratio becomes independent of the maximum operating pressure.

Given the importance of gradient elution conditions in 2D-LC, a significant portion of the chapter has been devoted to optimization of these conditions. The key takeaways from this discussion are:

* except for very fast separations (dead time of < 1 s) where column re-equilibration can be a limiting factor, higher peak capacities can be obtained under gradient conditions than with isocratic conditions;

- for reversed-phase gradient elution separations optimized using the three-parameter approach, peak capacity increases with the 1/4th power of the column dead time (t_0) or gradient time (t_g). Thus, doubling the peak capacity requires a 16-fold increase in the analysis time; and
- under the same conditions, peak capacity is proportional to $S \cdot \Delta\phi \cdot t_0^{1/4}$, and peak capacity is proportional to the to the 1/3rd power of the optimized column length.

APPENDIX

See the Glossary at the end of the book for definitions of all symbols in this Appendix.

A.1 RELATIONSHIPS BETWEEN PLATE HEIGHT AND MOBILE PHASE VELOCITY

Although there had been many important theoretical studies of peak broadening in continuous linear chromatography, most especially that of Lapidus and Amundson [110], van Deemter, Zuiderweg and Klinkenberg [3] were the first to produce a theoretical equation (see Eq. A.1) relating the height equivalent of a theoretical plate, denoted H, to u_e – the *interstitial* linear velocity of the mobile phase. Here H is the ratio of column length (L) to the number of plates (N) on the column. The Greek symbols in Eq. A.1 are variables independent of u_e but dependent of various analyte and column parameters including factors related to diffusion coefficients, retention, and column and packing material characteristics such as porosities. Such issues will be discussed in more detail below. Van Deemter *et al.* generated a three-term equation the form of which is now almost universally referred to as the van Deemter equation.

$$H = 2\lambda d_p + \frac{2\gamma D_m}{u_e} + \omega \frac{u_e d_p^2}{D_m} \tag{A.1}$$

Equation A.1 is frequently written in the simplified form of Eq. A.2a to emphasize the three different dependencies of H on the linear mobile phase velocity (u_0), defined as as L/t_0, where t_0 is the elution time of an unretained species.

$$H = A' + \frac{B'}{u_0} + C'u_0 \tag{A.2a}$$

However, in most recent work the interstitial velocity (u_e) is preferred to u_0. We will write the van Deemter equations as Eq. A.2b:

$$H = A + \frac{B}{u_e} + Cu_e \tag{A.2b}$$

Note that d_p and D_m in Eq. A.1 represent the packing material particle size, and the analyte's molecular diffusion coefficient in the mobile phase in the interstices between particles, respectively. Physically the first term (the "A" term which here has units of length) represents the longitudinal dispersion of the analyte due to variations in the interstitial mobile phase velocities – sometimes called the velocity biases – as well as the geometry of the packed bed. The A term is often called the "eddy" dispersion or longitudinal flow heterogeneity induced dispersion term. In van Deemter's work (Eqs. A.1 and A.2a) the A term, which has units of length as does H, is independent of both mobile phase velocity and the analyte's retention factor; however, later work by Giddings [111] and Desmet [15] showed that in fact this term does depend on the interstitial velocity and analyte retention. The second term (the "B" term which depends inversely on velocity and here has units of length^2 /time) accounts for longitudinal (axial) molecular diffusion. In liquid chromatography this process results from longitudinal analyte diffusion in the moving fluid outside the particles, in the stagnated mobile phase inside the particle pores, and in the stationary phase on the interior surface of the particles. This is quite different from the situation in gas-liquid chromatography where longitudinal diffusion in the stationary phase inside the particles – that is in the stationary phase liquid – is essentially negligible compared to the rate of longitudinal diffusion in the gaseous mobile phase. The third term (the "C" term, which is proportional to mobile phase velocity and has units of time) accounts for slow interphase mass transfer between the moving mobile phase and the mobile phase on the interior of the particles. Thus, this term has both convective and diffusive components. Van Deemter developed equations for A, B and C that related them to several factors: 1) the ratio of the analytes' effective diffusion coefficients in the mobile and stationary phases; 2) an empirical column packing homogeneity factor on the order of unity; 3) the relative amounts of mobile and immobile phases in the column; 4) the analyte equilibrium constant for transfer from the moving mobile phase to the stationary phase; 5) transfer rate coefficients in the mobile and stationary phases; and 6) the thickness of the stationary phases film. We will describe these relationships as they pertain to LC in more detail below.

For the purpose of optimizing performance and speed it is important to understand that there is a minimum H in plots of H versus velocity regardless of whether it is expressed as u_0 or u_e. It is easily shown that the velocity that minimizes H is:

$$u_{e,opt} = \sqrt{\frac{B}{C}} \qquad (A.3)$$

The minimum H, which maximizes N, is:

$$H_{min} = A + 2\sqrt{BC} \qquad (A.4)$$

Several groups, especially Giddings [112], have pointed out that when one substitutes dimensionless (also called *reduced*) variables for H and u, much simpler and more important universally applicable equations and plots result. The reduced plate height (h) and the reduced *interstitial* velocity (v_e) are defined as:

$$h \equiv \frac{H}{d_p} \qquad (A.5)$$

$$v_e \equiv \frac{u_e d_p}{D_m} \qquad (A.6)$$

Examination of the units show that both h and v_e have no dimensions. When Eqs. A.5 and A.6 are substituted in Eq. A.2 we get:

$$h = A + \frac{B}{v_e} + C v_e \qquad (A.7)$$

As shown in Figure A1.A [112], plots of H vs u_e for columns packed with different size particles are not at all the same, but as we see in Figure A1.B, when one plots the same data after they are

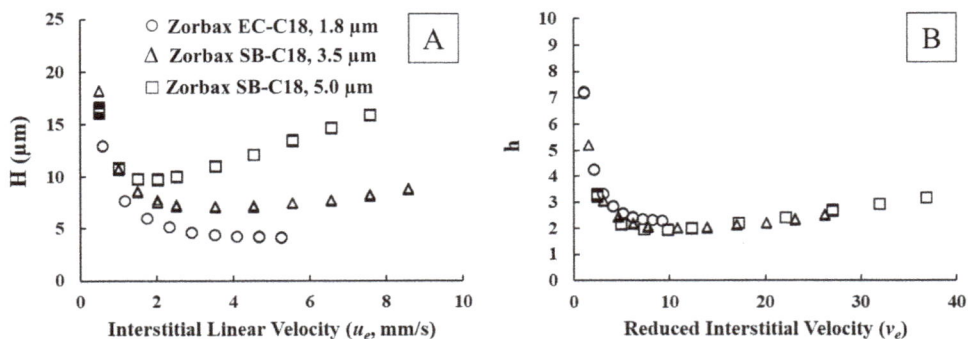

FIGURE A.1 Experimental plate height versus mobile phase velocity curves for 1.8, 3.5, and 5.0 μm particles plotted using A) non-reduced (i.e., H vs. u_e) coordinates, and B) reduced (h vs. v_e) coordinates. Republished with permission of Elsevier, from M. Dittmann, X. Wang, New materials for stationary phases in LC/MS, in *Handbook of Advanced Chromatography / Mass Spectrometry Techniques*, C. Byrdwell (Ed.), 2017; permission conveyed through Copyright Clearance Center, Inc.

transformed into reduced variables (h and v_e) the plots are close to superimposable. Perhaps one of the most important insights is that the velocity that is most fundamental in understanding the various dynamic processes that are responsible for the longitudinal dispersion of the analyte is the interstitial velocity (u_e), which corresponds to the speed at which analyte molecules *move past the solid particles* while such molecules are in the moving phase. The interstitial velocity and the linear mobile phase velocity (u_0), sometimes called the percolation velocity, are related by the ratio of the total porosity of the column ($\varepsilon_T = \varepsilon_e + \varepsilon_i \cdot (1 - \varepsilon_e)$) to the porosity outside the particles, that is the porosity of the column *external* to the particles (ε_e, where ε_i is the internal porosity of the particles), thus:

$$\frac{u_e}{u_0} = \frac{\varepsilon_e + \varepsilon_i (1 - \varepsilon_e)}{\varepsilon_e} \tag{A.8}$$

Clearly, u_e is always higher than u_0 unless totally non-porous particles (NPPs) are used.

A.2   DETAILS RELATED TO THE "A" TERM

Giddings [113] soon recognized on both empirical and theoretical grounds that the first term in Eqs. A.2a and A.2b, the so-called "eddy dispersion" term, could not be independent of velocity. This led to his famous "coupling theory" and the resulting "coupling equation" for the eddy dispersion process.

$$h_{eddy,Gidd} = \left[\left[\frac{1}{A_{Gidd}} \right] + \left[\frac{1}{C_{Gidd} \, v_e} \right] \right]^{-1} \tag{A.9}$$

We note that the high velocity limit of Eq. A.9 is:

$$h_{eddy,Gidd} = A_{Gidd} \ \text{as} \ v_e \to \infty \tag{A.10}$$

This corresponds to the simple van Deemter form for eddy dispersion, where plate height is a constant (A_{Gidd}) independent of v_e. In contrast, the low velocity limit (Eq. A.11) shows that the eddy dispersion plate height is linearly dependent on velocity:

$$h_{eddy,Gidd} = C_{Gidd} v_e \ as \ v_e \rightarrow 0 \qquad (A.11)$$

Obviously, at intermediate velocities the coupling theory shows that h_{eddy} vs. v_e is a curve with less than a first power dependence on velocity. In accord with this idea as well as experimental measurements, Knox [114] showed that for liquid chromatography h_{eddy} as a function of velocity is well approximated by an empirical equation of simpler form than Eq. A.9.

$$h_{eddy,Knox} = A_{Knox} v_e^\alpha \ with \ 0 \le \alpha \le 1 \qquad (A.12)$$

Knox [114] favored a value of α equal to approximately *1/3* based on his experimental results. Thus, in practice, his eddy dispersion equation has only a single adjustable parameter (A_{Knox}). Others have shown that the exponent (α) depends on the range of velocities used to experimentally determine α, and most of these studies found that α was between 0.5 and 1.0 [14].

In Giddings's coupling work [113] he pointed out that there are five different lateral length scales over which velocity biases, each having a different longitudinal velocity, might exist. He termed these five length scales *trans-channel*, *trans-particle*, *short-range*, *long-range*, and *trans-column*. A specific longitudinal velocity bias, denoted with the index *j*, is terminated by the solute moving from one flow stream by either lateral molecular diffusion or convectively by some local lateral velocity. Each of these two lateral transport processes (diffusion and convection) can be coupled using the formalism expressed in Eq. A.9. Each convective contribution results in its own $A_{Gidd,j}$ and each diffusive contribution has its own $C_{Gidd,j}v_e$. Both parts are coupled harmonically as in Eq. A.9. The sum of all five processes generates the final expression of Giddings's general coupling theory:

$$H_{eddy,Gidd} = \sum_{j=1}^{N} \left[\left[\frac{1}{A_{Gidd,j}} \right] + \left[\frac{1}{C_{Gidd,j} v_i} \right] \right]^{-1} \ with \ N = 5 \qquad (A.13)$$

Desmet and his group in an elegant series of papers [12–16] succeeded at clarifying and correcting a number of short comings of the Giddings eddy coupling theory. First, and perhaps most significantly, until Desmet's paper [13] eddy dispersion was almost always treated as if it did not depend on analyte retention. Considering the finite parallel zone (PFZ) approach based on the insightful work of Berdichevsky and Neue [115] and applied by Desmet *et al.* [13], the eddy dispersion term should be rewritten as follows:

$$h_{eddy,Desm} = \sum_{j=1}^{N} C_{Desm,j} v_e \left[1 - \frac{C_{Desm,j} v_e}{2A_{Desm,j}} \cdot \left(1 - \exp\left(\frac{-2A_{Desm,j}}{C_{Desm,j} v_e} \right) \right) \right] \qquad (A.14)$$

The exponential terms in Eq. A.14 account for the time dependences of the five local transport processes (trans-channel, short-range, etc.) that lead to the termination of a local velocity bias in Desmet's model regime. These transients do not show up in the Giddings model because Giddings assumed that the processes have essentially achieved complete equilibrium. Clearly, there is a

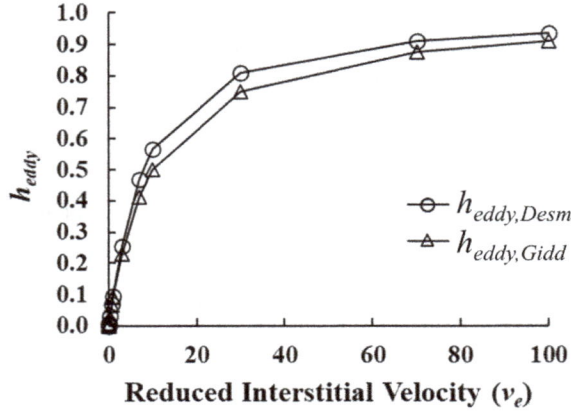

FIGURE A.2 h_{eddy} vs. v_e calculated using either Eq. A.9 ($h_{eddy,Gidd}$) or Eq. A.15 ($h_{eddy,Desm}$) with A = 1 and C = 0.1 in both cases.

marked similarity in both treatments of eddy dispersion. This is emphasized and clarified by writing an equation analogous to Eq. A.9 using a single term in Eq. A.14:

$$h_{eddy,Desm} = C_{Desm} v_e \left[1 - \frac{C_{Desm} v_e}{2A_{Desm}} \cdot \left(1 - \exp\left(\frac{-2A_{Desm}}{C_{Desm} v_e} \right) \right) \right] \tag{A.15}$$

Furthermore, it is easy to show that the high and low velocity limits of Eq. A.15 are the same as Eqs. A.10 and A.11. Desmet has shown [13] that using the value of A equal to 1.0 and the value of C equal to 0.1 in both his and Giddings' equations, the greatest difference in plots of h_{eddy} against v_e only amounts to 12%. The two plots converge at both the low and high limits of v_e (see Figure A.2 for more details).

Equation A.14 is the general form including all five types of velocity biases contributing to eddy dispersion, including the trans-column velocity dispersion term (see Eqs. 29 and 30 in reference [13]). Figure A.3 shows that Eqs. A.9 and A.15 actually give very similar results with maximal differences of 12% over a wide range in A and C values, not just their values given for Figure A.2.

The near numerical equivalence of these plots makes it impossible to empirically decide between the two forms. However, Desmet et al. [14] argue strongly for Eq. A.15 because it can be derived from the solution of the convective-diffusion mass balance equations for transport between two different velocity regimes. Furthermore, the same equation works for all five types of eddy dispersion including the trans-column process for which Eq. A.9 (the Giddings coupling equation) theoretically does not apply because the trans-column process is often so slow that it is not at equilibrium and the exponential term in Eq. A.15 has not gone to zero.

A.3   DETAILS RELATED TO THE "B" TERM

The most complete description of the total reduced plate height is [14]:

$$h = h_{eddy} + h_B + h_{Cm} + h_{Cs} \tag{A.16}$$

The terms h_{eddy}, h_B, h_{Cm}, and h_{Cs} represent the eddy dispersion, longitudinal molecular diffusion, and resistances to mass transfer in the mobile phase outside the particle, and in the non-moving phases

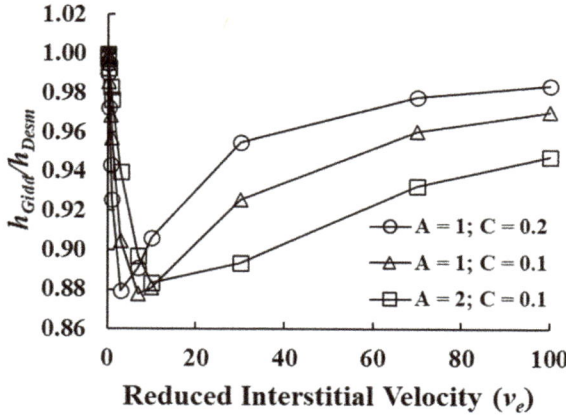

FIGURE A.3 Ratio of h_{eddy} values ($h_{eddy,Gidd}/h_{eddy,Desm}$) vs. v_e for different combinations of A and C values. h_{eddy} values were calculated using Eqs. A.9 and A.15.

inside the particle, respectively. The h_{eddy} component was discussed extensively in section A.2. The remaining terms are given as follows based on various papers by the Desmet group [15, 16, 116] and several others:

$$h_B = \frac{B}{v_e} = \frac{2}{v_e} \cdot \frac{D_{eff}}{D_m} \cdot (1+k") \tag{A.17}$$

$$h_{Cm} = \frac{1}{3} \cdot \left[\frac{k"}{1+k"}\right]^2 \cdot \frac{\varepsilon_e}{1-\varepsilon_e} \cdot \frac{v_e}{Sh} = C_m \cdot v_e \tag{A.18}$$

$$h_{Cs} = \frac{1}{30} \cdot f(\rho) \cdot \frac{k"}{(1+k")^2} \cdot \frac{D_m}{D_{pz}} \cdot v_e = C_s \cdot v_e \tag{A.19}$$

In the above equations D_{eff} is the effective coefficient accounting for longitudinal diffusion in the mobile phase outside the particles, plus molecular diffusion in both the non-moving eluent and the retentive stationary phases in the pores, D_m is the analyte's molecular diffusion coefficient in the free mobile phase. Additionally, $k"$ is the "zone retention factor" defined below (see Eq. A.20) and ε_e is the interstitial porosity of the packed bed in contrast to the internal porosity of the packing particles (ε_i). The total porosity of the column is denoted ε_T as described above.

$$k" = (1+k) \cdot \frac{\varepsilon_T}{\varepsilon_e} - 1 \tag{A.20}$$

Conceptually, the zone retention factor combines the fraction of the solute that is the stationary phase with the fraction of solute in stagnated mobile phase, thus $k"$ quantifies the mass fraction of the analyte that at equilibrium is not moving by convection (i.e., it is the fraction of mass inside the particle where transport only occurs by diffusion). The conventional phase retention factor (k) is given in Eq. A.22 in terms of the partition (phase equilibrium) constant (K), the total volume of stationary phase (V_s) and the total volume of mobile phase (V_m). Of course, V_m is sum of both the moving volume of eluent and the stagnated mobile phase inside the pores of the particles. The contribution to the reduced plate height from slow transport inside the stationary phase depends on the dimensionless diameter (ρ), which is the ratio of the impenetrable core (d_{core}) to the overall particle diameter (d_{part})

$$\varepsilon_T = \varepsilon_e + \varepsilon_i \cdot \left(1 - \varepsilon_e\right) \tag{A.21}$$

$$k = \frac{K \cdot V_s}{V_m} = \frac{t_r - t_0}{t_0} \tag{A.22}$$

$$f\left(\rho\right) = \frac{1 - 5\rho^3 + 9\rho^5 - 5\rho^6}{1 - \rho^3} \tag{A.23}$$

$$\rho = \frac{d_{core}}{d_{part}} \tag{A.24}$$

$$\text{FPP}: \rho = 0; f\left(\rho\right) = 1$$

$$\text{NPP}: \rho = 1; f\left(\rho\right) = 0$$

Another major contribution of Desmet and his coworkers to the understanding of peak broadening was their seminal work of the longitudinal diffusion term $(h_B;$ Eq. A.17) using effective medium theory (EMT) [117, 118]. Most prior work followed Knox and Scott's simpler intuitive approach and treated the interior of the particle as a single homogeneous medium comprised of the stagnant mobile phase and the actual stationary phase, and handled longitudinal diffusion as the time weighted average (TWA) of molecular diffusion in the mobile phase and inside the particles [112, 119]. Thus, to estimate the overall effective diffusion coefficient, Knox used the time weighted average (TWA) of three distinct diffusion coefficients:

1. Diffusion in the interstitial mobile phase (i.e., outside of the particles) with an effective diffusion coefficient equal to the product of a so-called obstruction or tortuosity factor (γ_e) and the free solution diffusion coefficient in the mobile phase liquid (D_m).
2. Diffusion in the stagnated mobile phase inside the pores, again as the product of an internal obstruction/tortuosity (denoted γ_i) and D_m.
3. Diffusion in the stationary phase on the surface of the mesopores of the porous particle (denoted D_s). The diffusion in the stagnated mobile phase and in the stationary phase were treated as if they took place in series. In contradistinction, EMT says that they are actually coupled in a more complex fashion.

However, subsequent experimental and theoretical studies [122, 123] showed that equations for diffusion based on EMT are superior to those based on the TWA approach. The elegant overview of column dynamics by Dittmann and Wang [112] should be consulted for details of the EMT approach and discussions of differences between SPP and FPP type particles. Readers interested in more detail on EMT are referred to [122]. For purposes of maximum development of peak capacity in 2D-LC it does appear that SPP type particles do allow higher peak capacities principally due to their lower pressure drops, thus longer or coupled columns can be used [112, 123, 124].

A.4   DETAILS RELATED TO THE "C" TERMS

The h_{Cs} and h_{Cm} terms represent broadening due to slow mass transfer inside the particles (that is, the non-convecting part of the eluent fluid and the stationary phase) and transport by convection and diffusion from the moving fluid to the outside boundary of the particle's surface. Note that for FPPs there is no core thus ρ (Eq. A.24) is equal to 0 and $f(\rho)$ becomes 1. Consequently, h_{Cs} (Eq. A.19) takes on its traditional form with the shape factor of 1/30 that applies for spherical particles. In contrast, for totally non-porous particles where the solid core has the same diameter as the entire particle there is no stationary phase or stagnated moving phase thus k'' and $f(\rho)$ are both 0 and it follows that there can be no stationary phase resistance to mass transfer, that is, h_{Cs} is also 0. The function $f(\rho)$ decreases monotonically as the core becomes a greater fraction of the whole particle. Equation A.23 is entirely consistent with Horvath and Lipsky's [125] work, as well as Desmet's [13] and Felinger's [126].

Another difficulty in using and understanding the Giddings formulation for h_{eddy} is that historically the A_j and C_j terms, which encompass a distance scale of at least a few particle diameters, have often been confounded with the C_s and C_m coefficients in Eqs. A.18 and A.19. In fact, C_m and C_s only pertain to mass transfer to and from a single particle diameter [14]. These and other problems in understanding the physical basis of eddy dispersion connected to the Giddings form of h_{eddy} were exhaustively described by the Desmet group [14]. One of the major conclusions of this paper and of references [13, 15, 16, 116] is that in contrast to virtually all earlier work, eddy dispersion must depend on analyte retention. This result is shown in Figure A.4 [14].

The eddy dispersion contribution to the reduced plate height was obtained by measuring the total reduced plate height and subtracting measured values of for h_B and calculated (i.e., theoretical, but empirically validated) values for h_{Cs} and h_{Cm}. As seen in Figure A.5 plots of the $\log(h_{eddy})$ vs. $\log(v_e)$ are good straight lines.

This means that the Knox equation (Eq. A.12) does a good job of fitting the data, but the values of A_{Knox} and α_{Knox} are both strongly dependent on the zone retention factor (k'') as shown by Desmet's fitting equations [14] (see Eqs. A.25 and A.26 below), which are plotted in Figure A.6:

$$\alpha_{Knox} = 0.55 - 0.26 \cdot k''^{-0.45} \tag{A.25}$$

$$A_{Knox} = 0.19 + 0.12 \cdot k''^{-0.93} \tag{A.26}$$

Equation A.18 gives the reduced plate height contribution (C_m) resulting from the slow mass transfer in the moving mobile phase to the outer surface of the particle that holds the non-moving phases. This equation is written in terms of the Sherwood number (Sh), which is a dimensionless measure of the kinetics of external transport due to the combined processes of convection and diffusion. In chromatographic practice, one of the most commonly used correlations for Sh is the Wilson-Geankoplis [127] equation.

$$Sh_{WG} = 1.09 \cdot \frac{\left(\varepsilon_e \cdot v_e\right)^{1/3}}{\varepsilon_e} = 2.07 \cdot v_e^{1/3}; \text{ with } \varepsilon_e = 0.38 \tag{A.27}$$

This is an empirical correlation that was determined from experimental data where the smallest reduced velocity was about 50, which is far above the reduced velocities encountered in liquid chromatography because in HPLC and UHPLC the particles used are much smaller than those used in typical chemical engineering operations. As a consequence, Eq. A.27 significantly underestimates

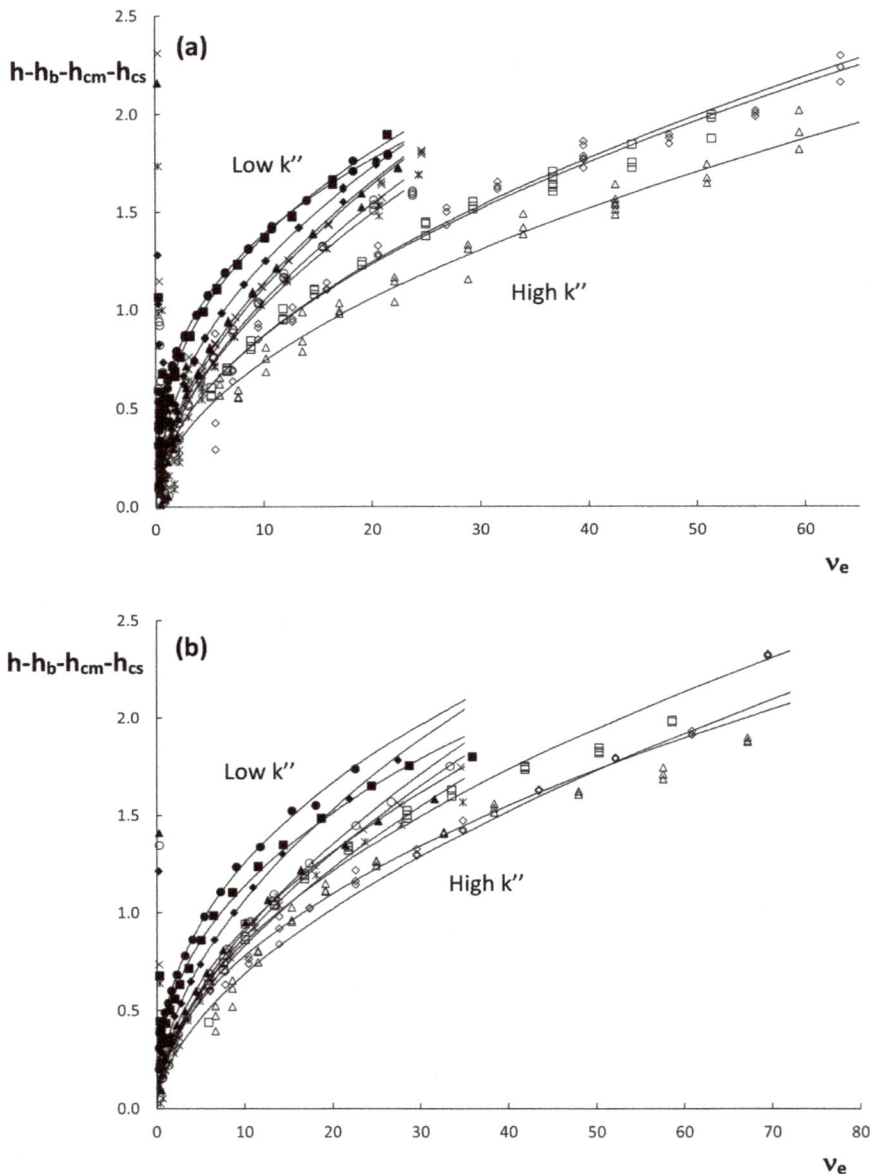

FIGURE A.4 Plots of $h_{eddy} = h - h_b - h_{cm} - h_{cs}$ vs. reduced interstitial velocity (v_e) for SPP (a) and FPP (b) columns. Symbols indicate data for alkylphenone homologs covering a range of k" from about 0.7 to 123. Reprinted from *Journal of Chromatography, A*, 1626, G. Desmet, H. Song, D. Makey, D. Stoll, D. Cabooter, Experimental investigation of the retention factor dependency of eddy dispersion in packed bed columns and relation to Knox's empirical model parameters, Copyright 2020, with permission from Elsevier.

Sh at the low reduced velocities encountered in analytical work and thus the Wilson-Geankoplis correlation overestimates h_{Cm}. Using the methods of computational fluid dynamics, Desmet and his group came up with a much improved correlation for Sh_{Desm} that can be used at lower reduced velocities more typical of HPLC.

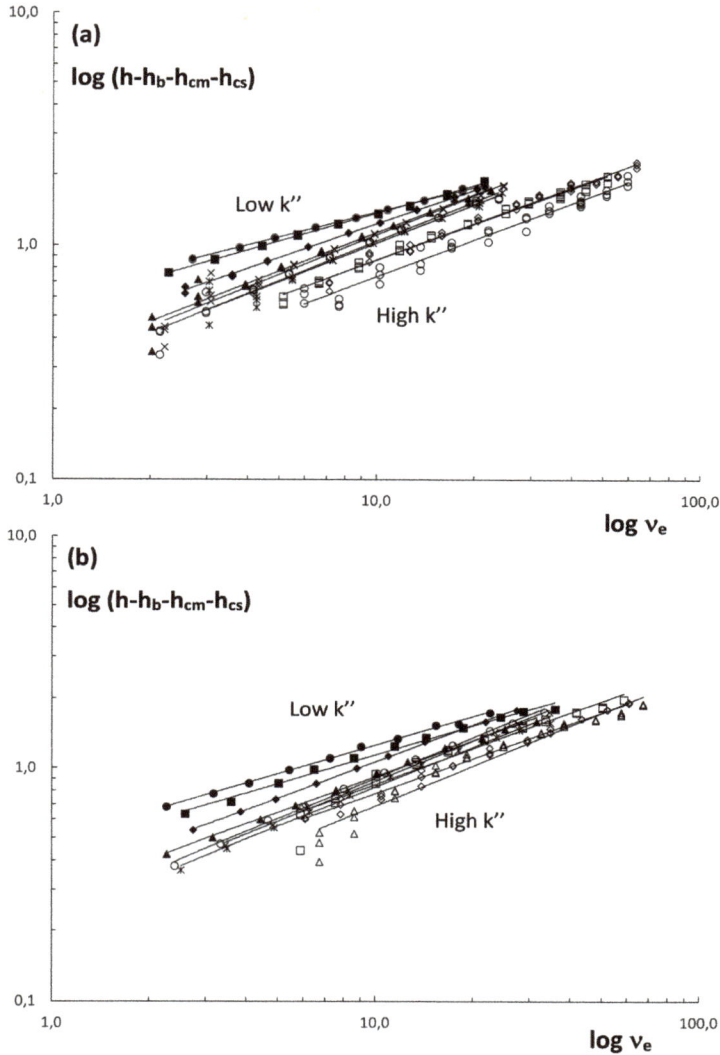

FIGURE A.5 Plots of h_{eddy} vs. v_e in log/log coordinates for SPP (a) and FPP (b) particles. These are the same data as in Figure A.4, plotted on different axes. Reprinted from *Journal of Chromatography, A*, 1626, G. Desmet, H. Song, D. Makey, D. Stoll, D. Cabooter, Experimental investigation of the retention factor dependency of eddy dispersion in packed bed columns and relation to Knox's empirical model parameters, Copyright 2020, with permission from Elsevier.

$$Sh_{Desm} = \frac{13}{1+2.1 \cdot v_e} + 8.6 \cdot v_e^{0.21} \tag{A.28}$$

Plots of Eqs. A.27 and A.28 over a range of v_e values used in analytical liquid chromatography are shown in Figure A.7. Even at a reduced velocity of 2.0, more than 30% of the external mass transport results from diffusion and not convection. The finite limit of this equation at zero velocity tells us that diffusion makes a significant contribution to external transport even at reduced velocities. This is a more physically sensible result than the zero-velocity limit of Eq. A.27, which indicates very

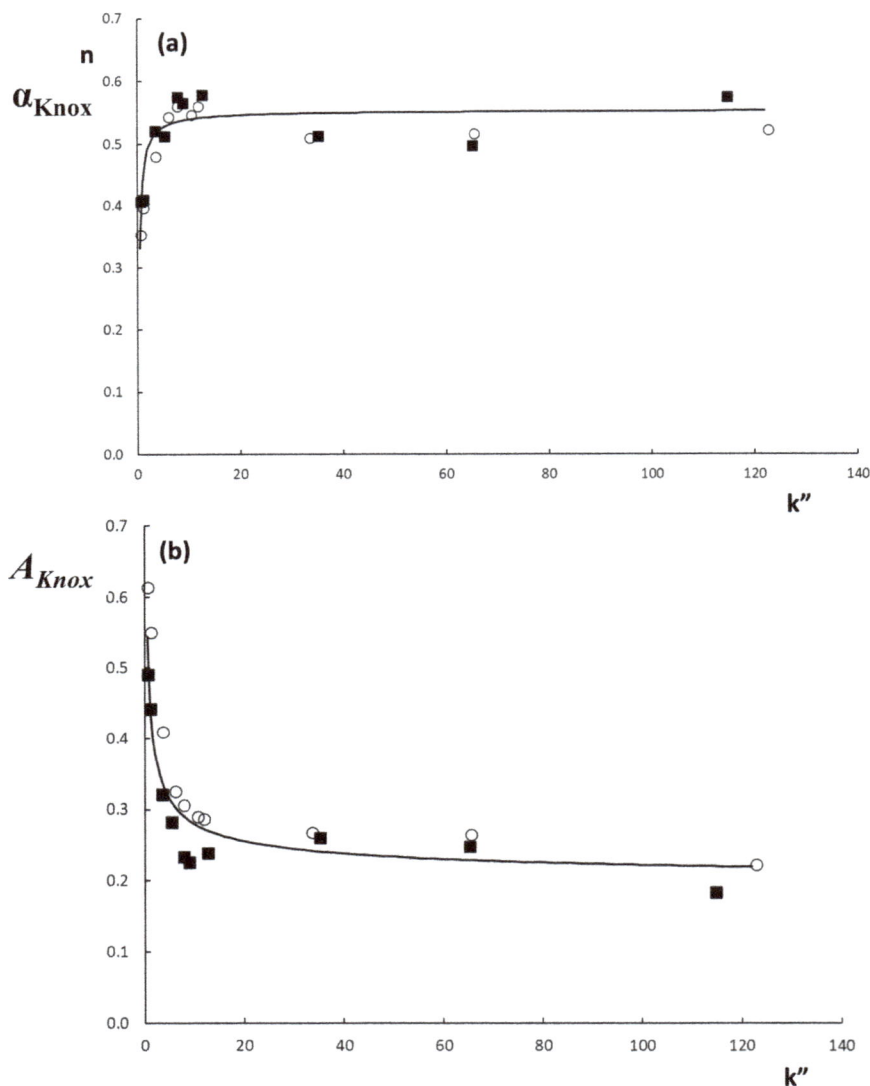

FIGURE A.6 Plots of the fitting coefficients α_{Knox} (a) and A_{Knox} (b) in Eq. A.12 vs. k" for the data shown in Figures A.4 and A.5. Reprinted from *Journal of Chromatography, A*, 1626, G. Desmet, H. Song, D. Makey, D. Stoll, D. Cabooter, Experimental investigation of the retention factor dependency of eddy dispersion in packed bed columns and relation to Knox's empirical model parameters, Copyright 2020, with permission from Elsevier.

little diffusive transport at zero reduced velocity. This physically unrealistic result stems from the long extrapolation of experimental data from rather high velocities to zero velocity to obtain Sh_{WG}.

A.5   RELATIONSHIPS RELATED TO MOBILE PHASE FLOW AND PRESSURE DROP ACROSS THE COLUMN

The fact that the resolving power, analysis time, and thus speed (N/t_0) depend so strongly on pressure (see Eqs. 2.18–2.22) requires that the relationship between pressure and the column variables must

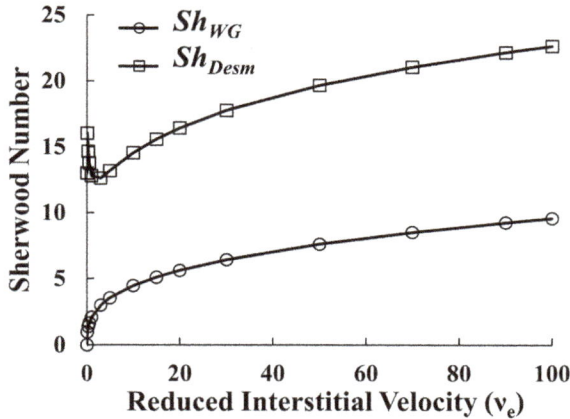

FIGURE A.7 Comparison of Sherwood numbers calculated using the Wilson-Geankopolis correlation ($\circ$, Eq. A.27) or the Desmet correlation ($\square$, Eq. A.28).

be understood. Experimentally Darcy's law (Eq. A.29) relates the pressure drop (ΔP, in Newtons/m^2 (= Pascals)) across a column to the interstitial linear velocity of the mobile phase (u_e, m/s), column length (L, m), mobile phase viscosity (η, Pascal·s)), and the column's permeability (K_c, m^2). The relationship is found to be correct only under laminar flow and isothermal conditions for incompressible, Newtonian fluids. The permeability (K_c) is obtained in a semi-empirical manner by use of the Kozeny-Carman equation (A.30, [112]).

$$\Delta P = \frac{u_e \cdot \eta \cdot L}{K_c} \tag{A.29}$$

For flow in a packed bed the constant that describes the column's permeability K_c (often called the Kozeny-Carman constant) is estimated by modeling a column packed with particles as a bundle of cylindrical capillaries each of which obeys the Poiseulle equation for a single capillary with the effective area available for flow [128, 129]. The Kozeny-Carman equation (A.30) appears in many different forms that differ based on which measure of velocity is used. It is vital that the form of this equation, specifically the dependence on porosity, be consistent with the type of velocity used in Eq. A.29. Unfortunately, this has not always been done properly. Because the truly physically relevant velocity is the interstitial velocity (u_e), we use the equation discussed by both Giddings [85] and Probstein [130], and thus K_c in Eq. A.29 is:

$$K_c = d_p^2 \cdot \left[\frac{1}{36 \cdot k_c} \cdot \frac{\varepsilon_e^2}{\left(1 - \varepsilon_e\right)^2} \right] = \frac{d_p^2}{180} \cdot \frac{\varepsilon_e^2}{\left(1 - \varepsilon_e\right)^2} = \frac{d_p^2}{\Phi} \tag{A.30}$$

The parameter k_c is obtained from experimental pressure data. For beds of packed spheres k_c is about 5. This yields a factor of about 180 in the denominator [85, 112]. Using an interstitial porosity of 0.38 gives a value of 479 for Φ. This is usually rounded to 500 which is the value used in the tables and figures of Chapter 2.

The Reynolds number (see Eq. A.31) must be less than about 10 to maintain laminar flow in a bed of well-packed spheres [129]. Here ρ and η are the mobile phase density and viscosity and u_s is the superficial velocity – that is, the linear fluid velocity in an open tube operated at the same volumetric flow rate as the packed tube.

$$Re = \frac{d_p \cdot \rho \cdot u_s}{\eta} \cdot \left(\frac{1}{1 - \varepsilon_e} \right) \leq 10 \tag{A.31}$$

A.6 COLUMN SELF-HEATING DUE TO VISCOUS FLOW AND OTHER THERMAL EFFECTS

The presence of axial and most especially radial temperature gradients in the column are due to a complex set of phenomena that can have a very deleterious impact on the peak shape and width as well as alter retention. The four processes responsible for these gradients include:

1. First, the mobile phase temperature going into the column may not be equal to that of the column wall. When the column is thermostated at about room temperature the eluent is usually warmer than the "set point" temperature of the column temperature controller because the mobile phase is stored at room temperature but it is warmed both by being compressed and by contact with a warm pump head. There must be an efficient heat exchanger placed between the pump and the column preferably located before the sample injection point to bring the mobile phase to the same temperature as the column entrance and column wall. The need for a highly effective exchanger increases as the flow rate is increased to speed up chromatography. Poppe and Kraak [24] noted that there was an optimum difference in the temperature of the eluent and the column wall which produces a minimum HETP. Subsequently, Thompson *et al.* [25] showed in their study of LC at elevated column temperatures that strongly retained solutes with larger enthalpies of retention were much more sensitive to peak distortion by thermal mismatch than were less retained solutes. When the temperature difference between the fluid entering 2.1 mm i.d. stainless steel columns immersed in a stirred hot oil bath was less than about 5 °C, peak broadening due to this difference became essentially negligible. It should be noted that when the mobile phase compression process is adiabatic the heat will be generated reversibly [131], consequently the temperature change will be given as:

$$\frac{\Delta T}{\Delta P} = \frac{\alpha T V}{C_P} \tag{A.32}$$

Where α is the mobile phase's coefficient of volume expansion per unit temperature, C_p is the fluid's heat capacity at constant pressure, and V is the molar volume of the fluid. For methanol this gives a distinctly non-negligible temperature increase of 1.8 K for only 100 bar of compression. Clearly, heating by compression effects in the pump especially in UHPLC can be a significant issue if the post-compression eluent temperature is not pre-conditioned before going into the column.

2. The second process that perturbs the fluid's temperature is its gradual decompression as it moves through the column. This is so slow that it is not adiabatic and thus is not simply the reverse of the compression in process in #1. This decompression event is the liquid analog to the well-known Joule-Thompson expansion of gases; however, for a liquid it could even be exothermic as the entropy change involved in decompression of a liquid is much smaller than for a gas. Lin and Horvath [132] considered this process negligible compared to the heat due to viscous friction but others consider this questionable.

3. Third, in principle the sorption and desorption of the solute zone could cause either temperature increases or decreases depending on the enthalpy of retention. However, under analytical
conditions, *i.e.* infinite analyte dilution, this process is entirely negligible.
4. Fourth, we must consider the heat generated by the frictional (viscous) forces between the
moving mobile phase and the column bed. Halasz *et al.* were the first to recognize that
these effects set the upper limit to the speed of LC and the size of the smallest particles that
could be used [35]. It is important to note that this heating effect is not important under the
conditions needed to do high resolution LC, that is, with large particles operated at low velocity while using the highest possible pressure (see Section 2.4.2.3 and Eq. 2.11). Poppe *et al.*
and Horvath *et al.* virtually simultaneously developed detailed solutions to the complex heat
transfer equations, albeit using somewhat different assumptions, to estimate the temperature
rise resulting from frictional heating [131, 132].

Although these processes lead to a time-dependent axial temperature gradient inside the column,
which impacts analyte retention, the major concern is that heat transfer (leakage) from the column
wall to the column's surroundings will induce a radial temperature gradient inside the column.
When the column's set point temperature is equal to the temperature of the incoming mobile phase,
the temperature at the center of the column, *i.e.* on the column axis, can be significantly higher than
that at the interior surface of the column wall and furthermore the warmer will be the fluid at the
column exit. If the column is both sufficiently long and the transit time of the fluid great enough,
the temperature on the axis will become independent of time and column length – that is, there is a
substantial thermal entrance length which is on the same order as the length of the column. The most
important things that references [132] and [133] teach are:

1. The magnitude of the radial temperature gradient increases with the flow rate and pressure
drop and decreases with the column diameter [133].
2. The radial temperature profile is essentially parabolic.
3. Assuming that viscosity is nearly linearly dependent on temperature, the resulting gradient in
mobile phase velocity is also radially parabolic and thus according to Guiochon [41] Aris's
[134] approach to peak broadening can be used. This leads to a sixth power dependence of
the friction-induced contribution to the HETP on the column diameter [41]. This means that
a two-fold decrease in column diameter decreases the contribution to *H* from viscous heating
by 64-fold.
4. The above only applies to the broadening of unretained species. The radial temperature gradient can be sufficiently large to produce an even larger non-parabolic radial retention gradient
especially for analytes having large enthalpies of retention [25]. Desmet [135] has shown that
the effect of retention enthalpy on the analyte's local velocity leads to at most a factor of 10
increased contribution to the HETP than does the actual radial gradient in eluent velocity. It
has not as yet been definitively shown that well retained solutes in 100 mm long by 2.1 mm i.d.
columns packed with sub-2 µm particles are free of viscosity induced broadening at optimum
operating velocities.

Desmet has shown [135], based on the work of Knox and Grant [136], that the temperature at
any point along the radius of a column of length (*L*) that exceeds the thermal entrance length,
and that is perfectly adiabatic so no heat is lost except that which leaves by flow out the column
exit, is:

$$T(r) = T_w + \frac{q \cdot R_{out}^2}{4\lambda} \cdot \left[1 - \left(\frac{r}{R_{out}} \right)^2 \right] = T_w + \Psi \cdot T_w \cdot \left[1 - \left(\frac{r}{R_{out}} \right)^2 \right] \qquad (A.33)$$

where T_w is the wall temperature, q the rate of heat production per unit volume (power/l^3), R_{out} is the inside radius of the packed column, λ is the thermal conductivity of the mobile phase or of the contents of the column, and r is radial position measured from the wall. Under all the above conditions q is given as:

$$q = \frac{u_e^2 \cdot \eta \cdot \varepsilon_e}{K_c} \tag{A.34}$$

Under conditions of K-S-H optimization we know that the optimum *reduced interstitial velocity* $v_e \; (= u_e \cdot d_p/D_m)$ is a constant for all t_o and fixed P_{max} when the interstitial velocity, particle size, and column length are simultaneously optimized. It follows that:

$$u_{e,opt} = \frac{v_{e,opt}}{d_p} \cdot D_m = \frac{\sqrt{\dfrac{B}{C}}}{d_p} \cdot D_m \tag{A.35}$$

Upon substitution for u_e and K_c in A.34 we get:

$$q = \frac{\left(\dfrac{B}{C}\right) \cdot D_m^2 \cdot \eta \cdot \varepsilon_e \cdot \Phi}{d_p^4} \tag{A.36}$$

It follows that at fixed P_{max} under K-S-H optimization the rate of heating and consequently the radial temperature gradient varies with the fourth power of the packing particle diameter. Clearly, there is a rather significant price to be paid for using smaller particles. Assuming a parabolic radial velocity distribution, Aris's method can be used to calculate the contribution to the plate height. For the case of zero retention enthalpy Desmet [135] gets the same results as Knox and Grant [136], and Poppe [24]:

$$H_{viscous} = \frac{1}{48} \cdot \frac{\left(T_w \cdot \Psi\right)^2}{\left(2 + \Psi \cdot T_w\right)} \cdot u_{Rw} \cdot \frac{R_{out}^2}{D_{rad}} \tag{A.37}$$

Here $H_{viscous}$ is the contribution of the total plate height (H) due solely to frictional heating for analytes with zero enthalpy of retention, u_{Rw} is the velocity of a retained solute near the wall at temperature equal to T_w, and D_{rad} is the radial mass dispersion coefficient. Clearly, the wider is the column the bigger is the contribution of $H_{viscous}$ to the total plate height.

In order to better understand the details of frictional heating, Desmet et al. [137] used computational fluid dynamics simulations to examine some idealized versions of column operating conditions. Consider a wall-less, or effectively infinitely thin, column so that there is no axial heat flux from the hot exit end of the column backwards along the column wall to the cooler column entrance. We then consider two extreme cases:

1. The column wall is assumed to be perfectly adiabatic. In this situation no heat is transported radially out of the wall thus there will be no *radial* thermal, velocity, or retention gradient and consequently there can be no additional band broadening due to the viscous heating. Of course, the temperature inside the fluid temperature varies axially as the eluent warms up as it moves through the column.

2. The entire length of the exceedingly thin column wall is assumed to be perfectly controlled by a thermostat held at the temperature of the eluent entering the column (denoted T_w). In this case there must be a radial gradient in temperature because the fluid immediately adjacent to the inner column wall is at T_w whereas the temperature at the center of the tube is warmer.

Of course, no real column can be operated as above. Any real column wall will allow some flux of heat backwards from the warmer column exit towards the cooler entrance. This will induce a radial thermal gradient whose amplitude will depend on how much heat leaks out of the column into its surroundings. Clearly, the better insulated the column wall, the smaller will be the radial temperature gradient. This has been recognized for some time and is the basis of attempts to improve the performance of conventional 2.1 mm diameter columns by insulating them from the column compartment thermostat or by encapsulating them in a highly adiabatic vacuum jackets [138–140]. At best the use of perfectly adiabatic surrounding can help by about a factor of two in reducing $H_{viscous}$; the remainder is due to the effect of backwards conduction of heat along the column wall [141] as described above.

As discussed by Desmet and his coworkers there are a number of additional ways in which one might decrease the radial thermal gradient:

1. A column of total length (L) can be divided into n sections each of length L/n. These subsections are insulated and connected by short lengths of efficiently cooled thermostated tubing. The pressure drop in each column is limited to $\Delta P/n$ and thus the radial thermal temperature gradient is n times smaller [142].
2. The Guiochon group [143] has suggested that columns packed with silica-based SPPs have higher thermal conductivities than FPP-based packed columns due to the higher thermal conductivity of the solid cores of SPPs. However, the computational fluid dynamic work of Desmet [141] shows that because the cores of separate SPPs particles are not in direct contact this will not help much and it is more effective to increase the thermal conductivity of the shell; however, such particles are not yet available. The best approach seems to be to make a monolithic packing material so that a single more highly thermally conductive structure extends the entire length of the column [144].
3. A third way to proceed is to use a 1 mm i.d. packed column. However, we do not know how to pack 1 mm columns as well as we do 2.1 mm columns. Additionally, instruments with even lower extra-column broadening than current UHPLC instruments would have to developed [144].

A.7   ESTIMATION OF MOLECULAR DIFFUSION COEFFICIENTS IN LIQUIDS

The solute molecular diffusion in free solution (D_m) is clearly an essential parameter for understanding peak broadening in all forms of chromatography. It is a vital factor as it is present in the very definition of the reduced velocity. Thus, we need ways to approximate this property. There are a number of semi-theoretical methods for estimating molecular diffusion coefficients in liquids [145]. By far the most popular in LC work is the Wilke-Chang equation (Eq. A.38) [146] and variants of it [145]. This correlation was developed by chemical engineers for use with low molecular weight solutes in organic solvents and water. In non-associated solvents, such as hexane, it is accurate to about 10% compared to experimental measurements. Use in associated solvents such as water and alcohols requires an empirical association factor. The Wilke-Chang model is certainly oversimplified as it does not include any solute-solvent interaction terms, nor solute shape dependent factors [147]. The estimates are not as accurate in associated solvents, especially in mixed water-organic solvents of interest in RPLC.

$$D_m = 7.4 \times 10^{-8} \cdot \frac{T \cdot \sqrt{\chi \cdot MW}}{\eta \cdot \left(\bar{V}_{solute} \right)^{0.6}} \tag{A.38}$$

D_m is the solute's diffusion coefficient in the solvent or solvent mixture, T is the absolute temperature in Kelvin, MW and η are the molecular weight (g/mol) and viscosity (centipoise) of the solvent respectively, and $\bar{V}$ is the molar volume of the solute at the normal boiling point (mL/g·mol). χ is an empirically determined association constant of the solvent (e.g. for water χ is 2.6). Subsequently Li developed a modified Wilke-Chang-like correlation (Eq. A.39) based on measurements of D_m for a homolog series of solutes in mixtures of water with either methanol or acetonitrile for use in RPLC [148]. For the majority of the measurements made with the homologs used in a number of solvent compositions, the correlations were accurate to better than 3–4% with only a few errors as large as 10%. It is thus 3–4 times as accurate as the Wilke-Chang equation for the range of solutes and RPLC solvent mixtures that Li studied.

$$D_m = \kappa \cdot \frac{T}{\eta^\beta} \cdot \frac{MW^\alpha}{\bar{V}^\gamma} \tag{A.39}$$

The exponents α, β and γ, and the coefficient κ are determined empirically by regression of Eq. A.40:

$$\ln\left(\frac{D_m}{T} \right) = \ln(\kappa) + \alpha \cdot \ln(MW) - \beta \cdot \ln(\eta) - \gamma \cdot \ln(\bar{V}) \tag{A.40}$$

Here, κ is a solvent dependent factor, but is presently taken as a constant. The constants α, β, and γ are adjustable exponents obtained by regression of Eq. A.40. The molecular weight of a mixed solvent (MW_{mix}) is calculated as the mole fraction weighted average for a binary water-organic mixture:

$$MW_{mix} = x_{org} \cdot MW_{org} + x_{water} \cdot MW_{water} \tag{A.41}$$

More recently Desmet's group has measured by means of the Aris-Taylor dispersion method a large body of data for 45 polar, nonpolar, and charged solutes at 10 water-acetonitrile mixtures ranging from 0 to 100% (v/v) acetonitrile. This encompasses a range in solvent composition appropriate for both RPLC and HILIC [149]. The reader is advised to use these meticulously measured data in future work because the data were scrupulously obtained under conditions that ruled out longitudinal diffusion, as well as secondary flow and extra-column broadening effects that can impact the Aris-Taylor method. For nine solutes the measurements had a mean and median difference from the literature data of 1.9 and 1.7 %, respectively, and in all cases were better than 4%.

REFERENCES

[1]   G. Guiochon, Chapter 1. Optimization in liquid chromatography, in: C. Horváth (Ed.), High-Performance Liquid Chromatography: Advances and Perspectives, Academic Press, New York, 1980: pp. 1–56.

[2]   J.A. Lewis, J.W. Dolan, L.R. Snyder, I. Molnar, Computer simulation for the prediction of separation as a function of pH for reversed-phase high-performance liquid chromatography, J. Chromatogr. A. 592 (1992) 197–208. https://doi.org/10.1016/0021-9673(92)85086-9.

[3]   J.J. van Deemter, F.J. Zuiderweg, A. Klinkenberg, Longitudinal diffusion and resistance to mass transfer as causes of nonideality in chromatography, Chem. Eng. Sci. 5 (1956) 271–289. https://doi.org/10.1016/0009-2509(56)80003-1.

[4] H. Poppe, Some reflections on speed and efficiency of modern chromatographic methods, J. Chromatogr. A. 778 (1997) 3–21. https://doi.org/10.1016/S0021-9673(97)00376-2.

[5] J.H. Knox, M. Saleem, Kinetic conditions for optimum speed and resolution in column chromatography, J. Chromatogr. Sci. 7 (1969) 614–622. https://doi.org/10.1093/chromsci/7.10.614.

[6] I. Halász, G. Görlitz, Optimal parameters in high speed liquid chromatography (HPLC), Angew. Chem., Int. Ed. 21 (1982) 50–61. https://doi.org/10.1002/anie.198200501.

[7] V.R. Meyer, How to generate peak capacity in column liquid chromatography, J. Chromatogr., A. 1187 (2008) 138–144. https://doi.org/10.1016/j.chroma.2008.02.019.

[8] A.J. Matula, P.W. Carr, Separation speed and power in isocratic liquid chromatography: Loss in performance of Poppe vs Knox-Saleem optimization, Anal. Chem. 87 (2015) 6578–6583. https://doi.org/10.1021/acs.analchem.5b00329.

[9] J.C. Giddings, Comparison of theoretical limit of separating speed in gas and liquid chromatography, Anal. Chem. 37 (1965) 60–63. https://doi.org/10.1021/ac60220a012.

[10] P.W. Carr, D.R. Stoll, Xiaoli. Wang, Perspectives on recent advances in the speed of high-performance liquid chromatography, Anal. Chem. 83 (2011) 1890–1900. https://doi.org/10.1021/ac102570t.

[11] P.W. Carr, X. Wang, D.R. Stoll, Effect of pressure, particle size, and time on optimizing performance in liquid chromatography, Anal. Chem. 81 (2009) 5342–5353. https://doi.org/10.1021/ac9001244.

[12] G. Desmet, D. Clicq, P. Gzil, Geometry-independent plate height representation methods for the direct comparison of the kinetic performance of LC Supports with a different size or morphology, Anal. Chem. 77 (2005) 4058–4070. https://doi.org/10.1021/ac050160z.

[13] G. Desmet, A finite parallel zone model to interpret and extend Giddings' coupling theory for the eddy-dispersion in porous chromatographic media, J. Chromatogr. A. 1314 (2013) 124–137. https://doi.org/10.1016/j.chroma.2013.09.016.

[14] G. Desmet, H. Song, D. Makey, D.R. Stoll, D. Cabooter, Experimental investigation of the retention factor dependency of eddy dispersion in packed bed columns and relation to Knox's empirical model parameters, J. Chromatogr. A. 1626 (2020) 461339. https://doi.org/10.1016/j.chroma.2020.461339.

[15] G. Desmet, B. Huygens, W. Smits, S. Deridder, The checkerboard model for the eddy-dispersion in laminar flows through porous media. Part I: Theory and velocity field properties, J. Chromatogr. A. 1624 (2020) 461195. https://doi.org/10.1016/j.chroma.2020.461195.

[16] G. Desmet, W. Smits, S. Deridder, The checkerboard model for the eddy-dispersion in laminar flows through porous media. Part II: Application to ordered and disordered 2-D flow systems, J. Chromatogr. A. 1624 (2020). https://doi.org/10.1016/j.chroma.2020.461196.

[17] S. Hjertén, J.-L. Liao, R. Zhang, High-performance liquid chromatography on continuous polymer beds, J. Chromatogr. A. 473 (1989) 273–275. https://doi.org/10.1016/S0021-9673(00)91309-8.

[18] F. Svec, J.M.J. Frechet, Continuous rods of macroporous polymer as high-performance liquid chromatography separation media, Anal. Chem. 64 (1992) 820–822. https://doi.org/10.1021/ac00031a022.

[19] N. Ishizuka, H. Kobayashi, H. Minakuchi, K. Nakanishi, K. Hirao, K. Hosoya, T. Ikegami, N. Tanaka, Monolithic silica columns for high-efficiency separations by high-performance liquid chromatography, J. Chromatogr. A. 960 (2002) 85–96. https://doi.org/10.1016/S0021-9673(01)01580-1.

[20] T. Hara, H. Kobayashi, T. Ikegami, K. Nakanishi, N. Tanaka, Performance of monolithic silica capillary columns with increased phase ratios and small-sized domains, Anal. Chem. 78 (2006) 7632–7642. https://doi.org/10.1021/ac060770e.

[21] J.E. MacNair, K.C. Lewis, J.W. Jorgenson, Ultrahigh-pressure reversed-phase liquid chromatography in packed capillary columns, Anal. Chem. 69 (1997) 983–989. https://doi.org/10.1021/ac961094r.

[22] J.W. Jorgenson, Capillary liquid chromatography at ultrahigh pressures, Annu. Rev. Anal. Chem. 3 (2010) 129–150. https://doi.org/10.1146/annurev.anchem.1.031207.113014.

[23] F. D. Antia, C. Horváth, High-performance liquid chromatography at elevated temperatures: examination of conditions for the rapid separation of large molecules, J. Chromatogr. A. 435 (1988) 1–15. https://doi.org/10.1016/S0021-9673(01)82158-0.

[24] H. Poppe, J.C. Kraak, Influence of thermal conditions on the efficiency of high-performance liquid chromatographic columns, J. Chromatogr. A. 282 (1983) 399–412. https://doi.org/10.1016/S0021-9673(00)91617-0.

[25] J.D. Thompson, J.S. Brown, P.W. Carr, Dependence of thermal mismatch broadening on column diameter in high-speed liquid chromatography at elevated temperatures, Anal. Chem. 73 (2001) 3340–3347. https://doi.org/10.1021/ac010091y.

[26] J. Li, Y. Hu, P.W. Carr, Fast separations at elevated temperatures on polybutadiene-coated zirconia reversed-phase material, Anal. Chem. 69 (1997) 3884–3888. https://doi.org/10.1021/ac9705069.

[27] H. Chen, Cs. Horváth, High-speed high-performance liquid chromatography of peptides and proteins, J. Chromatogr. A. 705 (1995) 3–20. https://doi.org/10.1016/0021-9673(94)01254-C.

[28] J.H. Knox, High speed liquid chromatography, Annu. Rev. Phys. Chem. 24 (1973) 29–49. https://doi.org/10.1146/annurev.pc.24.100173.000333.

[29] J.H. Knox, The speed of analysis by gas chromatography, J. Chem. Soc. (1961) 433. https://doi.org/10.1039/jr9610000433.

[30] J.H. Knox, Gas Chromatography, Wiley, New York, 1964.

[31] I. Halász, H. Schmidt, P. Vogtel, Particle size, pressure and analysis time in routine high-performance liquid chromatography, J. Chromatogr. A. 126 (1976) 19–33. https://doi.org/10.1016/S0021-9673(01)84060-7.

[32] J. Purnell, Gas Chromatography, John Wiley & Sons, Inc., New York, 1962.

[33] K.M. Grinias, J.M. Godinho, E.G. Franklin, J.T. Stobaugh, J.W. Jorgenson, Development of a 45kpsi ultrahigh pressure liquid chromatography instrument for gradient separations of peptides using long microcapillary columns and sub-2μm particles, J. Chromatogr. A. 1469 (2016) 60–67. https://doi.org/10.1016/j.chroma.2016.09.053.

[34] J.E. MacNair, K.D. Patel, J.W. Jorgenson, Ultrahigh-pressure reversed-phase capillary liquid chromatography: Isocratic and gradient elution using columns packed with 1.0-μm particles, Anal. Chem. 71 (1999) 700–708. https://doi.org/10.1021/ac9807013.

[35] I. Halász, R. Endele, J. Asshauer, Ultimate limits in high-pressure liquid chromatography, J. Chromatogr. A. 112 (1975) 37–60. https://doi.org/10.1016/S0021-9673(00)99941-2.

[36] M.J. Wirth, Sub-micron Plate Heights for Capillaries Packed with Silica Colloidal Crystals, HPLC 2010: Boston, MA, 2010.

[37] D.S. Malkin, B. Wei, A.J. Fogiel, S.L. Staats, M.J. Wirth, Submicrometer plate heights for capillaries packed with silica colloidal crystals, Anal. Chem. 82 (2010) 2175–2177. https://doi.org/10.1021/ac100062t.

[38] B.A. Rogers, Z. Wu, B. Wei, X. Zhang, X. Cao, O. Alabi, M.J. Wirth, Submicrometer particles and slip flow in liquid chromatography, Anal. Chem. 87 (2015) 2520–2526. https://doi.org/10.1021/ac504683d.

[39] F. Gritti, G. Guiochon, Measurement of the axial and radial temperature profiles of a chromatographic column, J. Chromatogr. A. 1138 (2007) 141–157. https://doi.org/10.1016/j.chroma.2006.10.095.

[40] F. Gritti, G. Guiochon, Complete temperature profiles in ultra-high-pressure liquid chromatography columns, Anal. Chem. 80 (2008) 5009–5020. https://doi.org/10.1021/ac800280c.

[41] G. Guiochon, The limits of the separation power of unidimensional column liquid chromatography, J. Chromatogr. A. 1126 (2006) 6–49. https://doi.org/10.1016/j.chroma.2006.07.032.

[42] A. Alvarez-Zepeda, B.N. Barman, D.E. Martire, Thermodynamic study of the marked differences between acetonitrile/water and methanol/water mobile-phase systems in reversed-phase liquid chromatography, Anal. Chem. 64 (1992) 1978–1984. https://doi.org/10.1021/ac00041a037.

[43] J.W. Jorgenson, K.D. Lukacs, Zone electrophoresis in open-tubular glass capillaries, Anal. Chem. 53 (1981) 1298–1302. https://doi.org/10.1021/ac00231a037.

[44] T. Teutenberg, High-Temperature Liquid Chromatography: A User's Guide for Method Development, RSC Pub, Cambridge, 2010.

[45] J.D. Thompson, P.W. Carr, A study of the critical criteria for analyte stability in high-temperature liquid chromatography, Anal. Chem. 74 (2002) 1017–1023. https://doi.org/10.1021/ac010917w.

[46] X. Li, D.R. Stoll, P.W. Carr, Equation for peak capacity estimation in two-dimensional liquid chromatography, Anal. Chem. 81 (2009) 845–850. https://doi.org/10.1021/ac801772u.

[47] C.G. Horvath, S.R. Lipsky, Rapid analysis of ribonucleosides and bases at the picomole level using pellicular cation exchange resin in narrow bore columns, Anal. Chem. 41 (1969) 1227–1234. https://doi.org/10.1021/ac60279a024.

[48] J.J. Kirkland, Controlled surface porosity supports for high-speed gas and liquid chromatography, Anal. Chem. 41 (1969) 218–220. https://doi.org/10.1021/ac60270a054.

[49] F. Gritti, A. Cavazzini, N. Marchetti, G. Guiochon, Comparison between the efficiencies of columns packed with fully and partially porous C18-bonded silica materials, J. Chromatogr. A. 1157 (2007) 289–303. https://doi.org/10.1016/j.chroma.2007.05.030.

[50] D. Cabooter, F. Lestremau, F. Lynen, P. Sandra, G. Desmet, Kinetic plot method as a tool to design coupled column systems producing 100,000 theoretical plates in the shortest possible time, J. Chromatogr. A. 1212 (2008) 23–34. https://doi.org/10.1016/j.chroma.2008.09.106.

[51] J.M. Cunliffe, T.D. Maloney, Fused-core particle technology as an alternative to sub-2 μm particles to achieve high separation efficiency with low backpressure, J. Sep. Sci. 30 (2007) 3104–3109. https://doi.org/10.1002/jssc.200700260.

[52] X. Wang, P.W. D.R. Stoll, Dwight, A simple approach to performance optimization in HPLC and its application in ultrafast separation development, LCGC North Am. 28 (2010) 932, 934–936, 938, 940, 942.

[53] J.J. DeStefano, T.J. Langlois, J.J. Kirkland, Characteristics of superficially-porous silica particles for fast HPLC: some performance comparisons with sub-2 μm particles, J. Chromatogr. Sci. 46 (2008) 254–260. https://doi.org/10.1093/chromsci/46.3.254.

[54] F. Gritti, D.S. Bell, G. Guiochon, Particle size distribution and column efficiency. An ongoing debate revived with 1.9 μm Titan-C18 particles, J. Chromatogr. A. 1355 (2014) 179–192. https://doi.org/10.1016/j.chroma.2014.06.029.

[55] Y. Zhang, X. Wang, P. Mukherjee, P. Petersson, Critical comparison of performances of superficially porous particles and sub-2 μm particles under optimized ultra-high pressure conditions, J. Chromatogr., A. 1216 (2009) 4597–4605. https://doi.org/10.1016/j.chroma.2009.03.071.

[56] F. Gritti, I. Leonardis, J. Abia, G. Guiochon, Physical properties and structure of fine core-shell particles used as packing materials for chromatography, J. Chromatogr. A. 1217 (2010) 3819–3843. https://doi.org/10.1016/j.chroma.2010.04.026.

[57] J.H. Knox, Band dispersion in chromatography – a universal expression for the contribution from the mobile zone, J. Chromatogr. A. 960 (2002) 7–18. https://doi.org/10.1016/S0021-9673(02)00240-6.

[58] F. Gritti, G. Guiochon, Mass transfer resistance in narrow-bore columns packed with 1.7 μm particles in very high pressure liquid chromatography, J. Chromatogr. A. 1217 (2010) 5069–5083. https://doi.org/10.1016/j.chroma.2010.05.059.

[59] S.A. Schuster, J.J. Kirkland, B.M. Wagner, B.E. Boyes, W. Johnson, T.J. Langlois, J.J. DeStefano, Fused-core Particles: Varying Shell Thickness and Pore Size, HPLC 2010: Boston, MA, 2010.

[60] G. Guiochon, F. Gritti, Shell Particles, Trials, Tribulations, and Triumphs, HPLC 2010: Boston, MA, 2010.

[61] F. Gritti, K. Horvath, G. Guiochon, How changing the particle structure can speed up protein mass transfer kinetics in liquid chromatography, J. Chromatogr. A. 1263 (2012) 84–98. https://doi.org/10.1016/j.chroma.2012.09.030.

[62] G. Guiochon, Monolithic columns in high-performance liquid chromatography, J. Chromatogr. A. 1168 (2007) 101–168. https://doi.org/10.1016/j.chroma.2007.05.090.

[63] F. Svec, Y. Lv, Advances and recent trends in the field of monolithic columns for chromatography, Anal. Chem. 87 (2015) 250–273. https://doi.org/10.1021/ac504059c.

[64] O. Núñez, K. Nakanishi, N. Tanaka, Preparation of monolithic silica columns for high-performance liquid chromatography, J. Chromatogr. A. 1191 (2008) 231–252. https://doi.org/10.1016/j.chroma.2008.02.029.

[65] N. Tanaka, H. Kobayashi, N. Ishizuka, H. Minakuchi, K. Nakanishi, K. Hosoya, T. Ikegami, Monolithic silica columns for high-efficiency chromatographic separations, J. Chromatogr. A. 965 (2002) 35–49. https://doi.org/10.1016/S0021-9673(01)01582-5.

[66] A. Vaast, K. Broeckhoven, S. Dolman, G. Desmet, S. Eeltink, Comparison of the gradient kinetic performance of silica monolithic capillary columns with columns packed with 3 μm porous and 2.7 μm fused-core silica particles, J. Chromatogr. A. 1228 (2012) 270–275. https://doi.org/10.1016/j.chroma.2011.07.089.

[67] K.K. Unger, R. Skudas, M.M. Schulte, Particle packed columns and monolithic columns in high-performance liquid chromatography-comparison and critical appraisal, J. Chromatogr. A. 1184 (2008) 393–415. https://doi.org/10.1016/j.chroma.2007.11.118.

[68] P. Gzil, N. Vervoort, G.V. Baron, G. Desmet, General rules for the optimal external porosity of LC supports, Anal. Chem. 76 (2004) 6707–6718. https://doi.org/10.1021/ac049202u.

[69] R.W. Brice, X. Zhang, L.A. Colón, Fused-core, sub-2 μm packings, and monolithic HPLC columns: A comparative evaluation, J. Sep. Sci. 32 (2009) 2723–2731. https://doi.org/10.1002/jssc.200900091.

[70] N. Tanaka, D.V. McCalley, Core-shell, ultrasmall particles, monoliths, and other support materials in high-performance liquid chromatography, Anal. Chem. 88 (2016) 279–298. https://doi.org/10.1021/acs.analchem.5b04093.

[71] K.D. Patel, A.D. Jerkovich, J.C. Link, J.W. Jorgenson, In-depth characterization of slurry packed capillary columns with 1.0 µm nonporous particles using reversed-phase isocratic ultrahigh-pressure liquid chromatography, Anal. Chem. 76 (2004) 5777–5786. https://doi.org/10.1021/ac049756x.

[72] N. Wu, J.A. Lippert, M.L. Lee, Practical aspects of ultrahigh pressure capillary liquid chromatography, J. Chromatogr. A. 911 (2001) 1–12. https://doi.org/10.1016/S0021-9673(00)01188-2.

[73] N. Wu, Y. Liu, M.L. Lee, Sub-2 µm porous and nonporous particles for fast separation in reversed-phase high performance liquid chromatography, J. Chromatogr. A. 1131 (2006) 142–150. https://doi.org/10.1016/j.chroma.2006.07.042.

[74] G. Desmet, Comparison techniques for HPLC column performance, LC-GC Eur. 21 (2008) 310–320.

[75] T.J. Causon, K. Broeckhoven, E.F. Hilder, R.A. Shellie, G. Desmet, S. Eeltink, Kinetic performance optimisation for liquid chromatography: Principles and practice: Liquid Chromatography, J. Sep. Sci. 34 (2011) 877–887. https://doi.org/10.1002/jssc.201000904.

[76] S. Eeltink, G. Desmet, G. Vivó-Truyols, G.P. Rozing, P.J. Schoenmakers, W.Th. Kok, Performance limits of monolithic and packed capillary columns in high-performance liquid chromatography and capillary electrochromatography, J. Chromatogr. A. 1104 (2006) 256–262. https://doi.org/10.1016/j.chroma.2005.11.112.

[77] S. Heinisch, G. Desmet, D. Clicq, J.-L. Rocca, Kinetic plot equations for evaluating the real performance of the combined use of high temperature and ultra-high pressure in liquid chromatography, J. Chromatogr. A. 1203 (2008) 124–136. https://doi.org/10.1016/j.chroma.2008.07.039.

[78] P.A. Bristow, J.H. Knox, Standardization of test conditions for high performance liquid chromatography columns, Chromatographia. 10 (1977) 279–289. https://doi.org/10.1007/BF02263001.

[79] F. Lestremau, A. de Villiers, F. Lynen, A. Cooper, R. Szucs, P. Sandra, High efficiency liquid chromatography on conventional columns and instrumentation by using temperature as a variable, J. Chromatogr. A. 1138 (2007) 120–131. https://doi.org/10.1016/j.chroma.2006.10.042.

[80] D.R. Stoll, T.J. Lauer, K. Broeckhoven, Where has my efficiency gone? impacts of extracolumn peak broadening on performance, Part IV: Gradient elution, flow splitting, and a holistic view, LCGC North Am. 39 (2021) 308–314.

[81] K. Broeckhoven, C. Gunnarson, D.R. Stoll, But why doesn't it get better? Kinetic plots for LC, Part III – Pulling it all together, LCGC North Am. 40 (2022) 111–115.

[82] A.S. Kaplitz, G.A. Kresge, B. Selover, L. Horvat, E.G. Franklin, J.M. Godinho, K.M. Grinias, S.W. Foster, J.J. Davis, J.P. Grinias, High-throughput and ultrafast liquid chromatography, Anal. Chem. 92 (2020) 67–84. https://doi.org/10.1021/acs.analchem.9b04713.

[83] J.M. Godinho, A.E. Reising, U. Tallarek, J.W. Jorgenson, Implementation of high slurry concentration and sonication to pack high-efficiency, meter-long capillary ultrahigh pressure liquid chromatography columns, J. Chromatogr. A. 1462 (2016) 165–169. https://doi.org/10.1016/j.chroma.2016.08.002.

[84] B. Yan, J. Zhao, J.S. Brown, J. Blackwell, P.W. Carr, High-temperature ultrafast liquid chromatography, Anal. Chem. 72 (2000) 1253–1262. https://doi.org/10.1021/ac991008y.

[85] J.C. Giddings, Flow transport and viscous phenomena, in: Unified Separation Science, Wiley, New York, 1991: pp. 62–65.

[86] N. Wu, A.M. Clausen, Fundamental and practical aspects of ultrahigh pressure liquid chromatography for fast separations, J. Sep. Sci. 30 (2007) 1167–1182. https://doi.org/10.1002/jssc.200700026.

[87] U.D. Neue, in: C.F. Poole, I.F. Wilson (Eds.), Encyclopedia of Separation Science, Elsevier Science Ltd, United Kingdom, 2007.

[88] F. Gritti, G. Guiochon, Kinetic performance of narrow-bore columns on a micro-system for high performance liquid chromatography., J. Chromatogr., A. 1236 (2012) 105–114. https://doi.org/10.1016/j.chroma.2012.03.007.

[89] L.R. Snyder, J.J. Kirkland, J.W. Dolan, Chapter 2. Basic concepts and the control of separation, in: Introduction to Modern Liquid Chromatography, 3rd ed, Wiley, Hoboken, NJ, 2010: p. 55.

[90] E. Grushka, Chromatographic peak capacity and the factors influencing it, Analytical Chemistry. 42 (1970) 1142–1147. https://doi.org/10.1021/ac60293a001.

[91] J.C. Giddings, Maximum number of components resolvable by gel filtration and other elution chromatographic methods, Anal. Chem. 39 (1967) 1027–1028. https://doi.org/10.1021/ac60252a025.

[92] J.C. Giddings, Generation of variance, theoretical plates, resolution, and peak capacity in electrophoresis and sedimentation., Separ. Sci. 4 (1969) 181–9. https://doi.org/10.1080/01496396908052249.

[93] C.G. Horvath, S.R. Lipsky, Peak capacity in chromatography, Anal. Chem. 39 (1967) 1893–1893. https://doi.org/10.1021/ac50157a075.

[94] J.C. Giddings, Two-dimensional separations: concept and promise, Anal. Chem. 56 (1984) 1258A–1270A. https://doi.org/10.1021/ac00276a003.

[95] P.W. Carr, D.R. Stoll, A study of peak capacity optimization in one-dimensional gradient elution reversed-phase chromatography: A memorial to Eli Grushka, in: N. Grinberg (Ed.), Advances in Chromatography, CRC Press, Boca Raton, FL, 2018: pp. 1–21.

[96] J.M. Davis, J.C. Giddings, Statistical theory of component overlap in multicomponent chromatograms, Anal. Chem. 55 (1983) 418–424. https://doi.org/10.1021/ac00254a003.

[97] L.R. Snyder, J.W. Dolan, High-performance Gradient Elution: The Practical Application of the Linear-Solvent-Strength Model, John Wiley, Hoboken, NJ, 2007.

[98] F. Gritti, G. Guiochon, Separations by gradient elution: Why are steep gradient profiles distorted and what is their impact on resolution in reversed-phase liquid chromatography, J. Chromatogr. A. 1344 (2014) 66–75. https://doi.org/10.1016/j.chroma.2014.04.010.

[99] U. Neue, Theory of peak capacity in gradient elution, J. Chromatogr. A. 1079 (2005) 153–161. https://doi.org/10.1016/j.chroma.2005.03.008.

[100] U. Neue, Peak capacity in unidimensional chromatography, J. Chromatogr., A. (2008) 107–130. https://doi.org/10.1016/j.chroma.2007.11.113.

[101] U. Neue, D. Marchand, L.R. Snyder, Peak compression in reversed-phase gradient elution, J. Chromatogr., A. 1111 (2006) 32–39. https://doi.org/10.1016/j.chroma.2006.01.104.

[102] F. Gritti, G. Guiochon, The distortion of gradient profiles in reversed-phase liquid chromatography, J. Chromatogr. A. 1340 (2014) 50–58. https://doi.org/10.1016/j.chroma.2014.03.004.

[103] H. Poppe, J. Paanakker, M. Bronckhorst, Peak width in solvent-programmed chromatography, J. Chromatogr. A. 204 (1981) 77–84. https://doi.org/10.1016/S0021-9673(00)81641-6.

[104] F. Gritti, M.-A. Perdu, G. Guiochon, Gradient HPLC of samples extracted from the green microalga botryococcus braunii using highly efficient columns packed with 2.6 µm Kinetex-C18 core–shell particles, J. Chromatogr. A. 1229 (2012) 148–155. https://doi.org/10.1016/j.chroma.2012.01.013.

[105] L.M. Blumberg, Theory of gradient elution liquid chromatography with linear solvent strength: Part 1. Migration and elution parameters of a solute band, Chromatographia. 77 (2014) 179–188. https://doi.org/10.1007/s10337-013-2555-y.

[106] L.M. Blumberg, Theory of gradient elution liquid chromatography with linear solvent strength: Part 2. Peak width formation, Chromatographia. 77 (2014) 189–197. https://doi.org/10.1007/s10337-013-2556-x.

[107] L.M. Blumberg, G. Desmet, Metrics of separation performance in chromatography: Part 3: General separation performance of linear solvent strength gradient liquid chromatography, J. Chromatogr. A. 1413 (2015) 9–21. https://doi.org/10.1016/j.chroma.2015.07.122.

[108] L.M. Blumberg, G. Desmet, Optimal mixing rate in linear solvent strength gradient liquid chromatography, Anal. Chem. 88 (2016) 2281–2288. https://doi.org/10.1021/acs.analchem.5b04078.

[109] X. Wang, D.R. Stoll, P.W. Carr, P.J. Schoenmakers, A graphical method for understanding the kinetics of peak capacity production in gradient elution liquid chromatography, J. Chromatogr. A. 1125 (2006) 177–181. https://doi.org/10.1016/j.chroma.2006.05.048.

[110] L. Lapidus, N.R. Amundson, Mathematics of adsorption in beds. VI. The effect of longitudinal diffusion in ion exchange and chromatographic columns, J. Phys. Chem. 56 (1952) 984–988. https://doi.org/10.1021/j150500a014.

[111] J.C. Giddings, Chapter 2: Dynamics of zone spreading, in: Dynamics of Chromatography: Principles and Theory, Marcel Dekker, New York, 1965: p. 52.

[112] M. Dittmann, X. Wang, New materials for stationary phases in LC/MS, in: C. Byrdwell, M. Holcapek (Eds.), Handbook of Advanced Chromatography / Mass Spectrometry Techniques, Academic Press – Elsevier, New York, 2017: pp. 179–218.

[113] J.C. Giddings, Nature of gas phase mass transfer in gas chromatography, Anal. Chem. 34 (1962) 1186–1192. https://doi.org/10.1021/ac60190a005.

[114] J.H. Knox, J.F. Parcher, Effect of the column to particle diameter ratio on the dispersion of unsorbed solutes in chromatography, Anal. Chem. 41 (1969) 1599–1606. https://doi.org/10.1021/ac60281a009.

[115] A.L. Berdichevsky, U.D. Neue, Nature of the eddy dispersion in packed beds, J. Chromatogr. A. 535 (1990) 189–198. https://doi.org/10.1016/S0021-9673(01)88944-5.

[116] G. Desmet, K. Broeckhoven, Equivalence of the different C_m – and C_s – Term expressions used in liquid chromatography and a geometrical model uniting them, Anal. Chem. 80 (2008) 8076–8088. https://doi.org/10.1021/ac8011363.

[117] S. Deridder, G. Desmet, Effective medium theory expressions for the effective diffusion in chromatographic beds filled with porous, non-porous and porous-shell particles and cylinders. Part II: Numerical verification and quantitative effect of solid core on expected B-term band broadening, J. Chromatogr. A. 1218 (2011) 46–56. https://doi.org/10.1016/j.chroma.2010.10.086.

[118] S. Deridder, G. Desmet, Calculation of the geometrical three-point parameter constant appearing in the second order accurate effective medium theory expression for the B-term diffusion coefficient in fully porous and porous-shell random sphere packings, J. Chromatogr. A. 1223 (2012) 35–40. https://doi.org/10.1016/j.chroma.2011.12.004.

[119] J.H. Knox, H.P. Scott, B and C terms in the Van Deemter equation for liquid chromatography, J. Chromatogr. A. 282 (1983) 297–313. https://doi.org/10.1016/S0021-9673(00)91609-1.

[120] K. Broeckhoven, D. Cabooter, F. Lynen, P. Sandra, G. Desmet, Errors involved in the existing B-term expressions for the longitudinal diffusion in fully porous chromatographic media, J. Chromatogr. A. 1188 (2008) 189–198. https://doi.org/10.1016/j.chroma.2008.02.058.

[121] G. Desmet, K. Broeckhoven, J. De Smet, S. Deridder, G.V. Baron, P. Gzil, Errors involved in the existing B-term expressions for the longitudinal diffusion in fully porous chromatographic media, J. Chromatogr. A. 1188 (2008) 171–188. https://doi.org/10.1016/j.chroma.2008.02.018.

[122] Effective medium approximations (n.d.). https://en.wikipedia.org/wiki/Effective_medium _ approximations.

[123] J. De Vos, C. Stassen, A. Vaast, G. Desmet, S. Eeltink, High-resolution separations of tryptic digest mixtures using core-shell particulate columns operated at 1200bar, J. Chromatogr. A. 1264 (2012) 57–62. https://doi.org/10.1016/j.chroma.2012.09.065.

[124] X. Wang, W.E. Barber, P.W. Carr, A practical approach to maximizing peak capacity by using long columns packed with pellicular stationary phases for proteomic research, J. Chromatogr. A. 1107 (2006) 139–151. https://doi.org/10.1016/j.chroma.2005.12.050.

[125] C.G. Horvath, S.R. Lipsky, Column design in high pressure liquid chromatography, J. Chromatogr. Sci. 7 (1969) 109–116. https://doi.org/10.1093/chromsci/7.2.109.

[126] A. Felinger, Diffusion time in core–shell packing materials, J. Chromatogr. A. 1218 (2011) 1939–1941. https://doi.org/10.1016/j.chroma.2010.10.025.

[127] E.J. Wilson, C.J. Geankoplis, Liquid mass transfer at very low Reynolds numbers in packed beds, Ind. Eng. Chem. Fund. 5 (1966) 9–14. https://doi.org/10.1021/i160017a002.

[128] Kozeny-Carman Equation (n.d.). https://en.wikipedia.org/wiki/Kozeny-Carman_equation (accessed February 27, 2022).

[129] R.B. Bird, W.E. Stewart, E.N. Lightfoot, Chapter 6. Interphase transport in isothermal systems, in: Transport Phenomena, John Wiley & Sons, Inc., New York, 1960.

[130] R.F. Probstein, Physicochemical Hydrodynamics: An Introduction, 2nd ed, Wiley-Interscience, Hoboken, NJ, 2003.

[131] H. Poppe, J.C. Kraak, J.F.K. Huber, J.H.M. van den Berg, Temperature gradients in HPLC columns due to viscous heat dissipation, Chromatographia. 14 (1981) 515–523. https://doi.org/10.1007/BF02265631.

[132] H. Lin, C.G. Horváth, Viscous dissipation in packed beds, Chem. Eng. Sci. 36 (1981) 47–55. https://doi.org/10.1016/0009-2509(81)80047-4.

[133] A. Moussa, S. Deridder, K. Broeckhoven, G. Desmet, Detailed computational fluid dynamics study of the parameters contributing to the viscous heating band broadening in liquid chromatography at pressures up to 2500 bar in 2.1 mm columns, J. Chromatogr. A. 1661 (2022) 462683. https://doi.org/10.1016/j.chroma.2021.462683.

[134] R. Aris, On the dispersion of a solute in a fluid flowing through a tube, Proc. R. Soc. Lond. A. 235 (1956) 67–77. https://doi.org/10.1098/rspa.1956.0065.

[135] G. Desmet, Theoretical calculation of the retention enthalpy effect on the viscous heat dissipation band broadening in high performance liquid chromatography columns with a fixed wall temperature, J. Chromatogr. A. 1116 (2006) 89–96. https://doi.org/10.1016/j.chroma.2006.03.024.

[136] J.H. Knox, I.H. Grant, Miniaturisation in pressure and electroendosmotically driven liquid chromatography: Some theoretical considerations, Chromatographia. 24 (1987) 135–143. https://doi.org/10.1007/BF02688476.

[137] K. Broeckhoven, G. Desmet, Considerations for the use of ultra-high pressures in liquid chromatography for 2.1 mm inner diameter columns, J. Chromatogr. A. 1523 (2017) 183–192. https://doi.org/10.1016/j.chroma.2017.07.040.

[138] F. Gritti, Designing vacuum-jacketed user-friendly columns for maximum resolution under extreme UHPLC and SFC conditions, LCGC North Am. 36 (2018) 18–23.

[139] F. Gritti, M. Gilar, J.A. Jarrell, Quasi-adiabatic vacuum-based column housing for very high-pressure liquid chromatography, J. Chromatogr. A. 1456 (2016) 226–234. https://doi.org/10.1016/j.chroma.2016.06.029.

[140] F. Gritti, M. Gilar, J.A. Jarrell, Achieving quasi-adiabatic thermal environment to maximize resolution power in very high-pressure liquid chromatography: Theory, models, and experiments, J. Chromatogr. A. 1444 (2016) 86–98. https://doi.org/10.1016/j.chroma.2016.03.070.

[141] A. Moussa, S. Deridder, K. Broeckhoven, G. Desmet, Computational fluid dynamics study of potential solutions to alleviate viscous heating band broadening in 2.1 millimeter liquid chromatography columns, J. Chromatogr. A. 1654 (2021). https://doi.org/10.1016/j.chroma.2021.462452.

[142] R.D. Pauw, B. Degreef, H. Ritchie, S. Eeltink, G. Desmet, K. Broeckhoven, Extending the limits of operating pressure of narrow-bore column liquid chromatography instrumentation, J. Chromatogr. A. 1347 (2014) 56–62. https://doi.org/10.1016/j.chroma.2014.04.056.

[143] J. Kostka, F. Gritti, K. Kaczmarski, G. Guiochon, Modified equilibrium-dispersive model for the interpretation of the efficiency of columns packed with core–shell particle, J. Chromatogr. A. 1218 (2011) 5449–5455. https://doi.org/10.1016/j.chroma.2011.06.019.

[144] S. Deridder, W. Smits, K. Broeckhoven, G. Desmet, A multiscale modelling study on the sense and nonsense of thermal conductivity enhancement of liquid chromatography packings and other potential solutions for viscous heating effects, J. Chromatogr. A. 1620 (2020). https://doi.org/10.1016/j.chroma.2020.461022.

[145] R.C. Reid, J.M. Prausnitz, B.E. Poling, Chapter 11, in: The Properties of Gases and Liquids, 4th ed, McGraw-Hill, New York, 1987.

[146] C.R. Wilke, P. Chang, Correlation of diffusion coefficients in dilute solutions, Am. Inst. Chem. Eng. J. 1 (1955) 264–270.

[147] A.J. Easteal, L.A. Woolf, Solute-solvent interaction effects on tracer diffusion coefficients, J. Chem. Soc., Faraday Trans. 1. 80 (1984) 1287. https://doi.org/10.1039/f19848001287.

[148] J. Li, P.W. Carr, Accuracy of empirical correlations for estimating diffusion coefficients in aqueous organic mixtures, Anal. Chem. 69 (1997) 2530–2536. https://doi.org/10.1021/ac961005a.

[149] H. Song, Y. Vanderheyden, E. Adams, G. Desmet, D. Cabooter, Extensive database of liquid phase diffusion coefficients of some frequently used test molecules in reversed-phase liquid chromatography and hydrophilic interaction liquid chromatography, J. Chromatogr. A. 1455 (2016) 102–112. https://doi.org/10.1016/j.chroma.2016.05.054.

3 Theoretical Guiding Principles for Two-Dimensional Liquid Chromatography

Dwight R. Stoll and Peter W. Carr

CONTENTS

3.1 INTRODUCTION

Early experimental demonstrations of what was possible in both the heartcut and comprehensive modes of two-dimensional liquid chromatography (2D-LC) were published in the late 1970s and early 1980s [1, 2]. Much of the theoretical work that forms the basis of how researchers approach 2D-LC today was carried out in the 15-year period from the late 1990s to the early 2010s. Given this timeline, most of the topics discussed in this chapter are considered settled issues by the 2D-LC research community, and patterns have emerged in the literature in recent years that reflect this understanding. Of course, there is still room for further improvement on these topics, but research on them has slowed in favor of other topics such as data analysis and method development.

Our emphasis at this point is that the ideas discussed in this chapter are pitched as concepts that should guide development of 2D-LC methods. They are not rigid rules strictly applicable to every situation where 2D-LC will be used. Rather, utilizing these concepts will generally increase the likelihood that method development will lead to a 2D-LC method that performs better (as measured by analysis time, resolution, sensitivity, etc.) than the best available 1D-LC method. As with all theoretical guidance, there will be practical situations where there are good reasons not to follow the prevailing guidance.

3.2 TWO-DIMENSIONAL PEAK CAPACITY

The concept of peak capacity is discussed in Section 2.12 for conventional one-dimensional (1D) separations. One of the attractive characteristics of comprehensive 2D-LC (LC×LC) separations is that they offer a tremendous potential increase in peak capacity compared to 1D separations. This increase is due to the multiplicative relationship between the 2D peak capacity ($n_{c,2D}$) and the peak capacities of the individual 1D separations (^1n_c and ^2n_c) that contribute to the LC×LC separation, as expressed in Eq. 3.1.

$$n_{c,2D} = {}^1n_c \times {}^2n_c \tag{3.1}$$

DOI: 10.1201/9781003090557-3

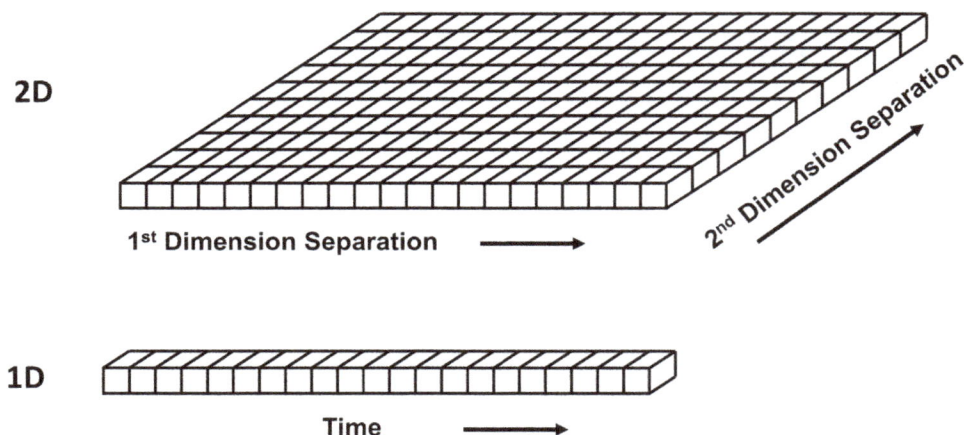

FIGURE 3.1   Illustration comparing the peak capacities of 1D and comprehensive 2D separations. Each bin corresponds to one unit of peak capacity.

Source: Adapted from [5].

This idea, and its potential impact are illustrated in Figure 3.1. One view of this expression is that the resolving power of the ^{2}D separation acts as a multiplier of the resolving power of the first dimension. Adding a ^{2}D separation with a modest peak capacity of 10 to an existing 1D separation with a peak capacity of 200 would ideally result in a 2D peak capacity of 2,000. Achieving such a peak capacity for a 1D-LC separation of any molecules in any mode is literally impossible within any practically useful analysis time.

We refer to Eq. 3.1 as the "product rule" for the peak capacity of comprehensive 2D separations. The notion of this multiplicative relationship was first discussed by Karger, Snyder, and Horvath [3], and later by Giddings [4]. The legitimacy of Eq. 3.1 depends on the fulfillment of the following two criteria, stated by Giddings [5]. These can never be perfectly fulfilled in practice, thus 2D peak capacities calculated in this way represent an ideal upper bound. Effective 2D peak capacities are always smaller – and sometimes much smaller – as is discussed below in Section 3.5.

1. The retention pattern for compounds in one separation mode must be independent of the retention pattern in the second separation mode. In most of the recent 2D-LC literature this "independence" is termed as "orthogonality" [6]. We prefer the term "complementary" when discussing how the two separation modes relate to each other, as is discussed more in Section 3.3 below.

2. Whenever two compounds are separated by one separation dimension, they must remain separated as they move through the other separation dimensions. In other words, any resolution gained in one separation should not be diminished by the other separation. This criterion has proven to be extremely challenging to fulfill in practice, and the biggest issue occurs when we encounter the combination of narrow ^{1}D peaks and slow ^{2}D separations. This leads to "undersampling", which is discussed in more detail in Section 3.4. In essence, the narrower the ^{1}D peaks, the more difficult it is to do the ^{2}D separation fast enough to avoid degrading the resolution, i.e. the peak capacity of the ^{1}D separation.

Fair and realistic comparisons of LC×LC separations to 1D ones, or comparisons of multiple 2D-LC separations to each other, must account for the inability to perfectly fulfill these criteria in real separations. The framework we used for doing so [7] is discussed in Section 3.5.

As a metric of separation performance, peak capacity is most useful for discussing LC×LC separations because LC×LC is rarely applied in cases where baseline resolution of all components in the sample is both possible and desired. Since peak capacity is related to the average resolution of components in a sample [8], it is a practically useful measure of the overall degree of separation for all sample components. The peak capacity concept can also be applied to heartcut separations, however we do not spend any time discussing this here because usually the more preferred metric of separation quality in heartcut separations is the resolution of the specific pairs of analytical interest.

3.3 SEPARATION ORTHOGONALITY AND COMPLEMENTARITY

The range of 2D retention patterns that can be observed in experimental 2D-LC separations is shown qualitatively in Figure 3.2. Panel A shows a retention pattern where the retention times of the ^{2}D separation are very highly correlated with those of the ^{1}D separation. 2D separations resulting in retention patterns like this are usually not very useful (though there are important exceptions, as in the case of resolving a small peak hiding in the front or tail of a large ^{1}D peak; e.g., see ref. [9]); in a case like this it would be better to invest experimental effort in an improved 1D separation (e.g., by using a longer column) than the 2D separation. Panel B shows a retention pattern with a very low correlation between the ^{1}D and ^{2}D retention times. This type of pattern is highly desirable in 2D-LC, and is becoming increasingly common as researchers find pairs of stationary and mobile phase combinations that are highly complementary for specific applications, and as "shifting gradients" are used more often to further spread the retention pattern (see Section 4.5 for details on the implementation of shifting gradients). Finally, the retention pattern in Panel C is an example of a moderate correlation between the ^{1}D and ^{2}D retention patterns, and one where the need to account for the departure from "independent" retention patterns is greatest. In a case like this, not properly accounting for the fact that a majority of the available 2D separation space is not optimally used will lead one to believe that the 2D separation is much higher performing than it actually is, and in the worst case it may not even be competitive with a high quality 1D separation.

Considerable effort has been invested by several research groups to develop ways of quantifying the extent to which ^{1}D and ^{2}D separations are complementary, and approach the "independence" spoken of by Giddings [5]. Most of this work took place in the period from the late 1990s to the mid-2010s, and has been critically assessed in two research articles that discuss the strengths and weaknesses of several of the approaches [11, 12]. Two of the approaches we used in our work are illustrated in Figures 3.3 and 3.4. We favor these methods because they are easy to understand and implement in practice, and they are being used by a number of other research groups as well. The first approach (Figure 3.3), which we refer to as a box or bin counting method, was initially

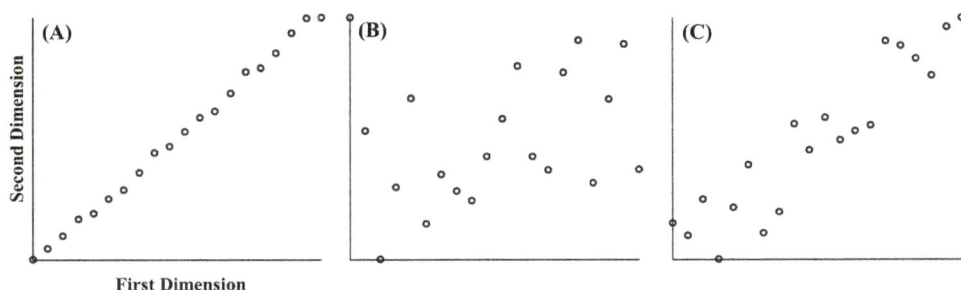

FIGURE 3.2 Illustration of different retention patterns observed in LC×LC separations, ranging from highly correlated (A) to uncorrelated (B). All patterns contain the same number of peaks.

Source: Adapted from [10].

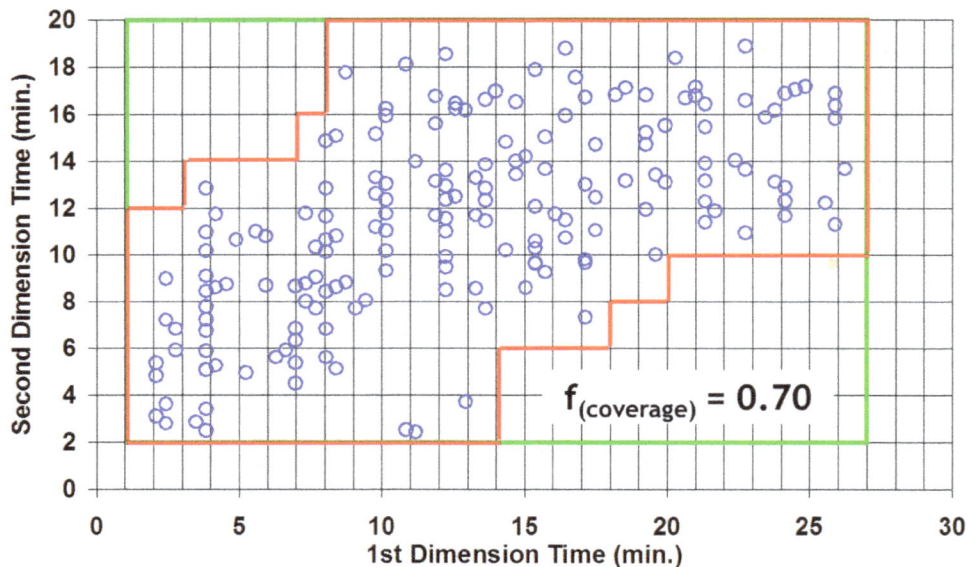

FIGURE 3.3 Illustration of a box-counting approach to quantifying the usage of a 2D separation space. Reprinted with permission from J. Davis, D. Stoll, P. Carr, Dependence of effective peak capacity in comprehensive two-dimensional separations on the distribution of peak capacity between the two dimensions, *Analytical Chemistry* 80 (2008) 8122–8134. Copyright 2008 American Chemical Society.

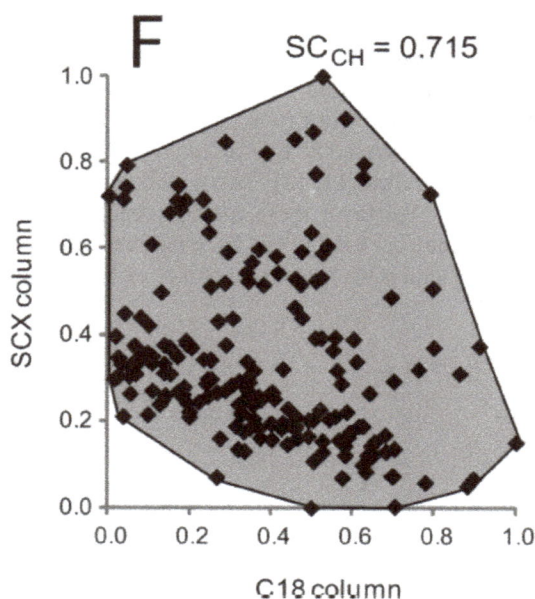

FIGURE 3.4 Illustration of the use of the convex hull to quantify the fraction of the 2D separation space occupied by peaks. Reprinted with permission from M. Gilar, J. Fridrich, M. Schure, A. Jaworski, Comparison of orthogonality estimation methods for the two-dimensional separations of peptides, *Analytical Chemistry* 84 (2012) 8722–8732. Copyright 2012 American Chemical Society.

described by Gilar *et al.* [13], and adapted by Davis *et al.* [14]. The goal of this approach is to quantify the fraction of the available 2D separation space that is occupied by peaks observed in experimental chromatograms. A grid is cast on the separation space, and the ratio of the number of occupied bins to the total number of bins is calculated. Variations on this theme either do or do not count bins that do not contain peaks, but are located between bins that do contain peaks. In the example shown in Figure 3.3, the fraction of the total number of bins that contains peaks (in this case some bins are included that do not contain peaks) is 0.7 (referred to as the "fractional coverage", f_{cov}). As is discussed below in Section 3.5, this fraction can be used to make a correction to the "product rule" (Eq. 3.1) that accounts for the fact that the available 2D separation space is not fully utilized. We note that this approach has several significant weaknesses, including a dependence of f_{cov} on the bin size, and the high sensitivity of f_{cov} to peaks that are relatively isolated and lie far away from the majority of the peak distribution. Other approaches, such as the so-called asterisk method developed by Camenzuli and Schoenmakers are less prone to these problems [15]. The second approach (Figure 3.4) uses the geometric concept of a "convex hull" to quantify the fraction of the 2D separation space that is occupied by peaks [16]. The convex hull for a chromatogram is established by drawing a polygon around the peak pattern where each vertex of the polygon is co-located with a chromatographic peak on the periphery of the distribution. This approach does not involve discrete bins, and thus is not sensitive to bin size, but is still quite sensitive to the location of outlying peaks. This sensitivity can be weakened by excluding from the hull the most outlying peaks [16].

Note that we have deliberately avoided the use of the term "orthogonality" here, in part because it has a strict mathematical definition that can be a distraction when discussing whether or not two separation modes are complementary enough to be useful in a 2D separation format. Many innovative ideas have been introduced over the years as ways to measure use of the 2D separation space. For practical separations the means to arrive at these estimates are rarely critically important, thus we have not discussed them extensively here. Readers interested in these details are referred to research articles that have critically compared many of the different approaches [11, 12].

3.4   UNDERSAMPLING

As stated above in Section 3.2, the second of Giddings' criteria regarding the product rule for 2D peak capacity is that whenever two compounds are separated by one separation dimension, they must remain separated as they move through the other separation dimensions. In many, if not most, real 2D-LC separations this is difficult to achieve because ^{2}D separations are slow relative to the temporal width of typical ^{1}D peaks. This results in a situation where a fraction of ^{1}D effluent that is transferred to the second dimension contains more than one peak, or a volume that corresponds to the volume of more than one peak. The compounds in those peaks get mixed together as they enter the ^{2}D column (sometimes described as "remixing"), thus undoing the resolution of those peaks that was achieved by the first separation, thereby decreasing the effective peak capacity of the ^{1}D separation. In other words, this is at least a partial step backward on the journey toward resolving components of the sample, not a full step forward. We refer to this scenario where the large volume of ^{1}D effluent fractions degrades the ^{1}D resolution as "undersampling"; its effect on ^{1}D resolution is illustrated qualitatively in Figure 3.5. When the sampling time becomes large relative to the width of ^{1}D peaks prior to sampling, the resolution of a ^{1}D peak from its neighbor is entirely lost.

The undersampling phenomenon as it manifests in 2D-LC has been studied in great detail by several groups [18–21], and summarized in multiple review articles and book chapters [10, 17, 22], thus it is only discussed rather briefly here. Readers interested in a more detailed perspective are referred to the extensive treatments published earlier. The first group to quantify the impact of undersampling in 2D separations was Murphy, Schure, and Foley [23]. The main finding of their work was that ^{1}D separations should be sampled at a rate corresponding to three to four samples

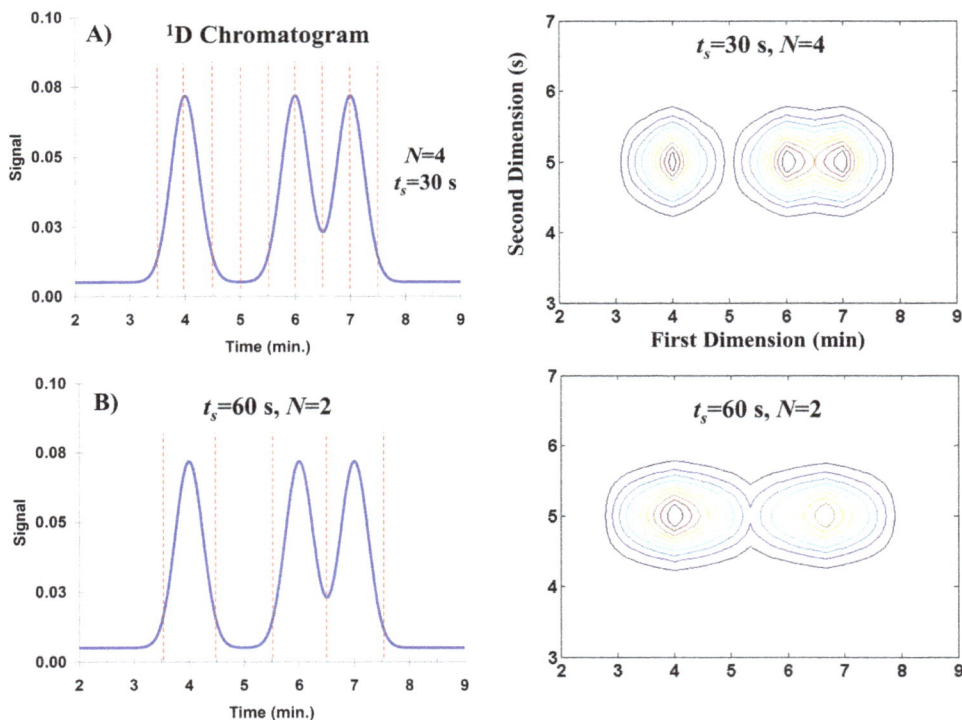

FIGURE 3.5   Illustration of the effect of increasing the sampling time on the effective ^{1}D resolution. Dashed red lines indicate the beginning and end of each sampling period. In this hypothetical example, all three sample components have the same ^{2}D retention time. N refers to the number of fractions taken over a period corresponding to $8 \cdot ^1\sigma$.

Source: Reprinted from [17].

over a time corresponding to eight times the standard deviation of a typical ^{1}D peak prior to sampling ($8 \cdot ^1\sigma$) to avoid losing a significant fraction of the native ^{1}D peak capacity. This finding has had a major impact on the way that the development of LC×LC methods is approached, as is discussed below in Section 3.5, but it was also a major driving force for the development of sLC×LC [24]. In our study of undersampling we used a numerical simulation and stochastic technique to establish a quantitative relationship between the sampling rate and the effective peak capacity of the ^{1}D separation. Davis *et al.* found the empirical relationship shown in Eq. 3.2, where $<\beta>$ is the factor that quantifies the effective broadening of a ^{1}D peak due to undersampling, t_s is the sampling time, and $^1\sigma$ is the standard deviation of the ^{1}D peak prior to sampling. For the conditions considered by Davis, the constant κ has a value of 0.21 that is applicable when $0.2 < \left(\dfrac{t_s}{^1\sigma} \right) < 16$ [25].

$$< \beta > = \sqrt{1 + \kappa \left(\frac{t_s}{^1\sigma} \right)^2} \qquad (3.2)$$

The broadening factor $<\beta>$ can then be used to calculate a corrected peak capacity for the ^{1}D separation (^1n_c) after it has been sampled, as in Eq. 3.3 [17].

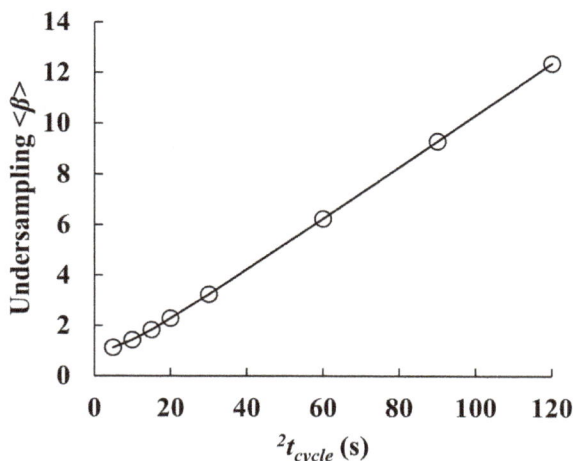

FIGURE 3.6 Effect of ^{2}D cycle time on the magnitude of the broadening factor $<\beta>$ (calculated using Eq. 3.2) for the case of sampling a 30-min ^{1}D separation with a peak capacity of 100.

$$^1n_c' = \frac{^1n_c}{<\beta>} \tag{3.3}$$

Figure 3.6 shows that the broadening factor $<\beta>$ can be very large even when sampling a 30-min. ^{1}D separation with a modest peak capacity of 100. In this case the $^1\sigma$ is 4.5 s; if a very short 5-s ^{2}D cycle time can be implemented, the $<\beta>$ value is just 1.1, and only 10% of the native ^{1}D peak capacity will be lost. However, this fast cycle time is difficult to achieve in practice, and much longer ^{2}D cycle times are usually used. Increasing $^2t_{cycle}$ to just 20 s leads to a $<\beta>$ of 2.3 (about a 55% loss in ^1n_c), and moving to a more comfortable $^2t_{cycle}$ of 1 min. leads to a $<\beta>$ of 6.3 and a devastating 84% loss of ^{1}D peak capacity.

The full implications of this correction to the ^{1}D peak capacity are discussed in more detail in the next section, but considering how serious the undersampling problem is, it is natural for one to raise the question of whether or not it is really worth the experimental effort to do 2D-LC. Indeed, this question motivated our 2008 study critically comparing the resolving power of 1D-LC and LC×LC [7].

3.5 OPTIMIZATION OF TWO-DIMENSIONAL PEAK CAPACITY

We have argued that a realistic estimate of the effective peak capacity of a LC×LC separation should include corrections for incomplete usage of the 2D separation space (discussed in Section 3.3), and undersampling (discussed in Section 3.4) [7]. Equation 3.4 shows the equation that incorporates these two corrections to the product rule (Eq. 3.1) to give a corrected 2D peak capacity ($n_{c,2D}^*$):

$$n_{c,2D}^* = \frac{^1n_c}{<\beta>} \cdot {}^2n_c \cdot f_{cov} \tag{3.4}$$

One of the most important decisions that must be made when developing a LC×LC method is to choose a sampling time. On one hand we are instinctively inclined to use long sampling times because this allows long ^{2}D separations to be used to generate higher values of ^2n_c and thus of $n_{c,2D}^*$. However, as was discussed in Section 3.4, increasing the sampling time also increases the likelihood that compounds separated by the ^{1}D column will be remixed during the sampling process, thus reducing the effective peak capacity of the ^{1}D separation. This situation – where increasing the sampling time leads to higher ^{2}D peak capacities, but lower effective ^{1}D peak capacities – leads to an optimum sampling time that is neither too long, nor too short. Indeed, this optimum has been predicted using theoretical arguments [26, 27], and its impact on real LC×LC separations has been verified by studying the number of peaks observed in LC×LC chromatograms obtained as the sampling time has been varied [27].

Figure 3.7 quantifies the impact of the sampling time on the effective 2D peak capacity, where we assume that the time available for each ^{2}D separation cycle ($^2t_{cycle}$) is the same as the sampling time (t_s), as is usually the case in LC×LC. Panel B shows that the peak capacity contributed by the ^{2}D separation increases as longer cycle times are used, as expected (see Section 2.10 for the dependence of n_c on analysis time). Based on this dependence alone, one is motivated to use longer ^{2}D cycle times. However, Panel A shows the penalty for increasing the cycle time, as the corrected ^{1}D peak capacity decreases according to Eq. 3.3, with increasing cycle time. When $^1n_c'$ is multiplied by ^2n_c (and assuming for the sake of discussion here that $f_{cov} = 1$) we get the dependence of $n^*_{c,2D}$ on $^2t_{cycle}$ shown in Panel C. One of the most interesting aspects of this curve is that the optimum occurs around a ^{2}D cycle time of about 25 s. The optimum certainly is not more than a minute, and thus this has been a major motivation for our work on the development of LC×LC separations with very fast ^{2}D cycle times [28–30]. We note here that the position of the optimum with respect to time is dependent on retention characteristics of the molecules under study and the time required to re-equilibrate the ^{2}D column, because these both affect the shape of the curve in Panel B.

Finally, we need to comment on several statements in the literature that one way to mitigate the effects of undersampling is to deliberately slow down the ^{1}D separation and/or deliberately broaden the ^{1}D peaks [31–35]. While this may seem warranted from a mathematical point of view (i.e., as $^1\sigma$ increases, $t_s/^1\sigma$ decreases, leading to smaller β values), from a practical point of view we find little value in this strategy. Deliberately broadening the ^{1}D peaks while holding the analysis time constant will lead to loss of resolution provided by the ^{1}D separation. So, even if the broadening due to undersampling is reduced by broadening the ^{1}D peaks prior to sampling, more ^{1}D resolution will have been lost already anyway before only a fraction of that loss will be regained by the slight increase in ^2n_c. Obtaining ^{1}D peaks that are broad enough to drive β close to 1 by extending the time

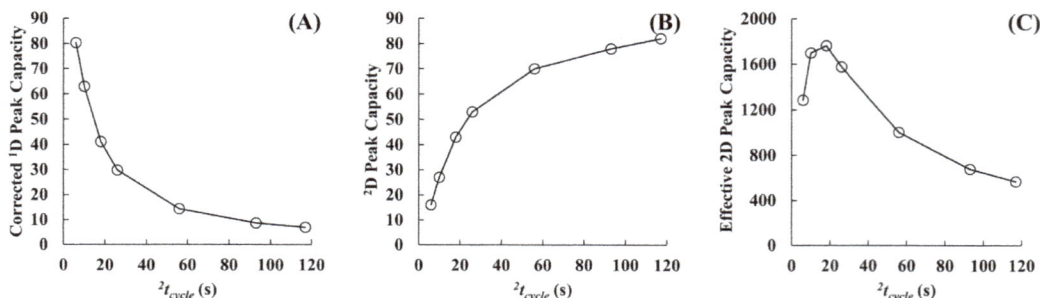

FIGURE 3.7 Illustration of the impact of the ^{2}D cycle time ($^2t_{cycle}$) on ^{1}D, ^{2}D, and 2D peak capacities for LC×LC separations. These are calculated trends that assume typical behavior for small molecule reversed-phase separations, and were made as follows: the native ^{1}D analysis time and peak capacity (i.e., without undersampling) were assumed to be 30 min. and 100, respectively; the corrected ^{1}D peak capacity (A) was calculated using Eq. (3.4); the dependence of ^{2}D peak capacity (B) on $^2t_{cycle}$ is the same as that reported previously [36]; and, the effective 2D peak capacity (C) was calculated using Eq. 3.4.

Source: Adapted from [26].

of the ^{1}D separation (e.g., to several hours) is a different scenario, but this is not a viable solution for most practical applications. Our view is that the only justifiable reason to deliberately broaden ^{1}D peaks is to facilitate data analysis using chemometric tools that rely on the appearance of a compound in more than one ^{2}D chromatogram (see further discussion in Sections 8.1 and 8.4). However, at this point we are not aware of any definitive guidance on what constitutes an optimal number of samples per ^{1}D peak in terms of facilitating data analysis using chemometric methods. In our own work we generally strive for high quality ^{1}D separations, without deliberate peak broadening, and let the framework discussed above (Eq. 3.4) account for the effect of undersampling on resolving power.

3.6   DETECTION SENSITIVITY

For many years detection sensitivity was not considered as a primary performance metric in most publications describing the development of 2D-LC methods. However, over the last decade or so it has gained more attention, probably because it became increasingly evident that poor detection sensitivity was a general weakness of LC×LC methods that had to be improved for 2D-LC to be more widely adopted. The fundamental problem is that – generally speaking – the local concentration of analyte decreases as an analyte band moves from the column inlet to the column outlet in chromatographic separations due to broadening of the analyte zone, unless something is deliberately done to overcome this phenomenon. In 2D-LC the consequences of this broadening and dilution of the analyte zone are amplified because it happens twice – once as the zone proceeds through the ^{1}D column, and then again as the band moves through the ^{2}D column [37]. If the ^{1}D separation of a 2D-LC system is comparable to a 1D separation, then this is the reason that the detection sensitivity of a 2D separation is generally worse than that for a 1D separation. Schure was the first to quantitatively describe the seriousness of this problem, and for a variety of potential couplings of different separation techniques in multi-dimensional formats [38].

The extent to which adding the ^{2}D separation to a 1D separation dilutes the analyte zone relative to its concentration at the ^{1}D column outlet is given by the dilution factor (^2DF) in Eq. 3.5, provided that mass is conserved from the point of injection into the ^{2}D column through to the point of detection [39–41]. Here $^2C_{inj}$ and $^2C_{det}$ are the analyte concentrations at the point of injection and detection (i.e., at the peak apex) in the second dimension, respectively, and $^2\sigma_{v,col}$ is the standard deviation of a ^{2}D peak in volume units. When the calculated ^2DF is larger than one, the second dimension acts to dilute the analyte, and when ^2DF is less than one, it acts to concentrate the analyte. Clearly, to avoid significant dilution of the analyte by the ^{2}D separation, conditions must be chosen that minimize the widths of ^{2}D peaks in volume units while increasing the volume of ^{1}D effluent injected into the ^{2}D column as much as possible without compromising the resolving power of the separation.

$$^2DF = \frac{^2C_{inj}}{^2C_{det}} = \frac{^2V_{det}}{^2V_{inj}} = \frac{\sqrt{2\pi} \cdot ^2\sigma_{v,col}}{^2V_{inj}} \tag{3.5}$$

In some applications it is straightforward to achieve $^2DF < 1$, thereby increasing the detection sensitivity of the method by adding the ^{2}D separation. One of the most well-known examples of this is in IEX×RP separation of peptides. In this case peptides elute from the ^{1}D separation in a water-rich effluent. Very large volumes of this effluent can be injected into the ^{2}D RP column (i.e., injection volumes can be several multiples of the dead volume of the ^{2}D column) without seriously impacting the peak capacity of the ^{2}D separation. In most other pairings of modes of 2D-LC separations, however, mobile phase mismatch between the two dimensions prevents injection of such large volumes of ^{1}D effluent into the ^{2}D column because doing so will lead to serious broadening of ^{2}D peaks (see Section 4.4.4). In these cases there is a tradeoff between ^{2}D resolving power and detection sensitivity, and the user must decide which of these performance metrics to prioritize [41]. Advanced modulation techniques can help mitigate this problem and make the compromise less serious [42]. Readers interested in this possibility are referred to Section 4.4 for more details.

REFERENCES

[1] R.E. Majors, Multidimensional high performance liquid chromatography., J. Chromatogr. Sci. 18 (1980) 571–579.

[2] F. Erni, R.W. Frei, Two-dimensional column liquid chromatographic technique for resolution of complex mixtures, J. Chromatogr. A. 149 (1978) 561–569. https://doi.org/10.1016/S0021-9673(00)81011-0.

[3] B. Karger, L. Snyder, C. Horvath, Multistep separation schemes for complex samples, in: An Introduction to Separation Science, 1st ed., John Wiley & Sons, Inc., New York, 1973: pp. 558–562.

[4] J.C. Giddings, Two-dimensional separations: Concept and promise, Anal. Chem. 56 (1984) 1258A–1270A. https://doi.org/10.1021/ac00276a003.

[5] J.C. Giddings, Concepts and comparisons in multidimensional separation, J. High Resol. Chromatogr. 10 (1987) 319–323. https://doi.org/10.1002/jhrc.1240100517.

[6] B.W.J. Pirok, D. Stoll R., P.J. Schoenmakers, Recent developments in two-dimensional liquid chromatography – Fundamental improvements for practical applications, Anal. Chem. 91 (2019) 240–263. https://doi.org/10.1021/acs.analchem.8b04841.

[7] D.R. Stoll, X. Wang, P.W. Carr, Comparison of the practical resolving power of one- and two-dimensional high-performance liquid chromatography analysis of metabolomic samples, Anal. Chem. 80 (2008) 268–278. https://doi.org/10.1021/ac701676b.

[8] X. Wang, D.R. Stoll, A.P. Schellinger, P.W. Carr, Peak capacity optimization of peptide separations in reversed-phase gradient elution chromatography: Fixed column format, Anal. Chem. 78 (2006) 3406–3416. https://doi.org/10.1021/ac0600149.

[9] C.J. Venkatramani, J. Girotti, L. Wigman, N. Chetwyn, Assessing stability-indicating methods for coelution by two-dimensional liquid chromatography with mass spectrometric detection: Liquid chromatography, J. Sep. Sci. 37 (2014) 3214–3225. https://doi.org/10.1002/jssc.201400590.

[10] D. Stoll, Introduction to two-dimensional liquid chromatography – Theory and practice, in: M. Holcapek, Wm.C. Byrdwell (eds.), Handbook of Advanced Chromatography /Mass Spectrometry Techniques, Elsevier, London, 2017: pp. 227–286.

[11] M.R. Schure, J.M. Davis, Orthogonal separations: Comparison of orthogonality metrics by statistical analysis, J. Chromatogr. A. 1414 (2015) 60–76. https://doi.org/10.1016/j.chroma.2015.08.029.

[12] M. Gilar, J. Fridrich, M.R. Schure, A. Jaworski, Comparison of orthogonality estimation methods for the two-dimensional separations of peptides, Anal. Chem. 84 (2012) 8722–8732. https://doi.org/10.1021/ac3020214.

[13] M. Gilar, P. Olivova, A.E. Daly, J.C. Gebler, Orthogonality of separation in two-dimensional liquid chromatography, Anal. Chem. 77 (2005) 6426–6434. https://doi.org/10.1021/ac050923i.

[14] J.M. Davis, D.R. Stoll, P.W. Carr, Dependence of effective peak capacity in comprehensive two-dimensional separations on the distribution of peak capacity between the two dimensions, Anal. Chem. 80 (2008) 8122–8134. https://doi.org/10.1021/ac800933z.

[15] M. Camenzuli, P.J. Schoenmakers, A new measure of orthogonality for multi-dimensional chromatography, Anal. Chim. Acta. 838 (2014) 93–101. https://doi.org/10.1016/j.aca.2014.05.048.

[16] G. Semard, V. Peulon-Agasse, A. Bruchet, J.-P. Bouillon, P. Cardinaël, Convex hull: A new method to determine the separation space used and to optimize operating conditions for comprehensive two-dimensional gas chromatography, J. Chromatogr. A. 1217 (2010) 5449–5454. https://doi.org/10.1016/j.chroma.2010.06.048.

[17] P.W. Carr, J.M. Davis, S.C. Rutan, D.R. Stoll, Principles of online comprehensive multidimensional liquid chromatography, in: E. Gruska, N. Grinberg (Eds.), Advances in Chromatography, CRC Press, Boca Raton, FL, 2012: pp. 140–222.

[18] R.E. Murphy, M.R. Schure, J.P. Foley, Effect of sampling rate on resolution in comprehensive two-dimensional liquid chromatography, Anal. Chem. 70 (1998) 1585–1594. https://doi.org/10.1021/ac971184b.

[19] K. Horie, H. Kimura, T. Ikegami, A. Iwatsuka, N. Saad, O. Fiehn, N. Tanaka, Calculating optimal modulation periods to maximize the peak capacity in two-dimensional HPLC, Anal. Chem. 79 (2007) 3764–3770. https://doi.org/10.1021/ac062002t.

[20] J. Seeley, Theoretical study of incomplete sampling of the first dimension in comprehensive two-dimensional chromatography, J. Chromatogr., A. 962 (2002) 21–27. https://doi.org/10.1016/S0021-9673(02)00461-2.

[21] J.M. Davis, D.R. Stoll, P.W. Carr, Effect of first-dimension undersampling on effective peak capacity in comprehensive two-dimensional separations, Anal. Chem. 80 (2008) 461–473. https://doi.org/10.1021/ac071504j.

[22] D.R. Stoll, P.W. Carr, Two-dimensional liquid chromatography: A state of the art tutorial, Anal. Chem. 89 (2017) 519–531. https://doi.org/10.1021/acs.analchem.6b03506.

[23] R.E. Murphy, M.R. Schure, J.P. Foley, Effect of sampling rate on resolution in comprehensive two-dimensional liquid chromatography, Anal. Chem. 70 (1998) 1585–1594. https://doi.org/10.1021/ac971184b.

[24] S.R. Groskreutz, M.M. Swenson, L.B. Secor, D.R. Stoll, Selective comprehensive multi-dimensional separation for resolution enhancement in high performance liquid chromatography, Part I–Principles and instrumentation, J. Chromatogr. A. 1228 (2012) 31–40. https://doi.org/10.1016/j.chroma.2011.06.035.

[25] J.M. Davis, D.R. Stoll, P.W. Carr, Effect of first-dimension undersampling on effective peak capacity in comprehensive two-dimensional separations, Anal. Chem. 80 (2008) 461–473. https://doi.org/10.1021/ac071504j.

[26] F. Bedani, P.J. Schoenmakers, H.-G. Janssen, Theories to support method development in comprehensive two-dimensional liquid chromatography – A review, J. Sep. Sci. 35 (2012) 1697–1711. https://doi.org/10.1002/jssc.201200070.

[27] Y. Huang, H. Gu, M. Filgueira, P.W. Carr, An experimental study of sampling time effects on the resolving power of on-line two-dimensional high performance liquid chromatography, J. Chromatogr., A. 1218 (2011) 2984–2994. https://doi.org/10.1016/j.chroma.2011.03.032.

[28] A. Schellinger, D. Stoll, P. Carr, High speed gradient elution reversed-phase liquid chromatography, J. Chromatogr., A. 1064 (2005) 143–156. https://doi.org/10.1016/j.chroma.2004.12.017.

[29] D. Stoll, J. Cohen, P. Carr, Fast, comprehensive online two-dimensional high performance liquid chromatography through the use of high temperature ultra-fast gradient elution reversed-phase liquid chromatography, J. Chromatogr., A. 1122 (2006) 123–137. https://doi.org/10.1016/j.chroma.2006.04.058.

[30] C. Seidl, D.S. Bell, D.R. Stoll, A study of the re-equilibration of hydrophilic interaction columns with a focus on viability for use in two-dimensional liquid chromatography, J. Chromatogr. A. 1604 (2019). https://doi.org/10.1016/j.chroma.2019.460484.

[31] G.M. Leme, F. Cacciola, P. Donato, A.J. Cavalheiro, P. Dugo, L. Mondello, Continuous vs. segmented second-dimension system gradients for comprehensive two-dimensional liquid chromatography of sugarcane (Saccharum spp.), Anal. Bioanal. Chem. 406 (2014) 4315–4324. https://doi.org/10.1007/s00216-014-7786-8.

[32] P. Jandera, M. Staňková, T. Hájek, New zwitterionic polymethacrylate monolithic columns for one- and two-dimensional microliquid chromatography: Liquid chromatography, J. Sep. Sci. 36 (2013) 2430–2440. https://doi.org/10.1002/jssc.201300337.

[33] Q. Yang, X. Shi, Q. Gu, S. Zhao, Y. Shan, G. Xu, On-line two dimensional liquid chromatography/mass spectrometry for the analysis of triacylglycerides in peanut oil and mouse tissue, J. Chromatogr. B. 895–896 (2012) 48–55. https://doi.org/10.1016/j.jchromb.2012.03.013.

[34] Z. Liu, D. Zhu, Y. Qi, X. Chen, Z. Zhu, Y. Chai, Elucidation of steroid glycosides in Anemarrhena asphodeloides extract by means of comprehensive two-dimensional reversed-phase/polyamine chromatography with mass spectrometric detection: Liquid chromatography, J. Sep. Sci. 35 (2012) 2210–2218. https://doi.org/10.1002/jssc.201200236.

[35] M. Kivilompolo, J. Pol, Tuulia. Hyotylainen, Comprehensive two-dimensional liquid chromatography (LC x LC): A review, LC-GC Eur. 24 (2011) 232, 234, 236, 238, 240–243.

[36] J.M. Davis, D.R. Stoll, Likelihood of total resolution in liquid chromatography: Evaluation of one-dimensional, comprehensive two-dimensional, and selective comprehensive two-dimensional

liquid chromatography, J. Chromatogr. A. 1360 (2014) 128–142. https://doi.org/10.1016/j.chr oma.2014.07.066.

[37] I. François, K. Sandra, P. Sandra, Comprehensive liquid chromatography: Fundamental aspects and practical considerations – A review, Anal. Chim. Acta. 641 (2009) 14–31. https://doi.org/10.1016/j.aca.2009.03.041.

[38] M.R. Schure, Limit of detection, dilution factors, and technique compatibility in multidimensional separations utilizing chromatography, capillary electrophoresis, and field-flow fractionation, Anal. Chem. 71 (1999) 1645–1657. https://doi.org/10.1021/ac981128q.

[39] M.R. Schure, Limit of detection, dilution factors, and technique compatibility in multidimensional separations utilizing chromatography, capillary electrophoresis, and field-flow fractionation, Anal. Chem. 71 (1999) 1645–1657. https://doi.org/10.1021/ac981128q.

[40] G. Vivó-Truyols, Sj. van der Wal, P.J. Schoenmakers, Comprehensive study on the optimization of online two-dimensional liquid chromatographic systems considering losses in theoretical peak capacity in first- and second-dimensions: A pareto-optimality approach, Anal. Chem. 82 (2010) 8525–8536. https://doi.org/10.1021/ac101420f.

[41] M. Sarrut, A. D'Attoma, S. Heinisch, Optimization of conditions in on-line comprehensive two-dimensional reversed phase liquid chromatography: Experimental comparison with one-dimensional reversed phase liquid chromatography for the separation of peptides, J. Chromatogr. A. 1421 (2015) 48–59. https://doi.org/10.1016/j.chroma.2015.08.052.

[42] D.R. Stoll, H.R. Lhotka, D.C. Harmes, B. Madigan, J.J. Hsiao, G.O. Staples, High resolution two-dimensional liquid chromatography coupled with mass spectrometry for robust and sensitive characterization of therapeutic antibodies at the peptide level, J. Chromatogr. B. 1134–1135 (2019). https://doi.org/10.1016/j.jchromb.2019.121832.

4 Instrumentation for Two-Dimensional Liquid Chromatography

Dwight R. Stoll and Gabriel Mazzi Leme

CONTENTS

DOI: 10.1201/9781003090557-4

4.1   CHALLENGES UNIQUE TO 2D-LC SEPARATIONS

The more we have studied 2D-LC, the more we have realized over the years that there are several aspects of the instrumentation involved that are unique to 2D-LC – that is, these aspects simply are not relevant in 1D-LC, or at a minimum are radically different in the second dimension of 2D separations compared to 1D-LC. For example, in 1D chromatography it is widely understood that the injection volume should be on the order of 1% (or less) of the column dead volume [1]. This rule-of-thumb helps avoid significant pre-column dispersion of analyte bands due to the sample volume itself [2, 3]. This becomes especially important when there is a significant mismatch between the sample solvent and the mobile phase used in the separation [4]. However, following this rule-of-thumb in the second dimension of 2D-LC generally leads to extensive additional dilution of the analyte band (due to dispersion of the band on the way from the column inlet to the outlet). In the worst case, this can result in signals for low concentration analytes falling below the noise level of the ^{2}D detector, rendering the 2D separation useless because the analytes of interest cannot be detected. As was discussed earlier in Section 3.4, and again below in Section 4.4.4, a great deal of research by several groups has been dedicated to addressing this detection sensitivity issue in 2D-LC. We now know how to effectively increase the detection sensitivity of the second dimension in many 2D-LC applications. Usually this involves injecting rather large volumes of ^{1}D effluent into small ^{2}D columns, such that the injected volume is on the order of the dead volume of the ^{2}D column. Obviously, this is an entirely different situation compared to the rule-of-thumb that guides choice of injection for 1D-LC, and thus requires special attention during method development for 2D-LC. In the rest of this Section 4.1 we will discuss several other aspects of 2D-LC instrumentation and methods that are very different from those in 1D-LC.

4.1.1   Each 2D Separation Involves Many ^{2}D Separations

Decades ago, when the average 1D-LC analysis time was on the order of an hour, operating an instrument continuously for seven days would require about 300 injector valve movements (one move to load the sample into a loop, and one move to the sample into the mobile phase flow path). At this rate, continuous operation of the instrument for a year would require about 17,500 valve movements. In most cases, changing the valve rotor seal during annual preventative maintenance would prevent an interruption in operation of the instrument due to a failure of the rotor seal. The situation in 2D-LC is quite different. It is important to understand that for each complete 2D-LC analysis there are more than one ^{2}D separation carried out – sometimes many more than one. The different modes of 2D-LC separation were discussed briefly in Section 1.5.2 and illustrated in Figure 1.4. Looking at these illustrations we see that for the mLC-LC, sLC×LC, and LC×LC separations, the numbers of ^{2}D separations per 2D analysis would be 3, 6, and 8, respectively. The only situation where there is exactly one ^{2}D separation per 2D analysis is the case of single heartcut 2D-LC (LC-LC) as shown in Figure 1.4. Whereas for mLC-LC and sLC×LC the numbers of ^{2}D separations per 2D-LC analysis are typically on the order of 2 to 10, for LC×LC separations these numbers can be very large. For example, consider a 2-hr LC×LC analysis that involves a ^{2}D modulation time of 30 seconds, a total of 240 ^{2}D separations will be carried out over the course of the 2D analysis. Moreover, if we extend this to 24-hr and 7-day periods of work, this number of ^{2}D separations rises to 2,880 and 20,160, respectively. These are large numbers of separations! Although this is an extreme case, even a modest 30-min. sLC×LC analysis running six ^{2}D separations per 2D analysis will result in 288 ^{2}D separations per day, and 2,016 per week.

The two most practically critical consequences of these large numbers of ^{2}D separations are: 1) the valve used as the interface between the two dimensions of separation makes as many movements (or more, depending on the design) as there are ^{2}D separations; and 2) each transfer of ^{1}D effluent to the ^{2}D column is similar to a typical injection in 1D-LC, and these injections have an impact on the lifetimes of ^{2}D columns. The impact of these ^{2}D separations on the interface valve is straightforward in principle. All rotary valves of the types used for 2D-LC (see Section 4.4 for details) have lifetimes that are primarily limited by wear of the rotor and stator surfaces as one rotates against the other during transfer of ^{1}D effluent to the ^{2}D column. After thousands of movements, the valve can fail as a result of leaks between internal flow paths (usually in the rotor) or restriction of some of these paths.

The impact of repeated ^{2}D separations on the lifetime of columns used in the second dimension is a complex matter, and not well understood. There is very little literature on this topic. In our own work we have examined the effect of pressure changes at the ^{2}D column inlet that occur during the injection step on the lifetimes of ^{2}D columns [5]. The key finding from this work was that the design of the interface valve itself can have a dramatic effect on ^{2}D column lifetime. This is an important example of how technologies designed from the ground up for the purpose of 2D-LC are positively impacting the performance of 2D-LC overall. In more recent work we have also investigated the impact of the nominal ^{2}D operating pressure on the lifetimes of ^{2}D columns [6] under LC×LC conditions. In this work we found that above an operating pressure of about 400 bar ^{2}D columns failed much more quickly than they did when the ^{2}D operating pressure was restricted to less than 400 bar. This observation was consistent across different flow rates, column lengths, particle sizes, and particle types. Much more work remains to be done here to understand the origin of this pressure threshold and how this can be improved, but for now this finding serves as a useful guide when developing methods that must be robust.

4.1.2 Requirement for Fast ^{2}D Separations

As discussed in Section 3.4, avoiding undersampling of ^{1}D peaks – and the attendant loss in ^{1}D resolution (peak capacity) – requires fast ^{2}D separations, particularly when the ^{1}D separation produces narrow peaks, and in the LC×LC mode of 2D-LC. By "fast" here we mean ^{2}D separations on the timescale of 30 seconds or less. For isocratic elution during the ^{2}D separation this is not so challenging. But doing such separations under gradient elution conditions is much more difficult, and requires special attention to the type of pump that is used, the use of mobile phase mixers, and details associated with the flow path between the ^{2}D pump and column. These details will be discussed in detail in Section 4.3.

4.1.3   INJECTION PROCESS

The injection processes for conventional 1D chromatography and the second dimension of 2D chromatography share few similarities, and thus optimizing 2D separations inherently requires a different mindset from that used when developing 1D methods. For example, in the introduction to this section we discussed how in the second dimension of 2D-LC separations we typically use much larger injection volumes than in 1D-LC separations, and that as a result we must pay special attention to the potential for negative effects of mismatch between the ^{1}D effluent that is injected into the ^{2}D separation. Second, whereas in 1D chromatography we have the option to significantly under- or over-fill the sample loop used to inject sample into the column, in the second dimension of 2D separations neither of these is a good option. Using conditions where a fraction of ^{1}D effluent significantly under-fills a loop for transfer to the second dimension either requires that a very small fraction of ^{1}D effluent is used with a typical interface loop volume, or that a very large sample loop is used with a typical fraction volume. In the former case, using a very small fraction volume will lead to significant dilution of the injected analyte and correspondingly low detection sensitivity. In the latter case, using a large sample loop volume can lead to a severely tailing injection profile, which may impact the peak that elutes from the ^{2}D column, and leads to a long delay in the time it takes for a solvent gradient to travel through the loop itself, which is important when gradient elution is used in the second dimension. On the other hand, significantly over-filling the interface loop with ^{1}D effluent is usually not desirable because doing so results in loss of analyte of interest to waste. In 1D chromatography this is acceptable because the sample flowed into the injection loop is homogeneous, and loss of the analyte of interest is only a problem when the sample is volume limited. In the case of a 2D separation, however, the analyte concentration in the ^{1}D effluent varies with time, and losing any portion of that effluent stream will negatively impact the quantitative performance of the method, unless the portion that is lost is very consistent from one analysis to the next. Guidance for loop filling based on recent research is given in Section 4.4.13.

4.1.4   PRESSURE SPIKING AT ^{1}D DETECTOR

During typical operation of instruments designed for 1D-LC, the pressure at the outlet of the detector is low (< 20 bar), and consistent over time. In 2D-LC however, most interfaces used to connect the first and second dimensions of the system produce transient changes in the flow through the ^{1}D and ^{2}D flow paths through of the interface. This in turn can often result in large short-term changes (pulses) in pressure at various points within the fluid paths. For some components of the system these pressure changes are insignificant and the operator would have no way of knowing that they even occur. For other components, however, these pressure changes can be significant, and may result in phenomena that affect the data quality of the 2D separation. For example, some types of flow cells used for UV absorbance detectors are sensitive to such pressure changes, resulting in signal spikes that are artifacts and have nothing to do with the chromatographic separation *per se*. This has led manufacturers to commercialize instrument components that avoid these artifacts.

4.2   2D-LC SYSTEM CONTROL

4.2.1   NATIVE SOFTWARE VS. MASTER CONTROLLER

Although today there are good commercially available options for fully integrated, complete systems for 2D-LC, historically the early development of 2D-LC technology and methods largely relied on "home-built" systems. In most cases researchers have used instrument components designed for use in 1D-LC and assembled them in a way that results in 2D-LC functionality. Given the high degree of functionality and sophistication of commercially available 2D-LC instruments today, it is best

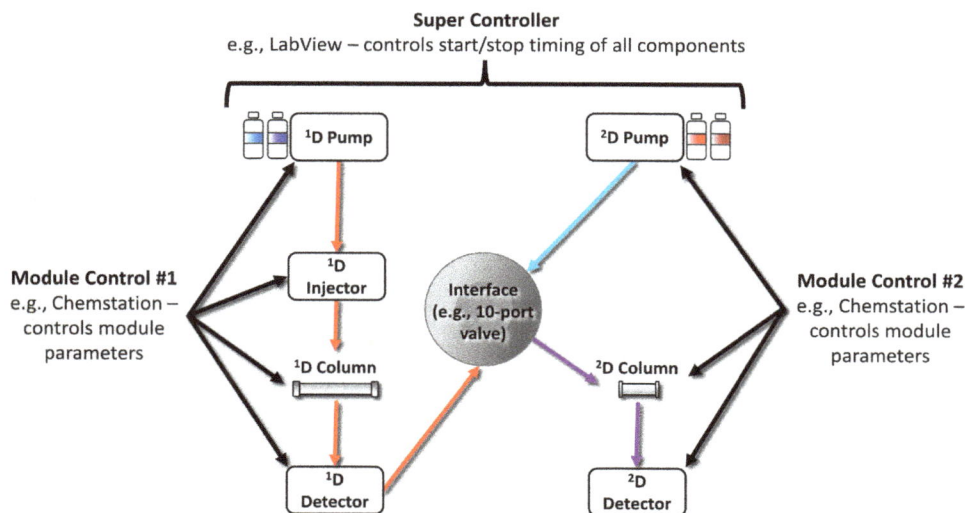

FIGURE 4.1 Illustration of an approach to control a 2D-LC system using "home-built" software to trigger instrument control and data acquisition.

to use these systems for routine applications, and in regulated environments. For research purposes however, building up a 2D-LC system from components used for 1D-LC, or augmenting a commercial 2D-LC system with customized components, remains a viable option. Currently this is being done frequently in the rapidly developing area of "automated characterization systems" for protein analysis, for example [7, 8]. In this case, a commonly used strategy is to use the sampler module that injects samples into the ^{1}D column to trigger a piece of controller software (e.g., LabView) that then sends commands to other components of the 2D-LC system. An illustration of such a strategy is shown in Figure 4.1.

In this case it is most common to use conventional 1D-LC software (e.g., Waters Empower, Agilent ChemStation) to control each individual instrument module (i.e., pumps, detectors, valves), but then use the controller software to dictate the timing of events that are critical to the core 2D-LC functionality. Foremost among these are switches of the interface valve connecting the two dimensions, and the mobile phase composition delivered by the ^{2}D pump, especially when gradient elution is used in the second dimension. For simple methods such as LC-LC or LC×LC with a single ^{2}D pump this type of control is pretty straightforward. Far more complex systems for mLC-LC and sLC×LC can also be implemented in this manner [9, 10], however programming such systems quickly becomes complicated and especially tedious when several different methods are needed that involve changes to event timing in the system.

4.2.2 PRECISION OF MODULE CONTROL

As the performance of 1D-LC has steadily improved, so too has the performance of ^{2}D separation in 2D-LC. State-of-the-art LC×LC separations are characterized by ^{2}D peaks with half-height widths on the order of 200 ms. At this level of performance, precision of ^{2}D retention time on the order of 50 ms or better is helpful to support grouping of peaks in adjacent modulations that belong to the same ^{1}D peak (i.e., the same analyte eluting from the ^{1}D separation) that has been sampled multiple times. Achieving this level of precision in ^{2}D retention requires precise switching of the interface valve, and tight synchronization with the execution of time-based operation of the ^{2}D pump (e.g., change in mobile phase composition over time as in gradient elution).

4.3 PUMPS

Modern LC pumps from multiple manufacturers have characteristics that are valuable for development of high performing 2D-LC methods. However, older pumps still in use in many laboratories have very different characteristics, and some of these may be detrimental to 2D-LC performance, depending on the aims and conditions of the 2D experiment. Differences between older and newer pumps, as well as different models of modern pumps, should be considered carefully when choosing equipment to assemble a 2D-LC system.

4.3.1 GRADIENT DELAY VOLUME

The gradient delay volume associated with a pump is arguably the single most important way that the characteristics can impact the capabilities and performance of a 2D-LC method. That is not to say that other characteristics such as mobile phase composition precision are not important, but using a pump with a delay volume that is not well matched with the aims of a particular 2D-LC method can really make the difference between success and failure. This point can be made most clearly by thinking through the impact of the delay volume on a comprehensive 2D-LC separation. Figure 4.2 shows a comparison between the expected solvent composition vs. time profiles for two

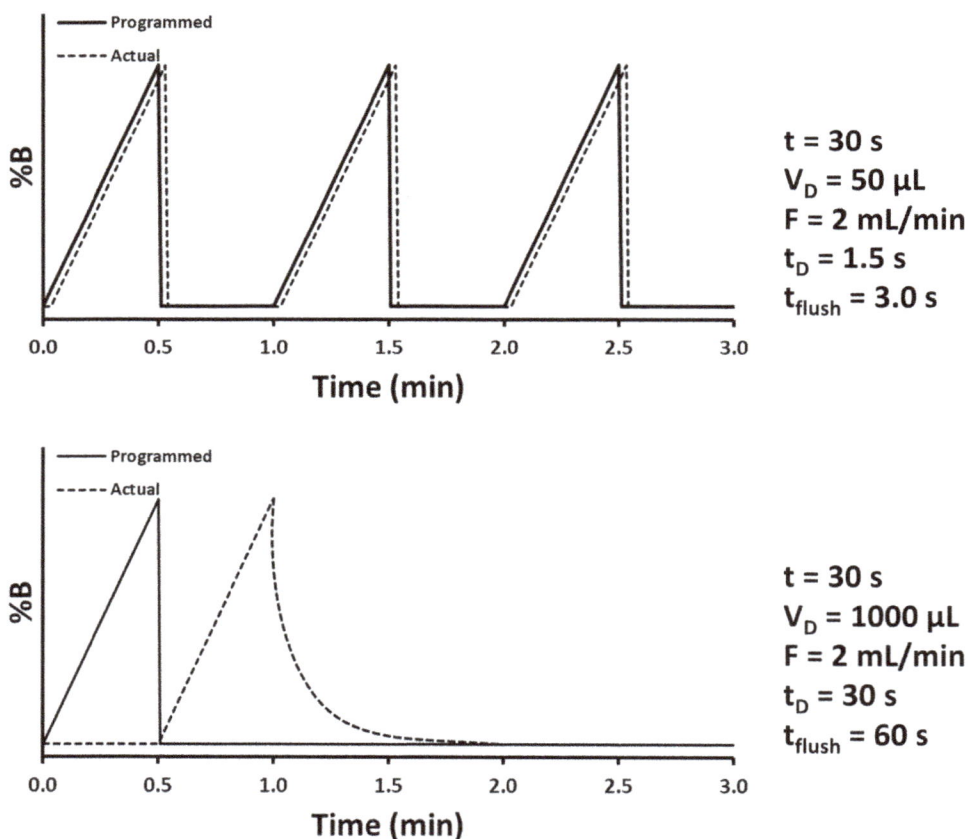

FIGURE 4.2 Illustration of the effect of gradient delay volume on the available time for actual separation of analytes in each separation cycle. The gradient delay volume of 1000 μL is typically of older pumps, whereas a delay volume of 50 μL is typical of state-of-the-art pumps used for analytical scale separations.

pumps with different delay volumes. In each case the programmed solvent profile (i.e., the mobile phase we would like the pump to deliver) is the same, and shown as the solid black trace. The mobile phase composition that is actually delivered to the LC system at the pump outlet is illustrated by the dashed black trace. There are two major differences between what is programmed, and what actually occurs. First, the entire profile is shifted later in time by an amount (t_d) that is related to the gradient delay volume of the pump (V_d), and the flow rate (F) ($t_d = \frac{V_d}{F}$). Here the flow rates are the same in each case, so the gradient delay time increases in proportion to the gradient delay volume. Second, when the pump is programmed to return the mobile phase back to the starting condition, it takes a while for this to actually occur and a kind of exponential decay curve is observed for the composition vs. time. In our own work we treat the "re-equilibration time" as two times the delay time (as defined immediately above). The net impact of these two differences between what is programmed and what happens is that when the delay time is large relative to the gradient time, a significant fraction of each ^{2}D cycle is wasted, essentially waiting for the pump.

The effects of the gradient delay volume discussed above are most acute in LC×LC separations, where each ^{2}D separation must be fast to minimize undersampling. The impact can also be significant in mLC-LC and sLC×LC separations as well, if many ^{2}D cycles are carried out in each 2D separation.

The impact of the delay time can also be serious in the first dimension. Again, this is of particular concern with LC×LC separations where ^{1}D flow rates tend to be low to avoid large fraction volumes. Table 4.1 shows the gradient delay times calculated for different combinations of 1D flow rate and gradient delay volume. Gradient delay volumes on the order of 1000 µL are typical for older pumps such as the Agilent 1100 and Waters Alliance systems. Volumes on the order of 100–200 µL are typical for newer binary pump designs (e.g., Agilent 1290, Waters Acquity). The 20 µL delay volume is not currently feasible for most analytical scale systems, however one can approach this level by optimizing the mixer volume and connecting capillaries. Obviously, gradient delay times of more than 10 minutes are undesirable for most 2D separations, especially if the total desired analysis time is less than one hour.

4.3.2 PRESSURE AND FLOW RATE RANGES

As discussed in Section 3.4, minimizing undersampling in LC×LC separations requires fast ^{2}D separations. On the one hand, such fast separations benefit from relatively high flow rates simply to get the sample into the ^{2}D column quickly. In cases where gradient elution is used, high flow rates enable delivery of the solvent gradient with a reasonable slope, and rapid flush-out of the pump and

TABLE 4.1
Gradient Delay Time (Min) for Different Combinations of ^{1}D Flow Rate and Gradient Delay Volume[a]

^{1}F (mL/min.)	^{1}D Gradient Delay Volume (µL)			
	20	100	200	1000
0.025	0.8	4	8	40
0.05	0.4	2	4	20
0.1	0.2	1	2	10
0.2	0.1	0.5	1	5
0.5	0.04	0.2	0.4	2
1.0	0.02	0.1	0.2	1

Note:
a) Shading indicates impact on throughput of the ^{1}D separation, ranging from minor (green) to serious (salmon).

TABLE 4.2
Pressure Drop Across Column [a, b] (Bar) for Different Combinations of Flow Rate, Particle Size, and Column Length

	Pressure Drop Across Column (bar)								
	1.8 μm			**2.7 μm**			**3.5 μm**		
Flow Rate (mL/min)	**20 mm**	**30 mm**	**50 mm**	**20 mm**	**30 mm**	**50 mm**	**20 mm**	**30 mm**	**50 mm**
0.5	92	138	229	41	61	102	24	36	61
1.0	184	275	459	82	122	204	49	73	121
1.5	275	413	688	122	184	306	73	109	182
2.0	367	551	918	163	245	408	97	146	243
2.5	459	688	1147	204	306	510	121	182	303
3.0	551	826	1376	245	367	612	146	218	364

Notes:
a) Assumes viscosity of water; Temperature, 60 °C; Interstitial porosity, 0.38; Column diameter, 2.1 mm.
b) Shading indicates pressure level relative to a 1500 bar limit, ranging from low (green) to high (salmon).

re-equilibration of the column after the gradient is complete. On the other hand, as discussed in Chapter 2, separations in the range of tens of seconds benefit greatly from the use of small particles and higher pressures and/or temperatures. This particular combination of high flow rate and small particles (which require high pressures to drive the mobile phase through the column) is not commonly encountered in 1D-LC, and thus is a somewhat unique challenge encountered in LC×LC separations. Table 4.2 shows the pressures required to operate different columns across a range of flow rates that are practically useful in the second dimension of contemporary LC×LC separations.

4.4   INTERFACE AND MODULATION

Without doubt the most important aspect of instrumentation for 2D-LC is the hardware used to connect the two dimensions of separation, which we refer to here as the "interface". This is the heart of any 2D-LC system, and thus it is not surprising that this is the area of 2D-LC instrumentation where most research has been focused over the last 40 years. In the section immediately below, we illustrate and discuss all of the different types of interfaces that have been studied and used for 2D-LC, along with comments about their advantages and disadvantages relative to the 8-port/2-position dual loop type interface that has been used most commonly in recent years.

4.4.1   DUAL LOOP INTERFACES

4.4.1.1   8-Port/2-Position Valves

The dual loop interface that has become the *de facto* standard, particularly for LC×LC separations, is shown in Figure 4.3, in the context of the other components of the 2D-LC system. This interface is composed of a rotary valve with eight ports that switches between two positions during 2D-LC operation. In the case of use for LC×LC, the valve is fitted with two nominally identical sample loop capillaries (labeled Loop 1 and Loop 2 here).

The position shown here is effectively the "load" position for Loop 1; effluent exiting the ^{1}D detector fills Loop 1 and displaces solvent in the loop to waste. The same position is effectively the "inject" position for Loop 2; ^{1}D effluent that was previously collected in the loop is displaced by the ^{2}D pump into the ^{2}D column where analytes can be further separated. The changing roles of Loop 1 and Loop 2 in the two positions of the valve are shown in Figure 4.4. Whereas in Position A ^{1}D effluent is collected in Loop 1, and previously collected effluent is injected from Loop 2 into the ^{2}D

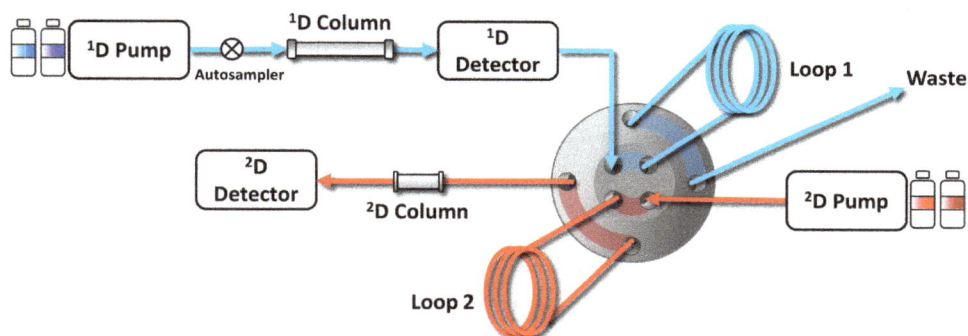

FIGURE 4.3 Illustration of an 8-port/2-position valve used as the interface between the first and second dimensions of a 2D-LC system.

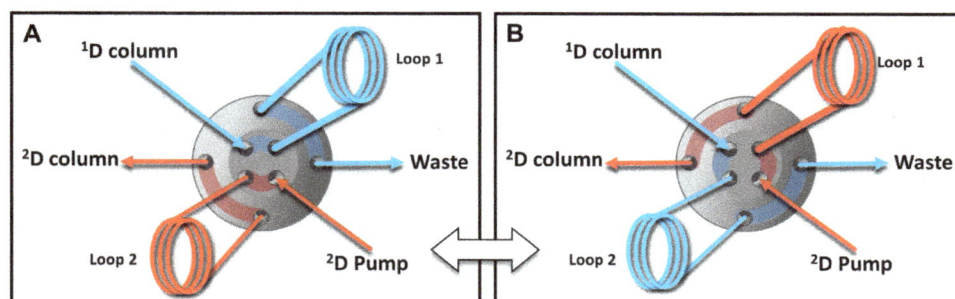

FIGURE 4.4 Detailed illustration of the fluid paths in the two positions of the 8-port/2-position valve shown in Figure 4.3. In LC×LC the valve switches between these two positions at regular intervals referred to as the "modulation time" or "sampling time".

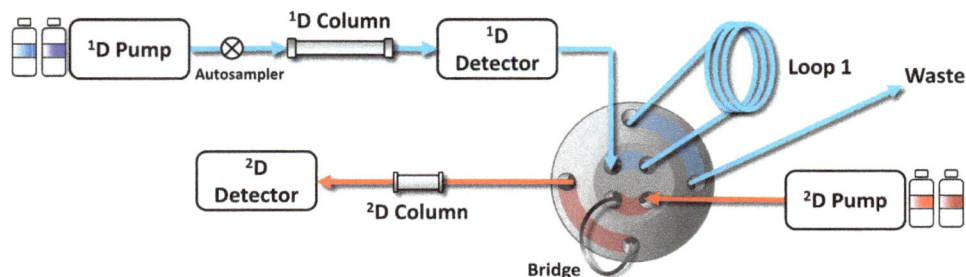

FIGURE 4.5 Detailed illustration of the fluid paths of an 8-port/2-position valve fitted with a single loop capillary used for LC-LC separations.

column, in Position B these roles are reversed – ¹D effluent is collected in Loop 2 while the material in Loop 1 is injected into the ²D column.

This same style of valve can be used for single heartcut 2D-LC as discussed in Section 1.5.2. In this case only one sample loop is really required, and the second loop can be replaced with a small volume capillary that only has the function to connect one side of the valve to the other. This configuration of the valve is shown in Figure 4.5. As in Figure 4.4, the loop is loaded with ¹D effluent

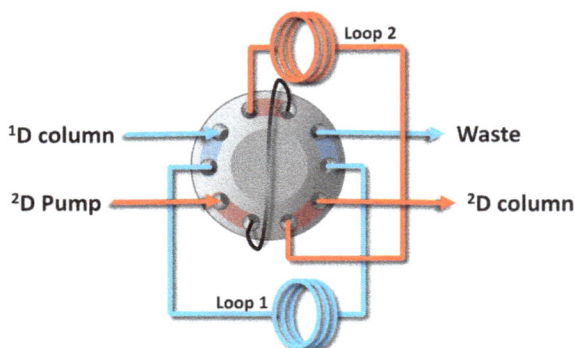

FIGURE 4.6 Illustration of a 10-port/2-position interface valve for 2D-LC. The bridge capillary drawn in black is external to the valve rotor/stator.

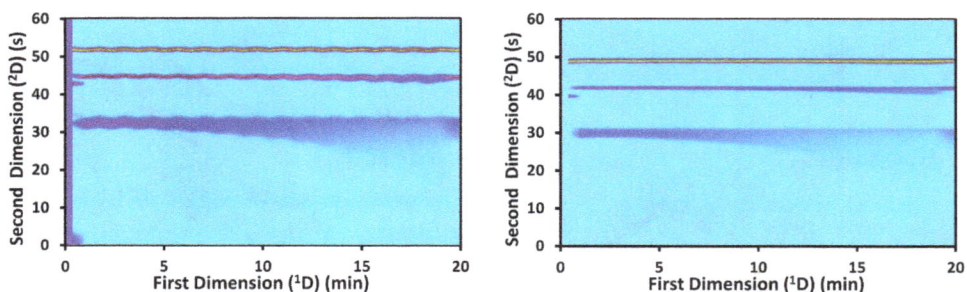

FIGURE 4.7 Comparison of ^{2}D retention consistency between modulations when using an asymmetric interface valve (A; as in Figure 4.6) or a symmetric valve (B; as in Figure 4.5). To collect these data a three-component mixture of neutral small molecules was infused directly into the interface with a syringe such that there was no ^{1}D separation. If the ^{2}D separation is perfectly repeatable we would expect to see smooth streaks from left to right in the images.

in Position A, and upon switching to Position B this material is displaced by the ^{2}D pump into the ^{2}D column (i.e., injected).

4.4.1.2   10-Port/2-Position Valves

The other commonly used type of dual loop interface is shown in Figure 4.5. This is a 10-port/2-position valve-based interface that is similar to that shown in Figure 4.2, except that the valve stator has two additional ports that must be connected with a "bridge" capillary. The function of this type of interface is similar to that discussed above, wherein the valve switches between "load" and "inject" positions for the loop capillaries fixed to the valve.

A significant disadvantage of this interface design compared to the 8-port/2-position design is that the volume of the bridge capillary causes a shift in ^{2}D retention time that depends on whether a fraction of ^{1}D effluent is injected from Loop 1 or Loop 2. When a fraction is injected from Loop 2, as shown in Figure 4.6, the bridge capillary is between the ^{2}D pump and the loop. On the other hand, when a fraction is injected from Loop 1, the bridge capillary is not involved in the flow path to the ^{2}D column. The difference in flow paths between the ^{2}D pump and ^{2}D column results in a different gradient delay volume, which can in turn affect the elution of fraction components from the ^{2}D column.

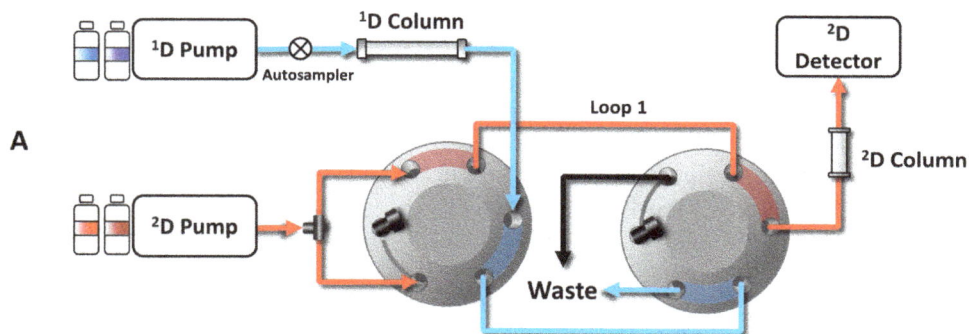

FIGURE 4.8   Illustration of a 2D-LC interface based on a pair of 6-port/2-position valves.

Source: Adapted from [11].

Experimental data that show the effect of the bridge capillary on the repeatability of ^{2}D retention time are shown in Figure 4.7. Whereas with the asymmetric interface valve design (Panel A) there is a clear, repeatable shift in ^{2}D retention time that depends on which loop the sample was injected from, *when using a symmetric valve (e.g.,* Figure 4.4) *no such shift is observed.*

4.4.1.3   Dual 6-Port/2-Position Valves

The basic functionality of the dual loop interfaces discussed above can also be achieved with pairs of 2-position valves rather than a single 8-, 10-, or 12-port valve. An example of such an interface that uses a pair of 6-port/2-position valves is shown in Figure 4.8. Perhaps the biggest advantage of this approach is simply convenience, especially when assembling a home-built 2D-LC system. Many labs have unused 6-port/2-position valves lying around because this has historically been the most often used valve design used in autosamplers and column-switching setups, and these can quickly be adapted for use in a 2D-LC interface. However, consistent operation of this type of interface requires very precise coordination of the movements of the two valves paired in this way. If one valve moves earlier than the other, there will be fluid moving in ways that are not intended.

Such interfaces have also been used with pairs of 10-port/2-position valves as in the work of François *et al.* [12]. These types of configurations provide a lot of flexibility, enabling the use of two ^{2}D pumping systems and two ^{2}D columns in parallel, for example.

4.4.2   Multi-Loop Interfaces

The dual-loop interfaces discussed above are well suited to LC×LC separations. However, they are also quite restrictive in the sense that the separation of a fraction injected from one loop must be completed before the next fraction from the other loop can be injected into the ^{2}D column (unless parallel ^{2}D columns are used; see Section 4.4.14 below). This restriction of dual-loop interfaces has motivated several groups to develop alternative approaches, referred to in this section as "multi-loop" interfaces. Although several different interfaces of this kind have been described in the literature, each with its own strengths and weaknesses, the principle they share in common is that the multi-loop array in the interface enables decoupling of two processes: 1) the sampling of the ^{1}D separation; and 2) the subsequent processing, or analysis, of fractions sampled from the first dimension by the second dimension [13]. This decoupling provides the analyst with valuable flexibility such that the ^{1}D separation can be carried out under conditions that are close to optimal for that separation, while the ^{2}D separation is carried out some time later (seconds to hours), under whatever conditions are needed to fully separate the analytes of interest that were not already separated by the first dimension. Sometimes these ^{2}D separations can be quite fast, on the order of tens of seconds, but sometimes they must be very slow, on the order of tens of minutes for each fraction. In principle, with these approaches there is no limit to the time used for each ^{2}D separation, no limit between the time of ^{1}D sampling and the time of ^{2}D separation, and no limit to the number of times the ^{1}D separation can be sampled in a single analysis. However, there are practical limits of course. For example, very long ^{2}D analysis times will preclude the collection of many fractions from the ^{1}D separation.

4.4.2.1 Multi-Loop Sampling With Serial ^{2}D Analysis

The first work we are aware of that describes collection of ^{1}D effluent fractions in a multi-loop interface followed by ^{2}D analysis of each fraction was described by Hamase *et al* [14]. The goal of this work was to separate fluorescently labeled amino acids from mammalian tissues and physiological fluids first by RP, and then by a ^{2}D chiral separation to resolve the D- and L- enantiomers of each of four aliphatic amino acids. Figure 4.9 shows an illustration of the setup used in this work. In this case a 6-port/2-position valve was used to direct either the ^{1}D effluent or the ^{2}D mobile phase to an array of up to ten sample loops. These loops were connected to a pair of selector valves, each of which had one inlet port and ten outlet ports. Synchronizing the positions of these two selector valves determines which loop is "active" and connected to either the ^{1}D or ^{2}D flow path.

It is important to note here that the type of setup shown in Figure 4.9 only allows the use of the multi-loop array for sampling the ^{1}D separation, or injection of collected fractions into the ^{2}D column. That is, the sampling and injection processes cannot be carried out in parallel. While this may not seem like a serious limitation, in practice it is not uncommon to encounter situations where this is a problem.

4.4.2.2 Multi-Loop Sampling With Multiple ^{2}D Column Chemistries

Zhang *et al.* extended the work of Hamase *et al.* by adding a second set of selector valves fitted with multiple ^{2}D columns with different chemistries, along with an additional 6-port/2-position valve to direct the ^{1}D or ^{2}D flow to the loop and column arrays [15]. This approach enables screening different column chemistries in the second dimension in an automated way as a means of optimizing the 2D separation. The setup used in this work is illustrated in Figure 4.10. A more detailed illustration of the interface at the heart of this setup is shown in Figure 4.11. Recent work by the Regalado group has extended this concept to include screening capabilities for both mobile and stationary phases in both dimensions [16]. The applications of this setup in pharmaceutical analysis by Zhang and coworkers are discussed in Sections 9.4 and 9.5.

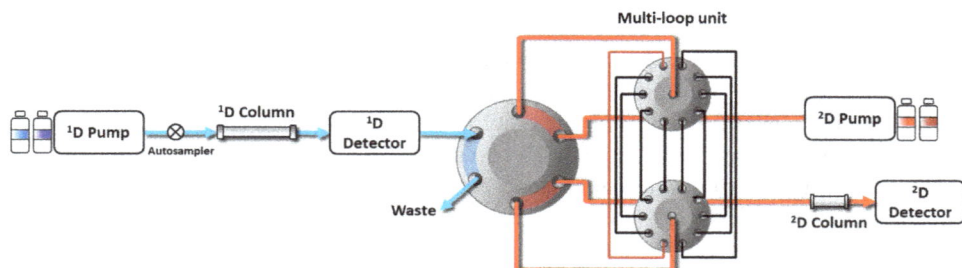

FIGURE 4.9 Illustration of the multi-loop interface used by Hamase *et al.* for the quantitation of D- and L-enantiomers of aliphatic amino acids found in mammalian tissues and physiological fluids.

Source: Adapted from [14].

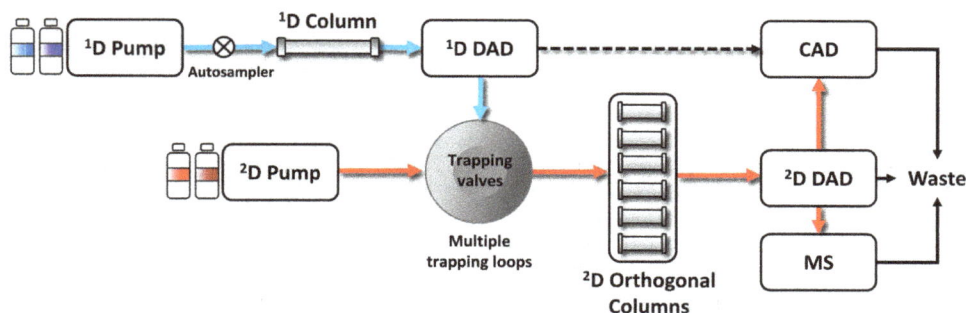

FIGURE 4.10 Illustration of a 2D-LC system with coordinated arrays of sampling loops and ^{2}D columns that enables automated screening of ^{2}D column chemistries as a means of optimizing a 2D separation.

4.4.2.3 Multi-Loop Sampling With Parallel ^{2}D Analysis, and Selective Comprehensive 2D Separation

In our own work in the early 2010s [10, 13, 17] we began developing multi-loop sampling interfaces that can be used in two ways that distinctly complement prior work. First, the interface enables sampling of the ^{1}D separation, and ^{2}D separation of previously collected fractions, in parallel. This minimizes the problems discussed in the preceding sections when ^{1}D sampling and ^{2}D separation must be carried out as serial processes. The ability to execute these processes in parallel provides the analyst with even more flexibility in method development and implementation, and allows ^{1}D separations to be carried out under conditions that are closer to conditions that would normally be used for 1D-LC. In other words, whereas in LC×LC the ^{1}D separation conditions typically involve lower flow rates and longer analysis times compared to 1D-LC, when we use a multi-loop interface that enables parallel ^{1}D sampling and ^{2}D separation we can usually simply add the second dimension of separation to an existing 1D-LC separation without having to make major adjustments to the 1D-LC conditions (i.e., flow rate, analysis time, etc.). Second, at this time we also introduced a mode of 2D-LC separation we refer to as selective comprehensive 2D separation (sLC×LC). This mode of 2D separation is illustrated conceptually in Figure 1.4. The multi-loop interface shown in Figure 4.12 enables both multiple heartcutting (mLC-LC, or MHC) and sLC×LC 2D separations. The early versions of this interface developed by Groskreutz *et al.* [13,17] and Larson *et al.* [10] involved up to eight valves and were quite complex. The version of the interface shown in Figure 4.12 is the most commonly used today for mLC-LC and sLC×LC.

4.4.3 OTHER INTERFACES

In 2003 Venkatramani and Zelechonok described one of the most innovative and unique 2D-LC interfaces published to date [18]. In this work a single 12-port/2-position valve was used as the interface, fitted with three open tubular sampling loops. A single pump was used for the entire system, with the flow split passively between the ^{1}D and ^{2}D flow paths. In the ^{2}D branch of the flow, the mobile phase was again split to go through two out of the three loops into two ^{2}D columns running in parallel, with the effluent from the two columns mixing back together before entering the ^{2}D detector. This approach has the advantage that it only requires a single pump for the 2D separation. However, this also complicates implementation because the ^{1}D and ^{2}D elution programs obviously cannot be run independently, and the multiple passive flow splitting steps makes retention times for all three columns (one first dimension, and two second dimension) prone to small changes in the flow restriction of various elements of the system.

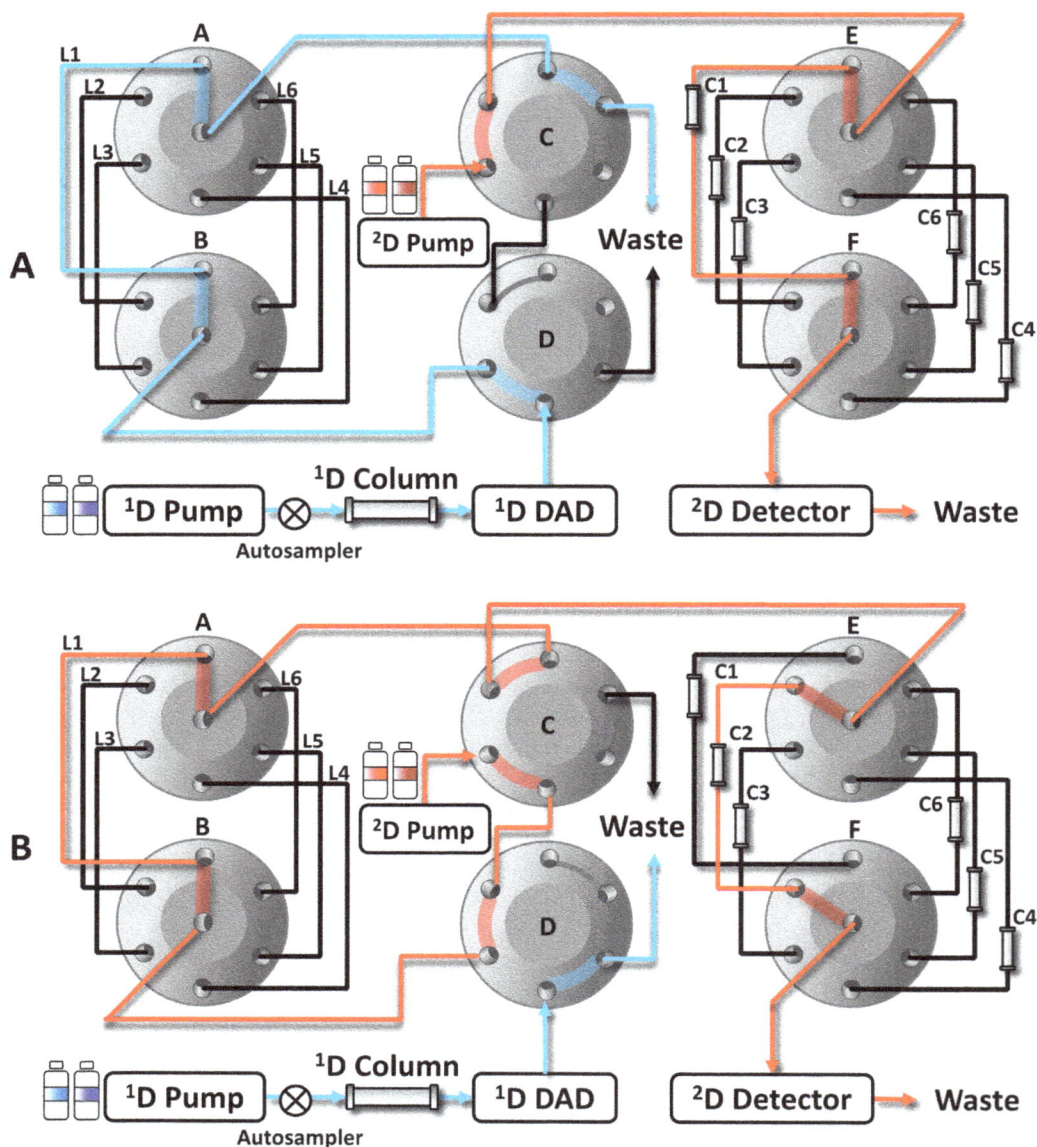

FIGURE 4.11 Detailed illustration of the multi-loop/multi-column interface shown in Figure 4.10 The left valves A/B are fitted with sampling loops, the right valves E/F are fitting with trapping cartridges, and valves C/D are used to direct the ^{1}D and ^{2}D flows to one or the other valve sets.

4.4.4 CHALLENGES ASSOCIATED WITH MOBILE PHASE MISMATCH IN MULTI-DIMENSIONAL SEPARATIONS

Historically one of the greatest challenges encountered in the development of effective 2D-LC methods has been what we refer to as mismatch between the mobile phases used in the two dimensions of a 2D-LC system. By "mismatch" we are not referring to differences in the properties of the two fluids *per se*. Rather, we are referring to the fact that the properties of the mobile phase used in the first dimension negatively interfere with the separation carried out in the second dimension. There

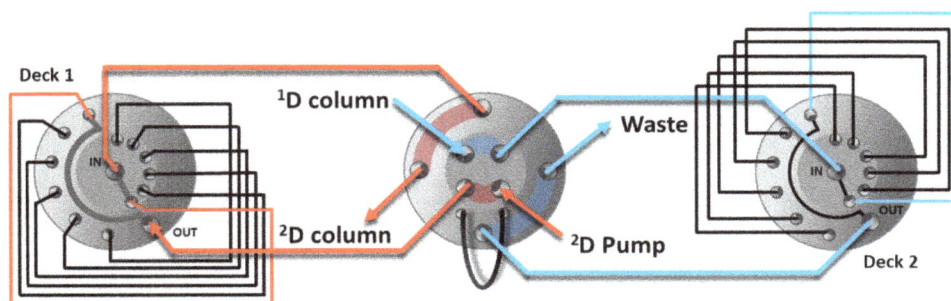

FIGURE 4.12 Illustration of a 2D-LC interface that supports mLC-LC and sLCLC with the possibility of parallel ¹D sampling and ²D separation.

most certainly are 2D-LC separations where this type of mismatch is hardly a concern (e.g., IEX-RP separation of proteins), however it is also true that there are many types of 2D-LC separation that are of great practical interest where this mismatch is a tremendous impediment to realizing the full potential of the 2D-LC method (e.g., HILIC-RP separations of peptides). This mismatch issue has led to the common use in the literature of the terms "incompatible" and "compatible" to describe combinations of separation mechanisms where mobile phase mismatch is or is not a significant challenge. Several previous review articles and book chapters have discussed potential solutions to this problem. Readers interested in this topic in particular are strongly encouraged to consult these publications because they are complementary in their perspective to the following discussion. Here, for the sake of completeness, we very briefly describe the commonly-discussed solutions shown in Table 4.3. The remainder of the chapter is primarily focused on the last two, because it seems the 2D-LC community is converging on these as the preferred options.

4.4.5 INTERFACES FOR SOLVENT-BASED MITIGATION OF MOBILE PHASE MISMATCH

In Chapter 5 we discuss the mismatch issue in broader terms, considering combinations of separation mechanisms ranging from IEX-RP to SEC-HILIC. In this section we focus on the specific case of coupling two RP separations, both because it is a commonly used combination in contemporary 2D-LC, and because the broad familiarity with the RP separation mechanism makes the challenge relatively easy to appreciate in this case. In this section we illustrate, by way of example, the fundamental issue, and then explain in the following subsections different approaches that have been developed to mitigate the problem. Readers interested in this topic are also referred to the recent review article of Chapel *et al.* [20] that complements this section with an up-to-date comparison of the strengths and weaknesses of different instrumental approaches to address the mismatch challenge.

Figure 4.13 shows a series of chromatograms obtained under 1D-LC conditions with the same sample and operating conditions that mimic conditions typical of a ²D RP separation in 2D-LC. In Panel A the sample matrix matches the starting point in the gradient (50/50 A/W) and a small volume of 0.2 μL is injected. In this case the peaks are symmetrical and narrow, as expected for these conditions. However, the peak height is low because such a small volume is injected – too low to be very useful, especially in real applications where some analytes may be present at much lower concentration. If we simply increase the volume of sample injected into the ²D column, with the goal of improving detection sensitivity and increasing peak height, we obtain the chromatogram shown in Panel B. Here we see that the peak height is indeed increased, however the early eluting peaks are broad and tailed because the analytes are injected in a large volume of sample matrix that matches the starting point in the RP gradient. The chromatogram in Panel C is obtained under conditions

TABLE 4.3

Approaches Used for Mitigation of Mobile Phase Mismatch between Dimensions of Multi-Dimensional LC Systems

Approach	References for Examples	Primary Limitations
Increase the ratio of column volumes used in the ^{2}D and ^{1}D separations	[19]	Decreases detection sensitivity [20]
Decrease the fraction of ^{1}D effluent delivered to the ^{2}D column	[21]	Decreases detection sensitivity [20]
Use a more retentive column in the second dimension compared to the first dimension	[22, 23]	Effective solution when it works, however more often than not the chemistry of the analytes makes this approach difficult
Deliberately use conditions that lead to "total breakthrough" of the analyte in the second dimension	[24, 25]	Main limitations of this approach are related to quantitation (i.e., only part of the analyte is retained, separated, and detected, and it is difficult to know what fraction of the mass has been retained for each compound)
Use temperature modulation during the injection of ^{1}D effluent into the ^{2}D column to help separate analytes of interest from the solvent components of the ^{1}D effluent	[26, 27]	Effect of temperature on retention is modest compared to the effect of solvent composition, and that the temperature modulation process is slow relative to the speed needed for 2D separations, particularly in the case of LC×LC
Use evaporation to remove volatile solvent components of the ^{1}D effluent prior to injection into the ^{2}D column	[28, 29]	Implementation is complex and there are no commercially available solutions, and that there must be clear differentiation between the volatility of the analytes and the ^{1}D effluent solvent components
Adjust the composition of ^{1}D effluent fractions by dilution with weak solvent prior to injection into the ^{2}D column	(see Section 4.4.5)	Requires additional instrument hardware (ranging from modest to expensive)
Use a solid phase sorbent to partially separate the analytes from the ^{1}D effluent solvent components prior to injection into the ^{2}D column	(see Section 4.4.6)	There are no "universal" trapping chemistries, which means that new trapping chemistries and conditions have to be worked out for each application

that mimic the worst-case scenario when coupling two RP separations in 2D-LC. This occurs when analytes elute from the ^{1}D column in a mobile phase that contains more organic solvent (i.e., the "strong solvent" for the ^{2}D separation) than is used for the ^{2}D mobile phase. Here we see that all of the peaks in the chromatogram are distorted and broadened to the point where all separation has been lost, and the chromatography is useless. On the other hand, if the solvent composition of the injected sample can be adjusted so that it contains less organic solvent than the mobile phase it is injected into, then good separation can be obtained even though a large injection volume is still used. This type of result is shown in Panel D, where the sample has been adjusted to contain 20% ACN less than the initial solvent used in the RP gradient. Now, because the analytes are focused into a small volume band at the column inlet, they elute as much sharper and well separated peaks. And, most importantly for the ^{2}D separation, the detection sensitivity is much better because we have both injected a large volume of sample, and observe tall, sharp peaks. The lesson here is that the combination of the composition of the injected sample, relative to the mobile phase composition, and the volume of sample that is injected is incredibly important to the performance of ^{2}D separations

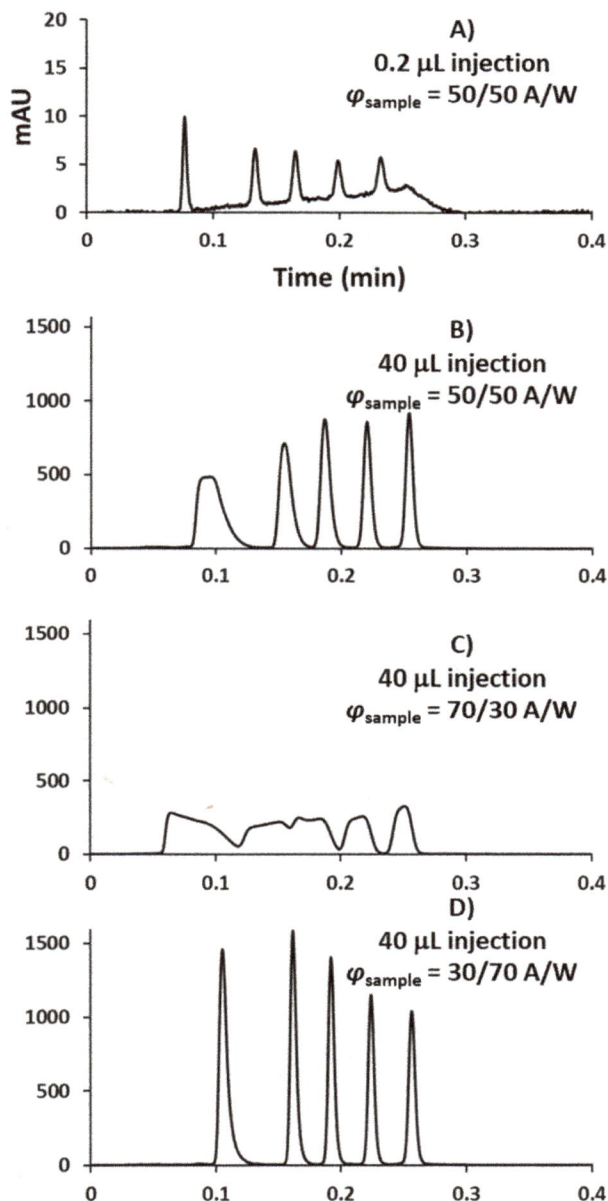

FIGURE 4.13 Demonstration of the mobile phase mismatch problem under conditions that mimic the use of RP separation in both dimensions of 2D-LC. Chromatographic conditions: Column, 30 mm x 2.1 mm i.d. SB C18 (3.5 micron); Flow rate, 2.5 mL/min.; Gradient elution from 50–90% B from 0-0.25 min., where A is water, and B is ACN; Temperature, 40 °C; Detection by absorbance at 210 nm; Sample, 10 μg/mL each of acetophenone (1), butyrophenone (2), valerophenone (3), hexanophenone (4), heptanophenone (5) in the sample solvents (φ_{sample} of acetonitrile (A) and water (W)) indicated; Injection volumes, A) 0.2 μL, B-D) 40 μL.

in 2D-LC. This fact has motivated many groups to investigate different approaches to address this challenge over the past few decades; these will be discussed in detail in the following sections.

4.4.5.1 Inline Dilution

The first report we are aware of in the literature that describes adjustment of the composition of the ^{1}D effluent prior to injection into a ^{2}D column in an online 2D-LC system was from Oda *et al.* in 1991 [30]. The particular instrument configuration used in this work was more complex than that used for simple inline dilution of the ^{1}D effluent, and will be discussed later in Sections 4.4.5.2 and 4.4.6. The next mention of a simple inline dilution approach was by Moore *et al.* in 1995 [31]. A generalized illustration of this approach is shown in Figure 4.14. In Moore's work the stated goal was to reduce the level of organic solvent in the samples injected into the ^{2}D column from 85% MeOH in the ^{1}D effluent to around 17%. This was close to the 6% ACN used as the initial mobile phase in each ^{2}D RP gradient elution program, and prevented serious breakthrough of analytes in the dead volume of the ^{2}D column.

This inline dilution approach has been used to great benefit in a variety of 2D-LC applications. It is simple to implement in practice, although it does typically require a third pumping system that is capable of precisely diluting the ^{1}D effluent. Since the dilution is implemented at the outlet of the ^{1}D column where the pressure is low, this pump does not need to be capable of producing high pressures. Indeed, the initial work of Moore *et al.* involved a low-pressure syringe pump. Inline dilution of the ^{1}D effluent has also been used in combination with sorbent-based traps in the 2D-LC interface; this approach will be discussed in more detail below in Section 4.4.6.

A common criticism of this kind of approach is that dilution of the ^{1}D effluent results in reduction of the analyte concentration in the fraction injected into the ^{2}D column, and that this can lead to poor detection sensitivity at the outlet of the ^{2}D column. If lowering the elution strength of the solvent in the injected fraction does not lead to focusing of the analyte molecules at the inlet of the ^{2}D column, then this criticism is justified. It is not possible to achieve this kind of focusing for all types of molecules under all chromatographic conditions, and thus the extent to which this criticism

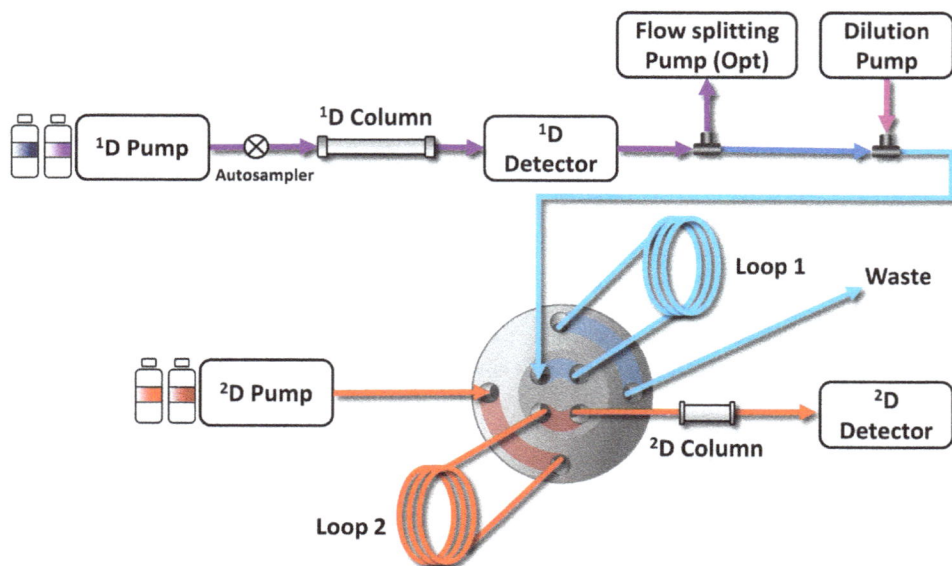

FIGURE 4.14 Illustration of inline dilution of ^{1}D effluent in online 2D-LC to minimize mobile phase mismatch between the ^{1}D and ^{2}D mobile phases. The use of a pump or other device to remove part of the ^{1}D effluent prior to dilution is optional, and has been used infrequently.

is meaningful depends strongly on the 2D-LC separation conditions. However, the inline dilution approach is typically used in cases where it is likely that lowering of the solvent strength in the injected sample will lead to analyte focusing, and thus the benefit of focusing the band by diluting the ^{1}D effluent outweighs the harm of lowering the analyte concentration in the injected fraction through the act of adding the diluent.

In our own work we have studied this compromise from a quantitative perspective to better understand the limits of the approach. The other common comment made when discussing the benefits of the inline dilution approach is that mobile phase mismatch in 2D-LC can be avoided by simply injecting smaller fractions (in volume terms) of the ^{1}D effluent into the ^{2}D column. Our 2014 study aimed to address both of these points [21]. The context we used for this study was the detection of low concentration degradants of a partially degraded active pharmaceutical ingredient (API). In this context, the second dimension of a 2D-LC separation must be sensitive enough to detect compounds that are present in the mixture at 0.05% (w/w) of the active ingredient. If we avoid exceeding a signal of 1500 mAU for a ^{1}D UV detector (so that we stay in the linear region of the detector response), then there is an upper limit to how much sample we can inject into the first dimension as a means of improving the signal for peaks detected in the second dimension. For this work we used the instrument configuration shown in Figure 4.14, which allows both splitting of the ^{1}D effluent as a means of varying the volume of each fraction of ^{1}D effluent injected into the ^{2}D column without altering the parameters of the ^{1}D separation. We then compared four scenarios. Scenarios 1–3 involved injecting fractions of ^{1}D effluent with volumes of 7, 20, or 40 μL, but no dilution of those fractions prior to injection into the ^{2}D column. The fourth scenario was similar to scenario #3 in that the ^{1}D effluent fraction was 40 μL, but then this was diluted 1:1 with water such that the total volume of the diluted fraction injected into the ^{2}D column was 80 μL. The most important view of the results from this comparison are shown in Figure 4.15, where we plot the limit of quantitation (in % w/w terms, relative to the API, naproxen) for four small test molecules of varying hydrophobicity, for each of the four 2D-LC interface scenarios. The dashed red line shows the 0.05% threshold above which quantitative results are desired. We see that for three of the four test molecules scenario #4 results in the lowest Limit of quantitation (LOQ). For naproxen it is the only condition that provides

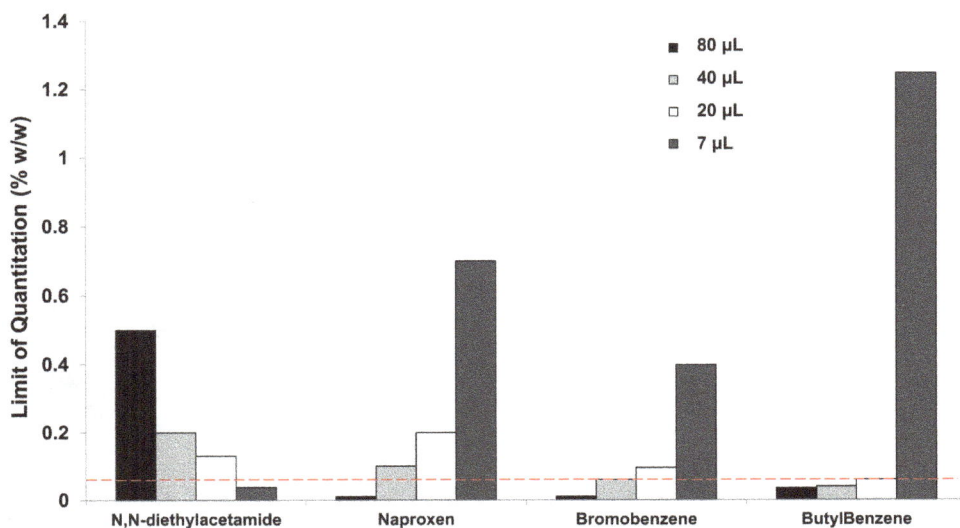

FIGURE 4.15 Limit of quantitation (LOQ) for four probe compounds under each interface condition. Reprinted by permission from Springer Nature: *Analytical and Bioanalytical Chemistry* (Evaluation of detection sensitivity in comprehensive two-dimensional liquid chromatography separations of an active pharmaceutical ingredient and its degradants, D. Stoll, E. Talus, D.C. Harmes, K. Zhang), Copyright (2014).

the sensitivity needed to yield quantitative results at the 0.05% threshold. The trend observed for the hydrophilic test analyte N,N-diethylacetamide opposes the trends for the other probes. Visual inspection of the chromatograms showed that this was mainly due to an interfering system peak eluting early in the mobile phase gradient.

The variant of the instrument configuration shown in Figure 4.14 without the splitting pump is frequently used for the inline dilution approach. In this implementation the ^{1}D and ^{2}D columns are never connected in the same flow path. A different approach to inline dilution that does connect the ^{1}D and ^{2}D flow paths is illustrated in Figure 4.16. This approach has been discussed by Rogatsky [32] and others [33], and is sometimes referred to as "at-column dilution" (ACD). The main advantage of this approach is that it is relatively simple to implement, particularly for LC-LC methods. As is shown in Figure 4.16 all that is required in addition to an existing 1D-LC system is a ^{2}D pump and column, and a 2-position/6-port valve to connect the first and second dimension flow paths. In one position of the valve the ^{1}D effluent goes to waste, while only the ^{2}D pump flow goes through the ^{2}D column. When it is time to sample a peak of interest from the

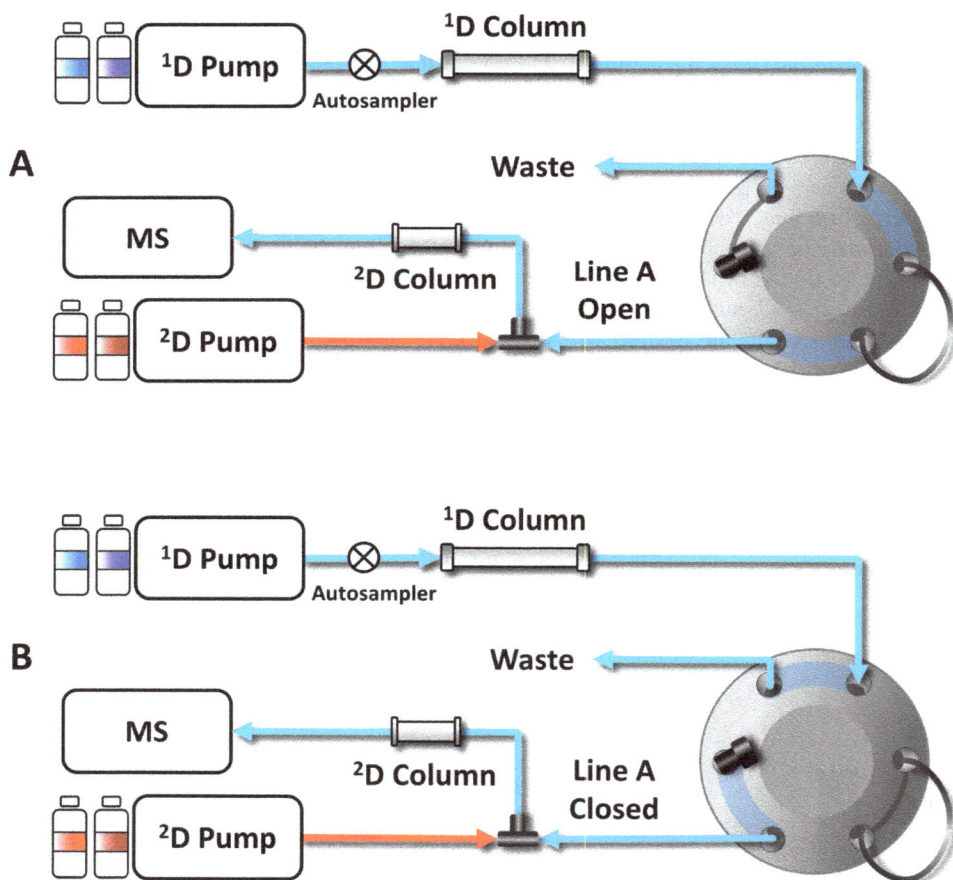

FIGURE 4.16   Instrument setup that enables inline dilution of ^{1}D effluent as it enters the ^{2}D column. A) ^{1}D effluent is directed to the ^{2}D column and combined with mobile phase from the ^{2}D pump, which acts as a diluent. In this position the ^{1}D and ^{2}D columns are connected in series. B) In this position the two columns are isolated; the ^{1}D effluent goes to waste, and only the ^{2}D mobile phase drives the ^{2}D separation.

Source: Adapted from Rogatsky *et al*. [32].

¹D separation, the valve is switched such that the ¹D effluent is mixed with the ²D mobile phase at the ²D column inlet. During this step the fraction of interest is transferred to the second dimension, and this fraction is diluted with ²D mobile phase; if the ²D mobile phase composition is chosen judiciously, very effective focusing of the analytes of interest can be achieved at the ²D column inlet. There are also significant disadvantages associated with this approach, however. Foremost among these are limitations on the ¹D and ²D flow rates that can be used effectively, and pressure fluctuations that the ²D column experiences when switching between positions of the valve. The flow rate limitation arises from the fact that during sampling the total flow from the ¹D and ²D pumps goes through the ²D column.

4.4.5.2   Fixed Solvent Modulation

Simple inline dilution of ¹D effluent during modulation to overcome solvent mismatch has significant disadvantages and/or limitations as discussed in the preceding section. Several groups have developed alternative approaches to address these shortcomings. One approach we refer to as Fixed Solvent Modulation (FSM), which is illustrated in Figure 4.17. A variant of this was first described by Oda and coworkers [30]; this particular setup also involved a sorbent-based trap between the first and second dimensions, and this will be discussed in more detail in Section 4.4.6. Subsequent implementations of FSM like that shown in Figure 4.17 were described by Gron [34], and by us [35]. The main principle in play here is that a bypass connection is installed that connects the outlet of the ²D pump to the inlet of the ²D column without going through the sample loop of the interface valve. This enables the use of the ²D pump itself as a source of the diluent for the sample exiting the sample loop, rather than having to add a third pump to the system. The primary advantage of this approach is that it is simple to implement, requiring only two tee pieces and the bypass connection capillary. The primary disadvantage of the approach is that the bypass is fixed – it is always in the flow path throughout the ²D separation cycle, and cannot be "turned off" without physically changing the hardware. The inability to turn off the bypass after the sample has been loaded into the ²D column has two serious consequences: 1) the two different flow paths for the ²D mobile phase result in a doubling of baseline features, particularly when using gradient elution; and 2) the slow re-equilibration of the ²D column due to slow flush-out of the sample loop at a flow rate lower than the total ²D flow. These features become particularly problematic when using the ²D chromatogram for

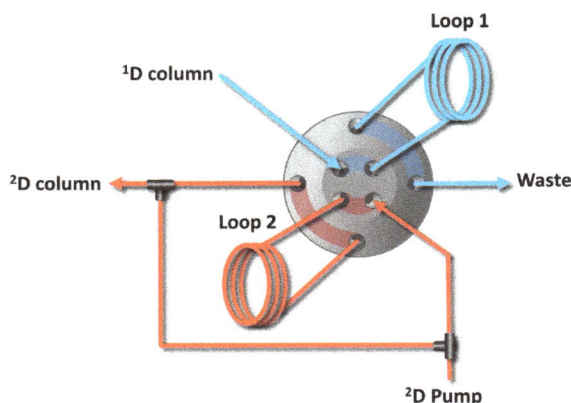

FIGURE 4.17   Example of a 2D-LC interface configured for Fixed Solvent Modulation (FSM). The flow restrictions through the sampling loops and the bypass connection determine the extent to which the fractions of ¹D effluent are diluted with ²D mobile phase prior to introduction into the ²D column.

Source: Adapted from [35].

quantitation of low abundance compounds (i.e., trace analysis), and when doing 2D-LC separations that require fast ^{2}D cycles (i.e., mainly in LC×LC).

4.4.5.3   Active Solvent Modulation

We have developed ASM to address the limitations and disadvantages of inline dilution and FSM discussed in the preceding sections. ASM is similar to FSM in the sense that a bypass connection enables flow between the ^{2}D pump and ^{2}D column inlet that does not go through the sample loop. The interface is illustrated in Figure 4.18 for the case where two loops are alternately used for ^{1}D effluent collection, and injection into the ^{2}D column, similar to the commonly used dual-loop interface shown in Figure 4.3. The principal difference between FSM and ASM is that ASM enables on/off control of flow through the bypass connection during each ^{2}D separation cycle. This is accomplished by adding two additional rotational positions to the two positions that are typically used with this type of dual-loop interface. During operation the bypass is only opened when a ^{1}D effluent fraction is being displaced from the sample loop so that it can be diluted with ^{2}D mobile phase. When this step is completed, the bypass is closed so that the doubling of baseline features that occurs with FSM is eliminated, and the ^{2}D column can be re-equilibrated faster using the full flow rate of the ^{2}D pump. An example of the changes in the mobile phase composition program during each ^{2}D elution cycle is shown in Figure 4.19. More detailed discussions of the impact of various ASM operational parameters (e.g., dilution ratio, loop flushing time, and injection volume) are given in Sections 4.4.5.5 and 6.7.

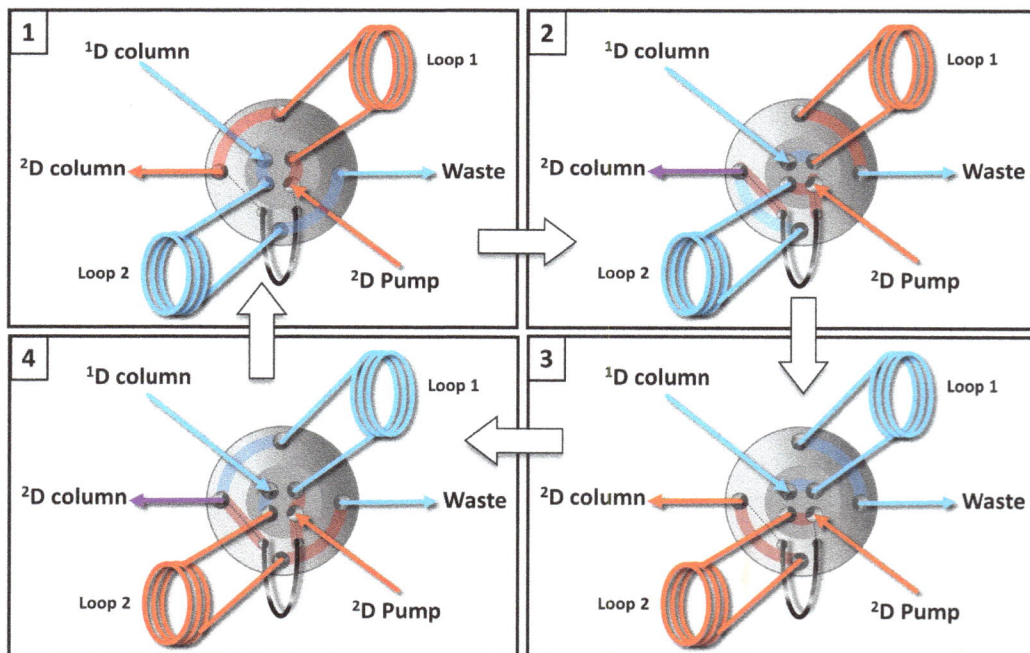

FIGURE 4.18   Dual-loop interface for Active Solvent Modulation. Position 1 enables collection of effluent from the ^{1}D column in Loop 2, and flow of ^{2}D mobile phase through the Loop 1 into the ^{2}D column. Position 2 enables displacement of the ^{1}D effluent from Loop 2 while also diluting this sample with ^{2}D mobile phase prior to injection of the mixture into the ^{2}D column. Positions 3 and 4 are similar to 1 and 2 except that the roles of Loops 1 and 2 are reversed.

Source: Adapted from [35].

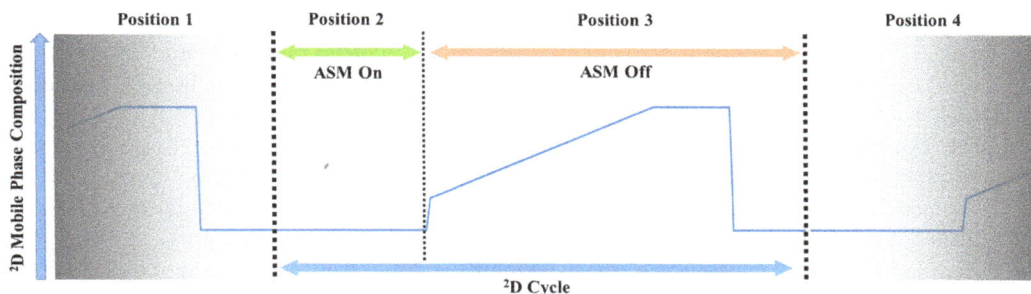

FIGURE 4.19 Illustration of the interaction between opening and closing of the ASM bypass connection and the mobile phase composition during a ^{2}D injection and elution cycle. During displacement of a ^{1}D effluent fraction from the sample loop, the bypass is open, and the ^{2}D pump is set to pump mobile phase at a composition (indicated by the blue trace) that promotes focusing of the analyte band at the inlet of the ^{2}D column. In the diagram shown here, RP separation is assumed, where water is the best possible diluent (i.e., the "weakest" solvent).

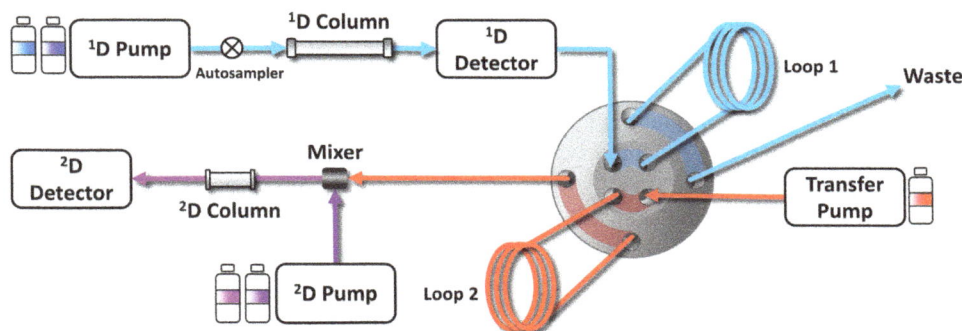

FIGURE 4.20 Illustration of the approach implemented by Schmitz *et al.* for flexible dilution of ^{1}D effluent fractions using an additional transfer pump. In the position shown, ^{1}D effluent is collected in Loop 1, while solvent from the transfer pump is mixed with mobile phase from the ^{2}D pump during elution of a fraction of ^{1}D effluent previously collected in and displaced from Loop 2. As with the other approaches discussed in the preceding three sections, the ^{2}D mobile phase can be chosen to act as an effective diluent (e.g., water if the ^{2}D separation is reversed phase) during the displacement of a fraction of ^{1}D effluent from the sample loop.

4.4.5.4 Inline Dilution Through Use of a Transfer Pump

In 2019 Schmitz's group introduced an approach to modulation that combines features of existing approaches in a way that enhances some of the advantages of these approaches while minimizing their weaknesses [36–38]. Their interface setup used for LC×LC is illustrated in Figure 4.20. The same type of 8-port/2-position valve shown in Figure 4.3 is used, but instead of routing the ^{2}D mobile phase flow from the pump through the sample loop, the ^{2}D pump is positioned downstream from the valve, and an additional transfer pump is added to displace ^{1}D effluent fractions from the sample loops. The primary advantages of this approach over others are:

- Precise control of the dilution ratio realized by mixing ^{2}D mobile phase with the fraction of ^{1}D effluent displaced from the sample loop. Using the ^{2}D mobile phase as a diluent has the benefit of FSM and ASM, with the added advantage that the dilution ratio can be changed via software (i.e., through the flow rate setpoints of the ^{2}D and transfer pumps); with FSM and ASM this ratio requires changing the bypass capillary to realize a change in the flow restrictions of the bypass and through-loop flow paths.

- The ^{2}D mobile phase never passes through the sample loop, thereby avoiding large ^{2}D gradient delay times that result from the use of low ^{2}D flow rates and large sampling loops. However, this could also be a disadvantage for "sticky" compounds, as they would not be thoroughly flushed from the loop by strong mobile phase.
- Modulation of the composition of the sample injected into the ^{2}D column is realized without connecting the ^{1}D and ^{2}D columns in series as in the approach shown in Figure 4.16. This should result in longer column lifetimes, and provides more flexibility with respect to the flow rates that can be used in the first and second dimensions.

The primary disadvantage of this approach compared to FSM and ASM is that a high-pressure capable pump is required to displace ^{1}D effluent from the sample loop and introduce it into the ^{2}D flow stream.

4.4.5.5 Implementation of Solvent-Based Modulation Approaches

The most commonly used approaches to solvent-based mitigation of mobile phase mismatch were illustrated and discussed in the preceding four sections. In our own work we have invested considerable effort in understanding when and to what degree these approaches solve the mobile phase mismatch problem. The harsh reality of 2D-LC is that there is no single approach that works very well for all applications. Some approaches are convenient and easy to implement, but are not as effective as others. Others are more effective, but may not be as robust, or may be more costly to implement for routine work. Understanding these compromises is valuable to those developing 2D-LC methods, because they can then focus their effort on the approaches that are most likely to be helpful for the application at hand.

It is well known for all types of LC that when the volume of the injected sample is small relative to the dead volume of the column (i.e., < 1%), the sample solvent composition (i.e., the sample matrix) does not have a major effect on resulting separations as measured by the width and shape of the peaks. However, as the volume of the injected sample increases, the particular nature of the sample solvent becomes more important [39, 40]. In the worst cases, bad combinations of sample volume and sample solvent can lead to wide, distorted, and even badly split peak shapes. Obviously, this can negatively impact the resolution of neighboring peaks, and detection sensitivity, since peak height and peak width are inversely related when the peak area is constant. These trends are observed in the series of chromatograms shown in Figure 4.13. When a small volume of sample is injected, the peaks are narrow and symmetrical, but they are short (see Figure 4.13A; i.e., sensitivity is low). If we increase the volume of the injected sample to increase sensitivity the peaks do get taller, but they also get wider, especially when the sample solvent is different from the mobile phase in an unfavorable way (see Figure 4.13C). In our view, then, it is this tradeoff between peak width (or resolution, peak capacity) and detection sensitivity that is the central issue, and measuring these values as a function of the 2D-LC separation conditions can give us a semi-quantitative sense for the nature of the problem, and the value of potential solutions.

Unfortunately, at this time there has not been any systematic experimental study of the effectiveness of the solvent-based modulation strategies discussed here over a wide range of analyte chemistries and chromatographic conditions. Such a study would be a highly valuable contribution to the field. However, there have been a few studies that have quantified the effectiveness of individual approaches within the context of specific applications. Such studies at least show us the kind of results we can expect from systematically varying modulation parameters (e.g., the extent of dilution in the case of inline dilution or ASM) that can guide method development. An example

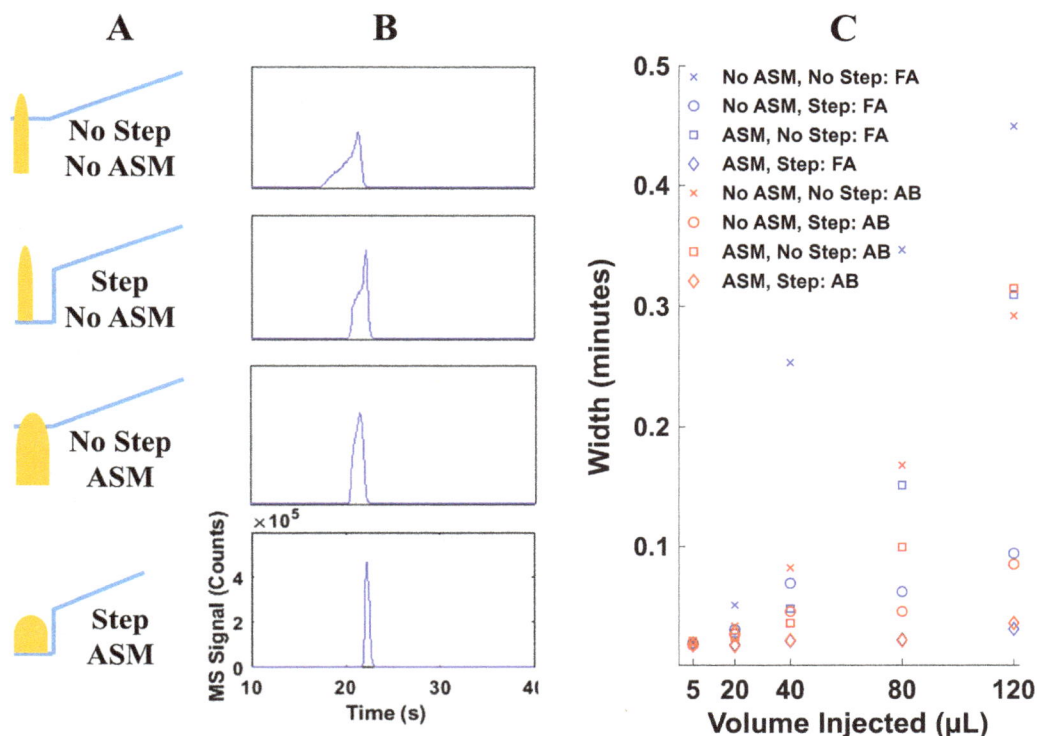

FIGURE 4.21 Comparison of peak shapes and widths obtained for a peptide eluting early in the second dimension of a RP×RP separation of tryptic peptides with different interface conditions (ASM-f = 3 in all cases). A) Illustration of different ways of operating the ASM interface. The gold plug represents the volume (width of the plug) and composition of the ¹D effluent injected into the 2D column (height represents ACN fraction). B) Chromatograms for the peptide obtained by MS detection. C) Peak width for the peptide observed for the four different modes, with several different loop volumes. FA indicates that the ¹D mobile phase additive was formic acid, and AB indicates that the additive was ammonium bicarbonate.

Source: Adapted from ref. [6].

of such results from our own work focused on the LC×LC separations of peptides is shown in Figure 4.21, where ASM was used. During method development, the loop volume and the ASM dilution ratio (ASM-f) can be varied to optimize peak capacity (or resolution) and detection sensitivity. This set of results shows us that when an ASM-f of 3 is used in this context, a loop volume of up to 120 μL can be used without sacrificing much peak capacity, because the peak width is similar with a 120 μL loop to the widths obtained with much smaller loops. Section 6.7 contains a more detailed discussion of factors to consider when developing a method using ASM. Interested readers are also referred to Table 1 of ref. [20] and our applications database (www.multidlc.org/literature/2DLC-Applications), both of which cite several papers describing methods that have been developed using ASM.

A more detailed understanding of the injection process in LC is emerging as a result of recent fundamental studies of cases where there is a significant mismatch between the sample solvent and the mobile phase, and the injection volumes are large relative to the column volume [41–44]. Additional work in this area may lead to the development of software tools that can simulate the injection process in 2D-LC with enough accuracy to support method development.

4.4.6 INTERFACES FOR SORBENT-BASED MODULATION

In the preceding section all of the modulation approaches discussed rely on adjustments to the composition of ^{1}D effluent fractions between the time they are collected from the ^{1}D column outlet and the time they are injected into the ^{2}D column to mitigate the negative impacts of mismatch between ^{1}D and ^{2}D mobile phases so often encountered in 2D-LC. While these approaches have proven very popular and useful over time [45], adjusting the solvent composition is not the only way to address the problem. Installation of a trapping device filled with a sorbent that will selectively retain only analytes of interest can be used to separate and possibly concentrate them from the solvent matrix that comes from the outlet of the ^{1}D column. Recently this approach has been referred to as Stationary Phase Assisted Modulation (SPAM). This strategy gained traction for biomolecule separations first [46, 47], and use for small molecules separations followed. Early demonstrations of the approach were published by Sweeney *et al.* [26], and later by Venkatramani *et al.* using the 12-port/2-position interface discussed in Section 4.4.3, but with sorbent-filled traps instead of open loops [48]. Early work relied mostly on 10-port/2-position valves [49], however most recent work has used 8-port/2-position interface valves. As of the 2019 review article by Pirok *et al.* [45], roughly 25% of recent applications of 2D-LC used some type of sorbent-based modulation strategy. A typical interface setup used for this approach is shown in Figure 4.22, where the simple sample loops in Figure 4.3 (i.e., open tubes) are replaced with some kind of trapping cartridge or column, such as a "guard column" normally used to protect the inlet of analytical columns in 1D-LC. The important principle is that when the sorbent material retains the analytes of interest more than the effluent components coming from the first dimension, then the analytes can be focused into a small volume while the majority of the solvent matrix in the ^{1}D effluent is flushed to waste. Then, when the trap is switched into the ^{2}D flow path, the trapped analytes are eluted from the trap into the ^{2}D analytical column for further separation, now in a volume of mobile phase that is typically smaller and more favorable for the ^{2}D column than would be the case if a simple open loop capillary were used in the interface. This approach is especially helpful in situations where the 1D effluent matrix is truly incompatible with the 2D mobile phase, potentially causing problems with miscibility and/or precipitation of the analyte. In this case the use of a trap between the first and second dimension separations to capture the analyte and facilitate a solvent exchange step is invaluable. For example, Pirok *et al.* developed a LC×LC separation of nanoparticles involving a ^{1}D separation based on hydrodynamic chromatography (HDC) and an aqueous buffer, and a ^{2}D separation based on SEC and THF as a mobile phase solvent [50]. In this case residual water from the ^{1}D separation would severely compromise the performance of the ^{2}D separation, and the use of sorbent-filled traps between the two dimensions was

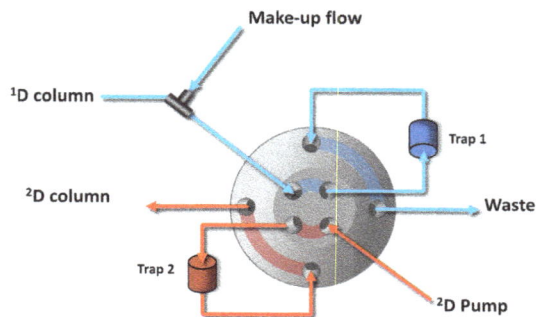

FIGURE 4.22 Illustration of the use of sorbent-filled traps in place of open sample loop capillaries in a standard 8-port/2-position 2D-LC interface valve. This approach can be used with or without the addition of the "make-up" flow, but typically the trapping efficiency of the sorbent is enhanced when the ^{1}D effluent can be diluted with weak solvent on its way to the trap.

critical for removing most of the water from the analyte band before injecting it into the ^{2}D column. A number of variations on this concept have been implemented. For example, Figure 4.22 illustrates the use of a diluent flow delivered by an additional pump, which when mixed with the ^{1}D effluent flow prior to the interface traps, can promote efficient trapping of the analytes in a small volume [46,51].

4.4.6.1 Implementation of Sorbent-Based Modulation Approaches

As was mentioned above for solvent-based modulation approaches, readers interested in developing a method involving sorbent-based modulation will find many examples in the literature (see Table 1 of ref. [20] and our applications database – www.multidlc.org/literature/2DLC-Applications), and can use these as starting points in their own work. Verstaeten and Desmet have used numerical simulations to explore the performance limits of traps used in conjunction with thermal modulation [52]. A highly generalized list of things to consider follows. The prioritization of these considerations will depend upon the goals of 2D-LC method (e.g., will the method prioritize throughput, sensitivity, or reproducibility?).

- *Trap chemistry and retention mechanism* – The success of a sorbent-based modulation approach depends largely on the ability of the sorbent to strongly retain the analytes of interest, but not too strongly. As with choosing separation mechanisms for the first and second dimensions (see Chapter 5), here too users must think carefully about the chemistry of their analytes, and choose a sorbent with chemistry that is likely to retain their analytes strongly enough to extract them from the ^{1}D effluent. The ability of the sorbent to retain them can be assessed experimentally by connecting a detector to the outlet of the trap to check for breakthrough during method development (i.e., connect the detector to the "waste" line at the right of Figure 4.22).
- *Trap volume and trapping conditions* – The ability to quantitatively extract analytes of interest from ^{1}D effluent fractions depends, of course, on the volume of the trap, and the extraction conditions. Increasing the trap volume will increase the volume of ^{1}D effluent that can be handled in a sampling period, particularly for weakly retained compounds, but there are practical limits to this. Increasing the trap volume also increases the time required to elute compounds from the trap into the ^{2}D column, and can also increase the gradient delay time for the ^{2}D separation. Generally speaking, diluting the ^{1}D effluent with weak solvent after the ^{1}D column and before the trap will improve trapping efficiency. The trap volume and dilution factor needed can either be determined by experiment, or by calculation if the retention behavior of the analytes of interest is known in detail from other work.
- *Impact on the ^{2}D separation* – After the analytes of interest have been extracted from the ^{1}D effluent, they must be completely eluted (i.e., desorbed) from the trap sorbent into the ^{2}D column. At this stage it is important to recognize that if the eluent needed to elute the analytes from the trap is strong (e.g., high organic content in the case of a RP mechanism) relative to the mobile phase needed for retention in the ^{2}D column that this can negatively affect the quality of the ^{2}D separation itself. Here again, the effect of these conditions on the ^{2}D separation must be determined by experiment, unless the retention behavior of the compounds is known from other work and can be used to predict the outcome.

4.4.7 Interfaces for Evaporation-Based Modulation

As is indicated by surveying different applications of 2D-LC, solvent- and sorbent-based approaches to improving modulation performance are by far the most widely used. Nevertheless, a few other innovative approaches have been described by academic laboratories. One of these is the use of vacuum to selectively remove the more volatile components of the ^{1}D effluent prior to injection of fractions into the ^{2}D column [28, 29]. Interest in this approach is primarily driven by the need to

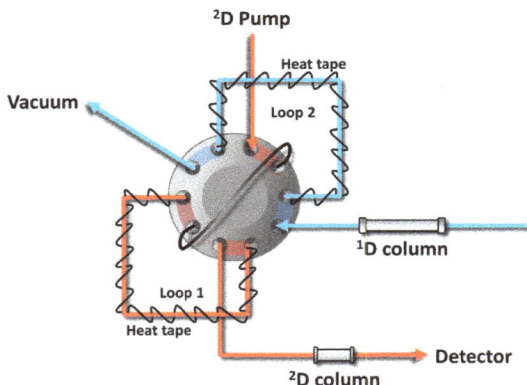

FIGURE 4.23   Illustration of an interface for 2D-LC that enables modulation of the composition of the solvent matrix of a fraction of ^{1}D effluent after it has been collected from the ^{1}D column, and before it is injected into the ^{2}D column, by applying vacuum to the loop outlet during fraction collection.

Source: Adapted from ref. [28].

mitigate problems associated with mismatch between the mobile phase solvents used in the first and second dimensions as discussed above in Section 4.4.4. A diagram of one type of interface that can be used in this way is shown in Figure 4.23. A dual-loop interface similar that shown in Figure 4.3 is used, but now the sample loop capillaries are heated, and vacuum is applied to the outlet of the loop being filled with ^{1}D effluent. Provided the ^{1}D effluent has a sufficiently low boiling point, the majority of the solvent can be evaporated from the loop by controlling the vacuum pressure and heating. In principle this leaves less-volatile analytes deposited on the inner walls of the loop, or in a much smaller volume of ^{1}D effluent than was flowed into the loop during the modulation period. These analytes are then dissolved in ^{2}D mobile phase when the loop is switched into the ^{2}D flow path. From a conceptual point of view the potential benefits of this approach are clear – ^{1}D effluent components that can cause broadening of ^{2}D peaks can be reduced or eliminated. However, in practice it seems this approach is likely limited to 2D-LC methods involving ^{1}D mobile phases composed entirely of volatile organic solvents [45], because removing water from this type of interface fast enough to be useful for rapid sampling of a ^{1}D separation would be difficult and lead to loss of analytes more volatile than water (which most are!). Even when using the approach with volatile NP solvents such as chloroform and hexane, Tian *et al.* found that recovery of analytes with boiling points above 400°C was only about 50% [28].

In a completely different approach, Fornells *et al.* have developed what they refer to as an Evaporative Membrane Modulation (EMM) interface for 2D-LC [53]; an illustration of the interface is shown in Figure 4.24. In this approach the ^{1}D effluent is brought into close proximity with a gas stream in a microfabricated device to facilitate removal of volatile constituents of the liquid, but separated by a thin, gas permeable hydrophobic membrane (e.g., polytetrafluoroethylene [PTFE]). The membrane assembly is heated using infrared light emitting diodes (LEDs) to control the rate of evaporation. Since the amount of heat required to produce a given evaporation rate is dependent on the composition of the ^{1}D effluent (e.g, the organic/water ratio), the flow rate of the liquid exiting the interface is measured using a calorimetric flow meter and used as a closed-loop feedback input to adjust the amount of heat provided by the LEDs and the flow rate of the gas at the inlet of the EMM device (i.e., the flow rate of the gas that sweeps away the evaporated solvent). In their initial investigation of this approach Fornells *et al.* found that the EMM interface was effective at removing ACN from a stream of effluent coming from a ^{1}D RP separation. This resulted in smaller ^{1}D peak volumes compared to when there was no evaporation. Also, the reduced ACN content of the ^{1}D

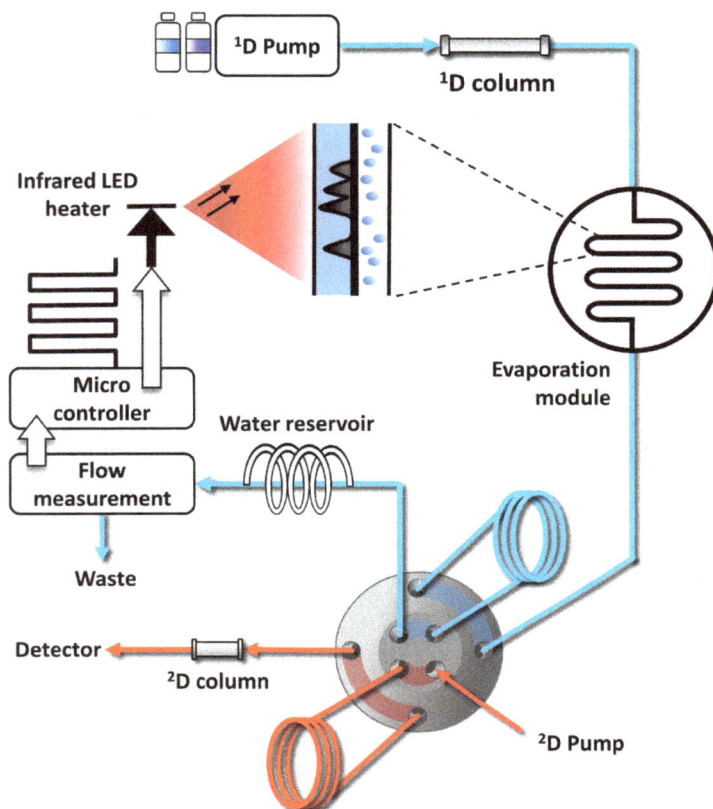

FIGURE 4.24 Illustration of an interface for 2D-LC enables modulation of the composition of ^{1}D effluent fractions as in Figure 4.23, but using a membrane-based module to control the selectivity and rate of evaporation of ^{1}D effluent components.

Source: Adapted from [53].

effluent fraction injected into the ^{2}D column resulted in ^{2}D retention times that were less dependent on the points in the ^{1}D ACN gradient from which the fractions were taken, compared to when there was no evaporation. However, the authors also note that the reduction in peak volume – and the expected attendant increase in peak height (i.e., concentration at the detector) – were not as high as expected based on the observed reduction in ^{1}D flow rate following evaporation. They attribute this to serious longitudinal broadening of ^{1}D peaks inside the EMM interface device. Thus, even though there is a reduction in peak volume, there is actually a significant increase in the width of ^{1}D peaks (in time units) as they move through the EMM device, resulting in a loss of ^{1}D resolution. Nevertheless, this is a very innovative approach, and perhaps future work will lead to optimization of this type of interface.

4.4.8 INTERFACES FOR TEMPERATURE-BASED MODULATION

Whereas in 2D-GC temperature-based modulation is commonplace, relatively little work has been reported related to the implementation of temperature-based modulation for 2D-LC. This is most likely due to the smaller influence of temperature on retention in LC compared to GC (i.e., enthalpy of transfer between mobile and stationary phases), and the relatively higher heat capacity of liquids compared to gases, which makes fast heating and cooling challenging in LC. There are three very different approaches that have been described for temperature-based modulation in 2D-LC, all of which rely on sorbent-based trapping in addition to temperature changes.

The first approach, described by Creese et al. [54] is functionally similar to temperature-based modulation in 2D-GC, as illustrated in Figure 4.25. In this case there is no modulation valve; the ^{1}D effluent flows into the interface at all times, and combined with ^{2}D mobile phase prior to a sorbent-based trap. Their approach is similar to prior work of Verstraeten et al., who used a low thermal mass metal sleeve fixed in place over the sorbent-filled trap to rapidly heat the trap [55]. In Creese's work, the resistively-heated metal sleeve is supported in a way that it can be moved along the long axis of the trap column. Creese et al. refer to this approach as Longitudinal on-column Thermal Modulation ($L^{OC}TM$). At the beginning of a modulation cycle (Figure 4.25A) the sleeve is positioned near the trap outlet and a cold air jet blows on the trap inlet to decrease the temperature of this zone, promoting retention of analytes entering the trap in ^{1}D effluent. After accumulating analytes for a fraction of the modulation cycle (e.g., tens of seconds), the hot sleeve is moved toward the trap inlet (Figure 4.25B). This warms the inlet, releasing previously trapped analytes to the outlet zone of the trap that has been cooled by a second cold air jet. This second trapping step has the effect of compressing the analyte zone. Finally, when the hot sleeve is returned to the outlet zone of the trap, analytes are released (eluted) to the inlet of the ^{2}D analytical column. In this way the modulator provides a kind of "catch-and-release" mechanism that delivers analytes to the ^{2}D column without

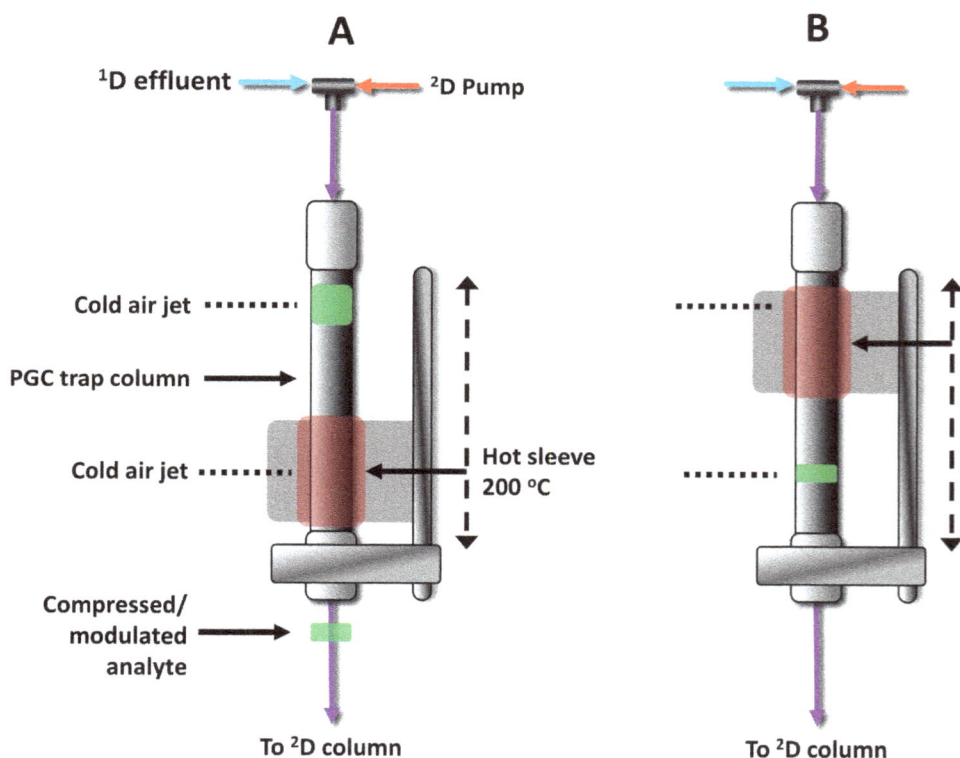

FIGURE 4.25 Illustration of the Longitudinal on Column Thermal Modulation ($L^{OC}TM$) approach developed for 2D-LC by Creese et al.

Source: Adapted from [54].

the need for a valve switch of any kind in the interface. In their proof-of-concept study that used RP separations in both dimensions for separations of an alkylphenone mixture and components of red wine, Creese et al. found that the L^{OC}TM approach reduced ^{2}D peak widths by about 60% relative to ^{2}D peaks obtained under the same conditions but with a standard valve-based interface like that shown in Figure 4.3. However, even with this improvement, the ^{2}D peak widths obtained with this system were on the order of 10 s. Given that optimized valve-based approaches routinely produce ^{2}D peak widths less than 1 s, the L^{OC}TM approach will have to be improved substantially for it to be competitive with other modulation approaches for 2D-LC.

In the second approach, described by van de Ven et al. [56], the authors focused on the principle of using a sorbent-based trap and temperature changes together to effect a significant change in the solvent that carries analytes from the modulation interface to the ^{2}D column. Briefly, ^{1}D effluent is loaded into a sorbent-filled trap when it is hot, and analytes are eluted from the trap when it is cold. In this case the eluting solvent can sweep out the analytes immobilized by the cold trap, resulting in both focusing of the analyte band and exchange of solvent so that the composition of the sample as it enters the ^{2}D column is dominated by the ^{2}D mobile phase rather than the ^{1}D effluent. Based on this initial proof-of-concept work it seems that this approach would be most useful in situations where the properties of the solvents used in the two dimensions would be dramatically different, as in coupling NP and RP separations, or perhaps HILIC and RP separations. At this time the concept has only been demonstrated by direct injection of analytes into the trap; we are not aware of any application of the approach in an actual 2D-LC separation.

Finally, very recently Niezen et al. have described the use of "cold trapping" to exploit the strong temperature dependence of the retention of polymers under specific conditions [27]. Their approach is similar to that described by Sweeney et al. two decades ago [26]. In this case a sorbent-filled trap is held at a temperature much lower (i.e., 75 °C in this case) than the ^{1}D and ^{2}D separation columns, which was shown to improve trapping efficiency, and thus recovery of the polymer analytes.

4.4.9 APPROACHES BASED ON SPLITTING THE ^{1}D FLOW PRIOR TO MODULATION

A significant challenge in the development of some 2D-LC methods is coping with the large volume of the fractions of ^{1}D effluent. This is particularly acute in the case where it is desirable to convert an existing 1D-LC method, which may be run at a flow rate of 1 mL/min or more, into a 2D-LC method. In a case like this, where the ^{1}D flow rate is 16 µL/s, taking a fraction over just five seconds produces a fraction with a volume of 80 µL. This may be manageable, but if we wish to collect an entire 30-s wide peak, then we have a fraction with a volume of 500 µL, which will be difficult to deal with in many situations.

In principle one can deal with this by passively splitting the ^{1}D effluent flow after the ^{1}D column outlet but before the sampling interface such that a portion of the ^{1}D flow is diverted to waste before the stream is sampled. This is simple to set up, as it only involves a tee-piece and a length of tubing to establish some resistance to flow to the waste container, but this approach can be unreliable and it is difficult to know when the flow rate to the sampling interface has changed. There are at least two literature reports that describe the use of an active flow splitting device for LC×LC that provides more control over the fraction of ^{1}D effluent that goes to the sampling interface [57, 58]. In these two reports the splitting device is essentially a pump operated "backwards" such that it pulls liquid out of the system (or the receiving piston allows liquid to flow into it at a controlled rate) rather than pushing liquid into it. A major focus of the work by Filgueira et al. was that such post-^{1}D flow splitting in LC×LC separations allows operation of the ^{1}D pump at flow rates that minimize gradient delay times to the point where they are a small fraction of the total analysis time [58]. This in turn increases the fraction of the analysis time that can actually be used for separation of the sample.

4.4.10   APPROACHES BASED ON DISCONTINUOUS ^{1}D ELUTION

In Section 1.5.1.3 we briefly discussed two approaches to 2D-LC that involve discontinuous execution of the ^{1}D separation – these are referred to as "Stop-Flow" and "Pulsed-Elution". The primary motivation for pursuing these implementations of 2D-LC is that they offer a means to provide more time to separate fractions of ^{1}D effluent in the second dimension.

4.4.10.1   Stop-Flow

As the name suggests, the principal idea at work in Stop-Flow 2D-LC is that the ^{1}D flow rate is modulated between on and off over the course of the 2D separation. When the ^{1}D flow is on, ^{1}D effluent is either transferred directly to a serially coupled ^{2}D column, or collected in an open loop or sorbent trap for subsequent transfer to the ^{2}D column. Once the portion of ^{1}D effluent of interest has been transferred, the ^{1}D flow is turned off and remains off until the ^{2}D separation of the transferred fraction is complete. In principle this could be as short as seconds, or as long as hours. A diagram of a system that would support Stop-Flow 2D-LC, as discussed by Bedani *et al.* [59] is shown below in Figure 4.26. Panel A shows an example of on-off modulation of the ^{1}D flow, and elution of sample constituents from the ^{2}D column while the ^{1}D flow is off. Panel B show an example of the system components that can be used to support a Stop-Flow implementation for 2D-LC. With valves in the positions shown in Panel A ^{1}D effluent is directly transferred to the ^{2}D column since the two columns are serially coupled through the valve. When the ^{1}D flow is turned off, both valves rotate such that the flow at the ^{1}D column outlet is blocked to maintain pressure on the column, and the ^{2}D pump is connected to the ^{2}D column for elution from the column. If there are regions in the ^{1}D chromatogram that are not transferred to the ^{2}D column, these can be diverted to waste by rotating the stop-flow valve when the ^{1}D flow is turned back on.

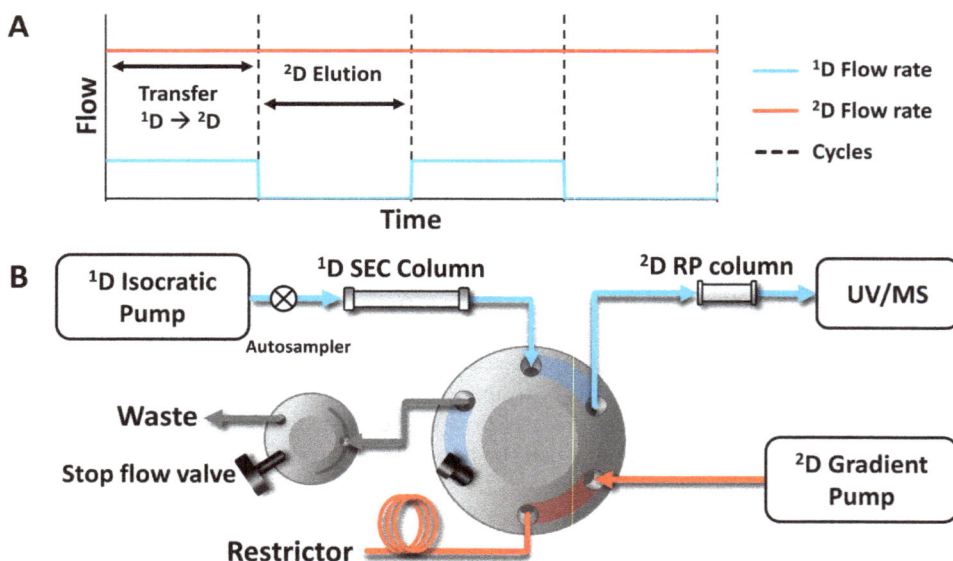

FIGURE 4.26   Illustration of the modulation of ^{1}D flow in a Stop-Flow 2D-LC system (A), and system components needed to implement this type of separation (B). In Panel B the ^{1}D effluent is transferred to the ^{2}D column. Switching both valves isolates the ^{1}D column, and allows elution of analytes from the ^{2}D column using the ^{2}D mobile phase while the ^{1}D flow is off.

Source: Adapted from [59].

As discussed above, the primary benefit of the Stop-Flow approach for 2D-LC is that it creates the opportunity to use very long ^{2}D analysis times without paying a significant undersampling penalty that would otherwise compromise the peak capacity of the ^{1}D separation. In principle there is no limit to the length of the ^{2}D separation that can be used. However, there is a practical limit because analytes remaining in the ^{1}D column during a period where the ^{1}D flow is off will diffuse longitudinally, thereby broadening the ^{1}D peaks and resulting in a loss of ^{1}D peak capacity. This compromise has been studied by several groups [60–62], with one of the most thorough studies by Xu *et al.* [63]. In this work they systematically quantified the additional broadening of ^{1}D peaks due to stopped flow periods for molecules ranging from 200 Da (small peptides) to 70 kDa (proteins). They found that for small molecules 50% of the efficiency (plate number) of the ^{1}D column is lost at a stop-flow time of 5,000 s (~1.4 hrs), but for large molecules only about 20% of the efficiency was lost after a stop-flow time of 10 hrs. This difference in behavior is due to the dependence of diffusion coefficients on the analyte's molecular weight. Given the square root dependence of ^{1}D peak capacity on plate number, these efficiency losses would translate into ^{1}D peak capacity losses of roughly 30 and 15%, respectively.

4.4.10.2 Pulsed-Elution

Recently Jakobson *et al.* introduced another approach that enables discontinuous elution in the first dimension of 2D-LC, which they refer to as "Pulsed Elution" [64]. This approach shares the same motivation with the Stop-Flow approach – that controlling the elution pattern in the first dimension provides more control over time spent on each ^{2}D separation – but accomplishes the objective by modulating the elution strength of the ^{1}D eluent over time, rather than modu-lating the ^{1}D flow rate over time. This concept is illustrated in Figure 4.27 where the volume fraction of organic solvent in the ^{1}D and ^{2}D eluents is plotted over time (a RP×RP separation is assumed in this case, but the concept is extensible to other types of separation). Whereas in Stop-Flow 2D-LC the ^{1}D flow rate is modulated while the elution strength is held constant (or continuously changing in the case of gradient elution), in Pulse Elution 2D-LC the ^{1}D flow rate is held constant while the elution strength of the ^{1}D eluent is modulated. The Pulsed Elution approach is on one hand attractive because modulating flow between on and off tends be a source of variability in ^{1}D retention times. On the other hand, effective execution of the Pulsed Elution approach requires the ability to rapidly modulate the ^{1}D eluent strength (i.e., a pump with very low gradient delay volume is required), and requires analytes that are sufficiently retained that they will not migrate very far through the ^{1}D column during a period when weak eluent is continuously pumped through the column.

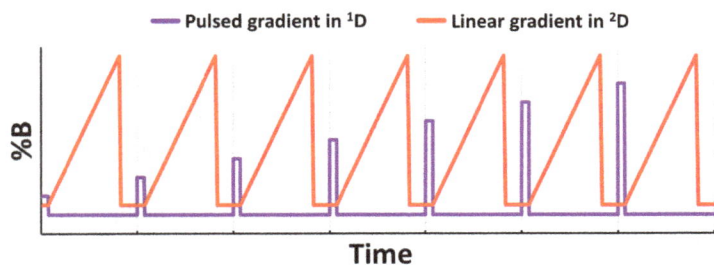

FIGURE 4.27 Illustration of the Pulsed Elution concept for used in the first dimension of 2D-LC. The elution strength of the ^{1}D eluent is increased over time, but as square pulses separated by periods of very low elution strength to decrease the migration velocity of analytes in the ^{1}D column.

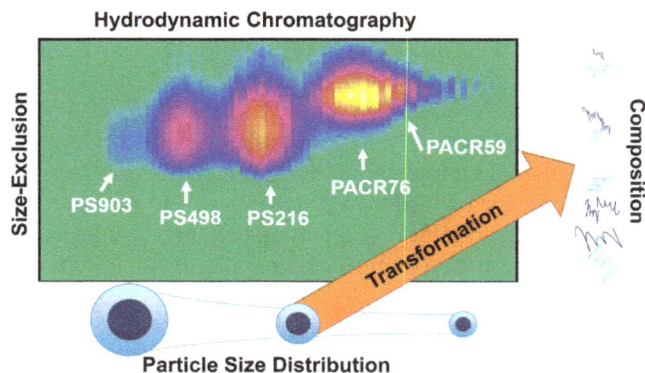

FIGURE 4.28 Illustration of the LC-transformation-LC concept. In this application nanoparticles are separated intact in the first dimension, which are then broken down into their constituent polymers in the interface, before finally separating the polymers by size in the second dimension. Reprinted with permission from [50] (https://pubs.acs.org/doi/10.1021/acs.analchem.7b01906). Permissions related to further use of this figure should be directed to the American Chemical Society.

4.4.11   LC – Transformation – LC

One interesting approach to multi-dimensional chromatography separations involves transformation of analytes after they have eluted from one column, but before they are loaded into the next column for further separation. The general concept is illustrated in Figure 4.28. The central idea is trace-able to so-called "diagonal electrophoresis", first demonstrated in 1966 [65]. In this initial work Brown and Hartley first separated a peptic digest of chymotrypsinogen A by paper electrophoresis. The partially resolved peptides where then oxidized using acidic hydrogen peroxide, the paper was rotated 90°, and electrophoresed again. Peptides that were not oxidized appear as spots on a diagonal in the resulting 2D electropherogram, but peptides that were chemically modified by oxidation migrate at different velocities, and thus appear as spots off the diagonal. A chromatographic version of this diagonal separation approach was developed many years later, and has mainly been used for selective isolation of peptides having specific functional groups. The approach, which evidently has only been applied using what we would refer to here as offline 2D-LC, has become known as combined fractional diagonal chromatography (COFRADIAC) [66].

Applications of this transformation approach in online 2D-LC have thus far been quite limited. Gilroy and Eakin demonstrated the use of an inline reduction step as part of an online heartcutting 2D-LC separation of antibody-drug conjugate (ADC) proteins [67]. Proteins were first separated by hydrophobic interaction (HIC), reduced using dithiothreitol (DTT) during the transfer of ^{1}D effluent to the ^{2}D column, and then the reduced ADC subunits were further separated by RP prior to MS detection. More recently there has been tremendous interest in this type of approach from researchers who sought to characterize variants of proteins at the amino acid level as a means of better understanding the dependence of protein function on amino acid sequence. For example, Bathke *et al.* have described the use of inline tryptic digestion of proteins following ^{1}D separation of intact proteins by mechanisms such as ion-exchange chromatography. After the proteins have been digested to peptides in the interface between dimensions, they are transferred to a ^{2}D RP column where they can be separated and ultimately sequenced by tandem mass spectrometery [8, 68].

In a completely different approach, Pirok *et al.* used aqueous hydrodynamic chromatography to first separate nanoparticles by size, transformed the nanoparticles to constituent polymers by dilution with a suitable diluent (tetrahydrofuran) in the 2D-LC interface, and finally separated the polymers in a ^{2}D separation based on size-exclusion (SEC) (see Figure 4.28) [50]. As this application shows, this LC-transformation-LC approach is amenable to many possible modes of transformation of analytes in the interface between two dimensions of separation. Readers interested in this concept are also referred to Section 10.2.3, which deals with transformation of protein analytes (e.g., digestion, reduction) between separation dimensions.

4.4.12   INJECTION LOOP MATCHING

Occasionally questions are raised about the importance of matching the volumes of loops used in the interface of a 2D-LC system. There are very few applications where the absolute volumes of the loops are important. Quantitation is usually achieved by comparison of peak area obtained for an unknown sample to that obtained for a reference standard. As long as the loops behave the same way for the standard as they do for an unknown sample, then the absolute volume of the loop should not affect quantitative accuracy and precision. The only study we are aware of that involved measurement of the reproducibility of loop volume in a commercially available system for mLC-LC was described by Iguiniz *et al.*, though the system was used under sLC×SFC conditions [69]. The authors reported that replicate injections from one set of five loops (the system was equipped with two sets of 10 μL loops) was more variable (9.5% RSD compared to 3.8%) than the other, but that the RSD below 10% was considered acceptable considering that the mLC-LC system had been adapted for sLC-SFC.

4.4.13   INJECTION LOOP FILLING

When using open tubular loops in the interface of a 2D-LC system, questions about the extent to which the loops should be filled – and in which direction – while sampling the ^{1}D separation are important during method development, and for the performance of the method. Qualitatively speaking, the loops should not be overfilled such that analytes of interest are lost from the loop; on the other hand, significantly underfilling the loop provides some safety margin to avoid analyte loss, but this usually requires a significant increase in the volume of the loop itself, which introduces other problems such as an increase in the gradient delay volume between the interface and the ^{2}D column. In this section we will summarize the primary topics to consider in this area. Although significant strides have been made developing a thorough understanding of the primary factors in play, more research is needed to deepen this understanding.

4.4.13.1   Analyte Losses Due to Breakthrough

When open tubular loops are used in the interface, there is the potential for analytes of interest to be lost during the sampling step. This is illustrated in Figure 4.29, where three stages of loop filling and emptying are shown. In the first stage, the sampling loop is full of mobile phase, and the analyte of interest approaches the loop from the outlet of the ^{1}D column. In the second stage, analyte flows into the loop. Analytes in the center of the tube move faster than the average velocity of the solvent due to the parabolic flow profile, and at some point, the "nose" of the parabola containing analyte will exit the loop and enter the waste line. Finally, when the interface valve is switched, any analyte material that exited the loop before the valve was switched is lost to the waste and will not appear in the 2D chromatogram. We refer to this as analyte "breakthrough" from the loop, which obviously can impact the quantitative accuracy of the method. The practical question, then, is – how much can I fill the loop before this breakthrough begins?

In the literature, a loop filling level of 80% is a commonly suggested guideline [70], though a few groups have suggested a more conservative level of 50% [71, 72]. However, to the best of our knowledge neither guideline has been studied thoroughly. This situation motivated us to do

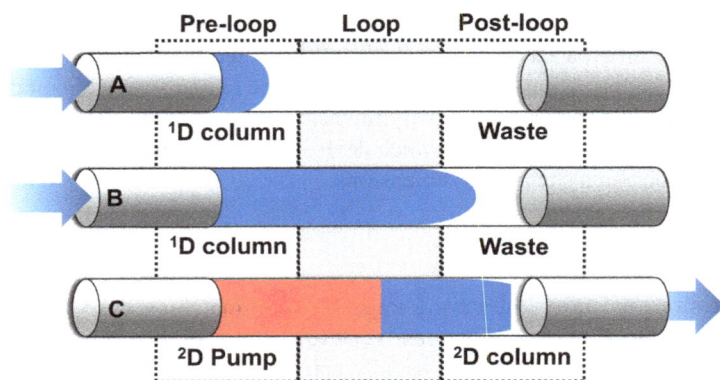

FIGURE 4.29 Illustration of analyte breakthrough from open tubular sampling loops due to the parabolic flow profile that develops under laminar flow conditions.

experiments and computational fluid dynamics simulations to more fully characterize what happens under conditions that are relevant to 2D-LC [70]. The most salient results from this study are shown in Figure 4.30. These trends are composed entirely of simulation results, but they were verified through selected experimental measurements. The x-axis in this plot is a dimensionless time axis (t^*) that represents the ratio of the time it takes to displace the liquid in the loop at a given flow rate (V_{loop}/F_{fill}) to the time associated with radial transport in the loop by diffusion (D_{mol}/R_{loop}^2). Here, V_{loop}, F_{fill}, D_{mol}, and R_{loop} are the loop volume, flow rate of solvent carrying analyte into the loop, diffusion coefficient of the analyte in the solvent, and the loop radius, respectively. When all other parameters are held constant (so the rate of radial transport is constant) and flow rate is increased, it takes less time to fill the loop with sample, allowing less time for radial transport. These conditions are reflected in smaller t^* values, and the left side of the plot. The y-axis represents the extent to which the loop is filled (V_{fill}/V_{loop}), and the blue lines indicate combinations of V_{fill}/V_{loop} and t^* that correspond to exactly 1% loss of analyte from the loop due to breakthrough. Conditions corresponding to combinations of (V_{fill}/V_{loop}) and t^* that lie above and to the left of these lines will result in analyte loss greater than 1%, and points that lie below and to the right of the lines will result in losses smaller than 1%. The fact that there are four different lines reflects that the fact that the amount of analyte that is lost depends on which part of a ^{1}D peak (or what type of concentration distribution) is sampled.

Discussion of a few examples helps one appreciate how these simulation results can be applied in practice. Consider first the dashed vertical line at $t^* \sim 1.8$, which corresponds to the conditions shown above the right-most purple arrow. This line intersects the blue line marked by crosses (+) at a V_{fill}/V_{loop} value near unity. This particular line is relevant to the case where an entire Gaussian peak is sampled (t_r +/- 3σ), where the analyte concentrations near the sampling boundaries are very low. This situation is very forgiving, and the loop can be filled nearly completely without losing much analyte. However, if we follow the t^*=1.8 line down to the blue line marked by x's, we see that it intersects at a V_{fill}/V_{loop} value of about 0.8. If the loop is filled more than 80%, greater than 1% of the analyte mass will be lost. This is because the leading edge of the ^{1}D peak being sampled contains a high concentration of analyte (i.e., the center of the Gaussian). Now, if we consider a different set of conditions (D_{mol}, V_{loop}, etc.) we see that the extent to which we can fill the loop is quite different. Consider the case indicated by the left vertical dashed line, which corresponds to conditions that are the same as the previous example, but with D_{mol} equal to that of a large protein. Now, under the same conditions, t^* is just 0.1, and the simulations predict that the loop can only be filled 55% without

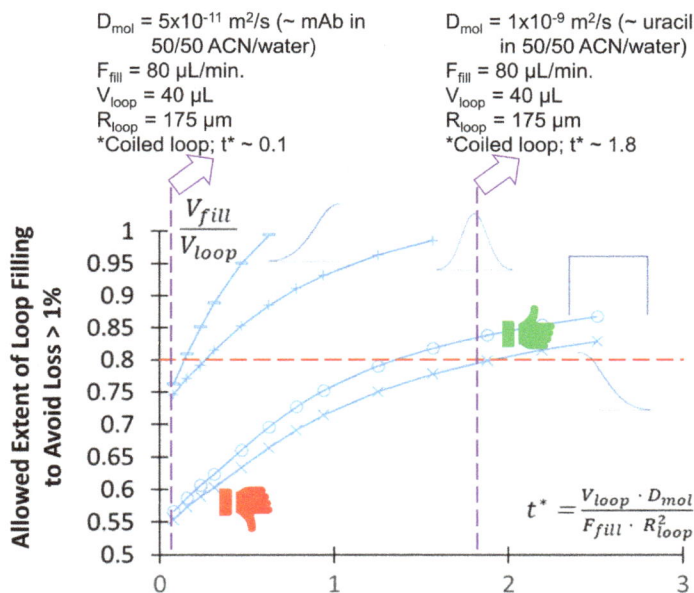

$D_{mol} = 5 \times 10^{-11}$ m²/s (~ mAb in 50/50 ACN/water)
$F_{fill} = 80$ µL/min.
$V_{loop} = 40$ µL
$R_{loop} = 175$ µm
Coiled loop; t ~ 0.1

$D_{mol} = 1 \times 10^{-9}$ m²/s (~ uracil in 50/50 ACN/water)
$F_{fill} = 80$ µL/min.
$V_{loop} = 40$ µL
$R_{loop} = 175$ µm
Coiled loop; t ~ 1.8

$$t^* = \frac{V_{loop} \cdot D_{mol}}{F_{fill} \cdot R_{loop}^2}$$

FIGURE 4.30 Curves showing the extent to which a sampling loop can be filled (V_{fill}/V_{loop}) without losing more than 1% of analyte mass from a fraction of ¹D effluent, as a function of the dimensionless loop filling time, t^*.

Source: Adapted from [70].

losing more than 1% of the analyte from the loop when sampling the peak such that the leading edge of the fraction is in the middle of the peak. This is obviously a very different outcome, which will have a significant impact on method development.

4.4.13.2 First-In-First-Out vs. First-In-Last-Out Modes of Operation

In addition to deciding how much to fill loops during sampling, one must decide *how* to fill the loops – that is, whether the loop will be filled in the same direction that it is emptied, or in the opposite direction. The two options are illustrated in Figure 4.31, and referred to as First-In-Last-Out (FILO; also referred to as "countercurrent") and First-In-First-Out (FIFO; also known as "concurrent"). Since movement of the collected fraction of ¹D effluent through the loop will lead to dispersion (broadening) and mixing of the fraction contents with other solvent inside the loop, the FILO mode will lead to less dispersion and mixing, and the FIFO mode will lead to more dispersion and mixing. Sometimes more dispersion of the collected fraction is desirable, and sometimes it is not. For example, it is desirable in cases where mixing of the collected fraction with the contents of the loop prior to sampling leads to a mixture that will promote focusing at the inlet of the ²D column, and thus narrow ²D peaks. This could occur, for example, if the loop were filled with water before collecting a fraction of ¹D effluent rich in organic solvent, prior to injecting that fraction into a ²D column under RP conditions. One example of a situation where dispersion of the fraction within the loop is not desirable is when SEC is used in the second dimension, because there is no opportunity for analyte focusing within the context of a SEC separation. A comparison of the ²D separation performance in this context when using either FILO or FIFO is shown in Figure 4.32. Here we see that the ²D peak obtained when using FILO is much narrower than when FIFO is used. Very recently, Moussa *et al.* studied analyte dispersion within the loop under FILO conditions using experiments

FIFO   FILO

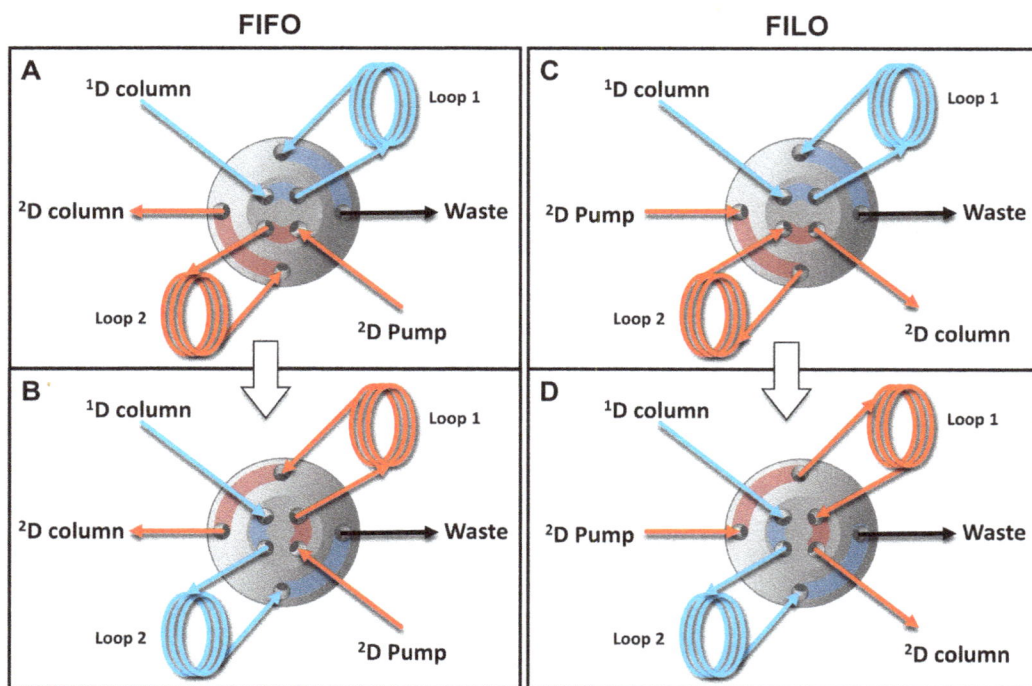

FIGURE 4.31 Illustration of the difference between use of open tubular sampling loops in the FIFO and FILO modes.

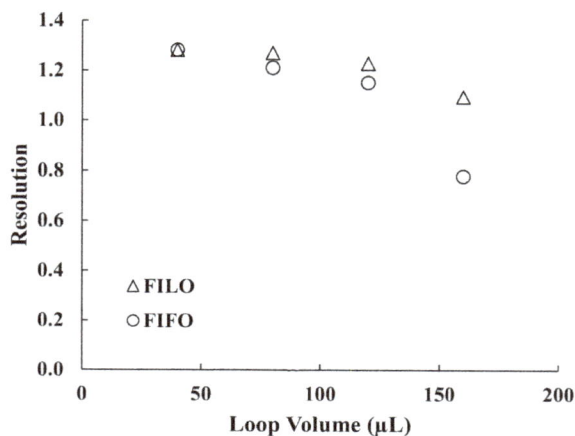

FIGURE 4.32 Comparison of resolution of mAb monomer and high molecular weight species peaks obtained in a LC-LC experiment involving SEC in the second dimension, and a sampling loop used in either FILO or FIFO mode. The dimensions of the ^{2}D SEC column were 150 mm x 7.8 mm i.d.

Source: Adapted from [75].

and computational fluid dynamics simulations [73]. Previously, Weatherbee *et al.* had studied injection profiles observed in the FIFO, and built a numerical model to predict these profiles over a range of loop volumes and mobile phase flow rates [74]. More research is needed to better understand the implications of using the FIFO and FILO modes.

4.4.13.3 Partial Filling

Usually, the open tubular loops of a 2D-LC setup like that shown in Figure 4.3 are used such that they are filled, or nearly filled, with ^{1}D effluent before injecting that material into the ^{2}D column. However, the idea of severely underfilling them (i.e., less than 50% fill) does have potential benefits. If the loop is filled with weak solvent prior to flowing ^{1}D effluent into it, and the loop is used in the FIFO configuration, the content of the loop prior to filling can act as a diluent for the ^{1}D effluent. Under these conditions the extent of dilution depends upon both the degree of underfilling, and the extent of mixing of the loop contents before it reaches in the ^{2}D column inlet. This benefit has been demonstrated experimentally by Regalado *et al.* [76], and briefly in our own work [77]. This effect is real and can be exploited in the context of real applications. However, at this time there is no way to predict *a priori* what extent of underfilling is optimal, and how much benefit (i.e., in ^{2}D resolution and sensitivity) could be realized in the best case; such details must be determined empirically on a case-by-case basis.

4.4.14 Use of Parallel ^{2}D Columns

Given the importance of high throughput in the second dimension of 2D-LC systems, particularly in LC×LC, it is reasonable to consider the benefits of using multiple ^{2}D columns in parallel. Fairchild *et al.* discussed the benefits and drawbacks of using parallel ^{2}D columns in online multi-dimensional separations from a theoretical perspective [78]. To the best of our knowledge, such an approach was first demonstrated experimentally by Opiteck *et al.* in 1998 [79], and then by Wagner *et al.* in the context of proteomics type applications using either two [80] or four [81] ^{2}D columns for 2D-LC separations of peptides and proteins. In the first of Wagner's studies, two full binary pumps were used to drive separations through each of the ^{2}D columns, in addition to the third pump used for the 1D separation. In the second case, a fourth isocratic pump was used to re-equilibrate one column in each pair of columns used in parallel in the second dimension. A few years later, Cacciola *et al.* demonstrated the use of two ^{2}D columns operated in parallel for 2D-LC separations of phenolic antioxidants [49]. Around 2010, François *et al.* first used parallel ^{2}D columns for the characterization of lemon oil extract and other small molecule analytes [82], and then for tryptic digests of proteins [12]. More recently, a system involving as many as 12 ^{2}D columns in parallel has been demonstrated for LC×LC separations of intact proteins [83]. While there is no question that the use of parallel ^{2}D columns can increase the peak capacity that can be produced per unit of analysis time [78], it is also true that such systems are more complicated than those that rely on a single ^{2}D column, and introduce other challenges such as ^{2}D retention shifts due to column-to-column variation in retention time. It is likely that these added complexities have impeded more widespread use of these systems, and as such we will not devote more space to their discussion here.

4.5 SECOND DIMENSION ELUTION SCHEMES

One of the most significant developments in 2D-LC methodology in the last decade has been the demonstration and widespread use of dynamic elution schemes in the second dimension of 2D separations. This has been especially important for LC×LC separations, where gradient elution is far more common in the second dimension than is isocratic elution. In the early years of LC×LC, the same ^{2}D elution pattern was always used over the entire set of all ^{2}D separations. This is illustrated

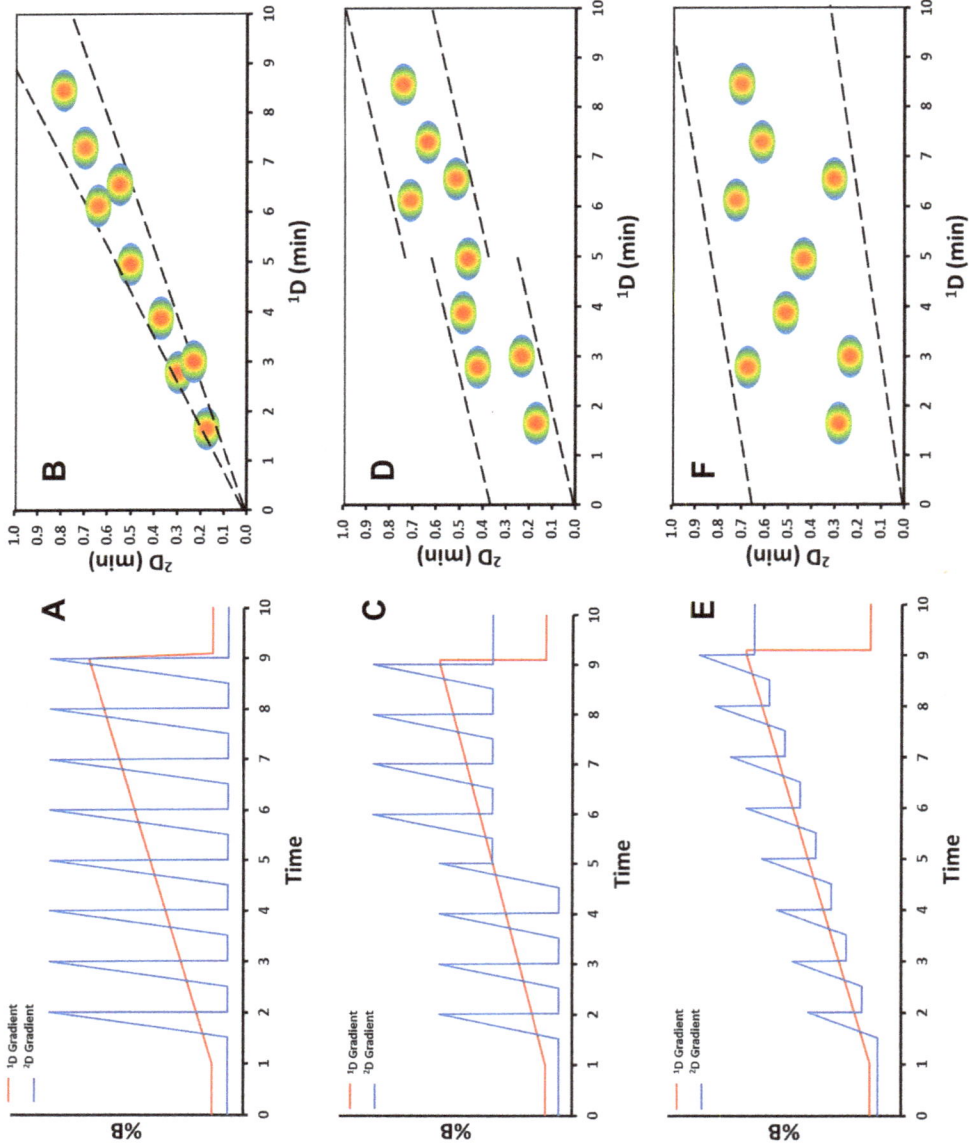

FIGURE 4.33 Illustration of different elution schemes used in the second dimension of LC×LC separations (A, C, E), and corresponding hypothetical chromatograms that emphasize the impact of the ^{2}D elution pattern on the distribution of peaks in the 2D separation space (B, D, F). Such schemes can be used for sLC×LC separations as well.

in Figure 4.33A, where each ^{2}D gradient starts and ends at the same %B (e.g., start at 10% and end at 90%B), independent of the ^{1}D time. In 2009, Bedani *et al.* demonstrated the value of using ^{2}D gradients whose starting and ending %B values evolved over the course of the 2D-LC experiment [84]. This type of scheme, which has become known as a continuously shifting gradient, is illustrated in Figure 4.33E. Comparing the corresponding hypothetical chromatograms in panels B and F we see that the value of the shifting gradient is that it provides the possibility of significantly spreading peaks out in the second dimension that would otherwise be clustered along the diagonal of the 2D plot. Over the next few years, the groups of Jandera, Schmitz, and Mondello continued to demonstrate the value of this new approach, and also explored variations on the theme [85–88]. Figure 4.33C illustrates what Jandera *et al.* referred to as a segmented approach to the ^{2}D elution scheme. Here, within each segment of ^{1}D time the ^{2}D elution pattern is the same, but the patterns are different for each segment of ^{1}D time. This approach has the advantage of more consistent ^{2}D retention times for compounds that are split across multiple fractions of ^{1}D effluent; however, we do not see this approach being used very often in practice today.

An example of the positive impact of the shifted gradient approach on a contemporary LC×LC separation of peptides is shown in Figure 4.34. Here we see that when the same "full" gradient is used consistently across the entire 2D separation, the majority of the peptide peaks only occupy a small fraction of the available separation space located near the diagonal in the 2D plot. However, with use of a shifted gradient that continuously evolves over the course of the ^{1}D separation, the area occupied by peaks is expanded dramatically, and practically fills the entire available space (note that in this case the first 10 s of each ^{2}D separation is committed to the sample injection step).

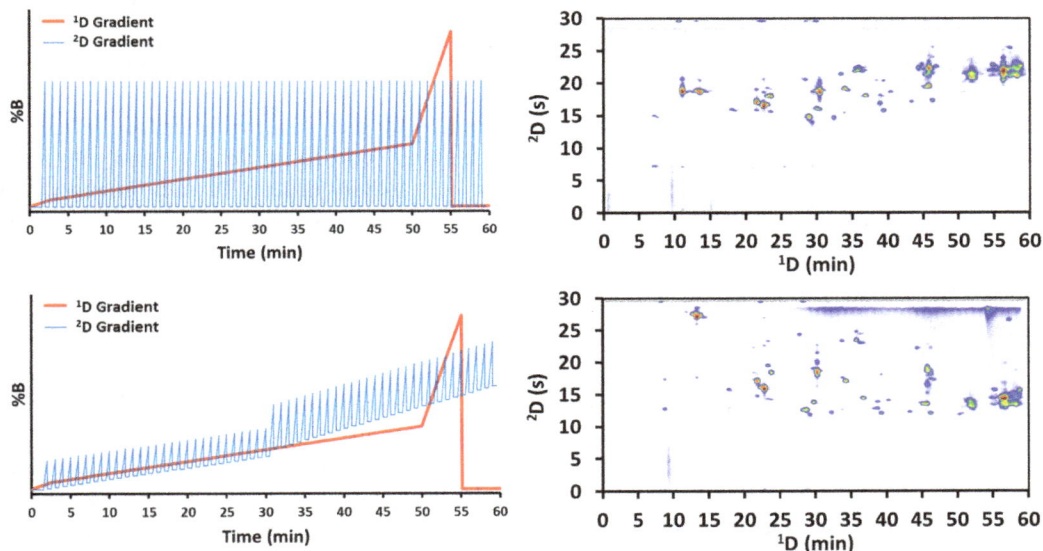

FIGURE 4.34 Unpublished results. Comparison of 2D-LC separations of a tryptic digest of a monoclonal antibody without (top row) with (bottom row) the use of shifted gradients in the second dimension. Chromatographic conditions are similar to those published in ref. [6]. Note that the gradients (left) are shown for a 60-s modulation time for clarity, whereas the actual separations shown (right) used a 30-s modulation time.

4.6 EXTRA-COLUMN PEAK DISPERSION

Dispersion (broadening) of peaks outside of the column (i.e., extra-column) in 2D-LC is gener-ally an important consideration just as it is in 1D-LC. Chromatographic resolution is expensive in 2D-LC just as it is in 1D-LC, and we should take all reasonable steps to avoid losing resolution to peak dispersion that happens outside of the LC columns. Many of the principles regarding extra-column dispersion applied in 1D-LC also apply in 2D-LC. For example, details related to injection of the sample into the ^{1}D column are important as they are for injection into the column in 1D-LC. However, there are some characteristics of 2D-LC instrumentation and separations that are espe-cially important to consider when making choices with the aim of avoiding serious extra-column dispersion.

4.6.1 DISPERSION DUE TO INJECTION INTO THE ^{2}D COLUMN

Dispersion of peaks as they enter the ^{2}D column deserves special attention both because relatively large volumes of ^{1}D effluent are injected into the ^{2}D column, and because the mobile phase compos-ition that is ideal for separation of sample components in the first dimension is often not the ideal sample solvent for the ^{2}D separation. This issue, and instrument-oriented solutions for mitigating the problem were discussed in Section 4.4, and are discussed in most of the following chapters in the context of different applications of 2D-LC.

4.6.2 FAST ^{2}D SEPARATIONS PRODUCE NARROW PEAKS

As is discussed in Chapter 3, fast ^{2}D separations are needed to avoid undersampling, particularly in LC×LC separations. In these situations, it is common to use short (20–30 mm), narrow (2.1 mm i.d.) columns. If a 20 mm x 2.1 mm i.d. column with 5,000 plates is used under gradient conditions where the local retention factor of the analyte at the column exit is 2, the volumetric variance of a peak exiting the column will be about 2 μL^2. This is extremely small for analytical scale separations in general, but not uncommon for LC×LC separations. Peaks with such small variances are extremely susceptible to dispersion in the tubing connecting the column outlet to the detector, and within the detector itself. Passing these peaks through a 120 μm i.d. capillary of just 200 mm in length will result in an increase in peak width corresponding to an apparent loss of 50% of the column effi-ciency [89]. Splitting the flow after the ^{2}D column to accommodate requirements of the ^{2}D detector makes this challenge even more serious, as discussed below in Section 4.7.1.

4.6.3 TURBULENT FLOW IN ^{2}D CONNECTING CAPILLARIES

As discussed in the preceding section, it is important to pay attention to the dimensions of connecting tubing to avoid serious extra-column dispersion, particularly in the second dimension where peaks can be very narrow. However, we must also be careful to avoid conditions where turbulent flow can develop, because this in turn can lead to much higher pressure drops across the connecting tubing than would be expected under conditions where laminar flow prevails. If this happens, it reduces the fraction of the pressure deliverable by the pump that can actually be used to drive the mobile phase through the column. Several variables determine when turbulent flow will develop, and a detailed discussion of them is out of the present scope. Generally speaking, beyond a crit-ical Reynold's number of about 1,500, turbulent flow conditions are more likely at high flow rate of mobile phases with low viscosities (e.g., due to solvent properties, or high temperatures). For example, the Reynold's number is about 2,400 when pumping acetonitrile at room temperature through a 70 μm i.d. capillary at 3.0 mL/min. Under these conditions, the pressure drop across the tubing is roughly double that expected when the flow is laminar (i.e., 140 bar compared to 70 bar). Readers interested in more detail are referred to our prior work on this topic [90].

4.7 DETECTION

A common question raised by those new to 2D-LC is whether or not a detector is needed in both dimensions of a 2D-LC system, particularly for LC×LC separations. In principle the answer is no, because by definition some fraction of all analytes that elute from the ^{1}D column will be injected into the ^{2}D column at some point, and thus will eventually be detected at the outlet of the ^{2}D column. However, we strongly recommend using detectors at the outlets of both the ^{1}D and ^{2}D columns. The second one is necessary, of course, but we find that being able to visualize what is happening at the outlet of the ^{1}D column is highly valuable, especially for troubleshooting. For example, a UV-Vis absorbance detector can not only detect analytes of interest, but also monitor pump performance in a way that is more direct than trying to diagnose problems based on information from the ^{2}D detector alone. Many mobile phase components absorb at least some UV light, and thus changes in the mobile phase composition in a gradient can be visualized this way. We have also explored the possibility that information from the ^{1}D detector can be used to enhance interpretation of 2D chromatograms produced using data from the ^{2}D detector [91].

Most of the detectors used for 1D-LC have been also used for 2D-LC at some point. The two most significant considerations when thinking about use of a particular detector in a 2D-LC method are their baseline characteristics, and data acquisition rate. The two most widely used detectors for 2D-LC are UV-Vis absorbance detectors, and mass spectrometers, and each of them is affected by both of these factors. In the case of UV-Vis detection, fast changes in the refractive index of the mobile phase (e.g., as in organic/water gradients in RPLC) cause larger baseline disturbances than those encountered in 1D-LC [92]. This is an especially important consideration in LC×LC where ^{2}D separations tend to be much faster than those used in 1D-LC, where gradient slopes can be quite large. This arguably is a challenge that must be addressed at the data analysis stage and not so much of a problem of the detector *per se*, but users should be aware that this is a new challenge they might encounter when developing 2D-LC methods with certain detectors. As was discussed in Section 4.6.2, some conditions used in the second dimension of 2D-LC methods can lead to 2D peaks that are exceptionally narrow in both volume and time units. In the time domain this can lead to a mismatch between the narrow widths of the peaks (e.g., < 200 msec at half-height) and the excessively slow detector data acquisition rate. In this case the slow acquisition rate can lead to apparent dispersion of the peaks as they exit the ^{2}D column. This is not so much of a problem with modern UV-Vis detectors based on diode-array technology, but it can be quite challenging for mass spectrometers used for LC-MS. Even with time-of-light (TOF) type detectors – which are some of the fastest in the industry – this is still a serious problem because of the compromise between S/N and acquisition rate. In our work we have observed that acquiring TOF mass spectra at rates above 10 spectra/s leads to serious compromises in detection limits and S/N.

4.7.1 POST-COLUMN FLOW SPLITTING

The combination of narrow peaks (in volume units) and high flow rates encountered in the second dimension of some 2D-LC separations, especially when using flow limited detectors such as MS, leads to unusual situations not normally encountered in 1D-LC. From the point of view of the 2D-LC separation itself, particularly in the case of LC×LC, high performing separations favor fast ^{2}D separations and flow rates that can be much higher than those tolerated by most interfaces designed for modern LC-MS (e.g., 2.5 mL/min. desired for the LC×LC compared to 0.5 mL/min. desired for the MS). This conflict is most commonly resolved by splitting off a portion of the flow as it exits the ^{2}D column using a tee-junction type passive splitter, before going into the MS. We now know, however, that this scenario can lead to serious extra-column dispersion of ^{2}D peaks – not due to the splitting process *per se*, but due to the splitting of an already small peak volume leading to an even smaller peak volume that is then highly susceptible to dispersion in connecting capillaries and the detectors themselves [93]. Figure 4.35 shows a snapshot of results obtained from our study aimed at quantifying the effect

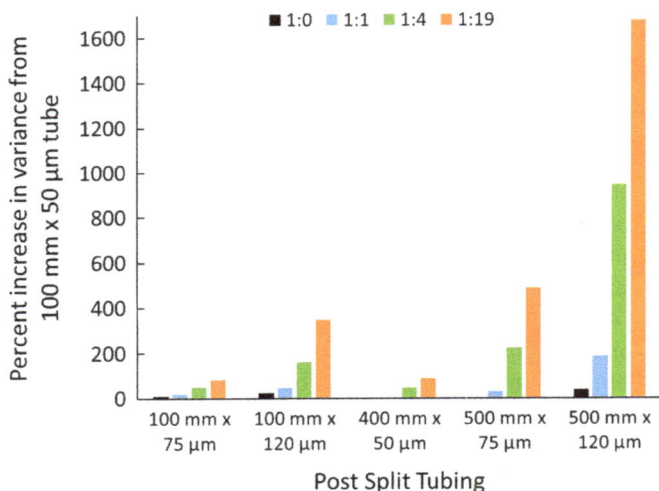

FIGURE 4.35   Effect of different dimensions of post-split tubing used with different split ratios (1:0, 1:1, 1:4, and 1:19) on percent increase in peak variance relative to the variance obtained from a small post-split tube of 100 mm x 50 μm i.d. (negligible contribution to the variance under these conditions).

Source: Adapted from [93].

of post-column flow splitting on the dispersion of narrow peaks like those often encountered in the second dimension of 2D-LC systems. For these experiments the column (30 mm x 2.1 mm i.d.) and conditions were chosen such that the variance of a peak entering the split point was relatively small at about 6.7 μL^2. This means that the peak volume as measured by the standard deviation of the peak distribution in volume units was about 2.6 μL. As discussed in the paper describing this work, a 1:1 post-column split results in a peak with a standard deviation of just 1.3 μL. Such peak volumes are rarely encountered in 1D-LC with "normal" analytical scale columns, and peaks with low volumes are extremely susceptible to dispersion in tubing and connections after the split point, and the detector itself. The results show that with a 1:1 split and a relatively short 100 mm x 120 μm i.d. tube between the split point and the detector leads to a roughly 150% increase in the peak variance! Post-column flow splitting is most commonly used in 2D-LC in conjunction with MS detection, where it is often difficult to locate the MS inlet very near the ^{2}D column outlet. This in turn leads to the use of long connecting tubing, which can lead to disastrous peak dispersion as the figure shows. The easiest ways to minimize this problem are to: 1) locate the split point as close to the detector as possible; and 2) use a narrow tube (e.g., 50 or 75 μm i.d.) to connect the outlet of the split point to the detector.

4.8   CLOSING REMARKS

For the first few decades of the development of online 2D-LC most instruments were "home-made", assembled by researchers using components repurposed from 1D-LC instrumentation. Currently, we clearly have moved beyond this stage and are in an era where instrument manufacturers are invested in development of both instrument hardware and software that is designed specifically for the purpose of multi-dimensional LC. Although great strides have already been made in this regard, the discussion in this chapter points to a variety of limitations of existing hardware that can be viewed as opportunities for further development. Moreover, the variety of implementations of multi-dimensional separation is expanding, not contracting, as indicated by the recent proliferation of online multi-dimensional systems for automated protein characterization (see Chapter 10 for more detail on this topic) [94]. It will be exciting to watch these and other technologies develop and mature in the years to come, and to see which approaches gain the most traction for routine use in industry.

REFERENCES

[1] J.W. Dolan, How much can I inject? Part I: Injecting in mobile phase, LCGC North Am. 32 (2014) 780–785.

[2] J. Sternberg, Extracolumn contributions to chromatographic band broadening, in: Advances in Chromatography, Marcel Dekker, New York, 1966: pp. 205–270.

[3] D.R. Stoll, K. Broeckhoven, Where has my efficiency gone? Impacts of extracolumn peak broadening on performance, Part II: Sample injection, LCGC North Am. 39 (2021) 208–213.

[4] D.R. Stoll, Mixing and mixers in liquid chromatography – Why, when, and how much? Part II, Injections, LCGC North Am. 36 (2018) 796–801.

[5] E.S. Talus, K.E. Witt, D.R. Stoll, Effect of pressure pulses at the interface valve on the stability of second dimension columns in online comprehensive two-dimensional liquid chromatography, J. Chromatogr. A. 1378 (2015) 50–57. https://doi.org/10.1016/j.chroma.2014.12.019.

[6] D.R. Stoll, H.R. Lhotka, D.C. Harmes, B. Madigan, J.J. Hsiao, G.O. Staples, High resolution two-dimensional liquid chromatography coupled with mass spectrometry for robust and sensitive characterization of therapeutic antibodies at the peptide level, J. Chromatogr. B. 1134–1135 (2019) 121832. https://doi.org/10.1016/j.jchromb.2019.121832.

[7] J. Camperi, L. Dai, D. Guillarme, C. Stella, Development of a 3D-LC/MS workflow for fast, automated, and effective characterization of glycosylation patterns of biotherapeutic products, Anal. Chem. 92 (2020) 4357–4363. https://doi.org/10.1021/acs.analchem.9b05193.

[8] C.J. Gstöttner, D. Klemm, M. Haberger, A. Bathke, H. Wegele, C.H. Bell, R. Kopf, Fast and automated characterization of antibody variants with 4D-HPLC/MS, Anal. Chem. 90 (2017) 2119–2125. https://doi.org/10.1021/acs.analchem.7b04372.

[9] S.W. Simpkins, J.W. Bedard, S.R. Groskreutz, M.M. Swenson, T.E. Liskutin, D.R. Stoll, Targeted three-dimensional liquid chromatography: A versatile tool for quantitative trace analysis in complex matrices, J. Chromatogr. A. 1217 (2010) 7648–7660. https://doi.org/10.1016/j.chroma.2010.09.023.

[10] E.D. Larson, S.R. Groskreutz, D.C. Harmes, I. Gibbs-Hall, S.P. Trudo, R.C. Allen, S.C. Rutan, D.R. Stoll, Development of selective comprehensive two-dimensional liquid chromatography with parallel first-dimension sampling and second-dimension separation – Application to the quantitative analysis of furanocoumarins in apiaceious vegetables, Anal. Bioanal. Chem. 405 (2013) 4639–4653. https://doi.org/10.1007/s00216-013-6758-8.

[11] D. Stoll, J. Cohen, P. Carr, Fast, comprehensive online two-dimensional high performance liquid chromatography through the use of high temperature ultra-fast gradient elution reversed-phase liquid chromatography, J. Chromatogr. A. 1122 (2006) 123–137. https://doi.org/10.1016/j.chroma.2006.04.058.

[12] I. François, D. Cabooter, K. Sandra, F. Lynen, G. Desmet, P. Sandra, Tryptic digest analysis by comprehensive reversed phase×two reversed phase liquid chromatography (RP-LC×2RP-LC) at different pH's, J. Sep. Sci. 32 (2009) 1137–1144. https://doi.org/10.1002/jssc.200800578.

[13] S.R. Groskreutz, M.M. Swenson, L.B. Secor, D.R. Stoll, Selective comprehensive multi-dimensional separation for resolution enhancement in high performance liquid chromatography, Part I–Principles and instrumentation, J. Chromatogr. A. 1228 (2012) 31–40. https://doi.org/10.1016/j.chroma.2011.06.035.

[14] K. Hamase, A. Morikawa, T. Ohgusu, W. Lindner, K. Zaitsu, Comprehensive analysis of branched aliphatic d-amino acids in mammals using an integrated multi-loop two-dimensional column-switching high-performance liquid chromatographic system combining reversed-phase and enantioselective columns, J. Chromatogr. A. 1143 (2007) 105–111. https://doi.org/10.1016/j.chroma.2006.12.078.

[15] K. Zhang, Y. Li, M. Tsang, N.P. Chetwyn, Analysis of pharmaceutical impurities using multi-heartcutting 2D LC coupled with UV-charged aerosol MS detection: Liquid chromatography, J. Sep. Sci. 36 (2013) 2986–2992. https://doi.org/10.1002/jssc.201300493.

[16] H. Wang, H.R. Lhotka, R. Bennett, M. Potapenko, C.J. Pickens, B.F. Mann, I.A.H. Ahmad, E.L. Regalado, Introducing online multicolumn two-dimensional liquid chromatography screening for facile selection of stationary and mobile phase conditions in both dimensions, J. Chromatogr. A. (2020) 460895. https://doi.org/10.1016/j.chroma.2020.460895.

[17] S.R. Groskreutz, M.M. Swenson, L.B. Secor, D.R. Stoll, Selective comprehensive multi-dimensional separation for resolution enhancement in high performance liquid chromatography, Part II – Applications, J. Chromatogr. A. 1228 (2012) 41–50. https://doi.org/10.1016/j.chroma.2011.06.038.

[18] C.J. Venkatramani, Y. Zelechonok, An automated orthogonal two-dimensional liquid chromatograph, Anal. Chem. 75 (2003) 3484–3494. https://doi.org/10.1021/ac030075w.

[19] P.J. Schoenmakers, G. Vivó-Truyols, W.M.C. Decrop, A protocol for designing comprehensive two-dimensional liquid chromatography separation systems, J. Chromatogr. A. 1120 (2006) 282–290. https://doi.org/10.1016/j.chroma.2005.11.039.

[20] S. Chapel, S. Heinisch, Strategies to circumvent the solvent strength mismatch problem in online comprehensive two-dimensional liquid chromatography, J. Sep. Sci. (2021). https://doi.org/10.1002/jssc.202100534.

[21] D.R. Stoll, E.S. Talus, D.C. Harmes, K. Zhang, Evaluation of detection sensitivity in comprehensive two-dimensional liquid chromatography separations of an active pharmaceutical ingredient and its degradants, Anal. Bioanal. Chem. 407 (2014) 265–277. https://doi.org/10.1007/s00216-014-8036-9.

[22] H. Gu, Y. Huang, M. Filgueira, P.W. Carr, Effect of first dimension phase selectivity in online comprehensive two dimensional liquid chromatography (LC×LC), J. Chromatogr. A. 1218 (2011) 6675–6687. https://doi.org/10.1016/j.chroma.2011.07.063.

[23] D.R. Stoll, X. Li, X. Wang, P.W. Carr, S.E.G. Porter, S.C. Rutan, Fast, comprehensive two-dimensional liquid chromatography, J. Chromatogr., A. 1168 (2007) 3–43. https://doi.org/10.1016/j.chroma.2007.08.054.

[24] S. Chapel, F. Rouvière, S. Heinisch, Pushing the limits of resolving power and analysis time in on-line comprehensive hydrophilic interaction x reversed phase liquid chromatography for the analysis of complex peptide samples, J. Chromatogr. A. (2019). https://doi.org/10.1016/j.chroma.2019.460753.

[25] S. Chapel, S. Heinisch, Strategies to circumvent the solvent strength mismatch problem in online comprehensive two-dimensional liquid chromatography, J. Sep. Sci. (2021). https://doi.org/10.1002/jssc.202100534.

[26] A. Sweeney, R. Shalliker, Development of a two-dimensional liquid chromatography system with trapping and sample enrichment capabilities, J. Chromatogr., A. 968 (2002) 41–52. https://doi.org/10.1016/S0021-9673(02)00782-3.

[27] L.E. Niezen, B.B.P. Staal, C. Lang, B.W.J. Pirok, P.J. Schoenmakers, Thermal modulation to enhance two-dimensional liquid chromatography separations of polymers, J. Chromatogr. A. 1653 (2021) 462429. https://doi.org/10.1016/j.chroma.2021.462429.

[28] H. Tian, J. Xu, Y. Guan, Comprehensive two-dimensional liquid chromatography (NPLC×RPLC) with vacuum-evaporation interface, J. Sep. Sci. 31 (2008) 1677–1685. https://doi.org/10.1002/jssc.200700559.

[29] H. Tian, J. Xu, Y. Xu, Y. Guan, Multidimensional liquid chromatography system with an innovative solvent evaporation interface, J. Chromatogr., A. 1137 (2006) 42–48. https://doi.org/10.1016/j.chroma.2006.10.005.

[30] Y. Oda, N. Asakawa, T. Kajima, Y. Yoshida, T. Sato, On-line determination and resolution of verapamil enantiomers by high-performance liquid chromatography with column switching, J. Chromatogr. A. 541 (1991) 411–418. https://doi.org/10.1016/S0021-9673(01)96013-3.

[31] A.W. Moore, J.W. Jorgenson, Comprehensive three-dimensional separation of peptides using size exclusion chromatography/reversed phase liquid chromatography/optically gated capillary zone electrophoresis, Anal. Chem. 67 (1995) 3456–3463. https://doi.org/10.1021/ac00115a014.

[32] E. Rogatsky, V. Tomuta, H. Jayatillake, G. Cruikshank, L. Vele, D.T. Stein, Trace LC/MS quantitative analysis of polypeptide biomarkers: Impact of 1-D and 2-D chromatography on matrix effects and sensitivity, J. Sep. Sci. 30 (2007) 226–233. https://doi.org/10.1002/jssc.200600250.

[33] B. Koshel, R. Birdsall, W. Chen, Two-dimensional liquid chromatography coupled to mass spectrometry for impurity analysis of dye-conjugated oligonucleotides, J. Chromatogr. B. 1137 (2020) 121906. https://doi.org/10.1016/j.jchromb.2019.121906.

[34] O. Gron, Evaluation of Different Sampling Approaches in 2D Multi Heart-Cut Applications, San Diego, CA, USA, 2015. www.facebook.com/CosmoSciences/

[35] D.R. Stoll, K. Shoykhet, P. Petersson, S. Buckenmaier, Active solvent modulation: A valve-based approach to improve separation compatibility in two-dimensional liquid chromatography, Anal. Chem. 89 (2017) 9260–9267. https://doi.org/10.1021/acs.analchem.7b02046.

[36] Y. Chen, L. Montero, J. Luo, J. Li, O.J. Schmitz, Application of the new at-column dilution (ACD) modulator for the two-dimensional RP×HILIC analysis of Buddleja davidii, Anal. Bioanal. Chem. (2020). https://doi.org/10.1007/s00216-020-02392-3.

[37] Y. Chen, L. Montero, O.J. Schmitz, Advance in on-line two-dimensional liquid chromatography modulation technology, Trends Analyt. Chem. 120 (2019) 115647. https://doi.org/10.1016/j.trac.2019.115647.

[38] Y. Chen, J. Li, O.J. Schmitz, Development of an at-column dilution modulator for flexible and precise control of dilution factors to overcome mobile phase incompatibility in comprehensive two-dimensional liquid chromatography, Anal. Chem. 91 (2019) 10251–10257. https://doi.org/10.1021/acs.analchem.9b02391.

[39] T.J. Lauer, D. Stoll R., Effects of buffer capacity in reversed-phase liquid chromatography, Part I: Relationship between the sample and mobile-phase buffers, LCGC North Am. 38 (2020).

[40] T.J. Lauer, D. Stoll R., Effects of buffer capacity in reversed-phase liquid chromatography, Part II: Visualization of pH changes inside the column, LCGC North Am. 38 (2020).

[41] S. Chapel, F. Rouvière, V. Peppermans, G. Desmet, S. Heinisch, A comprehensive study on the phenomenon of total breakthrough in liquid chromatography, J. Chromatogr. A. 1653 (2021). https://doi.org/10.1016/j.chroma.2021.462399.

[42] F. Gritti, M. Gilar, J. Hill, Mismatch between sample diluent and eluent: Maintaining integrity of gradient peaks using in silico approaches, J. Chromatogr. A. 1608 (2019). https://doi.org/10.1016/j.chroma.2019.460414.

[43] S.C. Rutan, L.N. Jeong, P.W. Carr, D.R. Stoll, S.G. Weber, Closed form approximations to predict retention times and peak widths in gradient elution under conditions of sample volume overload and sample solvent mismatch, J. Chromatogr. A. 1653 (2021). https://doi.org/10.1016/j.chroma.2021.462376.

[44] V. Pepermans, S. Chapel, S. Heinisch, G. Desmet, Detailed numerical study of the peak shapes of neutral analytes injected at high solvent strength in short reversed-phase liquid chromatography columns and comparison with experimental observations, J. Chromatogr. A. 1643 (2021). https://doi.org/10.1016/j.chroma.2021.462078.

[45] B.W.J. Pirok, D. Stoll R., P.J. Schoenmakers, Recent developments in two-dimensional liquid chromatography – Fundamental improvements for practical applications, Anal. Chem. 91 (2019) 240–263. https://doi.org/10.1021/acs.analchem.8b04841.

[46] A. Holm , E. Storbråten, A. Mihailova, B. Karaszewski, E. Lundanes, T. Greibrokk, Combined solid-phase extraction and 2D LC–MS for characterization of the neuropeptides in rat-brain tissue, Anal. Bioanal. Chem. 382 (2005) 751–759. https://doi.org/10.1007/s00216-005-3146-z.

[47] S.R. Wilson, M. Jankowski, M. Pepaj, A. Mihailova, F. Boix, G. Vivo Truyols, E. Lundanes, T. Greibrokk, 2D LC separation and determination of bradykinin in rat muscle tissue dialysate with in-line SPE-HILIC-SPE-RP-MS, Chroma. 66 (2007) 469–474. https://doi.org/10.1365/s10337-007-0341-4.

[48] C.J. Venkatramani, A. Patel, Towards a comprehensive 2-D-LC–MS separation, J. Sep. Sci. 29 (2006) 510–518. https://doi.org/10.1002/jssc.200500341.

[49] F. Cacciola, P. Jandera, E. Blahová, L. Mondello, Development of different comprehensive two dimensional systems for the separation of phenolic antioxidants, J. Sep. Sci. 29 (2006) 2500–2513. https://doi.org/10.1002/jssc.200600213.

[50] B.W.J. Pirok, N. Abdulhussain, T. Aalbers, B. Wouters, R.A.H. Peters, P.J. Schoenmakers, Nanoparticle analysis by online comprehensive two-dimensional liquid chromatography combining hydrodynamic chromatography and size-exclusion chromatography with intermediate sample transformation, Anal. Chem. 89 (2017) 9167–9174. https://doi.org/10.1021/acs.analchem.7b01906.

[51] R.J. Vonk, A.F.G. Gargano, E. Davydova, H.L. Dekker, S. Eeltink, L.J. de Koning, P.J. Schoenmakers, Comprehensive two-dimensional liquid chromatography with stationary-phase-assisted modulation coupled to high-resolution mass spectrometry applied to proteome analysis of Saccharomyces cerevisiae, Anal. Chem. 87 (2015) 5387–5394. https://doi.org/10.1021/acs.analchem.5b00708.

[52] M. Verstraeten, G. Desmet, Signal enhancement by trapping in microscale liquid chromatography: Numerical modelling, J. Sep. Sci. 34 (2011) 2822–2832. https://doi.org/10.1002/jssc.201100430.

[53] E. Fornells, B. Barnett, M. Bailey, E.F. Hilder, R.A. Shellie, M.C. Breadmore, Evaporative membrane modulation for comprehensive two-dimensional liquid chromatography, Anal. Chim. Acta. 1000 (2018) 303–309. https://doi.org/10.1016/j.aca.2017.11.053.

[54] M.E. Creese, M.J. Creese, J.P. Foley, H.J. Cortes, E.F. Hilder, R.A. Shellie, M.C. Breadmore, Longitudinal on-column thermal modulation for comprehensive two-dimensional liquid chromatography, Anal. Chem. 89 (2017) 1123–1130. https://doi.org/10.1021/acs.analchem.6b03279.

[55] M. Verstraeten, M. Pursch, P. Eckerle, J. Luong, G. Desmet, Thermal modulation for multidimensional liquid chromatography separations using low-thermal-mass liquid chromatography (LC), Anal. Chem. 83 (2011) 7053–7060. https://doi.org/10.1021/ac201207t.

[56] H.C. van de Ven, A.F.G. Gargano, Sj. van der Wal, P.J. Schoenmakers, Switching solvent and enhancing analyte concentrations in small effluent fractions using in-column focusing, J. Chromatogr. A. 1427 (2016) 90–95. https://doi.org/10.1016/j.chroma.2015.11.082.

[57] S. Stephan, C. Jakob, J. Hippler, O.J. Schmitz, A novel four-dimensional analytical approach for analysis of complex samples, Anal. Bioanal. Chem. 408 (2016) 3751–3759. https://doi.org/10.1007/s00 216-016-9460-9.

[58] M.R. Filgueira, Y. Huang, K. Witt, C. Castells, P.W. Carr, Improving peak capacity in fast online comprehensive two-dimensional liquid chromatography with post-first-dimension flow splitting, Anal. Chem. 83 (2011) 9531–9539. https://doi.org/10.1021/ac202317m.

[59] F. Bedani, W.Th. Kok, H.-G. Janssen, A theoretical basis for parameter selection and instrument design in comprehensive size-exclusion chromatography×liquid chromatography, J. Chromatogr. A. 1133 (2006) 126–134. https://doi.org/10.1016/j.chroma.2006.08.048.

[60] A.M. Striegel, Longitudinal diffusion in size-exclusion chromatography: A stop-flow size-exclusion chromatography study, J. Chromatogr. A. 932 (2001) 21–31. https://doi.org/10.1016/ S0021-9673(01)01214-6.

[61] B.Q. Tran, E. Lundanes, T. Greibrokk, The influence of stop-flow on band broadening of peptides in micro-liquid chromatography, Chromatographia. 64 (2006) 1–5. https://doi.org/10.1365/s10 337-006-0820-z.

[62] K.M. Kalili, A. de Villiers, Systematic optimisation and evaluation of on-line, off-line and stop-flow comprehensive hydrophilic interaction chromatography×reversed phase liquid chromatographic analysis of procyanidins, Part I: Theoretical considerations, J. Chromatogr. A. 1289 (2013) 58–68. https:// doi.org/10.1016/j.chroma.2013.03.008.

[63] J. Xu, D. Sun-Waterhouse, C. Qiu, M. Zhao, B. Sun, L. Lin, G. Su, Additional band broadening of peptides in the first size-exclusion chromatographic dimension of an automated stop-flow two-dimensional high performance liquid chromatography, J. Chromatogr. A. 1521 (2017) 80–89. https:// doi.org/10.1016/j.chroma.2017.09.025.

[64] S.S. Jakobsen, S. Verdier, C.R. Mallet, J.H. Christensen, N.J. Nielsen, Increasing flexibility in two-dimensional liquid chromatography by pulsed elution of the first dimension (Pulsed-elution 2D-LC): A proof of concept, Anal. Chem. 89 (2017) 8723–8730. https://doi.org/10.1021/acs.analc hem.7b00758.

[65] J. Brown, B. Hartley, Location of disulphide bridges by diagonal paper electrophoresis: The disulphide bridges of bovine chymotrypsinogen A, Biochem. J. 101 (1966) 214–228. https://doi.org/10.1042/ bj1010214.

[66] K. Gevaert, P.V. Damme, L. Martens, J. Vandekerckhove, Diagonal reverse-phase chromatography applications in peptide-centric proteomics: Ahead of catalogue-omics? Anal. Biochem. 345 (2005) 18–29. https://doi.org/10.1016/j.ab.2005.01.038.

[67] J.J. Gilroy, C.M. Eakin, Characterization of drug load variants in a thiol linked antibody-drug conjugate using multidimensional chromatography, J. Chromatogr. B. 1060 (2017) 182–189. https://doi. org/10.1016/j.jchromb.2017.06.005.

[68] A. Bathke, D. Klemm, C. Gstöttner, C. Bell, R. Kopf, Rapid characterization of protein modifications – A novel fully automated 2D-HPLC-MS approach, LC-GC Eur. 31 (2018) 10–21.

[69] M. Iguiniz, E. Corbel, N. Roques, S. Heinisch, On-line coupling of achiral reversed phase liquid chromatography and chiral supercritical fluid chromatography for the analysis of pharmaceutical compounds, J. Pharm. Biomed. Anal. 159 (2018) 237–244. https://doi.org/10.1016/j.jpba.2018.06.058.

[70] A. Moussa, T. Lauer, D. Stoll, G. Desmet, K. Broeckhoven, Numerical and experimental investigation of analyte breakthrough from sampling loops used for multi-dimensional liquid chromatography, J. Chromatogr. A. 1626 (2020) 461283. https://doi.org/10.1016/j.chroma.2020.461283.

[71] B.W.J. Pirok, A.F.G. Gargano, P.J. Schoenmakers, Optimizing separations in on-line comprehensive two-dimensional liquid chromatography, J. Sep. Sci. 41 (2017) 68–98. https://doi.org/10.1002/jssc.201700863.

[72] Y.Z. Baghdady, K.A. Schug, Online comprehensive high ph reversed phase×low ph reversed phase approach for two-dimensional separations of intact proteins in top-down proteomics, Anal. Chem. (2019). https://doi.org/10.1021/acs.analchem.9b01665.

[73] A. Moussa, T. Lauer, D. Stoll, G. Desmet, K. Broeckhoven, Modelling of analyte profiles and band broadening generated by interface loops used in multi-dimensional liquid chromatography, J. Chromatogr. A. (2021) 462578. https://doi.org/10.1016/j.chroma.2021.462578.

[74] S.L. Weatherbee, T. Brau, D.R. Stoll, S.C. Rutan, M.M. Collinson, Simulation of elution profiles in liquid chromatography – IV: Experimental characterization and modeling of solute injection profiles from a modulation valve used in two-dimensional liquid chromatography, J. Chromatogr. A. 1626 (2020). https://doi.org/10.1016/j.chroma.2020.461373.

[75] Z. Dunn, J. Desai, G.M. Leme, D. Stoll, D. Richardson, Rapid two-dimensional Protein-A size exclusion chromatography for determination of titer and aggregation for monoclonal antibodies in harvested cell culture fluid samples, MAbs. 12 (2020). https://doi.org/10.1080/19420862.2019.1702263.

[76] C.L. Barhate, E.L. Regalado, N.D. Contrella, J. Lee, J. Jo, A.A. Makarov, D.W. Armstrong, C.J. Welch, Ultrafast chiral chromatography as the second dimension in two-dimensional liquid chromatography experiments, Anal. Chem. 89 (2017) 3545–3553. https://doi.org/10.1021/acs.analchem.6b04834.

[77] D.R. Stoll, R.W. Sajulga, B.N. Voigt, E.J. Larson, L.N. Jeong, S.C. Rutan, Simulation of elution profiles in liquid chromatography – II: Investigation of injection volume overload under gradient elution conditions applied to second dimension separations in two-dimensional liquid chromatography, J. Chromatogr. A. 1523 (2017) 162–172. https://doi.org/10.1016/j.chroma.2017.07.041.

[78] J.N. Fairchild, K. Horváth, G. Guiochon, Theoretical advantages and drawbacks of on-line, multidimensional liquid chromatography using multiple columns operated in parallel, J. Chromatogr. A. 1216 (2009) 6210–6217. https://doi.org/10.1016/j.chroma.2009.06.085.

[79] G.J. Opiteck, S.M. Ramirez, J.W. Jorgenson, M.A. Moseley, Comprehensive two-dimensional high-performance liquid chromatography for the isolation of overexpressed proteins and proteome mapping, Anal. Biochem. 258 (1998) 349–361. https://doi.org/10.1006/abio.1998.2588.

[80] K. Wagner, Protein mapping by two-dimensional high performance liquid chromatography, J. Chromatogr., A. 893 (2000) 293–305. https://doi.org/10.1016/S0021-9673(00)00736-6.

[81] K. Wagner, T. Miliotis, G. Marko-Varga, R. Bischoff, K.K. Unger, An automated on-line multidimensional hplc system for protein and peptide mapping with integrated sample preparation, Anal. Chem. 74 (2002) 809–820. https://doi.org/10.1021/ac010627f.

[82] I. Francois, A. Devilliers, B. Tienpont, F. David, P. Sandra, Comprehensive two-dimensional liquid chromatography applying two parallel columns in the second dimension, J. Chromatogr. A. 1178 (2008) 33–42. https://doi.org/10.1016/j.chroma.2007.11.032.

[83] J. Ren, M.A. Beckner, K.B. Lynch, H. Chen, Z. Zhu, Y. Yang, A. Chen, Z. Qiao, S. Liu, J.J. Lu, Two-dimensional liquid chromatography consisting of twelve second-dimension columns for comprehensive analysis of intact proteins, Talanta. 182 (2018) 225–229. https://doi.org/10.1016/j.talanta.2018.01.072.

[84] F. Bedani, W.Th. Kok, H.-G. Janssen, Optimal gradient operation in comprehensive liquid chromatography×liquid chromatography systems with limited orthogonality, Anal. Chim. Acta. 654 (2009) 77–84. https://doi.org/10.1016/j.aca.2009.06.042.

[85] P. Jandera, T. Hájek, P. Česla, Comparison of various second-dimension gradient types in comprehensive two-dimensional liquid chromatography, J. Sep. Sci. 33 (2010) 1382–1397. https://doi.org/10.1002/jssc.200900808.

[86] P. Jandera, Programmed elution in comprehensive two-dimensional liquid chromatography – A review, J. Chromatogr., A. 1255 (2012) 112–129. https://doi.org/10.1016/j.chroma.2012.02.071.

[87] D. Li, O.J. Schmitz, Use of shift gradient in the second dimension to improve the separation space in comprehensive two-dimensional liquid chromatography, Anal. Bioanal. Chem. 405 (2013) 6511–6517. https://doi.org/10.1007/s00216-013-7089-5.

[88] G.M. Leme, F. Cacciola, P. Donato, A.J. Cavalheiro, P. Dugo, L. Mondello, Continuous vs. segmented second-dimension system gradients for comprehensive two-dimensional liquid chromatography of sugarcane (Saccharum spp.), Anal. Bioanal. Chem. 406 (2014) 4315–4324. https://doi.org/10.1007/s00216-014-7786-8.

[89] D.R. Stoll, T.J. Lauer, K. Broeckhoven, Where has my efficiency gone? Impacts of extracolumn peak broadening on performance, Part IV: Gradient elution, flow splitting, and a holistic view, LCGC North Am. 39 (2021) 308–314.

[90] J. Halvorson, A.M. Lenhoff, M. Dittmann, D.R. Stoll, Implications of turbulent flow in connecting capillaries used in high performance liquid chromatography, J. Chromatogr. A. 1536 (2016) 185–194. https://doi.org/10.1016/j.chroma.2016.12.084.

[91] D. Cook, S.C. Rutan, D.R. Stoll, P.W. Carr, 2D assisted liquid chromatography (2DALC) – A chemometric approach to improve accuracy and precision of quantitation in liquid chromatography using 2D separation, dual detectors, and multivariate curve resolution, Anal. Chim. Acta. 859 (2015) 87–95. https://doi.org/10.1016/j.aca.2014.12.009.

[92] M.R. Filgueira, C.B. Castells, P.W. Carr, A simple, robust orthogonal background correction method for two-dimensional liquid chromatography, Anal. Chem. 84 (2012) 6747–6752. https://doi.org/10.1021/ac301248h.

[93] C. Gunnarson, T. Lauer, H. Willenbring, E. Larson, M. Dittmann, K. Broeckhoven, D.R. Stoll, Implications of dispersion in connecting capillaries for separation systems involving post-column flow splitting, J. Chromatogr. A. (2021). https://doi.org/10.1016/j.chroma.2021.461893.

[94] J. Camperi, A. Goyon, D. Guillarme, K. Zhang, C. Stella, Multi-dimensional LC-MS: The next generation characterization of antibody-based therapeutics by unified online bottom-up, middle-up and intact approaches, Analyst. (2021) https://doi.org/10.1039/D0AN01963A.

5 Selecting Separation Modes and Selectivities for Multi-Dimensional LC

Bob W.J. Pirok and Dwight R. Stoll

CONTENTS

5.1 INTRODUCTION TO THE THOUGHT PROCESS OF CHOOSING SEPARATION MODES AND SELECTIVITIES

Having addressed the fundamental principles of multi-dimensional chromatography in Chapter 3, we will now see how we can combine these ideas with information about the strengths and weakness of different LC separation modes on the way to developing actual 2D-LC methods. In this chapter, we focus on selecting and combining two separation modes or selectivities that will increase the likelihood of an effective 2D separation. We discuss how to identify appropriate retention modes, and refine separations using the properties of both the stationary and mobile phases.

5.1.1 THE CONCEPT OF SAMPLE DIMENSIONALITY

Before we consider specific retention modes in the development of a 2D-LC method, we first have to learn as much as we can about the analyte mixture. In Chapter 3 we discussed the importance of the orthogonality or complementarity of the two separations used in a 2D method. But, of course, two separations that will be highly complementary for one set of molecules may not be complementary for a different set of molecules. In other words, the extent to which two separations are complementary depends on both the properties of the stationary and mobile phases used, and the properties of the analyte mixture. With this in mind, it is often useful to identify the molecular descriptors that are likely to influence retention (if they are known). These are sometimes referred to as "molecular handles". In chromatography this concept has been described by Giddings as identifying the "sample dimensionality" associated with the analyte mixture at hand [1].

At the most fundamental level the sample dimensionality is directly related to the number of different types of intermolecular forces by which the various analytes interact with either or both the mobile and stationary phases. In this regard LC and especially 2D-LC has a significant advantage over GC and 2D-GC in that interactions in the mobile phase can be used to adjust selectivity. These forces include: Coulombic, London (dispersion), dipole-dipole, dipole-induced dipole and all the various types of acid-base (donor/acceptor) interactions such as hydrogen bonding, and Lewis acid-base processes. The toolbox of the LC analyst offers a wide array of retention modes that can be used to exploit these different types of intermolecular interactions between analytes and mobile and stationary phases. The different retention modes that have been used in 2D-LC, their selectivities, and examples of typical stationary phases are shown in Table 5.1.

5.1.2 POTENTIAL FOR MIXED-MODE AND/OR UNINTENDED INTERACTIONS

In the case of a sample involving a mixture of components varying in hydrophobicity and charge, ideally we would have one dimension that separates exclusively on the basis of charge, whereas the other dimension would separate exclusively on the basis of analyte hydrophobicity. In practice, such an ideal distribution of selectivities is not very common. For example, in size exclusion (SEC) separations, chemical interactions can occur that influence the separation in ways that depend on analyte properties other than just size [2]. Another example of a single stationary phase exhibiting multiple selectivities is found in ion exchange (IEX) separations. In some cases, IEX stationary phase backbones may be quite low in polarity in nature [3, 4]. Such mixtures of retention mechanisms are sometimes deliberately exploited and the resulting phases and separations are typically referred to as "mixed-mode" (MM). Thus, depending on the stationary phase chemistry utilized, separation based on a pure retention mechanism can require adjustment of the mobile phase to counteract other interactions [5]. It is also possible for the mobile phase to either amplify or dampen the native selectivity of the stationary phase for a particular class of analytes. One example of such an effect is observed in the differential effect of the mobile phase modifiers acetonitrile and methanol on the retention of aromatic analytes by aromatic reversed-phase (RP) stationary phases (e.g., phenyl). The strongly dipolar and pi-electron rich acetonitrile dampens the selectivity of these phases for pi-electron rich analytes relative to that observed when using methanol [6, 7].

5.1.2 SUPPRESSION OF RETENTION MECHANISMS

Suppression of the retention mechanism that a stationary phase is designed for is also possible in extreme cases, such as the use of octadecylsilica columns for either RP or SEC separations of polymers. When used with aqueous-organic solvents as the mobile phase, polymers can be separated using solvent gradient elution, which reduces the dipolarity of the mobile phase as the fraction of organic solvent is increased. Yet, when the same column is used with the much stronger tetrahydrofuran as the mobile phase, analyte-stationary phase interactions are effectively suppressed, and the column essentially functions as a SEC column.

5.2 COMBINING SELECTIVITIES

The separation modes used in contemporary 2D-LC separations are summarized in Table 5.1, including acronyms, the chemical/physical basis of selectivity, and brief descriptions

TABLE 5.1

Overview of Separation Modes

Mode	Acronym	Selectivity	Common stationary phase (SP) selectors
Reversed-phase	RP	Hydrophobicity, Chain length, carbon skeleton	Alkyl (hydrocarbon: C1 to C30; most commonly C18), cyano (π-π)*, phenyl (π-π)*, carbon-clad zirconia (or graphitized carbon), PEG
Ion pairing	IP	Hydrophobicity, suppression of analyte ionization (acid/ bases)	Alkyl (hydrocarbon)
Hydrophobic interaction	HIC	Hydrophobicity	Short-chain alkyl hydrocarbons (C4 to C8)
Normal phase	NP	Polarity, Functional groups	Bare silica, Amino-propyl, diol, cyano
Argentation	AgLC	Degree of saturation, cis-trans isomers	IEX columns (*e.g.* sulfonic acid) or bare silica loaded with silver ions
Hydrophilic interaction	HILIC	Hydrophilicity, Polar character	Zwitterionic: Sulfobetaine, Phosphocoline; Basic: Amino propyl; Neutral: Diol, Amide
Ion exchange	IEX	Charge, Ionic interactions	Strong Cation Exchangers (SAX): Sulfonic Acid; Weak Cation Exchangers (WCX): Carboxylic Acid; Weak Anion Exchangers (WAX): Triethyl amine; Strong Anion Exchangers (SAX): Quaternary Amine
Size exclusion	SEC	Molecular size, Molecular weight	Crosslinked poly(styrene – divinyl-benzene) or methacrylate porous beads (SEC organic solvents); Polar-functionalized porous silica (SEC aqueous)
Mixed mode	MM	Combination of retention mechanisms	Anion-exchange/reversed-phase (AEX/RP), Cation exchange/reversed-phase (CEX/RP), Anion-exchange/cation-exchange/reversed-phase (AEX/CEX/RP); AEX/HILIC, CEX/HILIC, AEX/CEX/HILIC
Chiral	Chiral	Selector-specific chirality	Variety of selectors depending on the application. Most common are based on polysaccharide derivatives, antibiotics, and Pirkle phases
Affinity	Affinity	Selector-specific affinity	Stationary phases prepared for specific phase-analyte interactions (e.g., boronate-cis-diol; antibody-antigen)

Source: Adapted from [8].

of commonly-used stationary phase chemistries. Just because two different selectivities can be identified as ones that will both provide selectivity for analytes in the sample, and be complementary to each other, does not necessarily mean it will be easy to combine them in a 2D separation format. For many applications the number of possible combinations of stationary phase chemistries and mobile phase compositions is large, however some combinations have clear benefits and/or drawbacks compared to others. Thus, the challenge for the analyst is to narrow the list of possible combinations of selectivities to a small number to be evaluated experimentally. A list of factors for consideration is summarized in Table 5.2.

Some of the aspects listed in Table 5.2 may be surprising, simply because we do not have to consider them when developing 1D-LC methods. For example, slow column re-equilibration of columns

TABLE 5.2
Overview of Symbol Definitions as Used in Table 5.3

Symbol	Meaning	Relevant to	Description
A	Adsorption	^{1}D, ^{2}D	Lengthening of elution time due to injection solvent. Applies exclusively to SEC.
B	Breakthrough/Peak distortion	^{1}D, ^{2}D	Anomalous early elution in the second dimension. See Section 4.4.4 for more information.
E	Easy to modulate	2D-LC	Ease of developing active-modulation methods (e.g., trap columns or solvent dilution).
F	Fast separation	^{2}D	Method with short analysis times (e.g. <1 min).
H	High-resolution separation	^{1}D, ^{2}D	Method capable of high peak capacity.
I	Isocratic	^{1}D, ^{2}D	Possibility of (easily) running isocratic methods, reducing the complexity of the setup.
M	MS compatible	^{2}D	Possibility of using volatile mobile-phase additives and achieving good MS sensitivity.
O	Orthogonal/Complementary	2D-LC	Degree of independence of two separation mechanisms, assuming that the analyte mixture exhibits sample dimensions targeted by the two dimensions.
P	Applicability	2D-LC	Usefulness of the resulting separation.
Q	Column re-equilibration	^{2}D	Speed of column re-equilibration.
⇄	Reversed-order recommended	LC×LC	Recommended to consider the reversed order of the modes.
S	Selectivity/Specificity	^{1}D, ^{2}D	Capability of the separation method to separate based on chemical characteristics of sample components (e.g. shape, orientation, composition/ sequence).
X	Solvent compatibility	LC×LC	Extent of (in)compatibility of ^{1}D effluent and ^{2}D eluent.
	Polymers		Suitable/Unsuitable for separations of polymers.
	Proteins		Suitable/Unsuitable for separations of proteins.

Source: Adapted from [8].

following gradient elution is typically viewed as an inconvenience in 1D-LC, but not a factor that threatens the viability of the method altogether. However, in 2D-LC – and particularly LC×LC where fast ^{2}D separations are required – use of a separation that requires long re-equilibration times is simply not viable in the second dimension. Thus, fast separations (indicated in Table 5.2 with *F*) and short column re-equilibration times (indicated by *Q*) are useful in the second dimension of 2D-LC methods. This also explains why isocratic experiments may be desired in the second dimension in some cases (indicated in Table 5.2 by *I*). It is also good to realize that ultimately, all ^{1}D effluent components will enter the ^{1}D and ^{2}D detectors unless appropriate measures have been taken to avoid this when it is expected to be a problem. This is most important is when mass spectrometric (MS) detection is used in the second dimension.

Taking into account all of the potential benefits and drawbacks of each separation type, we can assess the potential effectiveness of each possible combination of separation modes. This information is organized in Table 5.3, which provides an overview of the strengths and weaknesses of possible combinations, utilizing the symbols listed in Table 5.2 to communicate where the strengths and weaknesses lie. While this table provides a one-stop overview, we cannot emphasize strongly enough

TABLE 5.3

Overview of the Possible Combinations of Separation Modes for Online 2D-LC

	²RP	²NP	²HILIC	²HIC	²IEX	²SEC-Aq	²SEC-Or	²Ag	²Chiral	²Affinity	²SFC
¹RP	E O⁺ P⁺ X⁺	B O²⁺ X²⁻	B O²⁺ X⁺	B E O⁺ X⁺	O⁺	A E O⁺ P⁺ X⁺	A E O⁺	B O²⁺ X⁻	O²⁺	O²⁺ X⁺	B O²⁺ X⁻
¹NP	B O²⁺ X²⁻	O⁺ P⁻ X⁻	O⁻ P⁻ X⁻	B O²⁺ P X²⁻	O²⁺	O²⁺ X²⁻	O²⁺ P⁺ X⁺	O⁺ X⁻	O²⁺	O⁺ X²⁻	O⁻ X²⁺
¹HILIC	B O²⁺ P⁺ X⁺	B O⁻ X⁻	O⁻ X⁻	B O²⁺ P X⁻	O⁺ X⁻	O²⁺ P⁻	A O⁺ X⁺	B O⁺ X⁻	O²⁺	X⁻	X⁺
¹HIC	E O⁺ X²⁺	B O²⁺ X²⁻	B O²⁺ X⁻	B O⁺ P X²⁺	B O⁺ P X²⁺	O²⁺ P X²⁻	A O⁺ P X⁻	B O²⁺ P X²⁻	O²⁺ P²⁻	O⁺ X⁻	O⁺ P² X²⁻
¹IEX	E O⁺ P⁺ X²⁻	B O²⁺ X²⁻	B O⁺ X⁻	B O⁺ P	B X⁻	O⁺ X²⁺	A O⁺ P X⁻	B O⁺ X⁻	O²⁺	O⁺ X⁻	O⁺ X²⁻
¹SEC-Aq	E O⁺ P⁺ X²⁻	B O²⁺ X²⁻	B O²⁺ X⁻	B O²⁺ P X²⁻	O²⁺ X⁻	O²⁺ P²⁻	A O²⁺ P²⁻ X²⁻	O²⁺ X²⁻	O²⁺ P⁻	O²⁺ X⁺	E O²⁺ P X⁻
¹SEC-Or	B²⁻ O⁺ X⁻	B O²⁺ X⁺	B O²⁺ X⁻	B O²⁺ P X²⁻	O²⁺ X⁻	O²⁺ P²⁻	O²⁻ P²⁻	O²⁺ X⁺	O²⁻ P⁻	O²⁺ P²⁻ X⁻	O⁺ P⁻ X⁺
¹Ag	B O²⁺	O²⁺	B O²⁺ P⁻	O²⁺ P²⁻	O²⁺ X⁻	O²⁺ P²⁻	O²⁺ P⁻	O²⁺ P²⁻	O²⁺	O²⁺ X²⁻	O⁺ X⁺
¹Chiral	O²⁺	O²⁺	O²⁻	O²⁺ P⁻	O⁺	O²⁺ P	O²⁺ P⁻	O²⁺	O²⁺	O⁻ P²⁻	O²⁺
¹Affinity	O²⁺ P⁻ X⁺	B O²⁺ P⁻ X⁻	B O²⁺ P⁻	O²⁺ P²⁻	O⁺ P X⁺	O²⁺ P X⁺	A O²⁺ P²⁻ X²⁻	B O²⁺ P X⁻	O²⁺ P²⁻	O²⁺ X²⁻	O⁺ P X²⁻
¹SFC	E O²⁺ X⁺	O⁺ X⁻	E O⁻	O²⁺ P³⁻	O²⁺ X⁻	O²⁺ P²⁻ X²⁺	O²⁺ X²⁻	O⁺ X⁺	O²⁺	O²⁺ X⁻	E O⁺ X²⁺

Source: Reprinted from [8].

that the strengths and weaknesses of each combination of selectivities depends strongly on the application at hand and the detector used. As such the table should be viewed as a first approximation – a starting point with which to begin thinking about which combinations should be considered first in method development, and then refine the perspective of the table based on nuances related to the specific application. Moreover, many of the challenges highlighted in Table 5.3 are not as serious when developing non-comprehensive 2D-LC separations – in short, more time in the second dimension helps mitigate many problems. It is also good to realize that the color does not indicate usefulness. Indeed, seemingly challenging combinations may be highly useful for certain applications. This has driven researchers to explore the limits of these challenges to develop practically useful methods.

In the following sub-sections we briefly discuss each of the different separation modes. In each case we briefly discuss the basis for separation, strengths, and weaknesses in the context of 2D-LC separations, and provide a few examples of 2D-LC applications involving that particular separation mode. Our intent is not to list all 2D-LC applications in this chapter; readers interested in a comprehensive list of all 2D-LC applications are referred to our free, online database (www.multidlc.org/lit erature/2DLC-Applications). Reversed-phase LC is frequently used in 2D-LC methods in a number of application areas, and thus the final section of this chapter is entirely devoted to the selection of RP phases for 2D-LC.

5.2.1 NORMAL-PHASE LIQUID CHROMATOGRAPHY

The oldest separation mode in our toolkit, normal-phase liquid chromatography (NPLC) utilizes a sorbent that is more polar than the eluent as the stationary phase. In combination with a low polarity mobile phase, compounds can be separated based on differences between their dipolarity and hydrogen bonding characteristics, and locations of the polar groups within the analyte structure. Retention is decreased by increasing the fraction of the polar component of the mobile phase. Whereas NPLC is usually thought of as involving mobile phases containing only organic solvents (or very little water), hydrophilic interaction chromatography (HILIC) is somewhat similar to NPLC but usually involves 1–30% (v/v) water in the mobile phase.

Strengths – Selectivity for regioisomers and amenable to separation of analytes that are only soluble in 100% organic solvent mobile phases.

Weaknesses – Application of NPLC within two-dimensional separation systems can be challenging due to incompatibility of the non-polar organic solvents employed in NPLC and the polar (especially water) solvents frequently used with other separation modes. This is the single biggest reason that combinations involving NPLC in Table 5.3 receive low scores. For example, in the case of RPLC×NPLC, the amount of water in the RPLC mobile phase typically makes it immiscible with many NPLC mobile phase. One approach to deal with this is to remove the volatile organic solvent from the ^{1}D NPLC effluent by evaporation [9] (see Section 4.4.7). When used in combination with RPLC, it is preferred to use the NPLC separation in the first dimension because active modulation techniques are more effective with this configuration compared to the opposite one (i.e., RPLC×NPLC). This can be avoided by using RPLC in the non-aqueous mode (NARP), but this is not useful for many classes of compounds. One exception to the poor compatibility of NPLC with other separation modes is the combination NPLC with SEC separations that use mobile phases composed of a solvent (i.e., "organic SEC").

The other significant weakness of NPLC is that equilibration of the stationary phase following a change in mobile phase conditions (e.g., as in gradient elution) is perceived to be slow. To minimize the impact of this limitation, most applications involving NPLC use it either in the first dimension, or under isocratic [10, 11] or pseudo-critical conditions [12] in the second dimension.

Representative applications – SEC×NPLC has been applied to the analysis of complex polymer mixtures [13]. NPLC×RPLC separations have been described for the analysis of cold-pressed lemon oil [14], alcohol ethoxylates [15], and oligomers [16].

5.2.1.1   Argentation (Silver-Ion) Normal-Phase Chromatography

A special form of NPLC is argentation (silver-ion) chromatography (AgLC). This separation mode utilizes a silica-based stationary phase that is treated with an aqueous solution of silver nitrate. When used in conjunction with an organic mobile phase, this mechanism is quite selective for π-π interactions between the double bonds of unsaturated analytes and the silver ions. This gives rise to selectivity that is based on differences in the extent and location(s) of unsaturation within different molecules, and thus AgLC has been mainly applied to the analysis of lipids. The silver ions bound to the stationary phase can easily be disturbed by the presence of small quantities of non-suitable solvents, and the relative instability of AgLC systems renders them mainly useful as ^{1}D separation. Example applications of AgLC×RP include the determination of lipids in samples of rice oil [17], soybean oil [18], peanut oil and mouse tissue [19].

5.2.1.2   Hydrophilic-Interaction Liquid Chromatography (HILIC)

Although there is ongoing study of the retention mechanism of HILIC separations [20] they can be thought of conceptually as a specific form of NPLC. The use of mobile phases rich in organic solvents and a small amount of water is thought to promote the development of a water-rich layer at the surface of the highly polar stationary-phase sorbents used for HILIC [21]. Hydrophilic analytes can interact with, and partition into, this aqueous layer. A wide range of stationary phase chemistries have been commercialized for HILIC separations, which sometimes incorporate ionogenic elements (e.g., carboxylate or amino groups). Thus, the ability to choose from very different HILIC selectivities for applications involving different analyte mixtures has attracted a lot of interest for its use in 2D-LC. The latest developments in HILIC research have been summarized in recent review articles [22, 23].

Strengths – The ability of HILIC to separate molecules with subtle differences in dipolarity and structure can be remarkable, and highly complementary to other separation modes such as RPLC. For example, we found that glycosylated proteins that coeluted under RPLC conditions were separated by tens of minutes in a HILIC separation [24, 25].

Weaknesses – HILIC has most of the same weaknesses as conventional NPLC, however they are not quite as serious. Several groups have demonstrated that active modulation techniques can be used to effectively couple HILIC and RPLC separations [26], and it has been shown that re-equilibration of HILIC phases is sufficiently repeatable even with short re-equilibration times (e.g., 3 s) to allow their use in the second dimension of 2D-LC even with gradient elution [27, 28]. A number of authors have employed HILIC in the second dimension. D'Attoma and Heinisch compared RPLC×RPLC to RPLC×HILIC and found the latter combination to suffer from injection effects [29]. Holčapek *et al.* analyzed lipidomic samples using RPLC×HILIC-MS [30].

Representative applications – Most 2D-LC applications involving HILIC use it in the first dimension, along with RPLC in the second dimension. This configuration leverages the strengths of these separations, and minimizes their weaknesses. Examples include separations of cocoa procyanidins [31], anthocyanins in red wine [32], phosphatidylcholine isomers [33], and surfactants [34]. Finally, the assortment of commercially available HILIC selectivities is making it increasingly practical to use two HILIC phases in a 2D-LC system [28]. For example, Wang *et al.* developed an online HILIC×HILIC method and applied it to separate the saponins from *Quillaja saponaria* [35].

5.2.2   Ion-Exchange Chromatography

Ion-exchange separations employ stationary phases functionalized with ionogenic groups that enable the separation of analytes in the charge state that is opposite of that of the stationary phase. Elution is facilitated by increasing the concentration of counterions in the mobile phase. So-called "strong" ion-exchange phases contain functional groups that remain charged independent of the mobile phase pH (e.g., sulfonate groups for cation-exchange, and quaternary ammonium groups for anion-exchange). In contrast, so-called "weak" ion-exchange phases carry functional groups whose ionization state depends strongly on the mobile phase pH *e.g.* diethylaminoethyl groups for anion-exchange (WAX), or carboxylate groups for weak cation-exchange (WCX).

Strengths – IEX is a very powerful approach for separation of analytes that have different charges in solution. This is particularly evident for polyvalent ions such as oligonucleotides, peptides, and proteins where IEX is an indispensible LC separation mode. IEX can also be highly complementary to other commonly used LC modes (e.g., RPLC). Yet, especially for small molecules or polymers with charged moieties as end-groups, much attention is required to achieve good usage of the separation space in IEX×RPLC systems. When the ^{1}D IEX separation is based primarily on the charge state of the analyte, ideally the charge should play no role in retention in the ^{2}D RPLC separation. This is difficult to realize because the charged groups increase solubility in the RPLC mobile phase. One way to address this is to use ion-pairing conditions in the ^{2}D separation. Pirok and coworkers used this strategy in their development of IEX×RPLC methods for the characterization of synthetic [5] and natural dyes in cultural heritage samples [36]. Since multi-valent analytes require a great deal more ion-pairing agent to fully suppress the charged character of the compound, the strategy appeared to be useful only when the number of ionogenic groups on the analytes was limited [5].

Weaknesses – Elution of ionogenic analytes from IEX phases requires "salty" mobile phases, which is a serious limitation when considering coupling with some detectors including MS and ELSD. In the area of protein separation by IEX this situation is improving slowly as several groups develop methods (i.e., stationary phases and mobile phase conditions) that are effective even with modest concentrations of volatile mobile phase buffers [37]. Nevertheless, the need for salty mobile phases in IEX separations is the major reason that IEX is primarily used as a first dimension in 2D-LC. This allows diversion of salts in the ^{1}D effluent to waste as they elute in the dead volume when RPLC is used in the second dimension, for example. Finally, IEX mobile phases are usually entirely aqueous, which makes coupling with LC modes running with organic-rich mobile phases challenging due to mobile phase mismatch, and the potential for precipitation of buffer salts from the mobile phase.

Representative applications – One of the most widely used applications of 2D-LC combines IEX and RPLC separations for the characterization of peptides using MS. One well-known variant of this approach is the so-called multi-dimensional protein-identification technology (MudPIT) [38]. Readers interested in this and other 2D-LC applications involving IEX are referred to Chapters 9, 10, and 13, which are focused on separations of small and large molecules in the pharmaceutical context, and chiral separations.

5.2.3 SIZE-EXCLUSION CHROMATOGRAPHY AND HYDRODYNAMIC CHROMATOGRAPHY

Out of all the LC separation types discussed in this book only size-exclusion chromatography (SEC) does not involve any chemical interaction between the analytes and stationary phase. In principle, separation by SEC is entirely entropic in nature, however it is not uncommon to observe non-specific analyte-phase interactions that influence the observed elution volume, and the mobile phase conditions are typically chosen to minimize these effects (e.g., tetrahydrofuran for organic SEC, and pH buffered aqueous solution for aqueous SEC of proteins and other bio-macromolecules). In the ideal case, SEC elution volume is entirely determined by steric exclusion of analytes from the pore volume. Molecules that are too large to enter the pores will only explore the interparticular volume, and thus elute faster through the column relative to smaller molecules that can (partially) permeate the pores. In some cases, large analytes excluded by the pores may be separated according to the mechanism of hydrodynamic chromatography [39]. Within the context of polymer separations, another interesting alternative is critical chromatography, formally known as "liquid chromatography at the critical conditions" (LCCC). In LCCC, the mobile-phase conditions are chosen to decouple retention from the molecular weight of the polymers. This mechanism can be regarded as a niche form of RPLC or NPLC and is not featured in Table 5.3.

Strengths – While the resolution obtained in SEC is rather limited relative to other modes of LC, calibration using standards allows estimation of the distribution of analyte molecular weights based on the elution volume. SEC is by its nature an isocratic technique, which makes it somewhat attractive as a second dimension because no re-equilibration is needed after elution, and overlapped injections can be used in the second dimension to make more efficient use of the available ^{2}D time [40].

Weaknesses – Generally speaking, large volume columns (e.g., 300 mm x 7.8 mm i.d.) are used for SEC. If these columns are used in the first dimension of a 2D-LC system this can lead to very large fraction volumes that must be transferred to the second dimension for high sensitivity. The generally poor resolving power of SEC limits its utility in LC×LC separations, however recent advances in SEC carried out under UHPLC conditions [41, 42], or with superficially porous particles [43, 44], are changing this situation. Aqueous SEC can work well when coupled with other separations such as RPLC and IEX. Coupling organic SEC with other LC separations can be more difficult due to the use of organic solvents for the mobile phase such as tetrahydrofuran, and the potential mobile phase mismatch that can arise. Organic SEC can also be quite sensitive to small amounts of water that lead to adverse adsorption effects, so care must be taken when coupling organic SEC as a second dimension with another LC mode that uses water in the mobile phase.

Representative applications – 2D-LC separations involving SEC are focused nearly exclusively on the analysis of large molecules such as polymers or proteins. Readers interested in these applications are referred to Chapter 11 (polymers) and Chapter 10 (proteins). In both cases one can find examples where SEC is used in the first dimension, the second dimension, or even both dimensions.

5.2.4 Hydrophobic Interaction Chromatography

Also known as salting-out chromatography, hydrophobic interaction chromatography (HIC) is exclusively applied to the separation of proteins [45]. High concentrations of salt (e.g. 2 M ammonium sulphate) in the mobile phase promote adsorption of the protein onto a moderately hydrophobic stationary phase [46]. Retention is decreased by decreasing the mobile phase salt concentration (i.e., the opposite of IEX), without the need for organic solvents (i.e., different from RPLC).

Strengths – Given that mobile phases used for HIC are almost always entirely aqueous, the native structure of protein analytes is preserved in the mobile phase environment, and the separation selectivity is different from that obtained with RPLC [47,48]. This in turn makes the coupling of these two separations attractive.

Weaknesses – Use of mobile phase salt concentrations in excess of 1 M is typically a pre-requisite for adequation retention in HIC, which prevents its use in the second dimension when MS is used as the 2D detector.

Representative applications – To date, the most active application area for 2D-LC separations involving HIC has been the characterization of antibody-drug conjugates (ADCs). In this case the HIC separation resolves different protein species according to the number of hydrophobic small molecule drugs bonded to the protein, and the ^{2}D RPLC separation effectively separates the protein analytes from the salt present in the ^{1}D effluent prior to MS detection [49, 50]. Interested readers are referred to Section 10.4 for further discussion of these applications.

5.2.5 Chiral Chromatography

The entirety of Chapter 13 of this book is devoted to the use of chiral separations in one or both dimensions of 2D-LC systems. Readers interested in this topic are directed to Chapter 13, and the strengths and weaknesses of this separation mode in the context of 2D-LC are only discussed briefly here. Chiral chromatography utilizes an array of column chemistries in which chiral selectors (e.g., small molecules, oligosaccharides, and proteins) are immobilized on a porous substrate (e.g., silica) to separate chemical compounds on the basis of their stereoconfigurations. Naturally, this separation mechanism is particularly useful for the separation of molecules having one or more chiral centers, including pharmaceuticals and their metabolites, agrochemical compounds, and amino acids. Method development for chiral separations typically involves lengthy studies to screen for chiral selectors that exhibit stereoselectivity for particular analytes of interest.

Strengths – The primary strength of chiral separations in the context of 2D-LC is that they can separate enantiomers and other structurally similar molecules that are difficult to separate any other way. As such, they are highly complementary to all other LC separation modes. Historically, much of the development of chiral separations was focused on NPLC conditions. However, in recent years there has been tremendous growth in the development and use of chiral separations that rely on other separation modes including HILIC, IEX, RPLC, and SFC [51,52]. These developments have created numerous opportunities to couple chiral separations with other achiral separation modes in 2D-LC formats.

Weaknesses – The two biggest weaknesses of chiral separations in the context of 2D-LC are the slow kinetics of desorption from the chiral support [53, 54], and unpredictable selectivity. The slow separation kinetics have hindered the use of chiral separations in the second dimension of LC×LC separations in particular, though this is beginning to change as discussed in Chapter 13. The inability to predict which chiral stationary phase is best suited for the separation of a particular pair of enantiomers means that many phases are often evaluated experimentally, and in some cases more than one chiral stationary phase is used in a 2D-LC separation (i.e., one phase is used in the second dimension to separate particular pairs of enantiomers that coelute in the first dimension, while other pairs are separated in the second dimension with a different phase; see Section 9.5.2).

Representative applications – Currently the most active application areas for 2D-LC involving one or more chiral separation are pharmaceutical analysis (e.g., analysis of molecules with one or more chiral centers, and their metabolites) and bioanalysis (e.g., separation of D-and L- amino acids). Please see Chapter 13 for a complete discussion of these and other applications.

5.2.6 AFFINITY CHROMATOGRAPHY

Affinity chromatography comprises a group of separations that on one hand covers a wide array of chemistries, and on the other hand can offer incredibly high selectivity. Stationary phases are often prepared by immobilizing proteins that exhibit strong interactions with a small number of molecules with particular three-dimensional structures. Prevalent examples include antibodies and protein receptors that are ubiquitous in biological systems [55–58]. Other stationary phases used for this purpose include phenylboronate [59, 60].

Strengths – Generally speaking, the selectivities of affinity-based separations are unparalleled. With sufficient effort, a stationary phase bearing a selector ligand can be made to be highly selective for almost any target analyte. Most affinity-based separations function well in highly aqueous mobile phases, making them highly compatible with most other LC separation modes.

Weaknesses – The extremely high selectivity (approaching absolute specificity in some specific instances) of affinity-based phases is also their greatest weakness. Any given stationary phase is only applicable to small number of target analytes. Desorption of the analyte from the stationary phase can also be slow [61], which limits their potential for use in ^{2}D separations. Sometimes affinity separations require elution conditions that can be detrimental to the following ^{2}D separation. This appears to be the case when separating antibodies using an ^{1}D affinity separation based on Protein A and a ^{2}D SEC separation (i.e., ProA-SEC). In this case the acidic mobile phase required for elution of the protein from the Protein A may induce protein aggregation during the SEC separation [62].

This is an area that should be looked at in more detail using class-selective affinity ligands such as boronates, lectins, and more general affinity ligands such as co-factor emulators.

Representative applications – While the number of published applications is small, they are significant. The use of ProA-SEC for the determination of titer and purity of therapeutical monoclonal antibodies (mAbs) has been demonstrated by several groups [57, 62, 63]. Hu *et al.* have used affinity chromatography and RPLC in the first and second dimensions of a LC×LC system for the characterization of traditional Chinese medicines [64].

5.2.7 SUPERCRITICAL-FLUID CHROMATOGRAPHY

While technically not part of the LC portfolio, supercritical-fluid chromatography (SFC) is increasingly being used in multi-dimensional separation systems. In SFC, the mobile phase is generally composed of carbon dioxide (CO_2) at sub- or supercritical conditions. Organic solvent modifiers are used to change retention and selectivity in ways similar to those used in NPLC under high pressure conditions [65].

Strengths – The advantages of SFC over LC are lower mobile-phase viscosity (and correspondingly lower pressure drops and higher diffusion coefficients of analytes) [66, 67], and NPLC-like selectivities without large quantities of hazardous organic solvents characteristic of NPLC [65]. The latter has been particularly important for chiral separations. The low viscosity of SFC mobile phases may enable faster ^{2}D separations than what is achievable with RPLC, however this has not been demonstrated in practice to date.

Weaknesses – When SFC is used in the second dimension, it is very susceptible to mobile phase mismatch effects when a first dimension is used that relies on a significant concentration of water in the mobile phase. Also, the scope of applications is relatively limited compared to RPLC (e.g., SFC is not broadly applicable to separation of biopolymers).

Representative applications – SFC is routinely applied in a variety of areas ranging from hydrocarbon analysis to lipids and pharmaceutical analysis. Several groups have demonstrated the use of SFC in the second dimension in 2D separations where RPLC is used in the first dimension [68, 69]. SFC is also potentially interesting as ^{1}D separation, because the mobile phase is compatible with ^{2}D RPLC, as demonstrated by François *et al.* [70, 71]. SFC×SFC using packed (capillary) columns has been demonstrated by Hirata [72, 73]. Open-tubular SFC×SFC [74] is an amazing technological achievement, but not a robust approach ready for routine practical use with current technology. Readers interested in these applications are also referred to Chapters 9 and 13.

5.3 CHOOSING REVERSED-PHASE SELECTIVITIES

Before discussing strategies for selecting RP stationary phases for use in 2D-LC, it is worth noting that in some situations adjusting mobile phase conditions can be just as effective as changing stationary phases. This point is illustrated in detail in Chapter 6 by way of a method development case study. Some contemporary applications, such as RP×RP separations of peptides where varying the mobile phase pH is particularly effective for inducing a selectivity difference between two similar RP phases [75, 76], have exploited the power of the mobile phase in 2D-LC to great effect.

5.3.1 SELECTING AN RP COLUMN FOR ONE DIMENSION OF A 2D-LC SEPARATION

When considering an RP separation for use in one dimension of a 2D-LC system, usually choosing the RP column is relatively straightforward. In most cases the particular RP selectivity (e.g., C8 vs. C18 vs. phenyl) that is used is not so important, because of the large differences in selectivity between the RP mode and other modes, in a general sense. For example, an ion-exchange separation will typically be highly complementary to a RP separation, so long as there is no strong correlation between the charge state of analytes in solution and their hydrophobicites. In our work in situations like this, we prioritize other characteristics of RP columns when choosing what to use in a 2D-LC system. These include chemical (i.e., pH and temperature) and physical (i.e., robustness of the particle bed) stability of the column, commercial availability of the phase in multiple column dimensions, and peak shape in simple mobile phases (e.g., dilute formic acid for MS detection). In other words, we prioritize ease-of-use over selectivity of the RP column in these situations.

5.3.2 SELECTING RP COLUMNS FOR BOTH DIMENSIONS OF A 2D-LC SEPARATION

Using two RP columns in a 2D-LC method is attractive from the point of view of compatibility of the ^{1}D and ^{2}D separations. Note that the upper-left corner of Table 5.3 is very green, indicating many benefits of the ways that RP separations work well together when used in both dimensions. However, one of the most prominent criticisms of this pairing is that it is not immediately obvious how the selectivities of two RP phases could be as complementary as RP paired with a very different separation mode such as SEC or HILIC. The good news is that there are more than 1,000 commercially available RP phases, which provides ample opportunity to discover RP phases that might be complementary enough to solve a given separation problem – the challenge, of course, is finding them quickly.

This challenge has inspired a number of groups to explore different strategies to narrow the number of phases that should be tried when screening different phase combinations. The Hydrophobic Subtraction Model (HSM) of RP selectivity developed by Snyder, Dolan, Carr, and coworkers has been used more than any other framework for addressing this problem. Readers interested in how the model was formulated and its uses are referred to other resources [77, 78]. In addition, a recent, comprehensive review article on modern models of RP selectivity is also a valuable resource [79]. Briefly, the HSM asserts that the selectivity of a given stationary phase (defined as the ratio of the retention factor of an analyte [k_x] to the retention factor of ethylbenzene [k_{EB}]) can be quantified using a linear combination of five stationary phase – analyte property pairs as shown in Eq. 5.1:

$$\ln\left(\frac{k_x}{k_{EB}}\right) = H \cdot \eta - S^* \cdot \sigma + A \cdot \beta + B \cdot \alpha + C \cdot \kappa \qquad (5.1)$$

Here, the H, S*, A, B, and C parameters quantify the contributions of the stationary phase to selectivity and represent hydrophobicity (H), resistance of the phase to penetration by bulky analytes (S*), hydrogen bond acidity (A) or basicity (B), and ionic character of the stationary phase (C), respectively, and the corresponding η, σ, β, α, κ parameters represent the conjugate characteristics of analytes. The column parameters have been measured for about 760 columns and are freely available through an interactive web-based database (www.hplccolumns.org). Although the initial motivation for the development of the HSM was focused on a strategy for identifying columns with similar selectivities, the data can also be used to identify columns that exhibit very different, or even complementary selectivities. All models of RP selectivity have strengths and weaknesses. One of the primary strengths of the HSM in the context of column selection for 2D-LC is the depth of the available data.

The HSM data has been used by a number of different groups, in a number of ways, for the purpose of identifying phases with selectivities different enough to be useful in 2D-LC separations. Zhang and Carr adapted Snyder's ternary plots [80] to visualize which columns lie at the periphery of the multi-dimensional space defined by the five parameters of the HSM [81]. We have further adapted this concept and developed an interactive tool for exploration of a three-dimensional space defined by any combination of three of the HSM parameters (www.hplccolumns.org). Lindsey,

Siepmann, and coworkers [82] used numerical simulations to predict RP×RP separations of a huge number of hypothetical small molecules using all possible combinations of RP columns in the HSM database, and then rank the resulting separations to identify the combinations of RP phases that looked most promising in a general sense. Interestingly, in spite of their very different approaches, both of these theoretical studies identified a common set of phases that appear to be good candidates for use in at least one of the two dimensions of 2D-LC separations that use RP in both dimensions. Moreover, it is interesting to note that none of these five phases is a traditional C18.

- Zorbax Bonus RP (polar embedded group)
- EC Nucleosil 100-5 Protect 1 (polar embedded group)
- BetaMax Acid (polar embedded group)
- Hypersil Prism C18 RP (polar embedded group)
- ZirChrom-PS (zirconia substrate with polystyrene coating)

It is important to recognize here that these conclusions represent the average selectivities of these phases toward small molecules of all kinds. As pointed out by Lindsey *et al.*, the identification of highly complementary column pairs is strongly dependent on the analyte set that is considered. It is entirely possible that for a given application the optimal column pair may not involve any of the members of the list above. So, the real value of this list is that it provides a rational starting point in method development; it does not guarantee that the final method must include one of these phases.

While the paragraphs above describe a theoretical basis for selecting RP phases for 2D-LC, there have also been some systematic experimental studies that provide support for the predictions from theory. Allen *et al.* used different combinations of six RP phases predicted by HSM to be highly complementary for RP×RP separations of a variety of small molecule extracts including those from corn seed, St. John's Wort, ginko biloba, and valerian root [83]. The stationary phases considered were C3, CN, PFP, C18, Bonus RP, and carbon-clad zirconia. Based on the experimental chromatograms the authors found that it was most important to have either the carbon-clad zirconia or the Bonus RP column in one of the two dimensions to obtain a good distribution of peaks across the 2D separation space. The carbon-clad zirconia column was not considered in the study of Lindsey *et al*, however this experimental work confirms the finding from the previous theoretical work that the Bonus RP phase is particularly attractive as a first choice to consider in method development.

The discussion in the preceding paragraphs was focused on RP selectivity selection for 2D-LC separations of small molecules. Over the past few years, Field, Euerby, and Petersson have initiated and continued a massive effort to characterize the selectivities of RP phases for (non-proteomic) separations of peptides [84–87]. Although it is still early in the development of this framework, and the knowledge has not yet been applied to 2D-LC separations, the framework looks very promising as a potential tool for selectivity selection in the case where a 2D-LC separation of peptides involving RP separation in both dimensions is desirable.

5.4 CLOSING REMARKS

Although much has been learned in recent years concerning the benefits and pitfalls associated with pairing different separation modes for 2D-LC methods (i.e., as summarized in Table 5.3), narrowing the list of viable options and selecting specific stationary phases is still a major impediment to 2D-LC method development. In some ways the same challenge is encountered in 1D-LC, but in 2D-LC it is considerably worse because there are two separations instead of one. Moreover, both must work well together, which is not always easy to arrange. In the years to come it will be important to develop more sophisticated frameworks to provide guidance in making these decisions that do not depend so strongly on user experience.

REFERENCES

[1] J.C. Giddings, Sample dimensionality: A predictor of order-disorder in component peak distribution in multidimensional separation, J. Chromatogr. A. 703 (1995) 3–15. doi:10.1016/0021-9673(95)00249-M

[2] A.M. Striegel, W.W. Yau, J.J. Kirkland, D.D. Bly, Modern Size-Exclusion Liquid Chromatography, 2nd ed., John Wiley & Sons, Inc., Hoboken, NJ, 2009. doi:10.1002/9780470442876

[3] T.H. Dean, J.R. Jezorek, Demonstration of simultaneous anion-exchange and reversed-phase behavior on a strong anion-exchange column, J. Chromatogr. A. 1028 (2004) 239–245. doi:10.1016/j.chroma.2003.11.068

[4] J. Xia, P.J. Gilmer, Organic modifiers in the anion-exchange chromatographic separation of sialic acids, J. Chromatogr. A. 676 (1994) 203–208. doi:10.1016/0021-9673(94)80461-3

[5] B.W.J. Pirok, J. Knip, M.R. van Bommel, P.J. Schoenmakers, Characterization of synthetic dyes by comprehensive two-dimensional liquid chromatography combining ion-exchange chromatography and fast ion-pair reversed-phase chromatography, J. Chromatogr. A. 1436 (2016) 141–146. doi:10.1016/j.chroma.2016.01.070

[6] K. Croes, A. Steffens, D.H. Marchand, L.R. Snyder, Relevance of π-π and dipole-dipole interactions for retention on cyano and phenyl columns in reversed-phase liquid chromatography, J. Chromatogr. A. 1098 (2005) 123–130. doi:10.1016/j.chroma.2005.08.090

[7] D.H. Marchand, K. Croes, J.W. Dolan, L.R. Snyder, R.A. Henry, K.M.R. Kallury et al., Column selectivity in reversed-phase liquid chromatography: VIII. Phenylalkyl and fluoro-substituted columns, J. Chromatogr. A. 1062 (2005) 65–78. doi:10.1016/j.chroma.2004.11.014

[8] B.W.J. Pirok, A.F.G. Gargano, P.J. Schoenmakers, Optimizing separations in online comprehensive two-dimensional liquid chromatography, J. Sep. Sci. 41 (2018) 68–98. doi:10.1002/jssc.201700863

[9] H. Tian, J. Xu, Y. Guan, Comprehensive two-dimensional liquid chromatography (NPLC×RPLC) with vacuum-evaporation interface, J. Sep. Sci. 31 (2008) 1677–1685. doi:10.1002/jssc.200700559

[10] A.M. Skvortsov, A.A. Gorbunov, Achievements and uses of critical conditions in the chromatography of polymers, J. Chromatogr. A. 507 (1990) 487–496. doi:10.1016/S0021-9673(01)84228-X

[11] R.T. Cimino, C.J. Rasmussen, Y. Brun, A. V. Neimark, Critical conditions of polymer adsorption and chromatography on non-porous substrates, J. Colloid Interface Sci. 474 (2016) 25–33. doi:10.1016/j.jcis.2016.04.002

[12] Y. Brun, P. Alden, Gradient separation of polymers at critical point of adsorption, J. Chromatogr. A. 966 (2002) 25–40. doi:10.1016/S0021-9673(02)00705-7

[13] K. Im, H. Park, S. Lee, T. Chang, Two-dimensional liquid chromatography analysis of synthetic polymers using fast size exclusion chromatography at high column temperature, J. Chromatogr. A. 1216 (2009) 4606–4610. doi:10.1016/j.chroma.2009.03.072

[14] P. Dugo, O. Favoino, R. Luppino, G. Dugo, L. Mondello, Comprehensive two-dimensional normal-phase (adsorption) – Reversed-phase liquid chromatography, Anal. Chem. 76 (2004) 2525–2530. doi:10.1021/ac0352981

[15] J.A. Raust, A. Bruell, P. Sinha, W. Hiller, H. Pasch, Two-dimensional chromatography of complex polymers, 8. Separation of fatty alcohol ethoxylates simultaneously by end group and chain length, J. Sep. Sci. 33 (2010) 1375–1381. doi:10.1002/jssc.200900775

[16] P. Jandera, J. Fischer, H. Lahovská, K. Novotná, P. Česla, L. Kolářová, Two-dimensional liquid chromatography normal-phase and reversed-phase separation of (co)oligomers, J. Chromatogr. A. 1119 (2006) 3–10. doi:10.1016/j.chroma.2005.10.081

[17] L. Mondello, P.Q. Tranchida, V. Stanek, P. Jandera, G. Dugo, P. Dugo, Silver-ion reversed-phase comprehensive two-dimensional liquid chromatography combined with mass spectrometric detection in lipidic food analysis, J. Chromatogr. A. 1086 (2005) 91–98. doi:10.1016/j.chroma.2005.06.017

[18] P. Dugo, T. Kumm, M.L. Crupi, A. Cotroneo, L. Mondello, Comprehensive two-dimensional liquid chromatography combined with mass spectrometric detection in the analyses of triacylglycerols in natural lipidic matrixes, J. Chromatogr. A. 1112 (2006) 269–275. doi:10.1016/j.chroma.2005.10.070

[19] Q. Yang, X. Shi, Q. Gu, S. Zhao, Y. Shan, G. Xu, On-line two dimensional liquid chromatography/mass spectrometry for the analysis of triacylglycerides in peanut oil and mouse tissue, J. Chromatogr. B. 895–896 (2012) 48–55. doi:10.1016/j.jchromb.2012.03.013

[20] D. V. McCalley, Hydrophilic interaction liquid chromatography: An update, LC-GC Eur. 37 (2019) 114–125.

[21] A.J. Alpert, Hydrophilic-interaction chromatography for the separation of peptides, nucleic acids and other polar compounds, J. Chromatogr. A. 499 (1990) 177–196. doi:10.1016/S0021-9673(00)96972-3

[22] P. Jandera, P. Janás, Recent advances in stationary phases and understanding of retention in hydrophilic interaction chromatography. A Review, Anal. Chim. Acta. 967 (2017) 12–32. doi:10.1016/j.aca.2017.01.060

[23] Y. Guo, Recent progress in the fundamental understanding of hydrophilic interaction chromatography (HILIC), Analyst. 140 (2015) 6452–6466. doi:10.1039/C5AN00670H

[24] M. Sorensen, D.C. Harmes, D.R. Stoll, G.O. Staples, S. Fekete, D. Guillarme et al., Comparison of originator and biosimilar therapeutic monoclonal antibodies using comprehensive two-dimensional liquid chromatography coupled with time-of-flight mass spectrometry, MAbs. 8 (2016) 1224–1234. doi:10.1080/19420862.2016.1203497

[25] D.R. Stoll, D.C. Harmes, G.O. Staples, O.G. Potter, C.T. Dammann, D. Guillarme, et al., Development of comprehensive online two-dimensional liquid chromatography/mass spectrometry using hydrophilic interaction and reversed-phase separations for rapid and deep profiling of therapeutic antibodies, Anal. Chem. 90 (2018) 5923–5929. doi:10.1021/acs.analchem.8b00776

[26] B.W.J. Pirok, D.R. Stoll, P.J. Schoenmakers, Recent developments in two-dimensional liquid chromatography: Fundamental improvements for practical applications, Anal. Chem. 91 (2019) 240–263. doi:10.1021/acs.analchem.8b04841

[27] J.C. Heaton, N.W. Smith, D.V. McCalley, Retention characteristics of some antibiotic and antiretroviral compounds in hydrophilic interaction chromatography using isocratic elution, and gradient elution with repeatable partial equilibration, Anal. Chim. Acta. 1045 (2019) 141–151. doi:10.1016/j.aca.2018.08.051

[28] C. Seidl, D.S. Bell, D.R. Stoll, A study of the re-equilibration of hydrophilic interaction columns with a focus on viability for use in two-dimensional liquid chromatography, J. Chromatogr. A. 1604 (2019). doi:10.1016/j.chroma.2019.460484

[29] A. D'Attoma, S. Heinisch, On-line comprehensive two dimensional separations of charged compounds using reversed-phase high performance liquid chromatography and hydrophilic interaction chromatography. Part II: Application to the separation of peptides, J. Chromatogr. A. 1306 (2013) 27–36. doi:10.1016/j.chroma.2013.07.048

[30] M. Holčapek, M. Ovčačíková, M. Lísa, E. Cífková, T. Hájek, Continuous comprehensive two-dimensional liquid chromatography-electrospray ionization mass spectrometry of complex lipidomic samples, Anal. Bioanal. Chem. 407 (2015) 5033–5043. doi:10.1007/s00216-015-8528-2

[31] K.M. Kalili, A. de Villiers, Systematic optimisation and evaluation of on-line, off-line and stop-flow comprehensive hydrophilic interaction chromatography×reversed phase liquid chromatographic analysis of procyanidins. Part II: Application to cocoa procyanidins, J. Chromatogr. A. 1289 (2013) 69–79. doi:10.1016/j.chroma.2013.03.009

[32] C.M. Willemse, M.A. Stander, A.G.J. Tredoux, A. de Villiers, Comprehensive two-dimensional liquid chromatographic analysis of anthocyanins, J. Chromatogr. A. 1359 (2014) 189–201. doi:10.1016/j.chroma.2014.07.044

[33] C. Sun, Y.Y. Zhao, J.M. Curtis, Elucidation of phosphatidylcholine isomers using two dimensional liquid chromatography coupled in-line with ozonolysis mass spectrometry, J. Chromatogr. A. 1351 (2014) 37–45. doi:10.1016/j.chroma.2014.04.069

[34] A.F.G. Gargano, M. Duffin, P. Navarro, P.J. Schoenmakers, Reducing dilution and analysis time in online comprehensive two-dimensional liquid chromatography by active modulation., Anal. Chem. 1 (2016) 1–16. doi:10.1021/acs.analchem.5b04051

[35] Y. Wang, X. Lu, G. Xu, Development of a comprehensive two-dimensional hydrophilic interaction chromatography/quadrupole time-of-flight mass spectrometry system and its application in separation and identification of saponins from Quillaja saponaria, J. Chromatogr. A. 1181 (2008) 51–59. doi:10.1016/j.chroma.2007.12.034

[36] B.W.J. Pirok, M.J. den Uijl, G. Moro, S.V.J. Berbers, C.J.M. Croes, M.R. van Bommel et al., Characterization of dye extracts from historical cultural-heritage objects using state-of-the-art comprehensive two-dimensional liquid chromatography and mass spectrometry with active modulation and optimized shifting gradients, Anal. Chem. 91 (2019) 3062–3069. doi:10.1021/acs.analchem.8b05469

[37] D.R. Stoll, Pass the salt – Evolution of coupling ion-exchange separations and mass spectrometry, LCGC North Am. 37 (2019) 405–409.

[38] D.A. Wolters, M.P. Washburn, J.R. Yates, An automated multidimensional protein identification technology for shotgun proteomics, Anal. Chem. 73 (2001) 5683–5690. doi:10.1021/ac010617e

[39] A.M. Striegel, A.K. Brewer, Hydrodynamic chromatography, Annu. Rev. Anal. Chem. 5 (2012) 15–34. doi:10.1146/annurev-anchem-062011-143107

[40] B.W.J. Pirok, N. Abdulhussain, T. Aalbers, B. Wouters, R.A.H. Peters, P.J. Schoenmakers, Nanoparticle analysis by online comprehensive two-dimensional liquid chromatography combining hydrodynamic chromatography and size-exclusion chromatography with intermediate sample transformation, Anal. Chem. 89 (2017) 9167–9174. doi:10.1021/acs.analchem.7b01906

[41] E. Uliyanchenko, Size-exclusion chromatography-from high-performance to ultra-performance, Anal. Bioanal. Chem. 406 (2014) 6087–6094. doi:10.1007/s00216-014-8041-z

[42] E. Uliyanchenko, P.J. Schoenmakers, S. van der Wal, Fast and efficient size-based separations of polymers using ultra-high-pressure liquid chromatography, J. Chromatogr. A. 1218 (2011) 1509–1518. doi:10.1016/j.chroma.2011.01.053

[43] B.W.J. Pirok, P. Breuer, S.J.M. Hoppe, M. Chitty, E. Welch, T. Farkas et al., Size-exclusion chromatography using core-shell particles, J. Chromatogr. A. 1486 (2017) 96–102. doi:10.1016/j.chroma.2016.12.015

[44] M.R. Schure, R.E. Moran, Size exclusion chromatography with superficially porous particles, J. Chromatogr. A. 1480 (2017) 11–19. doi:10.1016/j.chroma.2016.12.016

[45] A. Tiselius, Adsorption separation by salting out, Ark. Kemi, Miner. Geol. 26 (1948) 1–5.

[46] J.A. Queiroz, C.T. Tomaz, J.M.S. Cabral, Hydrophobic interaction chromatography of proteins, J. Biotechnol. 87 (2001) 143–159. doi:10.1016/S0168-1656(01)00237-1

[47] L. Xiu, S.G. Valeja, A.J. Alpert, S. Jin, Y. Ge, Effective protein separation by coupling hydrophobic interaction and reverse phase chromatography for top-down proteomics, Anal. Chem. 86 (2014) 7899–7906. doi:10.1021/ac501836k

[48] B. Chen, Y. Peng, S.G. Valeja, L. Xiu, A.J. Alpert, Y. Ge, Online Hydrophobic interaction chromatography – Mass spectrometry for top-down proteomics, Anal. Chem. 88 (2016) 1885–1891. doi:10.1021/acs.analchem.5b04285

[49] M. Sarrut, A. Corgier, S. Fekete, D. Guillarme, D. Lascoux, M.-C. Janin-Bussat et al., Analysis of antibody-drug conjugates by comprehensive on-line two-dimensional hydrophobic interaction chromatography x reversed phase liquid chromatography hyphenated to high resolution mass spectrometry. I – Optimization of separation conditions, J. Chromatogr. B. 1032 (2016) 103–111. doi:10.1016/j.jchromb.2016.06.048

[50] M. Sarrut, S. Fekete, M.C. Janin-Bussat, O. Colas, D. Guillarme, A. Beck et al., Analysis of antibody-drug conjugates by comprehensive on-line two-dimensional hydrophobic interaction chromatography x reversed phase liquid chromatography hyphenated to high resolution mass spectrometry. II- Identification of sub-units for the characteri, J. Chromatogr. B. 1032 (2016) 91–102. doi:10.1016/j.jchromb.2016.06.049

[51] C. West, Enantioselective separations with supercritical fluids-review, Curr. Anal. Chem. 10 (2014) 99–120. doi:10.2174/1573411011410010009

[52] K. De Klerck, D. Mangelings, Y. Vander Heyden, Supercritical fluid chromatography for the enantioseparation of pharmaceuticals, J. Pharm. Biomed. Anal. 69 (2012) 77–92. doi:10.1016/j.jpba.2012.01.021

[53] F. Gritti, G. Guiochon, Possible resolution gain in enantioseparations afforded by core–shell particle technology, J. Chromatogr. A. 1348 (2014) 87–96. doi:10.1016/j.chroma.2014.04.041

[54] K. Schmitt, U. Woiwode, M. Kohout, T. Zhang, W. Lindner, M. Lämmerhofer, Comparison of small size fully porous particles and superficially porous particles of chiral anion-exchange type stationary phases in ultra-high performance liquid chromatography: Effect of particle and pore size on chromatographic efficiency and kinetic, J. Chromatogr. A. 1569 (2018) 149–159. doi:10.1016/j.chroma.2018.07.056

[55] X. Zheng, Z. Li, S. Beeram, M. Podariu, R. Matsuda, E.L. Pfaunmiller et al., Analysis of biomolecular interactions using affinity microcolumns: A review, J. Chromatogr. B. 968 (2014) 49–63. doi:10.1016/j.jchromb.2014.01.026

[56] W. Cho, K. Jung, F.E. Regnier, Screening antibody and immunosorbent selectivity by two-dimensional liquid chromatography-MS/MS (2-D LC-MS/MS), J. Sep. Sci. 33 (2010) 1438–1447. doi:10.1002/jssc.200900860

[57] A. Williams, E.K. Read, C.D. Agarabi, S. Lute, K.A. Brorson, Automated 2D-HPLC method for characterization of protein aggregation with in-line fraction collection device, J. Chromatogr. B. 1046 (2017) 122–130. doi:10.1016/j.jchromb.2017.01.021

[58] S. Han, J. Huang, J. Hou, S. Wang, Screening epidermal growth factor receptor antagonists from Radix et Rhizoma Asari by two-dimensional liquid chromatography, J. Sep. Sci. 37 (2014) 1525–1532. doi:10.1002/jssc.201400236

[59] M. Pepaj, P.M. Thorsby, Analysis of glycated albumin by on-line two-dimensional liquid chromatography mass spectrometry, J. Liq. Chromatogr. Relat. Technol. 38 (2015) 20–28. doi:10.1080/10826076.2013.864980

[60] B. Zhang, S. Mathewson, H. Chen, Two-dimensional liquid chromatographic methods to examine phenylboronate interactions with recombinant antibodies, J. Chromatogr. A. 1216 (2009) 5676–5686. doi:10.1016/j.chroma.2009.05.084

[61] A.F. Bergold, D. Hanggi, A. Muller, P.W. Carr, High-performance affinity chromatography, in: C. Horvath (Ed.), High-Performance Liq. Chromatogr. Adv. Perspect. Vol. 5, Academic Press, New York, 1980: pp. 95–209.

[62] Z.D. Dunn, J. Desai, G.M. Leme, D.R. Stoll, D.D. Richardson, Rapid two-dimensional Protein-A size exclusion chromatography of monoclonal antibodies for titer and aggregation measurements from harvested cell culture fluid samples, MAbs. 12 (2020). doi:10.1080/19420862.2019.1702263

[63] K. Sandra, M. Steenbeke, I. Vandenheede, G. Vanhoenacker, P. Sandra, The versatility of heart-cutting and comprehensive two-dimensional liquid chromatography in monoclonal antibody clone selection, J. Chromatogr. A. 1523 (2017) 283–292. doi:10.1016/j.chroma.2017.06.052

[64] L. Hu, X. Li, S. Feng, L. Kong, X. Su, X. Chen et al., Comprehensive two-dimensional HPLC to study the interaction of multiple components in Rheum palmatum L. with HSA by coupling a silica-bonded HSA column to a silica monolithic ODS column, J. Sep. Sci. 29 (2006) 881–888. doi:10.1002/jssc.200500442

[65] E. Lesellier, C. West, The many faces of packed column supercritical fluid chromatography – A critical review, J. Chromatogr. A. 1382 (2015) 2–46. doi:10.1016/j.chroma.2014.12.083

[66] S. Jespers, K. Broeckhoven, G. Desmet, Comparing the separation speed of contemporary LC, SFC, and GC, LC-GC Eur. 30 (2017) 284–291.

[67] A. Grand-Guillaume Perrenoud, C. Hamman, M. Goel, J.-L. Veuthey, D. Guillarme, S. Fekete, Maximizing kinetic performance in supercritical fluid chromatography using state-of-the-art instruments., J. Chromatogr. A. 1314 (2013) 288–297. doi:10.1016/j.chroma.2013.09.039

[68] M. Sarrut, A. Corgier, G. Crétier, A. Le Masle, S. Dubant, S. Heinisch, Potential and limitations of on-line comprehensive reversed phase liquid chromatography×supercritical fluid chromatography for the separation of neutral compounds: An approach to separate an aqueous extract of bio-oil, J. Chromatogr. A. 1402 (2015) 124–133. doi:10.1016/j.chroma.2015.05.005

[69] C.J. Venkatramani, M. Al-Sayah, G. Li, M. Goel, J. Girotti, L. Zang et al., Simultaneous achiral-chiral analysis of pharmaceutical compounds using two-dimensional reversed phase liquid chromatography-supercritical fluid chromatography, Talanta. 148 (2016) 548–555. doi:10.1016/j.talanta.2015.10.054

[70] I. François, A. de Villiers, B. Tienpont, F. David, P. Sandra, Comprehensive two-dimensional liquid chromatography applying two parallel columns in the second dimension, J. Chromatogr. A. 1178 (2008) 33–42. doi:10.1016/j.chroma.2007.11.032

[71] I. François, K. Sandra, P. Sandra, P. Dugo, L. Mondello, F. Cacciola et al., Comprehensive Chromatography in Combination with Mass Spectrometry, John Wiley & Sons, Inc., Hoboken, NJ, 2011. doi:10.1002/9781118003466

[72] Y. Hirata, F. Ozaki, Comprehensive two-dimensional capillary supercritical fluid chromatography in stop-flow mode with synchronized pressure programming, Anal. Bioanal. Chem. 384 (2006) 1479–1484. doi:10.1007/s00216-006-0349-x

[73] D. Okomoto, Y. Hirata, Development of supercritical fluid extraction coupled to comprehensive two-dimensional supercritical fluid chromatography (SFE-SFC×SFC), Anal. Sci. 22 (2006) 1437–1440. doi:10.2116/analsci.22.1437

[74] J. Blomberg, Feasibility of comprehensive two-dimensional capillary supercritical fluid chromatography (SFC×SFC) on (heavy) petroleum fractions, in: 38th Int., Symp. Capill. Chromatogr., 2014.

[75] I. François, D. Cabooter, K. Sandra, F. Lynen, G. Desmet, P. Sandra, Tryptic digest analysis by comprehensive reversed phase×two reversed phase liquid chromatography (RP-LC×2RP-LC) at different pH's, J. Sep. Sci. 32 (2009) 1137–1144. doi:10.1002/jssc.200800578

[76] D.R. Stoll, H.R. Lhotka, D.C. Harmes, B. Madigan, J.J. Hsiao, G.O. Staples, High resolution two-dimensional liquid chromatography coupled with mass spectrometry for robust and sensitive characterization of therapeutic antibodies at the peptide level, J. Chromatogr. B. 1134–1135 (2019) 121832. doi:10.1016/j.jchromb.2019.121832

[77] J.W. Dolan, R. Snyder, Lloyd, The hydrophobic-subtraction model for reversed-phase liquid chromatography: A reprise, LC-GC North Am. 34 (2016) 730–741.

[78] L. Snyder, J. Dolan, D. Marchand, P. Carr, The hydrophobic-subtraction model of reversed-phase column selectivity, in Advances in Chromatography: 2012, pp. 297–376. doi:10.1201/b11636-8

[79] P. Žuvela, M. Skoczylas, J. Jay Liu, T. Bączek, R. Kaliszan, M.W. Wong et al., Column characterization and selection systems in reversed-phase high-performance liquid chromatography, Chem. Rev. 119 (2019) 3674–3729. doi:10.1021/acs.chemrev.8b00246

[80] L.R. Snyder, Classification of the solvent properties of common liquids, J. Chromatogr. A. 92 (1974) 223–230. doi:10.1016/S0021-9673(00)85732-5

[81] Y. Zhang, P.W. Carr, A visual approach to stationary phase selectivity classification based on the Snyder-Dolan Hydrophobic-Subtraction Model, J. Chromatogr. A. 1216 (2009) 6685–6694. doi:10.1016/j.chroma.2009.06.048

[82] R.K. Lindsey, B.L. Eggimann, D.R. Stoll, P.W. Carr, M.R. Schure, J.I. Siepmann, Column selection for comprehensive two-dimensional liquid chromatography using the hydrophobic subtraction model, J. Chromatogr. A. 1589 (2019) 47–55. doi:10.1016/j.chroma.2018.09.018

[83] R.C. Allen, B.B. Barnes, I.A. Haidar Ahmad, M.R. Filgueira, P.W. Carr, Impact of reversed phase column pairs in comprehensive two-dimensional liquid chromatography, J. Chromatogr. A. 1361 (2014) 169–177. doi:10.1016/j.chroma.2014.08.012

[84] J.K. Field, M.R. Euerby, K.F. Haselmann, P. Petersson, Investigation into reversed-phase chromatography peptide separation systems Part IV: Characterisation of mobile phase selectivity differences, J. Chromatogr. A. 1641 (2021) 461986. doi:10.1016/j.chroma.2021.461986

[85] J.K. Field, M.R. Euerby, P. Petersson, Investigation into reversed phase chromatography peptide separation systems part III: Establishing a column characterisation database, J. Chromatogr. A. 1622 (2020) 461093. doi:10.1016/j.chroma.2020.461093

[86] J.K. Field, M.R. Euerby, P. Petersson, Investigation into reversed phase chromatography peptide separation systems part II: An evaluation of the robustness of a protocol for column characterisation, J. Chromatogr. A. 1603 (2019) 102–112. doi:10.1016/j.chroma.2019.05.037

[87] J.K. Field, M.R. Euerby, J. Lau, H. Thøgersen, P. Petersson, Investigation into reversed phase chromatography peptide separation systems part I: Development of a protocol for column characterisation, J. Chromatogr. A. 1603 (2019) 113–129. doi:10.1016/j.chroma.2019.05.038

6 Method Development for Non-Comprehensive Two-Dimensional Liquid Chromatography

Stephan M.C. Buckenmaier, Sascha Lege, and Dwight R. Stoll

CONTENTS

6.1   INTRODUCTION

The simplest non-comprehensive mode of 2D-LC separation is heartcutting 2D-LC (LC-LC). The purpose of heartcutting is to further examine targeted sections of a 1D chromatogram. Different approaches that have been used to carry out LC-LC separations were discussed in detail in Chapter 4. Here we repeat schematic drawings of the most commonly used hardware configurations for ease of reference, but then go into more detail about how the hardware is used in practice as it relates to chromatograms and instrument operation. Figure 6.1A illustrates a simple LC-LC separation where two fractions (cut #1 and 2) are taken from regions of the ^{1}D separation that are unresolved; these fractions are then subjected to further separation in the second dimension.

The most commonly used approach to collect fractions from the first dimension and transfer them to the second dimension involves an interface valve, which connects the components of the separation hardware used in the two dimensions of the online 2D-LC system. The movements of the interface valve determine when the effluent from the ^{1}D column is collected and then injected into the ^{2}D column. Figure 6.1B shows an alternating dual-loop interface, which involves two independent flow paths. Each path can accommodate one fraction of the ^{1}D effluent (i.e., Loops 1 and 2). One loop is always connected to the first dimension for sampling, and the other loop is connected to the second dimension where the separation of a previously collected fraction occurs. Consequently, in Figure 6.1A, cut #1 is sampled in Loop 1 (B, pos-1) and upon switching the interface (B, pos-2) this fraction is injected into the ^{2}D column. Cut #2, which is collected in Loop 2, cannot be transferred to the second dimension until the separation of cut #1 in the second dimension has been completed.

A shortening of the ^{2}D separation cycle would enable collection and ^{2}D separation of additional fractions in a given region of a ^{1}D chromatogram. In the limit of very short ^{2}D times and collection of many fractions, this of course evolves into comprehensive 2D-LC (LC×LC). When a dual-loop interface like that shown in Figure 6.1B is used for LC×LC the time committed to sampling each fraction from the first dimension must be equal to the time committed to the separation of each fraction in the second dimension. Therefore, ^{2}D cycle times must be very short (typically less than 60 s), or the ^{1}D separation must be deliberately slowed down in order to increase the optimum ^{2}D

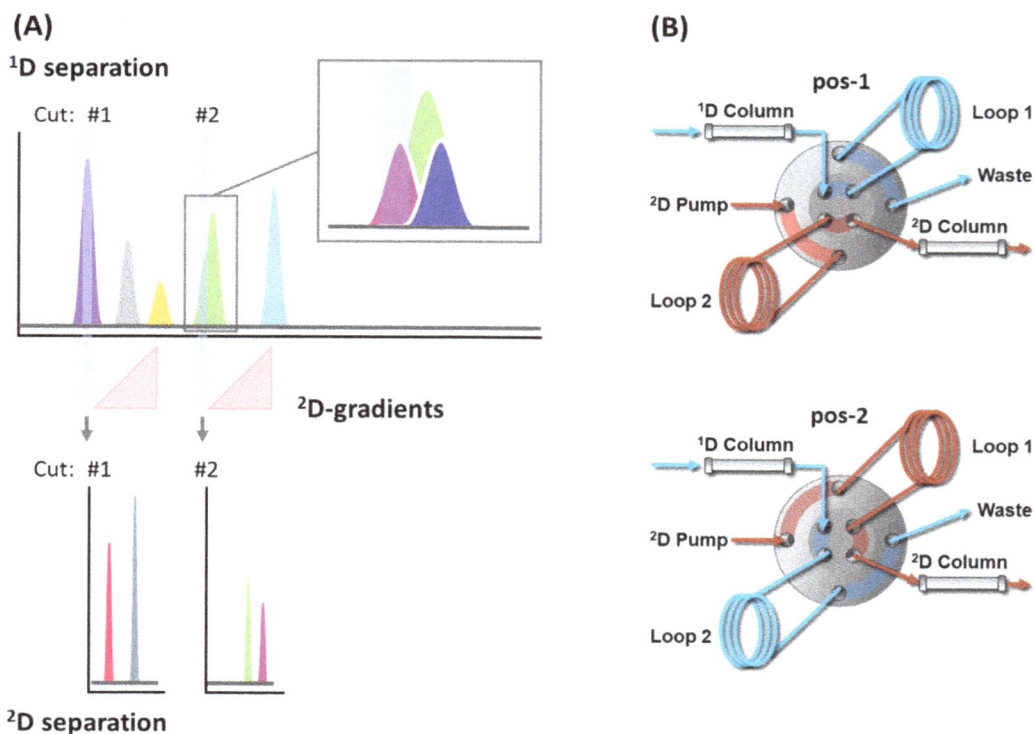

FIGURE 6.1 A) Illustration of heartcutting 2D-LC (LC-LC). B) Schematic of alternating dual-loop interface that connects the first and second dimensions of a 2D-LC system. Pos-1: Loop 1 samples ^{1}D effluent, Loop 2 connected for ^{2}D analysis; and Pos-2: roles of Loops 1 and 2 are reversed.

cycle time; the latter approach is certainly not desirable in situations where sample throughput is important.

An advanced format of LC-LC has been developed and become known as multiple heartcutting (MHC) 2D-LC (mLC-LC) [1–3]. MHC enables collection of more than one fraction of ^{1}D effluent while the second dimension is committed to separating previously collected fractions; that is, sampling fractions of the ^{1}D separation and ^{2}D separation of other fractions can occur in parallel. Figure 6.1A shows that the simple LC-LC approach does not allow sampling of ^{1}D peaks that elute close together in time; the MHC approach solves this problem. Figure 6.2A, where we assume that the same hypothetical ^{1}D separation is being done as in Figure 6.1A, shows how the MHC approach enables sampling of every peak of interest in the ^{1}D chromatogram without altering the ^{1}D separation time. This requires a special MHC interface connecting the first and second dimensions, which consist of two parking decks (A and B), each holding several sampling loops (Figure 6.2B). The decks are connected to the main 2D-LC interface valve that determines which deck is connected to the ^{1}D flow path, and which one is connected to the ^{2}D flow path. Figure 6.2B (pos-1) shows that cut #1, which was sampled into Loop 1 of Deck A, is immediately processed in the second dimension. During this ^{2}D separation cycle, cuts #2 and 3 become parked in Loops 1 and 2 of Deck B. As depicted in Figure 6.2B (pos-2), once the main interface valve switches, cuts #2 and 3 can be injected into the ^{2}D flow path, while cuts #4 to 7 are parked into Deck A.

The MHC approach illustrated in Figure 6.2 shows how it provides advantages of both LC-LC and LCxLC separations; it enables targeting many peaks of interest in the ^{1}D separation, but also provides tremendous flexibility, particularly in the time that can be committed to the ^{2}D separation

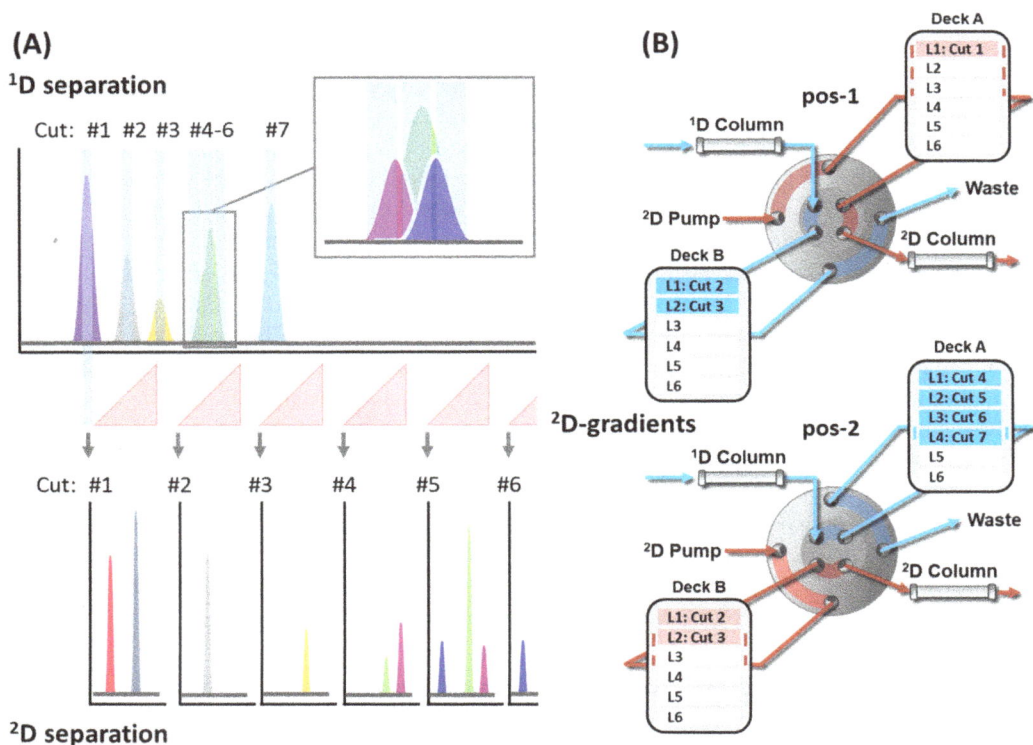

FIGURE 6.2 A) Illustration of multiple heartcutting (MHC) 2D-LC. (B) MHC interface with two parking decks (A and B) each holding six sampling loops. B) pos-1: Deck B samples ^{1}D effluent, while Deck A is connected to the ^{2}D flow path for separation of previously collected fractions and (B) pos-2: vice versa.

of each fraction of ^{1}D effluent. This in turn allows the user to operate both separation dimensions under nearly optimal conditions. The use of optimum flow rates, longer ^{2}D gradients, and columns with higher separation efficiency combined improve the overall quality of the 2D-LC separation. At the same time, the ability to accommodate many fractions over the width of a ^{1}D peak can help avoid information loss due to undersampling of the ^{1}D separation, as discussed in Chapter 3.

Heartcutting 2D-LC methods are now routinely applied for both qualitative and quantitative analyses. There are many literature reports describing applications in fields ranging from environmental and food/beverage science, to the chemical and (bio-) pharmaceutical industries [4]. The 2D-LC literature explicitly points to various advantages, including the enhanced level of confidence in chromatographic results that is gained by increasing the resolving power due to the addition of the ^{2}D separation, as well as the increase in sample throughput. A few representative examples are discussed briefly in the following sections.

Venkatramani et al. [5] demonstrated the use of 2D-LC to separate potential impurities/degradants in the front or tail of the peak for a low molecular weight active pharmaceutical ingredient. This particular study showed the benefit of deploying columns of the same type of stationary phase in both dimensions. The authors proposed that 2D-LC could be used as a method assessment tool in pharmaceutical and other industries to gain confidence in the correctness of results obtained from conventional 1D-LC methods.

Petersson et al. [6] used MHC 2D-LC to identify impurities related to biopharmaceutical molecules, which were separated under ^{1}D conditions when high salt concentrations are present in the mobile phase. Mass spectrometry (MS) analysis of peaks of interest in the first dimension was enabled by adding a second dimension operated with the same type of stationary phase used in the first dimension, but under MS friendly conditions. This 2D-LC approach provides an alternative to changing mobile phase conditions associated with the 1D-LC method to make them MS friendly, where a common challenge encountered when adjusting mobile phase conditions is that peaks of interest move around due to changes in selectivity, and tracking them can be difficult. Therefore, 2D-LC eliminates the guesswork associated with peak tracking during method development. Moreover, this work showed that 2D-LC enabled the differentiation of isomers that was not observed previously, quantification of low-level impurities, and a substantial enhancement in sample throughput with the time required for an impurity profiling workflow reduced from days to hours.

Other workers have used 2D-LC to increase throughput by measuring multiple quality attributes in a single assay [7, 8]. For instance, Dunn et al. [7] determined titre and aggregation level of monoclonal antibodies by directly analyzing clarified cell culture fluid in under 5 min. per sample, by combining affinity separation (Protein-A) and SEC in a LC-LC format. This compares favorably to 2D-LC methods previously developed for this purpose, which required 20 min. per analysis.

Another way to decrease analysis time is to use the additional selectivity and resolving power of a 2D-LC method rather than spending significant effort on optimizing a 1D-LC method for a particular sample (i.e. by screening stationary phases, mobile phases, gradient slopes, etc.). In other words, one can approach a novel sample type with a 2D-LC method that uses generic separation conditions in both the first and second dimensions. Each peak observed in the first dimension can be sampled and further analyzed in the second dimension for further separation of compounds co-eluting from the first dimension. With this approach, the development effort for a 1D-LC method, which could be time-consuming for complicated samples, can be dramatically reduced or possibly eliminated [9].

6.2   GENERAL THOUGHTS ON METHOD DEVELOPMENT FOR NON-COMPREHENSIVE 2D-LC

Comprehensive 2D-LC generally aims for the best coverage of the complete separation space and therefore requires orthogonal separation mechanisms in both dimensions. In contrast, non-comprehensive 2D-LC can be understood as a simpler case where differences in selectivity between the ^{1}D and ^{2}D separations have to be achieved, but *only for the regions of interest in the ^{1}D chromatogram*. Furthermore, since the timescales of the ^{1}D and ^{2}D separations are now decoupled, the traditional 1D-LC method development strategies (e.g., choice of flow rate, column dimensions) can be used, independent of which dimension is being modified.

The first step in any method development process is to consider the physico-chemical properties of the molecules that are to be separated. If the components of the sample vary by charge and hydrophobicity (e.g., peptides), one might choose to combine ion exchange (IEX) and reversed-phase (RP) separations. Other considerations include variations in size, chirality or H-bonding ability. This topic is discussed extensively in Chapter 5.

It is noteworthy that several successful 2D-LC applications have been developed that involve RP separations in both dimensions, in spite of the concept that two RP separation dimensions would not be generally considered orthogonal [1, 5, 6, 10–12]. The success of RP-RP separations is due in part to the fact that RP is the most frequently used mode for 1D-LC and analytical scientists are quite familiar with parameters that influence retention and selectivity of RP separations, including type of organic modifier, pH, gradient steepness, and temperature. There is a staggeringly wide variety of stationary phase chemistries available from many vendors, including C8, C18, phenyl-hexyl, and pentafluorophenyl (PFP). Moreover, marked differences in selectivity can arise from differences in the chemistry of the base silica particles used by different manufacturers. Finally, RP separations generally provide high plate numbers, and high peak capacities when gradient elution is used in both dimensions.

6.3 FOCUS OF THIS CHAPTER

This chapter outlines options and best practices for method development in non-comprehensive 2D-LC. To illustrate these using a real method development example, we have chosen to work with a moderately complex mixture of low molecular weight compounds, and demonstrate their resolution using RP separations in both dimensions.

Through this case study, we will demonstrate:

- that data from generic 1D-HPLC scouting runs can be used to develop successful 2D-LC methods.
- that parameters influencing chromatographic selectivity can be used to fine tune 2D-LC separations.
- strategies to optimize separation performance and robustness that are specific to 2D-LC.
- use of non-comprehensive 2D-LC for quantification purposes.

6.4 EXPERIMENTAL DETAILS

The following case study is focused on a sample of 15 low molecular weight compounds used commercially as herbicides; their structures are shown in Figure 6.3. Analytical solutions used in the experiments described below were prepared by dilution of a sample containing each compound at 1 mg/mL using a diluent similar to the mobile phase used as the starting point in gradient elution programs. Experimental details for the baseline conditions used in the case study are described in the following sections, and summarized in Table 6.1. The table is helpful for identifying which variables have been explored in each section. Other conditions described below illustrate the effects of changes in conditions relative to this baseline.

6.4.1 FIRST DIMENSION

The first dimension used 1290 Infinity II instrument modules (Agilent Technologies) including a binary pump, an autosampler, a multi-column thermostat, and a diode array UV/Vis absorbance

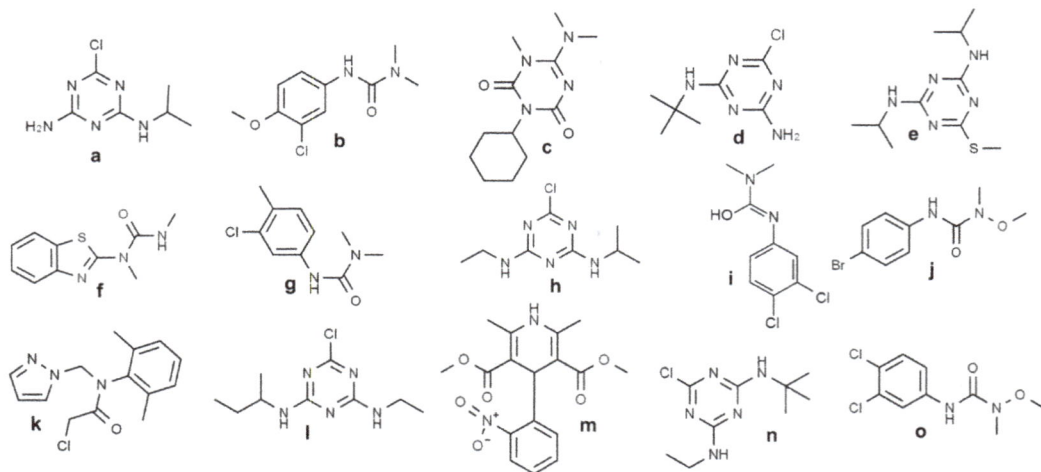

FIGURE 6.3 Chemical structures of the 15 compounds in the sample mixture. Atrazine-desethyl (a), metoxuron (b), hexazinone (c), terbuthylazine-desethyl (d), prometryn (e), methabenzthiazuron (f), chlorotoluron (g), atrazine (h), diuron (i), metobromuron (j), metazachlor (k), sebuthylazine (l), nifedipine (m), terbuthylazine (n), linuron (o).

detector equipped with 10 mm flow cell. The detector acquired the 254 nm UV absorbance trace at 20 pts/s. Columns were 50 mm×3.0 mm i.d. (1.8 µm) and held at 40 °C, unless otherwise noted. Gradients were 10–100 %B in 18 min (i.e. 5 %B/min) and the flow rate was held constant at 0.6 mL/min. Solvent A was water containing 0.2% formic acid (FA). Solvent B was acetonitrile (ACN) or methanol (MeOH), and the sample injection volume was 1 µL. Note that all solvent percentages indicated are (v/v) unless otherwise indicated.

6.4.2 SECOND DIMENSION

The second dimension used 1290 Infinity II instrument modules (Agilent Technologies) including a binary pump, a multi-column thermostat, and a diode array UV/Vis absorbance detector equipped with 10 mm flow cell. The detector tracked the 254-nm UV trace at 80 pts/s. Columns were 50 mm×3.0 mm or 2.1 mm i.d. (1.8 µm) and held at 40°C, unless otherwise noted. Gradients were 25–70 %B in 9 min (i.e., 5 %B/min) and the flow rate was held constant at 0.6 mL/min, unless otherwise noted. Solvent A was water containing 0.2% FA or 0.2% NH$_4$OH. Solvent B was ACN or MeOH. ASM experiments used an isocratic hold at 5 %B at the start of each ^{2}D gradient for the duration of sample displacement from the loop (here 3 x loop volume).

6.4.3 MULTIPLE HEARTCUTTING 2D-LC INTERFACE

The system was equipped with the multiple heartcutting (MHC) 2D-LC interface (Agilent Technologies; G4242A) including two parking decks each holding six sampling loops. Loops with 40, 80, 120, or 180 µL sampling capacity were used. A schematic is shown in Figure 6.2B, except that the data discussed below were collected with an active solvent modulation (ASM) valve as the central 2D-LC valve. The ASM functionality of the valve was only applied where explicitly stated. Its function has been described in detail in Chapter 4 of this book and will be briefly revisited in the context of Section 6.8 of this chapter.

TABLE 6.1

Experimental Conditions Used for 2D-LC Separation of the 15-Component Herbicide Mixture

Location		First Dimension			Second Dimension		
Section	Figure	Mobile phases	Column[a]	T (°C)	Mobile phases	Column[d]	T (°C)
6.5: Influence of mobile phase on selectivity	6.4A	A: Water (0.2 %FA), B: ACN	SB C18[b]	40	n.a.	n.a.	n.a.
	6.4B	A: Water (0.2 %FA), B: MeOH			n.a.	n.a.	n.a.
	6.4C	A: Water (0.2 %FA), B: MeOH			A: Water (0.2 %FA), B: ACN	SB C18	40
	6.5A-C	A: Water (0.2 %FA), B: MeOH	SB C18	40	A: Water (0.2 %FA), B: ACN	SB C18	25, 40, 55
	6.6	A: Water (0.2 %FA), B: MeOH	SB C18	40	A: Water (0.2 %FA), B: ACN	SB C18	40
	6.7A	A: Water (0.2 %FA), B: ACN	SB C18	40	A: Water (0.2 %NH$_4$OH), B: ACN	HPH C18	40
	6.7B	A: Water (0.2 %FA or NH$_4$OH), B: ACN	HPH C18[c]		n.a.	n.a.	n.a.
6.6: Influence of stationary phase on selectivity	6.8A 6.8B 6.8C	A: Water (0.2 %FA), B: ACN	SB C18	40	n.a. A: Water (0.2 %FA), B: ACN A: Water (0.2 %FA), B: MeOH	n.a. Bonus RP[e] Phenyl Hexyl[f]	n.a. 40 40
6.7: Optimizing gradient elution conditions for 2D	6.9A 6.9C	A: Water (0.2 %FA), B: MeOH	SB C18	40	n.a. A: Water (0.2 %FA), B: ACN	n.a. SB C18	n.a. 25
6.8: Active Solvent Modulation (ASM)	6.10A 6.10B	A: Water (0.2 %FA), B: MeOH	SB C18	40	A: Water (0.2 %FA), B: ACN	SB C18	25
	6.11A 6.11B 6.11C 6.11D	A: Water (0.2 %FA), B: MeOH	SB C18	40	A: Water (0.2 %FA), B: ACN	SB C18 SB C18[g] Bonus RP Bonus RP	25
6.10: Quantitation in 2D-LC	6.12A 6.12B 6.12C	A: Water (0.2 %FA), B: MeOH	SB C18	40	n.a. A: Water (0.2 %FA), B: ACN A: Water (0.2 %FA), B: ACN	n.a. SB C18 SB C18	n.a. 25 25

Notes:
a) Column dimensions: 50 mm x 3.0 mm i.d.
b) Zorbax SB-C18, 1.8 µm, unless noted otherwise
c) Poroshell HPH C18, 2.7 µm
d) Column dimensions: 50 mm x 3.0 mm i.d.
e) Zorbax Bonus RP, 1.8 µm
f) Zorbax Phenyl-Hexyl, 1.8 µm
g) Column dimensions: 50 mm x 2.1 mm i.d.

6.4.4 Mass Spectrometry

MS data were acquired using a 6550 iFunnel QTOF MS (Agilent Technologies) equipped with an ESI Jet Stream ion source, and operated in positive mode. Mass range, 100–3000 m/z; acquisition rate, 10 spectra/s; gas temperature, 250°C; gas flow, 14 L/min; nebulizer pressure, 35 psi; sheath gas temperature, 350°C; sheath gas flow, 11 L/min; Vcap, 3500 V; Nozzle, 500 V. Data were acquired using Mass Hunter LC/MS and OpenLAB CDS ChemStation (Agilent Technologies).

6.5 INFLUENCE OF THE MOBILE PHASE ON SELECTIVITY

The mobile phase in RP chromatography consists of an aqueous component and an organic solvent modifier. When ionizable compounds are analyzed, buffers and other additives may be present to control retention, peak shape, and selectivity. Selectivity is the most effective tool for optimizing methods in 1D-LC, as well as 2D-LC. Changes in mobile phase properties (e.g., composition, gradient steepness, pH etc.) would commonly be investigated during method development prior to changing either or both of the stationary phases to different chemistries [13].

6.5.1 Solvent Types

Solvent selection is one of the most important parameters in HPLC separations because of the effect it can have on the selectivity. Acetonitrile (ACN) and methanol (MeOH) are by far the most frequently used as the organic solvent modifiers in RP separations.

Figure 6.4 shows 1D chromatograms obtained for RP separations of the test mixture using either ACN (Panel A) or MeOH (Panel B) as the organic modifier (all other chromatographic conditions are described in Section 6.4.1).

Baseline separation of all analytes was not achieved by 1D-LC using either of the above two organic solvents. In the case of ACN, three regions of co-eluting compounds were observed (I to III in Figure 6.4A) containing compounds e-h, l+m, and n+o, respectively. When using MeOH, co-elution of analytes was observed in four regions of the 1D chromatogram (I to IV in Figure 6.4B), containing the compounds c+d, g+j, e+f+h+i, and l+o, respectively. However, what became obvious from the comparison of these chromatograms is the difference in chromatographic selectivity due to the change in organic solvents. For instance, analytes coeluting in the regions I, II, and IV in case of MeOH are well resolved in the 1D-LC separation with ACN. Only the resolution of the compound pair e+f might still remain critical under both conditions (region I in Figure 6.4A and region III in Figure 6.4B).

These observations led to initial 2D-LC experiments using the same stationary phase in both dimensions. Methanol was used as the organic modifier in the first dimension. It has lower elution strength than acetonitrile and thus facilitates analyte focusing at the inlet of the ^{2}D column; this was valuable since ASM (see Section 6.8) has not been applied in this experiment. Eleven 40 μL cuts were taken from the first dimension (grey highlights in Figure 6.4B) and re-analyzed in the second dimension using ACN as the organic modifier. The same gradient steepness was used in the second dimension as in the 1D-LC experiments. Figure 6.4C shows two-minute excerpts of the ^{2}D chromatograms obtained for cuts taken from the critical regions I to IV. Conclusions drawn from the selectivity differences shown by 1D chromatography were generally confirmed by the ^{2}D chromatograms. The peak pairs c+d (cut #3), g+j (cut #4), and l+o (cut #10) were well resolved. Satisfactory separation of the compounds h+i from each other and from the other two compounds (e+f) in the co-eluting region III was also achieved. Analytes e+f remained a critical pair. Additional cuts taken from non-critical regions of the chromatogram (e.g. cut #1) confirmed purity of the corresponding ^{1}D peaks (data not shown for brevity). Evaluation of the data revealed in addition, that the ^{2}D analysis time could be reduced by about 55%, since all peaks eluted within 5 min.

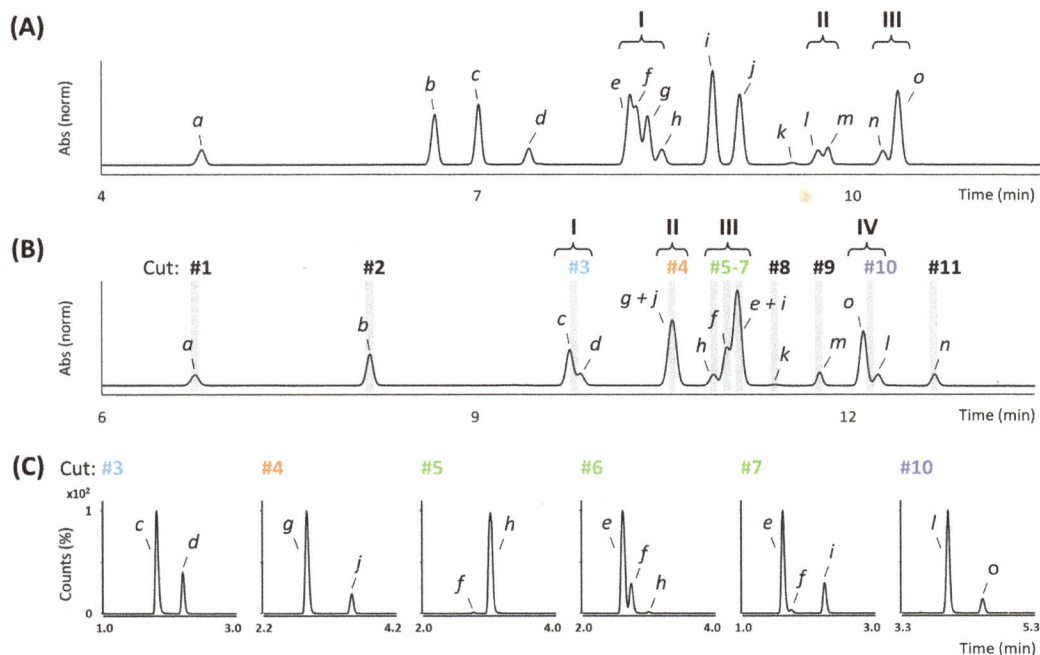

FIGURE 6.4 1D UV chromatograms obtained with ACN (A) or MeOH (B) as organic modifier, at low pH, using a Zorbax SB C18 (50 x 3.0 mm, 1.8 μm) column in both cases. Grey highlights in (B) indicate heartcuts of peaks taken for further separation by a second dimension column. (C) ^{2}D chromatograms shown as MS extracted ion chromatograms. The SB C18 column was used in both dimensions, with MeOH in the first dimension and ACN in the second dimension.

6.5.2 Temperature

Although the 2D-LC experiment shown in Figure 6.4, combining MeOH and ACN, gave superior results compared with 1D-LC separations, quantification would still require better peak resolution, in particular when using UV detection. As is well known [14,15], temperature is often useful for fine-tuning resolution in chromatography.

Figure 6.5B shows the superposition of the cuts #5 to 7 across ^{1}D region III in Figure 6.4B. This result was obtained at a column temperature of 40 °C. Figure 6.5A and C show the same results but with the ^{2}D column temperature held at 25 and 55°C, respectively. All compounds were baseline resolved at 25 °C. It is noteworthy that with only modest method development effort, merely combining two screening runs and adjustment of temperature in ^{2}D, full resolution for all compounds of this moderately complex sample was achieved by combining MeOH and a column temperature of 40°C in the first dimension with ACN and a column temperature of 25°C in second dimension (separation for cuts other than #5 to 7 at 25°C in ^{2}D not shown here).

Closer inspection of the dependencies of retention times on temperature reveal the reason temperature can be used as a tool to fine tune selectivity. Compounds f, h, and i all exhibit increased retention when the temperature is lowered as is expected from a van't Hoff type relationship [16]. The rate of retention change, however, is compound dependent. For instance, compounds h and i co-elute at 55°C and become progressively well separated at 40 and 25°C. Compound e stands out relative to the other compounds as it moved in opposite direction when temperature was decreased. Whereas compound f eluted prior to e at 55°C, the elution order is completely reversed at lower temperature.

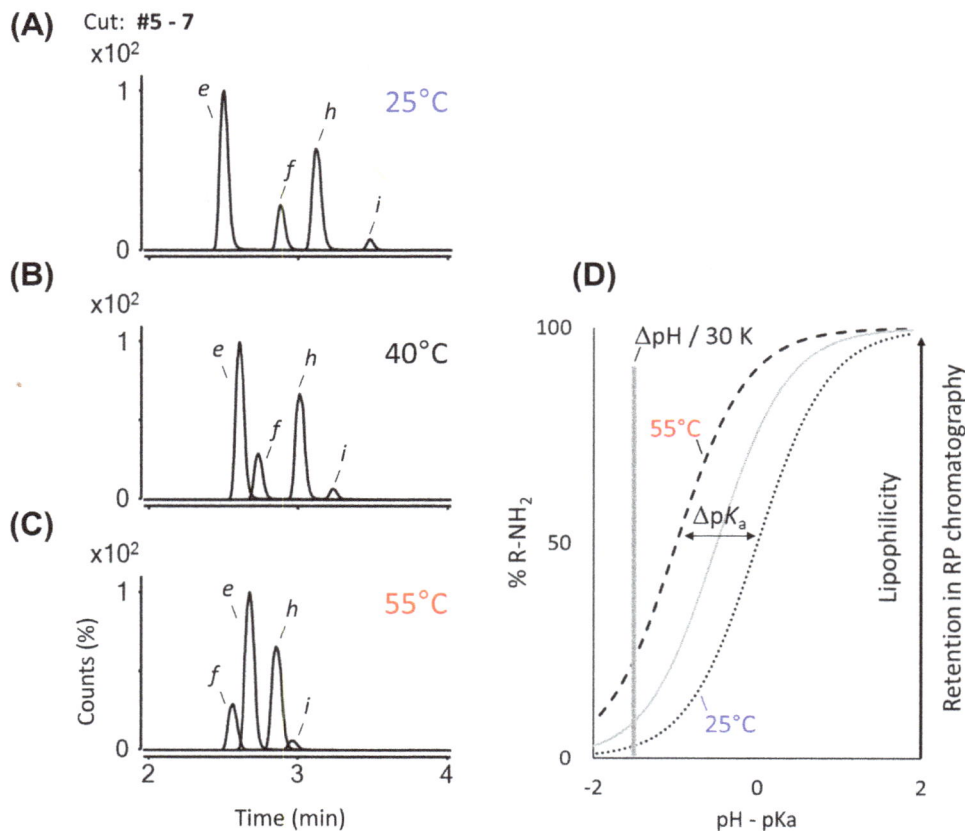

FIGURE 6.5   Second dimension extracted ion chromatograms (EIC) obtained with ACN as organic modifier at low pH and column temperatures of 25°C (A), 40 °C (B), or 55 °C (C) for cuts #5 – 7 of Figure 6.4B. All other conditions were the same as for Figure 6.4. Figure 6.5D: Schematic illustration of pK_a shift of a base and pH shift of formic acid with temperature as a rational for the anomalous retention behavior of compound e. %NH$_2$ – indicates the fraction of the base present in the deprotonated, neutral form.

A rationale for this behavior could be the change of pK_a with temperature, which could lead to a change in ionization state and thus hydrophobicity of the analyte [14, 17]. Compound e is a base with a pK_a of 4.1 at 25°C in aqueous solution, and the pH of 0.2% formic acid (mobile phase solvent A) was measured at 2.5. This base will be mostly protonated in mobile phase solvent A at 25°C as shown in the plot of the relative abundance of the ionized species against pH-pK_a in Figure 6.5D. It has been shown that with increasing temperature the pK_a of a protonated amine can shift at a rate of about -0.03 pKa-units/K [17]. In comparison, the shift for formic acid is only moderate with about +0.003 units/K [18]. Thus, if pH in the mobile phase remains about stable while the sigmoidal pH-pK_a curve shifts to the left on the x-axis at higher temperature, the base will be less protonated (i.e., a larger fraction present in the more neutral state) and thus more retained in RP chromatography. Of course, the situation in this analysis is more complicated because the pK_a (and thus also the pH) in aqueous solutions containing organic solvent is different from that in water and in addition, the concentration of formic acid changed dynamically during the gradient elution program [17, 19, 20].

A significant putative advantage in method development for 2D-LC is that there are two dimensions that can be altered in order to achieve the desired separation. Accordingly, an alternative to changing conditions in the second dimension would be to adjust conditions in the first dimension.

(A)

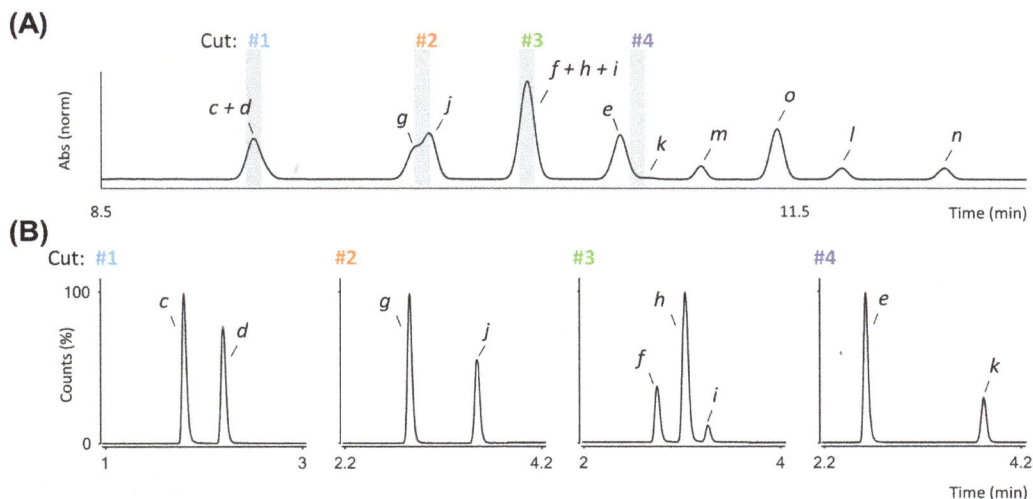

(B)

FIGURE 6.6 (A) ^{1}D UV chromatogram obtained with MeOH as organic modifier at low pH, 55°C; grey highlights indicate heartcuts. (B) ^{2}D EIC's obtained with ACN, low pH, 40 °C. All other conditions are as in Figure 6.4.

From screening experiments, we know that compound e is more retained at higher temperatures. Figure 6.6A shows an excerpt from 8.5 min to 12.0 min of the 1D chromatogram obtained with MeOH at 55°C. The elution pattern looks quite different compared to that in Figure 6.4B, which was obtained at 40°C.

For instance, co-elution region IV formerly observed at 40°C (containing the compounds l+o) is now well resolved, and compound e that was part of the e+f critical pair has moved to a later retention time where it now partially co-elutes with compound k. However, this co-elution of compound e+k might not pose a problem since ^{1}D scouting runs (Figure 6.4A) showed that they could be resolved in the second dimension using ACN as the organic solvent. Figure 6.6B shows ^{2}D chromatograms obtained using ACN at 40°C. Again, complete resolution of all compounds in the sample has been achieved, but this time by changing temperature in the first dimension. Predictions regarding the behavior of compound e were confirmed; at 55°C it moved out of region III of the ^{1}D separation (Figure 6.4B), and it was well resolved from compound k in the second dimension.

It is a best practice for development of robust methods to avoid conditions where separation performance (e.g., resolution) is very sensitive to small changes in conditions (e.g., temperature in this case) [21]. Here, the optimum scenario would be to use a mobile phase pH to push the protonation of the analyte away from the sigmoid part of the curve (i.e., to either fully protonated, or fully de-protonated) where small changes in temperature do not significantly shift the degree of protonation, and thus RP retention of the analyte. Nevertheless, complex mixtures may contain unknown analytes or analytes exhibiting a wide range of characteristics, where it can be difficult to find such conditions especially when MS detection is used.

6.5.3 pH

The pH of the mobile phase can have a great impact on selectivity for ionizable solutes because it influences their charge state. Briefly, the degree of ionization dictates the solutes' hydrophobicity and thus retention in reversed-phase LC. Figure 6.5D shows that when the mobile phase pH matches the pKa of a solute, the solute is 50% ionized; lowering pH increases protonation, and *vice versa*. A moderately strong base (e.g., benzylamine) will be most hydrophilic (thus showing the least retention in RP) at pH – pKa ≤ -2 and most hydrophobic (conversely exhibiting the most retention) at pH – pKa ≥ 2. Acids (e.g., benzoic acid) are affected in the opposite way (i.e., retention is higher at lower pH where the acid is protonated and neutral). Neutral and permanently charged solutes are mostly unaffected by mobile phase pH.

Figure 6.7A shows a ^{2}D chromatogram obtained at high pH (0.2% NH$_4$OH in water as solvent A) for an aliquot taken within region I from the ^{1}D separation shown in Figure 6.4A (Zorbax SB C18, ACN-gradient, low pH mobile phase) cutting compounds e and f. ACN was also used as the organic modifier in the second dimension, but with a Poroshell HPH C18 column (50 x 3.0 mm i.d.) that can withstand high pH conditions. The same gradients slopes were used in the first and second dimensions.

While compounds e and f co-eluted in the first dimension at low pH, they were well separated from each other at high pH, with compound e eluting more than 2 min later than f (Figure 6.7A). Compounds e and f are both bases with pKa values of 4.1 and 12.8 in aqueous solution, respectively. Setting aside the influence of the organic modifier on the pH of the mobile phase, at the low pH used in the first dimension both compounds are mostly protonated. At high pH (about 9.5), compound f remains in the protonated and positively charged form while compound e is mostly present in the deprotonated neutral form. Thus, compound e is in its more hydrophobic state at high pH and is therefore more retained using 0.2% NH$_4$OH as the mobile phase additive in the second dimension.

Surely, mobile phase pH is not the only factor influencing selectivity in these experiments since the stationary phases used in the first and second dimensions were different, even though both phases are referred to as "C18" (*Zorbax SB*: di-isobutyl n-octadecylsilane ligands, fully porous (80 Å) 1.8 μm silica support, non-endcapped; *HPH*: C18 ligands on superficially porous (100 Å) organically modified silica surface 2.7 μm particles, doubly endcapped). Figure 6.7B shows ^{1}D EICs for compounds e and f produced with the same Poroshell HPH C18 column that was operated with the same ACN gradient but at low pH (top) and high pH (bottom). In contrast to Figure 6.4A, the HPH column has resolved compound e from f under both conditions, which reflects the influence of the stationary phase on selectivity (both C18 phases were operated with the same chromatographic method). Nevertheless, the dominating effect of pH on retention discussed above has been

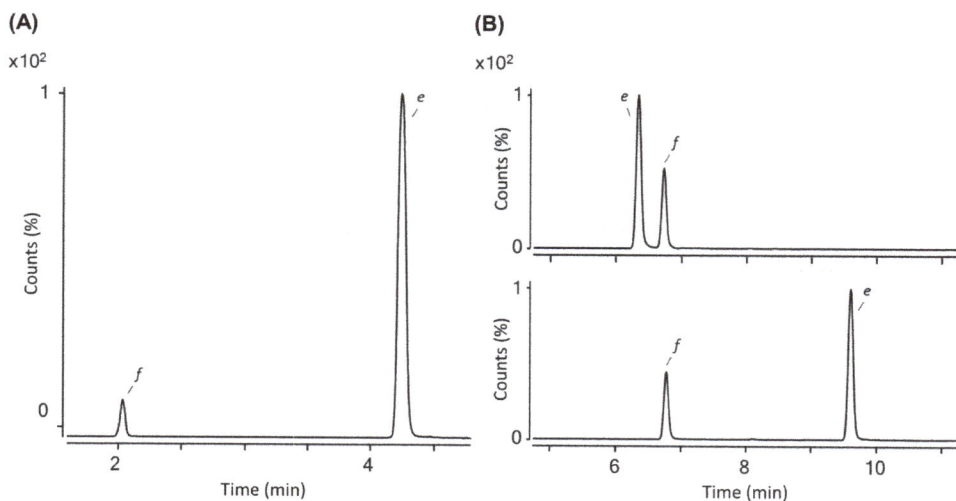

FIGURE 6.7 A) ^{2}D extracted ion chromatogram (EIC) obtained with ACN as organic modifier at high pH, 40 °C. Column, Poroshell HPH C18 (50 x 3.0 mm i.d., 2.7 μm). This result stems from a ^{1}D separation as in Figure 6.4A using low pH for mobile phase A, and ACN as mobile phase B, and cutting e+f in region I. B) ^{1}D EIC's obtained with ACN at low pH (top) and high pH (bottom), 40 °C. Column, Poroshell HPH C18 (50 x 3.0 mm i.d., 2.7 μm).

confirmed here as well. Compound f appears at around the same retention time for low and high pH conditions using the HPH column (Figure 6.7B), which suggests that its ionization state has not changed markedly with the variation of pH. Retention of compound e however, that eluted prior to f at low pH, increased significantly and to a similar extent as shown in panel A (Figure 6.7) using high pH conditions. The power of combining two RP separations – one at low pH and one at high pH – in a 2D-LC format has been described by other groups for the separation of biomolecules such as peptides [22–24].

6.6   INFLUENCE OF THE STATIONARY PHASE ON SELECTIVITY

The stationary phase can have a strong impact on selectivity in most modes of separation, as has been described in Chapter 5 of this book (see also www.hplccolumns.org). The selectivity of stationary phases used for reversed-phase separations can be compared using the Hydrophobic Subtraction Model (HSM) [25]. This model quantifies different types of analyte-stationary phase interactions including hydrophobic, hydrogen bonding, and electrostatic (i.e., ion-ion) interactions, as well as the influence of analyte bulkiness on selectivity. Recent work has shown that the model can be refined to account for strong π–π and dipolar interactions as well [26, 27].

When the HSM database (www.hplccolumns.org) is examined with the purpose of choosing two complementary RP phases, one can see that the Zorbax Bonus RP phase stands out as being highly complementary to most other commercially available RP phases. The Bonus RP phase has a diisopropyl-C14 alkyl chain covalently bound to the silica surface through an embedded amide moiety, and is triply endcapped. The reduction of strong electrostatic interactions between cationic analytes and ionized silanol groups renders this phase especially well-suited for separation of molecules with basic functional groups (e.g., amines).

Figure 6.8B shows the outcome of pairing a Bonus RP column (second dimension) with a SB-C18 column (first dimension) for the 2D-LC separation of the test mixture. The ^{1}D separation displayed in Figure 6.8A is the same as that shown in Figure 6.4A. The ^{1}D and ^{2}D separations were both carried out with low pH and ACN as the organic modifier, and both columns were held at 40°C. In contrast to the results shown earlier, these experiments used ASM (for details see Section 6.8) with an ASM factor of two and a slightly higher ^{2}D gradient slope (5.6 %B/min) to improve peak shape in the second dimension.

Excellent resolution was observed in the second dimension for all compounds transferred by the five cuts from regions of the ^{1}D separation where co-elution occurred (highlighted in panel A). Particularly interesting is the separation of compounds e and f (cuts #1 and 2), which were not well resolved by either dimension in the case where the same stationary phase was used in both dimensions, but different organic solvents and/or temperature were used to adjust selectivity.

The components of the test mixture contain aromatic moieties (see Figure 6.3). A column with phenyl-hexyl functionality was therefore evaluated in conjunction with MeOH as the organic modifier (ACN might attenuate the $\pi - \pi$ interactions between the analyte and stationary phase [26]), in order to leverage selectivity differences that arise from π-electron interactions. Aside from the change in stationary phase and organic modifier for the second dimension, all other conditions were the same as in Figure 6.8B. Figure 6.8C shows a promising initial result with many of the compounds being resolved. However, complete separation of compounds g and h would require additional method development.

These two stationary phases used in the second dimension are quite different in respect to their ligands compared to the Zorbax SB C18 used in the first dimension. It has, however, been demonstrated that even two more similar phases, both bearing C18 ligands, can bring about some difference in selectivity. While the SB C18 phase operated with ACN, at low pH, and 40°C did not resolve compounds e and f (Figure 6.4A) these were well resolved when using the HPH C18 phase under exactly the same conditions (Figure 6.7B, top).

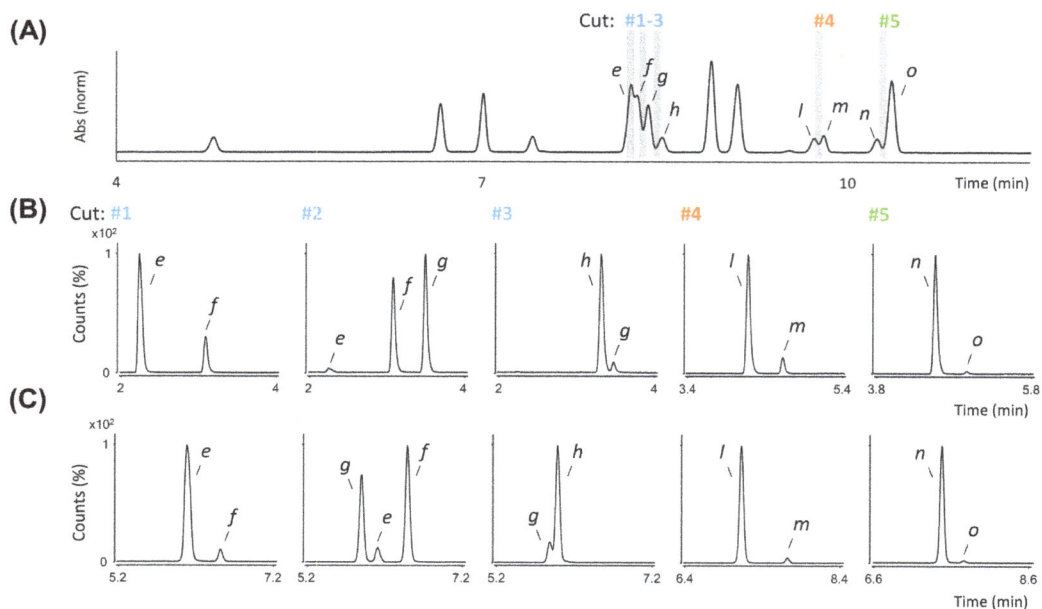

FIGURE 6.8 A) 1D UV chromatogram obtained with SB-C18, ACN as organic modifier at low pH (Figure 6.4A); grey highlights indicate heartcuts. B) ^{2}D chromatograms shown as MS EICs obtained with Zorbax Bonus RP (50 mm x 3.0 mm i.d., 1.8 μm) with ACN at low pH and 40 °C. C) 2D results as in (B) but obtained with Zorbax Phenyl-Hexyl (50 mm x 3.0 mm i.d., 1.8 μm) with MeOH at low pH and 40 °C. ^{2}D gradients were 25–75 %B in 9 min (i.e., 5.6 %B/min).

6.7   OPTIMIZING GRADIENT ELUTION CONDITIONS FOR THE ^{2}D SEPARATION

Optimization of the gradient elution conditions with the goal of maximizing resolution and/or speed can be achieved by adjusting the solvent strength at the start of the gradient and the gradient steepness. The gradient steepness is affected by many factors, including the change in solvent strength during the gradient, gradient time, flow rate, and column dimensions. Additionally, multiple segments within a gradient elution program can be used to improve selectivity in certain regions of the chromatogram [28].

In principle, 1D-LC experiments can be used to obtain the data needed to optimize elution conditions for the second dimension of a 2D-LC separation. These are often referred to as "scouting", "training", or "scanning" experiments. For example, one can prepare a sample containing only those compounds expected to co-elute in a certain region of the ^{1}D separation, and then vary elution conditions (i.e., gradient steepness) to determine the dependence of retention on these conditions. This is a viable approach for compounds that are known and for which standards are available. In cases where standards are not available, fraction collection can be used to obtain the compounds of interest for use in subsequent scouting experiments. Optimal separation conditions can then be predicted for the second dimension on the basis of these data, and finally combined with the ^{1}D separation to complete the 2D-LC method. This is effectively the approach demonstrated in the context of Figure 6.4.

Alternatively, we can carry out scouting experiments using actual 2D-LC separations. One advantage of doing so is that this avoids the complication of having to account for system parameters (e.g., gradient delay volume, pump-dependent gradient shape, 2D-LC loop volume) when translating results from 1D-LC scouting experiments to ^{2}D elution conditions. In other words, predicted ^{2}D separations will be more accurate if the scouting results are obtained directly from the second dimension of a 2D-LC system [29]. Multiple heartcutting 2D-LC is a convenient tool for this purpose because several small cuts can be taken across co-eluting regions in the first dimension, and then each cut can be subsequently analyzed in the second dimension under different conditions within a single 2D-LC analysis. In contrast, scouting experiments based on single heartcut 2D-LC require a full 2D-LC analysis for each individual change in ^{2}D gradient conditions. Fortunately, modern 2D-LC software provides the functionality (for instance "shifted gradients"; see Section 4.5) needed to make implementation of this type of method development approach relatively straightforward in practice.

Figure 6.9 shows results obtained from a method development approach involving full 2D-LC separations like that described above. Four cuts across the ^{1}D regions I and II were analyzed using different gradients for each cut in the second dimension. These gradient traces are shown in panel B. The same gradient slope of about 5% organic modifier/min. was used in both dimensions. Conditions for region I ranged from 15–35 to 30–50% ACN and from 20–40 to 35–55% ACN for region II. The

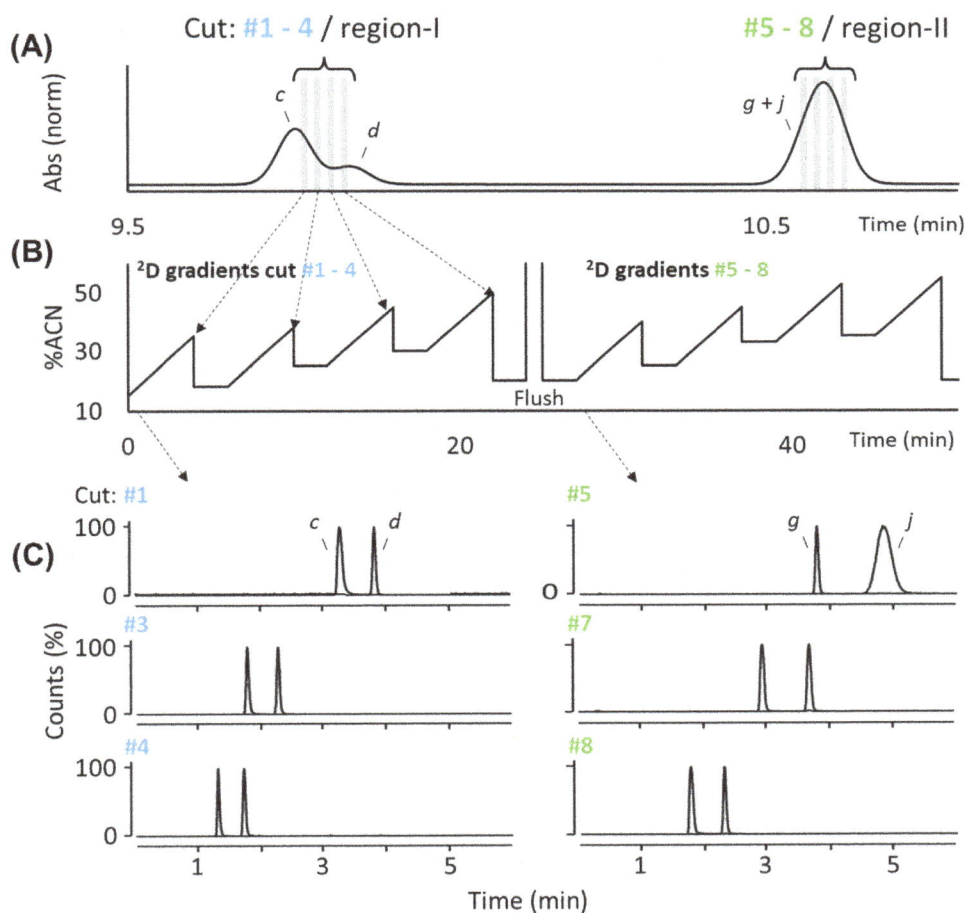

FIGURE 6.9 A) ^{1}D UV chromatogram obtained with MeOH as the organic modifier, using the same conditions as in Figure 6.4B; highlights indicate heartcuts. B) %B traces for ^{2}D gradients. C) ^{2}D chromatograms for ^{1}D region I (left) and region II (right) shown as EIC's obtained with Zorbax SB C18 (50 mm x 3.0 mm, 1.8 μm) with ACN as the organic modifier, at low pH, 25 °C, and 0.6 mL/min.

gradient duration was 4 min. in each case, and the time allowed for column re-equilibration following each gradient was 2 min. Figure 6.9 C shows representative ^{2}D chromatograms. For region II, a gradient of 20–40 %ACN (cut #5) was not a suitable condition, with the peak corresponding to compound j being broad and eluting outside of the gradient window. As expected, with increasing %ACN the retention of all compounds decreased. At the same time resolution was maintained, which means that the ^{2}D analysis time could have been reduced markedly under the optimized conditions. For instance, resolution of compounds g and j in cuts #7 and 8 was 10.3 and 9.0, respectively. Generally, the same trend was observed for the ^{2}D gradients applied to region I.

Further experiments were carried out to investigate the impact of varying gradient slopes, ranging from 5 to 10 %ACN/min, on the separation in the second dimension. All ^{2}D gradients for regions I and II started at 30 %ACN. Steeper gradients could decrease the time needed for the ^{2}D cycle, and possibly improve resolution due to a gradient focusing effect [30]. For instance, a gradient of 30–70% ACN in 4 min (10 %ACN/min) gave a resolution of 11 between compounds g and j (data not shown). The ^{2}D separation speed was further improved by doubling the flow rate from 0.6 to 1.2 mL/min. At the same time the duration of the gradient was halved to 2 min., thus maintaining a constant gradient steepness. In this way the ^{2}D cycle time was markedly reduced while still achieving a resolution of 8.0 and 12.2 for compound pairs c+d and g+j, respectively (data not shown). When mass spectrometry detection was used, however, care has to be taken to not exceed the maximum flow specifications of the ionization source. If the ^{2}D flow rate is too high for the MS, a post-column flow split can be used, but one should be careful to choose appropriate dimensions of post-split tubing to avoid unnecessary peak dispersion [31].

6.8   ACTIVE SOLVENT MODULATION

In the most common type of 2D-LC experiment portions of ^{1}D effluent are collected in sampling loops, which are subsequently transferred to a ^{2}D column for further separation. If the analytes of interest are weakly retained when ^{1}D effluent enters the ^{2}D column, broad, distorted peaks can result, leading to a loss in resolution and sensitivity. In the worst case scenarios the analytes are not retained and break through in the void volume of the ^{2}D column [8]. In the case of RP separation in the second dimension, this problem can occur when the ^{1}D effluent contains a high fraction of organic solvent. The effect becomes more pronounced with increasing loop size and decreasing volume of the ^{2}D column. It has been referred to as "solvent strength related incompatibility" and "mobile phase mismatch", and has been considered a major obstacle in method development [32, 33].

ASM provides a means to address this incompatibility problem. It is a valve-based approach that allows for dilution of the loop contents with weak solvent prior to its transfer to the ^{2}D column [34]. The weak solvent is delivered by the ^{2}D pump during the early phase of each ^{2}D cycle when a fraction of ^{1}D effluent is displaced from the sampling loop (i.e., the injection step). The ^{2}D pump is connected both to the sampling loop and to an additional flow path (ASM path) in parallel during this injection step. The two paths meet at the outlet of the ASM valve so that dilution with weak solvent occurs prior to arrival of the mixture at the ^{2}D column inlet. In many cases this results in significantly improved peak shapes, because compounds are retained at the ^{2}D column inlet under weak elution conditions before starting to migrate when the actual solvent gradient reaches a composition that promotes elution. In principle, any dilution factor (ASM factor, or "ASM-f") could be established, but the range used for most applications is between 1.5 and 5. For further information on ASM, please refer to Section 4.4.5.3 of this book.

6.8.1   Choosing the Right ASM-f

To demonstrate the effect of the ASM-f, different loop sizes were used to sample co-eluting compounds c+d from region I of the ^{1}D chromatography shown in Figure 6.4B. A first experiment used a Zorbax SB C18 (50 x 3.0 mm and 2.1 mm ID) column in ^{2}D, operated at low pH, 25°C, 0.6 mL/min., and ACN as the organic modifier. The gradient elution program was 25–45% ACN in 4 minutes. When ASM was activated, an isocratic hold at 5% ACN was used at the start of the ^{2}D cycle for the duration of sample displacement (3 x loop volume); 5% ACN was also used for column equilibration at the end of each cycle. The ^{2}D chromatograms obtained without the application of ASM were compared to separations performed using ASM factors 2, 3, and 5. Figure 6.10 shows 2-min excerpts of ^{2}D chromatograms containing compounds c and d obtained using the 3.0 mm i.d. column. The volumes of ^{1}D effluent transferred to the second dimension were 40 and 180 μL in panels A and B, respectively. The chromatograms in the top row were obtained without ASM (i.e., ASM-f 0); those in the bottom row were obtained with ASM-f 2.

Using 40 μL loops, hardly any difference was observed in peak shape, resolution and sensitivity when comparing the chromatograms obtained with ASM-f 2 or ASM-0. Hence, the ASM process is not needed for these 2D-LC conditions. The situation was different with 180 μL loops in that focusing was less effective when ASM was not used. Although there was no sign of breakthrough, the peak widths for both compounds were wider than expected, considering the results with the 40 μL loops, and thus peak heights were lower than expected. This is not entirely surprising because when using a loop of 180 μL the injected sample volume is about equal to the void volume of the 50 mm x 3.0 mm i.d. column (~200 μL), which really means that the 180 μL aliquot of the ^{1}D effluent containing about 50% MeOH acts as the ^{2}D mobile phase, at least temporarily, during the period when this material moves through the ^{2}D column. However, when using an ASM-f 2, peak shapes, resolution and sensitivity improved substantially so that the peak widths were similar to those obtained with 40 μL loops. As a consequence of the larger volume injected, the peaks are about

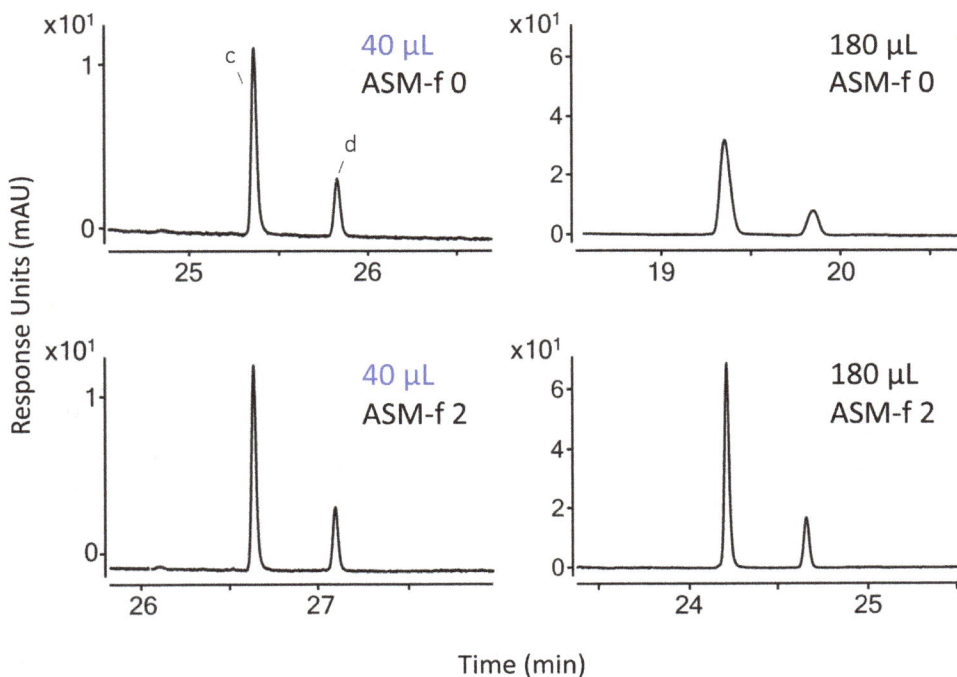

FIGURE 6.10 ^{2}D UV chromatograms (displayed on 1D timescale) obtained from separations of fractions from region I in Figure 6.4B, using different loop volumes and ASM factors. 2D conditions: Zorbax SB C18 (50 mm x 3.0 mm i.d.) ACN-gradient, low pH, 25 °C.

five -fold taller than when using the 40 µL, thereby providing a more sensitive analysis. Thus, in the case of the 180 µL loops, use of ASM provides significantly better results.

Figure 6.11A shows the peak widths at half-height corresponding to the ^{2}D chromatograms shown in Figure 6.10 for compounds c and d using different ASM factors. It confirms that with 40 µL loops the effect of ASM on the peak width of the selected compounds is minor. On the other hand, with 180 µL loops the peak widths decreased by about 50% when moving from ASM-f 0 to ASM-f 2. Furthermore, it is shown that peak widths were very similar for 40 and 180 µL loops when using ASM-f 2, and that larger ASM factors did not lead narrower peaks.

We recommend using the lowest ASM-f that gives satisfactory results. This is because the ASM period adds to the ^{2}D cycle time and the ASM period increases with increasing ASM-f. For instance, at ASM-f 0, the entire ^{2}D flow goes through the sampling loop. To flush out a 180 µL loop three times with the ^{2}D flow used here (600 µL/min) requires 0.9 min. At ASM-f 2, the ^{2}D flow is split in equal parts (i.e., 1:1) between the loop and the ASM bypass path. This means that the flow rate through the loop during the ASM step is reduced to 300 µL/min. and the time required to flush the loop three times doubles to 1.8 min. Similarly, using an ASM factor of 5 requires an ASM period of 4.5 min., which obviously adds significantly to the ^{2}D cycle time.

The plot in Figure 6.11B shows the results obtained for the 2.1 mm i.d. SB C18 column, which has a void volume (~100 µL) that is about half that of the 3.0 mm i.d. column. In this case, ASM was beneficial not only when using 180 µL loops, but also when using 40 µL loops. Peak widths decreased noticeably when moving from ASM-f 0 to 2, but again larger ASM factors of 3 or 5

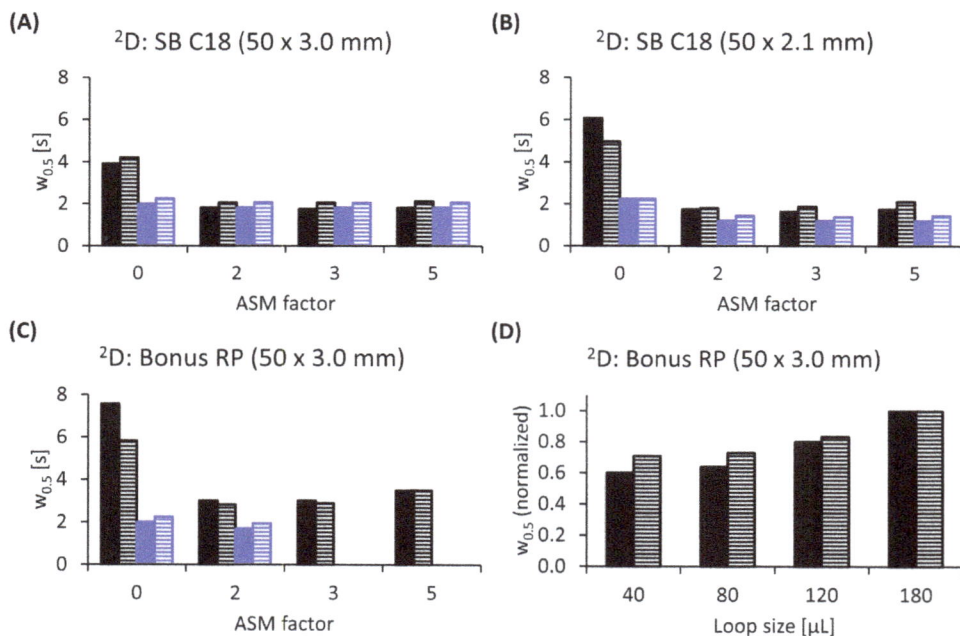

FIGURE 6.11 Plots corresponding to the chromatograms shown in Figure 6.10 showing peak widths at half-height ($w_{0.5}$) of ^{2}D peaks obtained for the compounds c (filled bars) and d (striped bars) with different ASM factors and different loop sizes (black: 180 µL loop, blue: 40 µL loop). Columns: (A) Zorbax SB C18 50 mm x 3.0 mm i.d.; (B) Zorbax SB C18 50 mm x 2.1 mm i.d.; (C) Zorbax Bonus RP 50 mm x 3.0 mm i.d.. Panel (D) is a plot of normalized $w_{0.5}$ for different loop sizes with ASM-f 3 and the 50 mm x 3.0 mm i.d. Bonus RP column. The column was operated at 40 °C with an ACN gradient from 25–45% ACN in 4 min. An isocratic hold at 5% ACN was used for the duration of the ASM period.

produced only minor changes in peak width. With 180 μL loops, use of ASM-f 2 led to a 64% improvement in peak width compared to ASM-f 0.

In contrast to the comparison of ASM factors when using the 3.0 mm i.d. column, the peak widths obtained for 180 μL loops never reached the values obtained with 40 μL loops when using the 2.1 mm i.d. column. For example, peak widths of 1.4 s and 1.8 s were observed for compound d when using ASM-f 2 and 40 μL and 180 μL loops, respectively. Nevertheless, resolution between compounds c and d was still satisfactory when using 180 μL loops and the 2.1 mm i.d. column. From a practical point of view a minor increase in peak widths could probably be accepted, if a 2.1 mm i.d. column would be preferred over a 3 mm i.d. column. For example, method development objectives may favor higher detection sensitivity (e.g. in the case of concentration-sensitive detectors such as UV absorbance), or a lower ^{2}D flow rate to improve compatibility with MS detection (e.g., [35]).

A second experiment was performed to investigate the impact of the ASM factor and loop volume on the peak width in the case where a less hydrophobic stationary phase was used in the second dimension. Figure 6.11C shows results from the use of a Bonus RP column (50 mm x 3.0 mm i.d.). These data confirmed the expectation that the less hydrophobic column would be more susceptible to the impact of injecting a large fraction of ^{1}D effluent containing a high level of organic solvent. The pattern of peak widths was similar to that observed with the 2.1 mm i.d. SB C18 column (Figure 6.11B). A small improvement in peak width was observed even for 40 μL loops when moving from ASM-f 0 to ASM-f 2. With 180 μL loops the improvements were remarkable using ASM-f 2 or 3 compared to ASM-f 0, and a slight increase in widths was observed when using ASM-f 5 (probably related to the long period of sample displacement). However, the peak widths observed with 180 μL loops never reached those obtained with 40 μL loops.

6.8.2 Using the Right Loop Size

The results in Figure 6.11A showed that with the Zorbax SB C18 column (50x3.0 mm i.d.) the volume of sample transferred to the second dimension had little effect on the widths of ^{2}D peaks when ASM was used. In other cases, for example when less retentive phases are used in the second dimension, focusing analytes at the column inlet can be more challenging. Figure 6.11C shows that despite the narrower peak widths observed with ASM-f 2 or 3 when using 180 μL loops, the widths were never as narrow as those obtained with the smaller (40 μL) loops. In cases where peak width must be decreased to achieve satisfactory ^{2}D resolution, a more detailed view of the effect of loop volume on ^{2}D peak width might be helpful for optimizing the separation.

Figure 6.11D shows the dependence of peak width on loop volume. The observed peak widths for compounds c and d were normalized to the widths observed with the 180 μL loops. Peak widths steadily improved, changing from 180 to 80 μL loops, with the latter giving about the same value for peak widths as 40 μL loops. Under these conditions, and for the best performance in the second dimension, the largest loop size that should be used is 80 μL.

6.9 USING A DIVERTER VALVE IN COMBINATION WITH MS

When separating charged compounds by RP LC, mobile phases with high ionic strength can be used to improve peak shapes [6, 36]. Some other separation modes require high ionic strength to control elution, such as, ion exchange (IEX) or hydrophobic interaction chromatography (HIC). Mass spectrometry is not compatible with the high ionic strength buffers used in these cases, because it can increase spectral complexity by introducing ion clusters and adducts, reduce sensitivity, for example by suppressing compound ionization, and shorten instrument maintenance intervals due to precipitation of non-volatile salts in the ionization source of the MS instrument. 2D-LC can be used to bridge MS-incompatible 1D-LC methods with MS detection by adding a second dimension of separation that can act as both a desalting step, and further separate compounds that co-elute from the first dimension (e.g., [6]). However, we cannot forget in this scenario that each fraction taken from the first dimension and transferred to the second dimension contains salt; the sampling frequency and volume determines the amount of salt transferred to the ^{2}D column.

In conventional 1D-LC-MS experiments it is common to use a diverter valve between the LC column outlet and the inlet of the MS to allow diversion of the column effluent to waste during elution periods where high concentrations of undesirable components elute from the column (i.e., during and around the column dead time). The same concept can be applied in 2D-LC-MS, where a diversion valve can be plumbed between the ^{2}D column outlet and the inlet to the MS. The difference here, of course, is that whereas in 1D-LC this valve is typically only switched once or a few times per analysis, in 2D-LC the valve must be switched at least once for each ^{2}D separation (i.e., tens or hundreds of times per 2D-LC analysis). Fortunately, contemporary 2D-LC software now fully automates the action of the ^{2}D diversion valve, making it much easier to use. The factors associated with timing the switching of the diversion valve are the same as with 1D-LC/MS; these include injection volume, extra-column volume, column dead volume, and flow rate used. For diversion of most of the unretained salt, we recommend to divert the column effluent for a time equivalent to the sum of these volumes (i.e., injection volume + extra-column volume + column dead volume). For applications requiring more thorough removal of the salt, diversion of the flow for two to three of these volumes is better. For example, for a system equipped with 40 μL loops (assuming full loop sampling), a ^{2}D extra-column volume of 30 μL (including a 10 μL UV flow cell and capillaries from the flow cell to the diverter valve and into the MS), a column dead volume of 210 μL, and a flow of 600 μL/min, the ^{2}D effluent should be diverted to waste for at least 0.5 minutes after the start of a modulation period. The longer the flow is diverted to the waste, the more careful one must be to avoid losing early eluting compounds from detection. When using RP in the second dimension, one should consider using an isocratic hold at low organic solvent during the diversion period. Furthermore, ASM can be helpful in this context for increasing the retention of hydrophilic analytes.

6.10   QUANTIFICATION IN 2D-LC

When 2D-LC is used for quantification purposes (e.g., ref. [10]), an important part of method development is to establish the quantitative performance of the method as determined using metrics such as instrumental variability, response linearity, and recovery (i.e., comparison of responses from ^{1}D and ^{2}D detectors). The case study described in this chapter provides an opportunity to discuss aspects of quantitation in 2D-LC that are not encountered in 1D-LC. As shown in Figure 6.12, compounds c and d co-elute from the ^{1}D column, making accurate quantitation with a UV detector impossible. Adding the ^{2}D separation rapidly provides a complete resolution of the pair, making quantitation possible.

6.10.1   PRECISION

The precision of quantitation was determined from six replicate analyses of the test mixture. Two cases are compared as shown in Figure 6.12. First, 40 μL sampling loops were installed in the MHC 2D-LC interface. High Resolution Sampling (a.k.a. the selective comprehensive mode or sLCxLC) was used to collect five fractions of ^{1}D effluent covering the elution window containing compounds c and d as shown in Figure 6.12A, each time filling the loop to 80% in order to prevent analyte loss [37]. Figure 6.12B shows the resulting five ^{2}D chromatograms.

The separation of compounds c and d was excellent in each case, and the typical "walking through the ^{1}D peak" pattern was observed, where the abundance of the earlier eluting compound c first increases from cut #1 to 2 and then decreases in consecutive cuts. The same occurs for compound d across cut #3–5. Table 6.2 summarizes the precision data calculated for six replicate measurements using peak areas obtained in each single cut and those for the total area. The total area

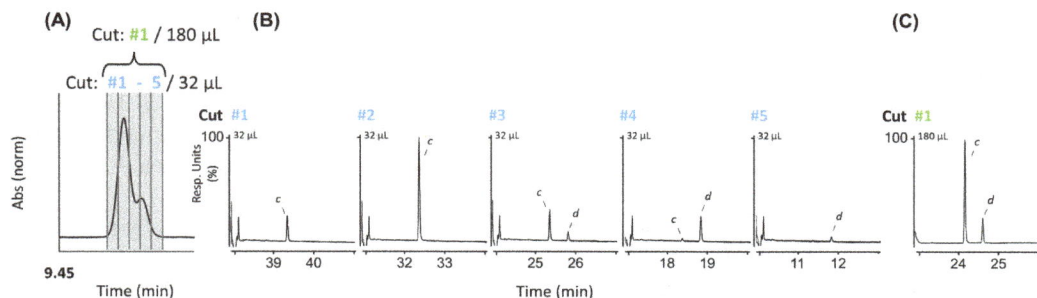

FIGURE 6.12 Chromatograms from 2D-LC separations illustrating different approaches to quantitation. All chromatograms are overlays from six replicate analyses. (A) Excerpt from ^{1}D UV chromatogram as in Figure 6.4B. Grey highlights show sampling scheme using either High Resolution Sampling with 40 μL loops (80% loop fill), or a single heartcut with 180 μL loop (full loop sampling). (B) ^{2}D chromatograms for cuts #1–5 made in the case of High Resolution Sampling. (C) ^{2}D chromatogram obtained from a single 180 μL heartcut. All ^{2}D separations used ASM with ASM-f 3.

TABLE 6.2
Precision of Individual and Cumulative Peak Areas Obtained for Six Replicate Analyses of the Test Mixture

Cut #	Compound c		Compound d	
	Average Peak Area (mAU·s)	RSD [%]	Average Peak Area (mAU·s)	RSD [%]
Measurements performed with 40 μL loops				
1	22.4	2.9	n.d.[a]	
2	86.7	0.4	n.d.	
3	25.5	3.4	9.2	2.7
4	2.6	4.8	24.0	0.6
5	n.d.		5.0	2.0
Total area	137.3	0.3	38.2	0.8
Measurements performed with 180 μL loops				
1	136.8	0.5	36.5	0.5

a) n.d.: not detected

for a compound is calculated by adding the peak areas from multiple ^{2}D chromatograms in a single measurement, for example, by adding areas for compound c across cuts #1 to 4.

In the first case where High Resolution Sampling was used, good area precision was obtained for individual cuts when they were made at the apex of the respective ^{1}D peaks for compounds c and d (i.e., cut #2 for compound c, and cut #4 for compound d). Other cuts before or after the apex of the ^{1}D peak exhibited larger RSD values. This may be caused by small retention shifts in the ^{1}D separation [10]. However, when the total ^{2}D peak area is considered by adding up all of the individual ^{2}D peak areas associated with a given compound, the area precision is excellent (less than 1% RSD), in spite of the variation observed for individual cuts.

In the second case where a single 180 μL fraction was transferred to the second dimension using a single large loop (Figure 6.12C), satisfactory area precision of 0.5% RSD was obtained for the co-eluting pair, which is close to the area precision observed for both compounds in ^{1}D (0.3 % RSD; calculation based on co-eluting region in Figure 6.12A).

6.10.2 Linearity

Linearity was evaluated using duplicate injections of the test mixture in the range from 0–160 μg/mL and 40 (High Resolution Sampling) or 180 (single heartcut) μL loops. Figure 6.13A provides plots of total peak areas obtained for ^{2}D peaks vs. analyte concentration for compounds c and d, from both experiments superimposed. Panel (B) shows a subset of the data focused on the low concentration samples.

Total ^{2}D peak areas obtained for both experiments were very similar, as indicated by the overlap of data points obtained with 40 and 180 μL loops across the entire concentration range. Good linearity is exhibited by R^2 values from regression analysis equal or greater than 0.9999. Analyte carryover was not an influencing factor in the experiments, as no analyte peaks were observed in analyses of blank samples, which were analyzed immediately after the samples with the highest test mixture concentration.

6.10.3 Recovery

Recovery must be demonstrated in the validation of a quantitative method analyzing drug substances post a sample extraction process [38]. To assess method robustness in quantitative 2D-LC analyses we recommend determining the recovery in 2D relative to the "extraction process" in 1D.

In a first step, the relative ^{1}D/^{2}D detector response should be compared. For UV detectors, this may be done by simply placing them in sequence while running a 1D-LC experiment. It is advisable to install the shortest, narrowest capillary possible to connect detector cells in order to reduce band spreading between the ^{1}D and ^{2}D detectors. Care must be taken so that the maximum pressure which the detector cell can withstand is not exceeded. In our UV measurements the ^{2}D detector gave 102% (± 1%) of the area obtained in ^{1}D. Second dimension results should therefore be multiplied by factor 0.98 to account for this subtle difference.

When measuring recovery in 2D-LC, it is advisable to use the same conditions in both dimensions. This allows neglecting solvatochromic effects if UV detection is used and for mass spectrometry, this would allow neglecting differences in ionization efficiency. Sampling of the entire ^{1}D peak is recommended by either using several cuts in the High Resolution Sampling (HiRes) mode or a large enough loop size for a single heartcut. In HiRes mode it is important to ensure that there is no sample loss due to loop overfill because this would deteriorate recovery. In most of our reversed-phase HPLC experiments we have seen that a loop fill of 80% was a good value. However, this may vary with the different flow profiles induced by using different solvents used in ^{1}D. Provided

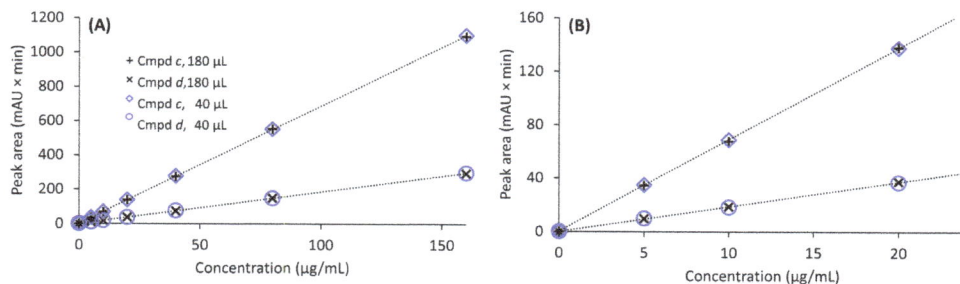

FIGURE 6.13 Total ^{2}D peak areas for compounds c and d using either 40 (High Resolution Sampling) or 180 (heartcut) μL loops. The full calibration range is shown in (A), while (B) shows a subset of the data focused on the low concentration samples. Injection volume in the first dimension was 1 μL, Plotted points are averages of areas obtained from two analyses. Other conditions were the same as those for Figure 6.12. R^2 values from regression analysis were always ≥ 0.9999.

a functional 2D-LC system, experiments with small molecular weight compounds typically gave 100% recovery, independent of whether HiRes or heartcutting are used. Sample loss during the transfer process from the first to the second dimension could thus be excluded.

It should, however, be noted that other types of analytes may interact with internal surfaces of the sampling loop, which could negatively influence recovery. These interactions may vary from analyte to analyte, thus the importance of such recovery measurements [7].

6.11 CLOSING REMARKS

First, whereas in LC×LC optimizing the ^{1}D separation for high peak capacity is not so important because of undersampling (see Section 3.4), when developing non-comprehensive 2D-LC methods we recommend starting with a good ^{1}D separation, and adapting the ^{2}D conditions accordingly, and as needed to complete the separation. Second, while much of the 2D-LC literature stresses the importance of column selection, we encourage readers to not lose sight of the power of the mobile phase for adjusting selectivity. Selectivity screening systems like those discussed in Sections 4.4, 9.5, and 9.6 are useful for identifying promising selectivities. But, once a candidate stationary phase is identified, either empirically or using guidance from the literature, we find that mobile phase organic modifier, pH, temperature, and gradient slope are all powerful levers for adjusting selectivity in 2D-LC just as they are in 1D-LC. Moreover, these variables are more convenient to adjust because they are software controllable. Finally, at this point in time multiple groups have convincingly shown that non-comprehensive 2D-LC methods can be used for high quality quantification, and even validated for use in regulated environments [39]. This is a very important development, because it opens the field to broader use of 2D-LC for more than just research activities.

REFERENCES

[1] S.M.C. Buckenmaier, Agilent 1290 Infinity 2D-LC solution for multiple heart-cutting. Technical overview, Agilent Technologies, n.d. www.agilent.com/cs/library/technicaloverviews/public/5991-5615EN.pdf (Accessed August 16, 2021).

[2] S.R. Groskreutz, M.M. Swenson, L.B. Secor, D.R. Stoll, Selective comprehensive multi-dimensional separation for resolution enhancement in high performance liquid chromatography, Part II – Applications, J. Chromatogr. A. 1228 (2012) 41–50. https://doi.org/10.1016/j.chroma.2011.06.038.

[3] S.R. Groskreutz, M.M. Swenson, L.B. Secor, D.R. Stoll, Selective comprehensive multi-dimensional separation for resolution enhancement in high performance liquid chromatography, Part I – Principles and instrumentation, J. Chromatogr. A. 1228 (2012) 31–40. https://doi.org/10.1016/j.chroma.2011.06.035.

[4] Multi-Dimensional Liquid Chromatography Applications Database, http://multidlc.org/literature/2DLC-Applications (Accessed June 5, 2022).

[5] C.J. Venkatramani, J. Girotti, L. Wigman, N. Chetwyn, Assessing stability-indicating methods for coelution by two-dimensional liquid chromatography with mass spectrometric detection: Liquid chromatography, J. Sep. Sci. 37 (2014) 3214–3225. https://doi.org/10.1002/jssc.201400590

[6] P. Petersson, K. Haselmann, S. Buckenmaier, Multiple heart-cutting two dimensional liquid chromatography mass spectrometry: Towards real time determination of related impurities of biopharmaceuticals in salt based separation methods, J. Chromatogr. A. 1468 (2016) 95–101. https://doi.org/10.1016/j.chroma.2016.09.023.

[7] Z. Dunn, J. Desai, G.M. Leme, D. Stoll, D. Richardson, Rapid two-dimensional Protein-A size exclusion chromatography for determination of titer and aggregation for monoclonal antibodies in harvested cell culture fluid samples, MAbs. 12 (2020). https://doi.org/10.1080/19420862.2019.1702263.

[8] M. Pursch, A. Wegener, S. Buckenmaier, Evaluation of active solvent modulation to enhance two-dimensional liquid chromatography for target analysis in polymeric matrices, J. Chromatogr. A. 1562 (2018) 78–86. https://doi.org/10.1016/j.chroma.2018.05.059.

[9] M. Pursch, Personal communication (2021).

[10] M. Pursch, S. Buckenmaier, Loop-based multiple heart-cutting two-dimensional liquid chromatography for target analysis in complex matrices, Anal. Chem. 87 (2015) 5310–5317. https://doi.org/10.1021/acs.analchem.5b00492.

[11] M. Pursch, P. Lewer, S. Buckenmaier, Resolving co-elution problems of components in complex mixtures by multiple heart-cutting 2D-LC, Chromatographia. 80 (2017) 31–38. https://doi.org/10.1007/s10337-016-3214-x.

[12] D.R. Stoll, X. Li, X. Wang, P.W. Carr, S.E.G. Porter, S.C. Rutan, Fast, comprehensive two-dimensional liquid chromatography, J. Chromatogr., A. 1168 (2007) 3–43. https://doi.org/10.1016/j.chroma.2007.08.054.

[13] L.R. Snyder, J.J. Kirkland, J.W. Dolan, Reversed-phase chromatography for neutral samples, in: Introduction to Modern Liquid Chromatography, 3rd ed, Wiley, Hoboken, NJ, 2010: pp. 265–270.

[14] S.M.C. Buckenmaier, D.V. McCalley, M.R. Euerby, Rationalisation of unusual changes in efficiency and retention with temperature shown for bases in reversed-phase high-performance liquid chromatography at intermediate pH, J. Chromatogr. A. 1060 (2004) 117–126. https://doi.org/10.1016/j.chroma.2004.04.019.

[15] Y. Mao, P.W. Carr, Adjusting selectivity in liquid chromatography by use of the thermally tuned tandem column concept, Anal. Chem. 72 (2000) 110–118. https://doi.org/10.1021/ac990638x.

[16] W. Melander, D.E. Campbell, C. Horváth, Enthalpy – Entropy compensation in reversed-phase chromatography, J. Chromatogr. A. 158 (1978) 215–225. https://doi.org/10.1016/S0021-9673(00)89968-9.

[17] S.M.C. Buckenmaier, D.V. McCalley, M.R. Euerby, Determination of ionisation constants of organic bases in aqueous methanol solutions using capillary electrophoresis, J. Chromatogr. A. 1026 (2004) 251–259. https://doi.org/10.1016/j.chroma.2003.11.007.

[18] J.M. Padró, A. Acquaviva, M. Tascon, L.G. Gagliardi, C.B. Castells, Effect of temperature and solvent composition on acid dissociation equilibria, I: Sequenced pssKa determination of compounds commonly used as buffers in high performance liquid chromatography coupled to mass spectroscopy detection, Anal. Chim. Acta. 725 (2012) 87–94. https://doi.org/10.1016/j.aca.2012.03.015.

[19] S.M.C. Buckenmaier, D.V. McCalley, M.R. Euerby, Determination of pKa values of organic bases in aqueous acetonitrile solutions using capillary electrophoresis, J. Chromatogr. A. 1004 (2003) 71–79. https://doi.org/10.1016/S0021-9673(03)00717-9.

[20] D. Sýkora, E. Tesařová, M. Popl, Interactions of basic compounds in reversed-phase high-performance liquid chromatography influence of sorbent character, mobile phase composition, and pH on retention of basic compounds, J. Chromatogr. A. 758 (1997) 37–51. https://doi.org/10.1016/S0021-9673(96)00691-7.

[21] L.R. Snyder, J.J. Kirkland, J.W. Dolan, Basic concepts and the control of separation, in: Introduction to Modern Liquid Chromatography, 3rd ed, Wiley, Hoboken, NJ, 2010: pp. 69–71.

[22] I. François, D. Cabooter, K. Sandra, F. Lynen, G. Desmet, P. Sandra, Tryptic digest analysis by comprehensive reversed phase×two reversed phase liquid chromatography (RP-LC×2RP-LC) at different pH's, J. Sep. Sci. 32 (2009) 1137–1144. https://doi.org/10.1002/jssc.200800578.

[23] M. Gilar, P. Olivova, A.E. Daly, J.C. Gebler, Orthogonality of separation in two-dimensional liquid chromatography, Anal. Chem. 77 (2005) 6426–6434. https://doi.org/10.1021/ac050923i.

[24] D.R. Stoll, H.R. Lhotka, D.C. Harmes, B. Madigan, J.J. Hsiao, G.O. Staples, High resolution two-dimensional liquid chromatography coupled with mass spectrometry for robust and sensitive characterization of therapeutic antibodies at the peptide level, J. Chromatogr. B. 1134–1135 (2019). https://doi.org/10.1016/j.jchromb.2019.121832.

[25] L.R. Snyder, J.W. Dolan, P.W. Carr, A new look at the selectivity of RPC columns, Anal. Chem. 79 (2007) 3254–3262. https://doi.org/10.1021/ac071905z.

[26] K. Croes, A. Steffens, D.H. Marchand, L.R. Snyder, Relevance of π-π and dipole-dipole interactions for retention on cyano and phenyl columns in reversed-phase liquid chromatography, J. Chromatogr., A. 1098 (2005) 123–130. https://doi.org/10.1016/j.chroma.2005.08.090.

[27] D.R. Stoll, T.A. Dahlseid, S.C. Rutan, T. Taylor, J.M. Serret, Improvements in the predictive accuracy of the hydrophobic subtraction model of reversed-phase selectivity, J. Chromatogr. A. (2020). https://doi.org/10.1016/j.chroma.2020.461682.

[28] L.R. Snyder, J.W. Dolan, High-performance Gradient Elution: The Practical Application of the Linear-Solvent-Strength Model, John Wiley, Hoboken, NJ, 2007.

[29] I.A. Haidar Ahmad, D.M. Makey, H. Wang, V. Shchurik, A.N. Singh, D.R. Stoll, I. Mangion, E.L. Regalado, *In Silico* multifactorial modeling for streamlined development and optimization of two-dimensional liquid chromatography, Anal. Chem. (2021). https://doi.org/10.1021/acs.analchem.1c01970.

[30] P. Petersson, A. Frank, J. Heaton, M.R. Euerby, Maximizing peak capacity and separation speed in liquid chromatography, J. Sep. Sci. 31 (2008) 2346–2357. https://doi.org/10.1002/jssc.200800064.

[31] C. Gunnarson, T. Lauer, H. Willenbring, E. Larson, M. Dittmann, K. Broeckhoven, D.R. Stoll, Implications of dispersion in connecting capillaries for separation systems involving post-column flow splitting, J. Chromatogr. A. (2021). https://doi.org/10.1016/j.chroma.2021.461893.

[32] B.W.J. Pirok, A.F.G. Gargano, P.J. Schoenmakers, Optimizing separations in on-line comprehensive two-dimensional liquid chromatography, J. Sep. Sci. 41 (2017) 68–98. https://doi.org/10.1002/jssc.201700863.

[33] S. Chapel, S. Heinisch, Strategies to circumvent the solvent strength mismatch problem in online comprehensive two-dimensional liquid chromatography, J. Sep. Sci. (2021). https://doi.org/10.1002/jssc.202100534.

[34] D.R. Stoll, K. Shoykhet, P. Petersson, S. Buckenmaier, Active solvent modulation – A valve-based approach to improve separation compatibility in two-dimensional liquid chromatography, Anal. Chem. 89 (2017) 9260–9267. https://doi.org/10.1021/acs.analchem.7b02046.

[35] S. Buckenmaier, C.A. Miller, T. van de Goor, M.M. Dittmann, Instrument contributions to resolution and sensitivity in ultra high performance liquid chromatography using small bore columns: Comparison of diode array and triple quadrupole mass spectrometry detection, J. Chromatogr. A. 1377 (2015) 64–74. https://doi.org/10.1016/j.chroma.2014.11.086.

[36] D.V. McCalley, Rationalization of retention and overloading behavior of basic compounds in reversed-phase HPLC using low ionic strength buffers suitable for mass spectrometric detection, Anal. Chem. 75 (2003) 3404–3410. https://doi.org/10.1021/ac020715f.

[37] A. Moussa, T. Lauer, D. Stoll, G. Desmet, K. Broeckhoven, Numerical and experimental investigation of analyte breakthrough from sampling loops used for multi-dimensional liquid chromatography, J. Chromatogr. A. 1626 (2020) 461283. https://doi.org/10.1016/j.chroma.2020.461283.

[38] Bioanalytical Method Validation – Guidance for Industry, United States Department of Health and Human Services, Food and Drug Administration, 2018.

[39] S.H. Yang, J. Wang, K. Zhang, Validation of a two-dimensional liquid chromatography method for quality control testing of pharmaceutical materials, J. Chromatogr. A. 1492 (2017) 89–97. https://doi.org/10.1016/j.chroma.2017.02.074.

7 Method Development for Comprehensive Two-Dimensional Liquid Chromatography Separations

Bob W.J. Pirok and Dwight R. Stoll

CONTENTS

DOI: 10.1201/9781003090557-7

7.1 INTRODUCTORY REMARKS

The goal of most comprehensive 2D-LC (LC×LC) separations is to gain as much information about the sample as possible. The higher peak capacities afforded by LC×LC methods generally improve the odds of observing more "pure" peaks, which can be translated into accurate quantitative information about complex samples. However, successful use of LC×LC requires striking a balance between the conditions used in the individual dimensions. This significantly complicates method development for LC×LC because the parameters of each dimension are strongly interdependent, and also depend on the modulation process and its parameters. To make things more challenging, there are additional complications, including undersampling (Section 3.4), sensitivity loss due to dilution (Section 3.6), mobile phase mismatch (Section 4.4.4), and poor complementarity (Section 3.3), any one of which can be detrimental to the chromatographic results.

Fortunately, years of fundamental work by many groups have resulted in a body of literature that holds lessons about how to mitigate these threats through decision-making during method development. Moreover, recent advances in novel modulation approaches (Section 4.4) give us new options to better deal with mobile phase mismatch and improve detection sensitivity. Nevertheless, even though this information is in the literature, translating this knowledge into functional and high-performing LC×LC methods can still feel like a daunting task.

In this chapter we first propose a set of generic workflows as starting points in method development, which leverage theoretical concepts discussed in Chapters 2–5, along with our own experiences. We then go on to demonstrate the use of these workflows in practice using two case studies of method development for two real LC×LC methods, and then close with a discussion of how software tools can be used to help develop and optimize these methods.

7.2 A GENERIC APPROACH TO METHOD DEVELOPMENT

In Figure 7.1 we propose a flow chart that reflects a generic approach to method development for LC×LC. This approach involves three main phases, which are discussed in turn in the following sections.

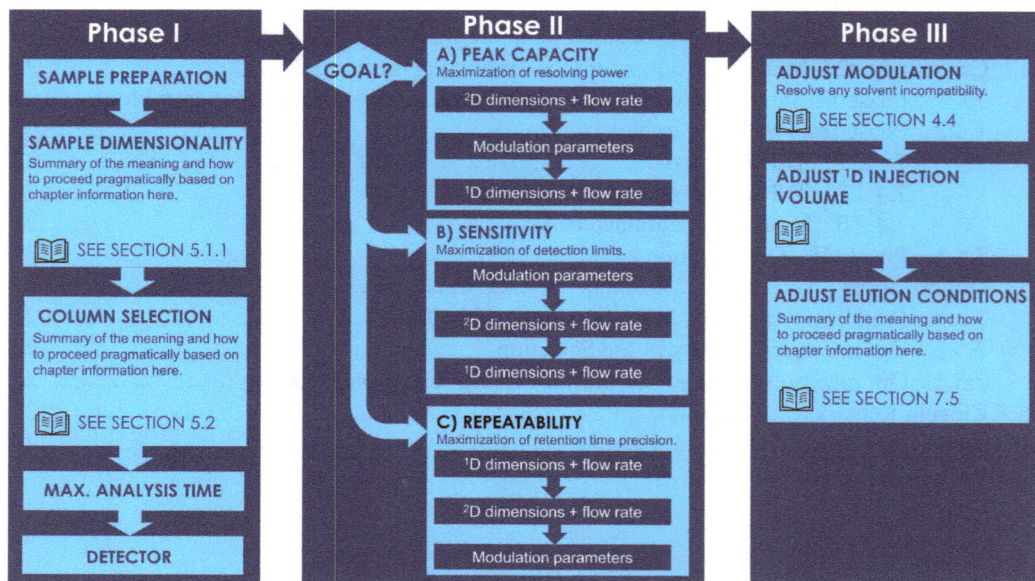

FIGURE 7.1 Generic approach to method development for LC×LC separations.

7.2.1 PHASE I – SAMPLE PREPARATION, SAMPLE DIMENSIONALITY, AND SEPARATION MODES

All methods should start with at least a consideration about whether or not sample preparation is needed prior to the 2D-LC separation. On one hand one of the perceived benefits of 2D-LC is that it alleviates the need for extensive, often time-consuming, sample preparation [1, 2]. On the other hand, LC×LC methods that target certain compound classes can benefit tremendously from a sample preparation step that removes compounds not targeted by the method, thus making the work of the 2D-LC method easier. Next, the dimensionality of the sample should be considered, which in turn guides the selection of appropriate separation modes (e.g., IEX, RP, etc.) and stationary phases; these were both discussed extensively in Chapter 5. At this point, decisions need to be made about the total analysis time that is desired, and the type of detector that will be used, as these decisions will strongly affect subsequent decisions made in Phase II of the flow chart.

7.2.2 PHASE II – PHYSICAL PARAMETERS

As was discussed in Section 1.9, method development goals that prioritize peak capacity (resolving power), detection sensitivity, throughput (analysis time), and robustness often conflict with each other (see also Figure 7 of [3]). Thus, the analyst has to decide which of these is most important, and then choose the appropriate workflow in Phase II of the flow chart.

As discussed in Section 3.2, for a 2D-LC method to be considered comprehensive, all ^{1}D effluent (or a consistent fraction of it) must be subjected to a ^{2}D separation. This means that:

1. the ^{2}D method must be sufficiently fast to be able to process the incoming ^{1}D fractions
2. the ^{1}D flow rate must be adapted to adhere to the requirements of the modulation interface and ^{2}D cycle time as well as the overall desired analysis time per sample
3. the sampling loops of the interface must be large enough to contain a full ^{1}D effluent fraction (or a portion of it if the ^{1}D flow is split (see Section 4.4.9))

The interdependencies between method parameters that result from these requirements can be complex, as discussed at length by Sarrut *et al.* [4]. Our emphasis here is that some of these dependencies must be assigned more weight when developing a LC×LC method with certain goals in mind. These nuances are discussed in the following Sections 7.2.2.1–7.2.2.3.

7.2.2.1 Prioritizing Peak Capacity

As discussed in Section 3.5, the most influential factor controlling the effective 2D peak capacity is the productivity of the 2D separation (i.e., $^2n_c/^2t_c$). This is why the flow chart in Phase IIA begins with decisions about the ^{2}D column and flow rate. High ^{2}D productivities generally favor short, narrow columns packed with small particles, operated at relatively high flow rates. Once these parameters are set, then decisions about the interface and modulation can be made. The volume of ^{1}D effluent injected into the ^{2}D column should not be so large that it compromises the performance of the ^{2}D separation. Of course, this will also be affected by the type of modulation that is used, with active modulation techniques tending to increase the volume of ^{1}D effluent that can be injected. Finally, once the interface loop volume and the ^{2}D cycle time are fixed, then the ^{1}D flow rate can be calculated, and the dimensions of the ^{1}D column can be chosen such that the dead time will be reasonable. Additional secondary considerations at this stage are the extent to which the loops are filled during each ^{2}D cycle (see Section 4.4.13), and whether or not the ^{1}D flow is split prior to entering the loops (see Section 4.4.9).

7.2.2.2 Prioritizing Detection Sensitivity

When prioritizing detection sensitivity of a LC×LC method, decisions about the interface and modulation parameters will have the largest impact on the performance of the method. Simply put, detection sensitivity improves as the mass of analyte injected into the ^{2}D is increased, and the volumes of peaks eluting from the ^{2}D column are decreased. Unfortunately, we cannot optimize sensitivity by simply increasing the loop size used in the interface, because in most cases there is a tradeoff between injection volume and ^{2}D performance (i.e., resolution or peak capacity). Active modulation techniques (see Section 4.4.4) can be very helpful for influencing this tradeoff by enabling injection of larger volumes without completely compromising the ^{2}D performance. Once the interface and modulation parameters are set, then the column dimensions and flow rate for the second dimension can be set. Here narrow columns and low flow rates generally favor small peak volumes (which increase sensitivity), but the flow rate cannot be so low that the ^{2}D cycle times become excessively long, as this will cause undersampling and a serious diminution of the effective 2D peak capacity. Finally, the ^{1}D flow rate can be calculated, and the dimensions of the ^{1}D column can be chosen. At this point the considerations are similar to those discussed in the preceding section.

7.2.2.3   Prioritizing Retention Repeatability

When comparing a large number of LC×LC chromatograms it is very important that analyte retention times be consistent from chromatogram to chromatogram so that the peak for an analyte is in the same place each time. Methods have been developed to deal with shifting peak locations across the chromatograms in a dataset (see Section 8.3), but generally speaking better results will be obtained when retention times are more consistent, even if alignment algorithms are used. While there have been several studies that have reported retention precision statistics for LC×LC [5–8], at this point in time we are not aware of any studies that have systematically assessed the impact of multiple LC×LC method parameters on retention repeatability in both dimensions. We do, however, have a lot of practical experience with this aspect, and we have observed that the ^{1}D flow rate has a particularly strong influence on the repeatability of ^{1}D retention time, with good precision favoring higher ^{1}D flow rates. There are many factors that could contribute to this observation, including temperature control of the ^{1}D column, and the accuracy and precision of the solvent gradient formation. Some of these have been discussed in detail elsewhere [9]. For these reasons, in our work we generally aim to use ^{1}D flow rates in the range of 100–200 µL/min. when prioritizing ^{1}D retention repeatability. Once this flow rate is set, and a tentative ^{2}D cycle time chosen, then the ^{2}D column dimensions and flow rate can be decided, followed by the interface and modulation parameters. Most of the considerations for the last two steps have been discussed in the preceding two sections, and are similar here.

7.2.2.4   Isocratic vs. Gradient Elution in the Second Dimension

A large majority of the LCLC methods published to date involve gradient elution in the second dimension. As discussed in Chapter 2, gradient elution generally yields higher peak capacities per unit of 2D time compared to isocratic elution. Moreover, using gradient elution reduces the time needed to elute all analytes from the ^{2}D column; this facilitates the use of short ^{2}D cycle times, resulting in high 2D peak capacities. These are the main reasons gradient elution is used so frequently in the second dimension of LC×LC methods. However, use of gradient elution requires that the column is re-equilibrated following one ^{2}D cycle before the next one can begin. Thus, gradient elution is not always feasible in the second dimension when using LC modes that require long equilibration times (see Section 5.2). Fortunately, studies have shown that just one column volume of re-equilibration is often sufficient for RPLC [10], and even for HILIC [11]. Finally, we note that most LC×LC methods use gradient elution that involves changes in the organic/aqueous composition of the mobile phase over time. In some cases, this can lead to differences in the severity of mobile phase mismatch at different points during the analysis. For example, in the case of RP×RP separations it is typical that the organic content of the ^{1}D effluent will be high near the end of the separation, and this causes problems when injected into a ^{2}D gradient that begins with a water-rich mobile phase. Recent work by Wicht *et al.* on so-called "temperature responsive" stationary phases provides a potential solution to this problem, as it enables gradient elution by changing the retentivity of the stationary phase over time, rather than through changes in the mobile phase properties [12].

7.2.3 PHASE III – OPTIMIZATION

Phases I and II of the flow chart are focused on initializing a method – making the decisions needed to run a method for the first time. Once initial results are obtained, we move to Phase III where adjustments are made to improve performance as demanded by the application at hand. Two of the major levers to improve performance involve adjusting the modulation parameters as well as the elution conditions to maximize usage of the 2D separation space. Varying the modulation parameters has been discussed in some detail in Sections 4.4.5, 4.4.6, and 6.7. Shifted gradient elution is a powerful tool for improving the usage of the 2D separation space. This was discussed conceptually in Section 4.5, and will be discussed in more detail in the context of the case studies in the following sections. Shifting gradients are especially useful and relatively easy to implement in those cases where there is partial correlation between the retention patterns in the first and second dimensions (e.g., in RP×RP, as discussed below in Section 7.4). For systems with uncorrelated ^{1}D and ^{2}D retention patterns shifting gradients can still be helpful, but they can be more difficult to develop as discussed in Section 7.3.

7.3 CASE STUDY #1 – METHOD DEVELOPMENT FOR SMALL MOLECULE DYES

Herein, we draw on the concepts introduced in earlier chapters and demonstrate how they can be applied to quickly establish a method to separate a mixture of natural and synthetic dyes. Dyes are generally small molecules, but span a large range in properties, including their hydrophilicity/ hydrophobicity and acidity/basicity. Whereas natural dyes typically do not feature permanently charged moieties, synthetic dyes can be cationic or anionic, often featuring functionalities such as sulphonates and quaternary amines. Indeed, this very diverse set of compounds provides opportunities to illustrate many of the method development steps discussed in this chapter.

The application context here is that a cultural-heritage institute seeks a method to obtain the most separation possible for mixtures of dyes present in extracts of historical objects. The primary analytical goal is to identify as many dyes as possible in these samples. The dyes might be degraded into similar components and/or contain contaminants. Since the identity or age of the object and its dyes is not always known *a priori*, the method ideally should be able to address all compounds within a particular molecular class that might be encountered in these samples. Given that thousands of dyes have already been identified, and that an even greater number of degradation products and impurities may be present, these mixtures are challenging even for the most powerful LC×LC separations.

7.3.1 PHASE I – SAMPLE PREPARATION, SAMPLE DIMENSIONALITY, AND SEPARATION MODES

In Phase I of method development we need to consider potential benefits of a sample preparation step, determine the sample dimensionality, and select separation modes.

7.3.1.1 Sample Dimensionality

Figure 7.2 shows the structures of several dye molecules, with different functional groups highlighted. The colors indicate relatively hydrophobic (purple), hydrophilic (blue), and charged (yellow) components of these molecules. From this coarse sketch, we can already recognize that HILIC, RPLC, and IEX separation modes could all be useful. Clearly, a number of potential pairings of these modes are worth considering, including RP×RP, HILIC×RP, and IEX×RP.

A number of factors now determine the combination of separation modes to use as a starting point. The first concerns identification of the chemical properties for which we expect the largest

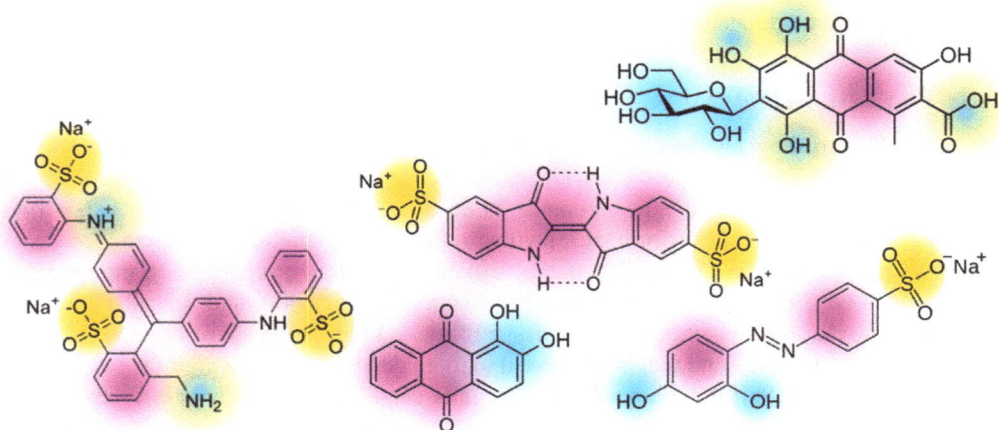

FIGURE 7.2 Example chemical structures of dye molecules. Relative chemical characteristics are highlighted with color: purple (hydrophobic), blue (hydrophilic) and yellow (charged).

Source: Reproduced from [13].

variety in the total population of analytes imaginable for this method. Dyes indeed exhibit a wide variability in the number of charges, which in turn may also be influenced by method parameters such as the pH of the mobile phase. Thus, a charge-based separation is attractive.

On the other hand, we also expect a significant variation in hydrophobic/hydrophilic character of these molecules, influenced by variation in the number of aromatic rings, for example. Given the numerous advantages of RPLC over other separation modes (see Chapter 5, Table 5.3), it is logical to include it in at least one of the dimensions of an initial method.

7.3.1.2 Column Selection

At this point a feasibility study is typically conducted to evaluate columns as candidates for a ¹D separation. In this case, we could indeed assess RPLC, IEX, and HILIC using different chemistries in each case. For example, for RPLC variations in mobile phase (e.g., pH and/or organic modifier) and stationary phase (e.g., alkyl vs. aromatic phases) can be investigated. Similarly, different HILIC selectivities could be studied, as well as mixed-mode (i.e. IEX and RP mechanisms with the same phase) phases. The depth of this feasibility study largely depends on the time available and the resources, the experience of the analyst with the analyte class, and the anticipated complications in combining different separation modes. The retention times obtained from these 1D separations can be plotted against each other to simulate a LC×LC chromatogram; this can help assess the complementarity of any combination of two separations. Based on these results, we chose IEX and RPLC for an initial method. Keeping in mind the benefits and limitations of the two possibilities for the separation order (i.e., IEX×RP, or RP×IEX) shown in Table 5.3, IEX×RP is clearly the best choice as a starting point.

For our IEX separation we must decide between weak or strong IEX stationary phases, and between cation- or anion-exchange. Again, we must draw on any prior knowledge regarding the analyte characteristics. In our case, the dyes are expected to mainly feature negatively charged moieties, which would benefit most from anion-exchange chromatography.

Figure 7.3 displays an example of a strong anion-exchange (SAX) separation of our dyes. The method involves a salt gradient up to 50 mM of ammonium sulphate. In addition, both mobile phase A and B contain 40% acetonitrile to suppress hydrophobic interaction between analyte molecules and the stationary

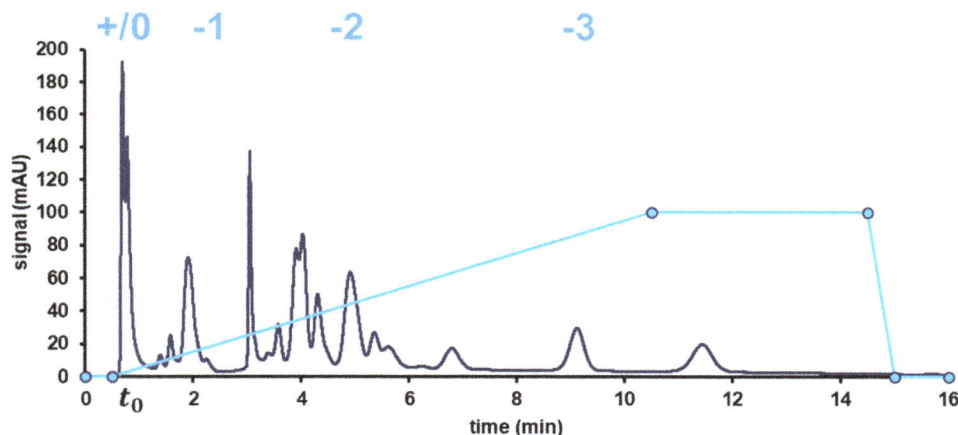

FIGURE 7.3 Example of a strong anion-exchange separation of dyes using a mobile phase composition of 40% acetonitrile in water, with a gradient running from 0 to 50 mM ammonium sulphate (shown as the blue line). The numbers at the top indicate the relative charge of the dominant analytes eluting around this retention time. Reprinted from *Journal of Chromatography, A*, 1436, B. Pirok, J. Knip, M. van Bommel, P. Schoenmakers, Characterization of synthetic dyes by comprehensive two-dimensional liquid chromatography combining ion-exchange chromatography and fast ion-pair reversed-phase chromatography, 141–146, Copyright (2016), with permission from Elsevier.

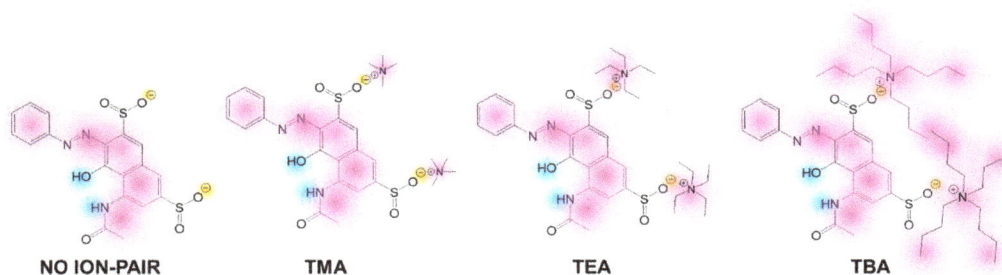

FIGURE 7.4 Examples of dye molecules and their interactions with tetramethylammonium (TMA), tetraethylammonium (TEA) and tetrabutylammonium (TBA) IPRs. Similar to Figure 7.1, relative hydrophobic character is roughly colored as pink, hydrophilic as blue and charged as yellow. Using IPRs neutralizes the charges on the anlaytes, but using an IPR that is too bulky may result in additional RP retention.

Source: Reproduced from [13].

phase. This is necessary for the ^{1}D retention to be dictated primarily by the charge state of the analyte, and so that the retention pattern is complementary to that obtained with the ^{2}D RP separation.

For the RPLC separation we could consider several stationary phase selectivities. However, we should first consider the impact of the charged functional groups of the dyes on their interactions with RP phases. As the number of charges on a molecule increases, its solubility in the polar mobile phase increases, and retention decreases. This is detrimental in this context because this trend is exactly opposite of that for the ^{1}D IEX separation (i.e., as charge increases, retention increases), which will compromise the orthogonality expected for the IEX and RP separations.

To mitigate this, an ion-pairing reagent (IPR) can be used to neutralize the effect of the charged groups on solubility in the mobile phase. Common IPRs used in RPLC include the

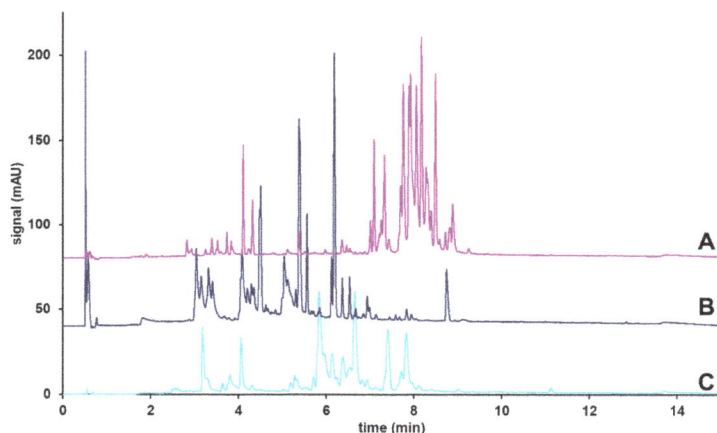

FIGURE 7.5   Offset overlay of RP separations using a C18 column and 5mM (A) tetrabutylammonium (TBA); (B) tetramethylammonium (TMA); or (C) sodium hydroxide in the mobile phase. Reprinted from *Journal of Chromatography, A*, 1436, B. Pirok, J. Knip, M. van Bommel, P. Schoenmakers, Characterization of synthetic dyes by comprehensive two-dimensional liquid chromatography combining ion-exchange chromatography and fast ion-pair reversed-phase chromatography, 141–146, Copyright (2016), with permission from Elsevier.

tetramethylammonium (TMA) and tetrabutylammonium (TBA) ions. The choice of IPR, of course, strongly affects the RP separation due to differences between the ways they interact with analytes, as illustrated in Figure 7.4. IPRs with large, bulky alkyl chains can be strongly retained by the RP stationary phase themselves, which creates a kind of dynamic IEX phase, further increasing the retention of highly charged analytes. In the extreme case this could again lead to correlation between the retention patterns in the first and second dimensions, which is not good for the LC×LC separation.

This also becomes apparent from Figure 7.5A where the bulky IPR TBA is used compared to Figure 7.5C where no IPR is used. Indeed, charged analytes that would not normally be well retained in RPLC are now highly retained together with the inherently hydrophobic compounds. TMA is thus more suitable for the IPR, balancing reduction in the effective charge of the analyte with additional retention due to the IPR.

7.3.1.3   Analysis Time and Detection

In this case the goal is to develop a separation method that allows characterization of dyes and their degradation products. The method would be used in conservation sciences, where the chromatographic fingerprint would be used qualitatively as evidence of the origin and history of the cultural heritage object. Thus, analysis time is not the highest priority, and the target analysis time is in the range of 1 to 5 hours.

As for detection, this is a case where UV-Vis detection will yield valuable information regarding the identity of the compound. Indeed, dyes have specific UV-Vis absorption spectra and relatively strong absorption such that sensitivity should not be a limitation. However, this may not be the case for degradation products. Thus, the use of MS detection is still desirable, although long-term screening methods would suffice with UV-Vis detection alone.

7.3.2   Phase II – Physical Parameters

Maximizing the peak capacity of the LC×LC method is the highest priority for this application, which is aimed at providing a qualitative fingerprint of the dyes and their degradation products found in cultural heritage object extracts. In practice, it may be prudent to test the 1D-LC version of the envisioned ^{2}D separation under the conditions intended for use in the LC×LC method. In this case, the chromatograms shown in Figure 7.5 were obtained at 1.0 mL/min. and a backpressure of roughly 230 bar (3300 psi). The pressure limit for this column stated by the manufacturer was 600 bar (8700 psi), thus a flow rate of 2.4 mL/min. was tested. To ensure sufficient time for equilibration, more experiments were conducted using shorter gradient times while holding the modulation time constant (and thus, steeper gradient slopes). Note that it is important to take into account the viscosity change of the mobile phase during a gradient when estimating the highest flow rate that can be used while staying within a pressure limit. Finally, the gradient was adapted to yield a total ^{2}D analysis time of 2 min. The mobile phase program ran from 100% A to 100% B in 1.5 min., followed by a 0.1-min. return to 100% A, leaving 0.4 min. for the re-equilibration time. The maximum backpressure under these conditions was roughly 560 bar (8100 psi).

At this point the two individual separations can be combined into one LC×LC method. At the heart of this lies the modulation process. While an experienced chromatographer might directly apply advanced modulation strategies (e.g., ASM) or sophisticated shifting gradients, we first explored a generic and simple approach using passive modulation. Initially we are simply interested in coupling both dimensions and do not worry about undersampling. With an analysis time of two minutes for the ^{2}D separation, we can set the modulation time to two minutes. Consequently, our sample loops should be able to contain at least the incoming ^{1}D effluent volume for this duration. Initially we set the ^{1}D flow rate to 10 µL/min. meaning that the interface loops should at least contain 20 µL, and preferably 30 to 40 µL if we want to avoid analyte breakthrough from the loops (see Section 4.4.13) [14, 15]. We finally adjust the ^{1}D gradient such that the gradient slope is similar to that used during the initial development of ^{1}D elution conditions.

7.3.3 Phase III

Having defined the modulation parameters, we then analyze a 20 µL sample using the LC×LC method. The resulting 2D separation is shown in Figure 7.6A. While there certainly is room for improvement at this stage, the initial separation is successful in the sense that charge-based separation does indeed occur in the first dimension, which appears to be highly complementary to the ^{2}D RP separation based on the peak pattern. One problem with this initial result is that there are several significant peaks eluting at the dead time of the ^{2}D column (indicated by orange line in Figure 7.6A). In the first dimension we expect to see unretained peaks due to neutral or cationic analytes, however in the second dimension we would not expect to see analytes so hydrophilic that they would elute from the RP column in a highly aqueous mobile phase. Inspection of the UV-Vis spectra for these peaks shows that they are similar to small peaks that elute later in the same ^{2}D cycle. Given that the ^{1}D mobile phase contains 40% ACN to suppress hydrophobic interactions between analytes and the IEX stationary phase, it is likely that some analytes are "breaking through" at the ^{2}D column dead time. Potential solutions to this include: i) decreasing the volume of ^{1}D effluent injected into the ^{2}D column; and ii) (actively) decreasing the ACN concentration in the ^{1}D effluent either by adjusting the ^{1}D mobile phase composition, or by using an active modulation technique (see Section 4.4.4).

Another observation from the initial result that requires attention is the variations in peak shape in the second dimension. While the chromatographic peaks for cationic, neutral, and even singly-charged anionic analytes are narrow, this is clearly not the case for the more highly charged analytes. Thanks to the LC×LC separation we can now see that this broadening in the RP separation is directly related to the anionic charge of the analyte. Evidently, the IPR concentration of 5 mM is not high enough to give good peak shapes for all of the analytes. The interactions between charged analyte functional groups and IPRs in the mobile phase are controlled by dynamic equilibria that can be imagined as a distribution of states; by increasing the IPR concentration, the equilibrium shifts in favor of neutralizing the analyte charge. For analytes with more charged functional groups, the number of possible types of IPR-analyte states is higher, which leads to broader peaks. Increasing

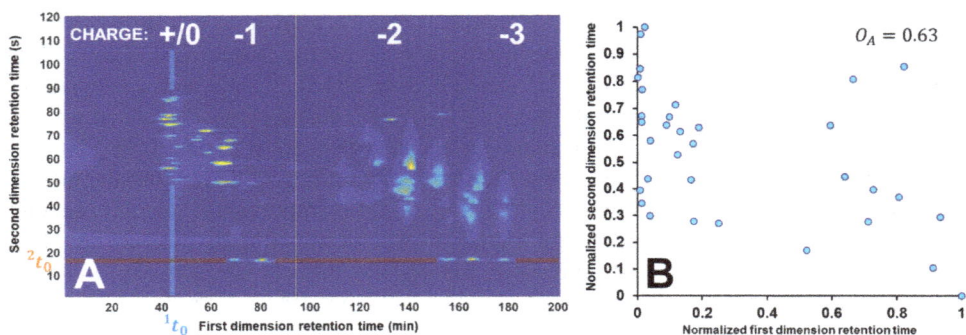

FIGURE 7.6   A: LC×LC-DAD separation of a mixture of synthetic dyes using SAX (x-axis) and IPRP (y-axis) for the ^{1}D and ^{2}D separations. ^{1}D: 150 mm×2.1 mm i.d. (8 μm) column operated at 10 μL/min; ^{2}D: 50 mm×4.6 mm i.d. (1.8 μm) column operated at 2.4 mL/min. Interface: Loops, 60 μL; Modulation time, 2 min. Chromatogram is plotted using absorbance at 400 nm. White numbers indicate the approximate distribution of analyte charge states across the ^{1}D separation. Approximate dead times for the ^{1}D and ^{2}D columns are indicated with vertical blue and horizontal orange lines, respectively. B: Orthogonality was calculated using the Asterisk method and found to be 0.63 [16]. Panel A is reprinted from *Journal of Chromatography, A*, 1436, B. Pirok, J. Knip, M. van Bommel, P. Schoenmakers, Characterization of synthetic dyes by comprehensive two-dimensional liquid chromatography combining ion-exchange chromatography and fast ion-pair reversed-phase chromatography, 141–146, Copyright (2016), with permission from Elsevier.

the IPR concentration improves peak shapes, but this does come at the cost of reducing compatibility with MS detection. At this point we can also adjust method conditions with an eye toward decreasing the total analysis time which was about four hours for the initial method.

Our first adjustment to the method is focused on the modulation strategy to reduce the occurrence of analyte breakthrough in the second dimension, and the potential for adding MS detection to the method, which was initially carried out using UV-Vis absorbance detection. In this case we chose to use the sorbent-based stationary phase-assisted modulation (SPAM) strategy (see Section 4.4.6 for more details on this type of approach) and replaced the open tubular loops with two RP guard columns. The approach relies on the retention of analytes on the stationary phase in these short columns, and provides a convenient means to "desalt" fractions of the ^{1}D effluent. Consequently, the ^{1}D flow rate now is 60 μL/min. utilizing the same column. The ^{2}D flow rate is 1.2 mL/min, now using a 50 mm×2.1 mm i.d. column packed with 1.8μm particles of the same stationary-phase chemistry as before. The modulation time is 0.75 min. Second, we replace tetramethylammonium (TMA) with triethylamine (TEA) as the IPR for better MS compatibility without affecting the separation selectivity too much.

Finally, we can modify the elution conditions to make better use of the 2D separation space. As shown in Figure 7.6 the peak pattern for this SAX×IPRP separation is already quite broad, even without using shifting gradients in the second dimension (see Section 4.5 for details on shifting gradients). Nevertheless, there are opportunities to improve the distribution of peaks by adjusting the ^{2}D elution conditions. LC×LC chromatograms obtained with two different shifting gradient programs are shown in Figure 7.7.

To ensure that analytes are retained by the traps, the ^{1}D effluent is actively diluted with water containing IPR as the diluent using a third pump. We evaluated two different dilution ratios as is shown in Figures 7.8A/B. Considering that the ^{1}D mobile phase contains 40% ACN ($^2\varphi = 0.4$), a dilution ratio of 40 would result in 1% ACN in the sample injected into the ^{2}D column, whereas a dilution ratio of 10 yields a concentration of 10% ACN. There is no breakthrough observed in the second dimension in either case, which suggests that the dilution ratio of 10 is adequate.

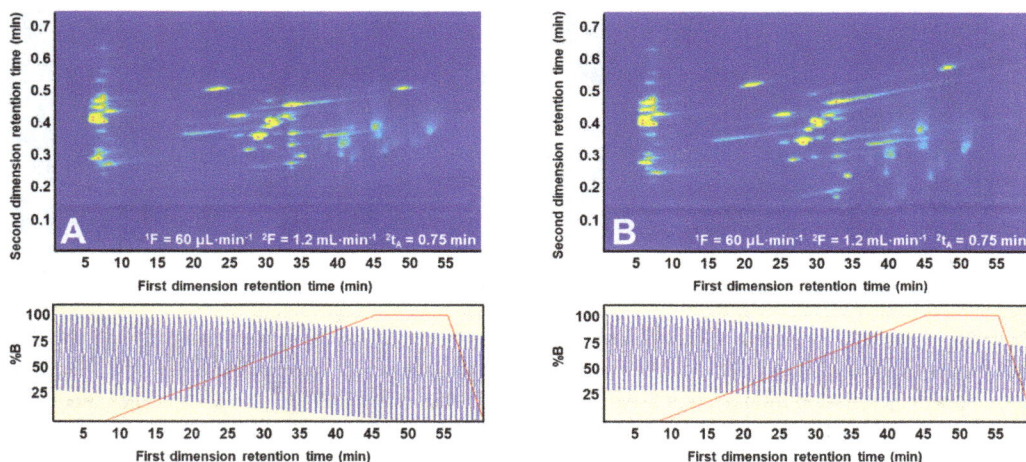

FIGURE 7.7 Comparison of two different shifting gradient programs and their influence on the use of the 2D separation space. Both methods use TEA as IPR. Reprinted with permission from [17] (https://pubs.acs.org/doi/10.1021/acs.analchem.8b05469). Permissions related to further use of this figure should be directed to the American Chemical Society.

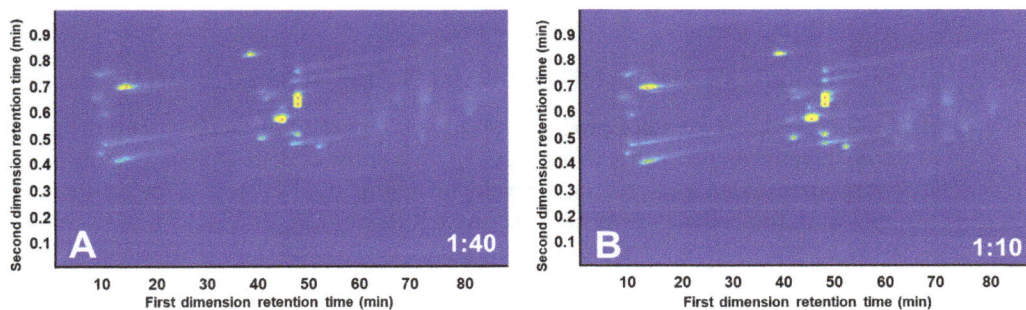

FIGURE 7.8 LC×LC separations of an expanded dye mixture using stationary phase-assisted modulation. The ^{1}D effluent is diluted by a factor of 40 (A) or 10 (B) prior to transfer to the trap columns. Compared to the previous figures a more complex sample was used containing twice as many dyes at more dilute concentrations. Reprinted with permission from Section S-2 of [17] (https://pubs.acs.org/doi/10.1021/acs.analchem.8b05469). Permissions related to further use of this figure should be directed to the American Chemical Society.

This is a good moment to emphasize that one of the advantages of using an active modulation strategy is that it makes the conditions used in the first and second dimensions much less interdependent. Thanks to the traps, the majority of ^{1}D effluent salts are removed and the analytes are concentrated into a much smaller volume before they are transferred to the ^{2}D column. Consequently, the ^{1}D flow rate can be increased as long as analytes remain retained on the trap column under these conditions. In addition, the injection volume into the ^{2}D column is now much lower, so that the ^{2}D column dimensions can also be reduced and the ^{2}D cycle time can be decreased. This is reflected in the change from the 4.6 mm (Figure 7.6) diameter column to the 2.1 mm diameter (Figures 7.7–7.10) column. As shown in Figure 7.9, moving to a higher ^{1}D flow rate, narrower ^{2}D column, and shorter modulation time all lead to a much shorter total analysis time of 45 min, albeit with a very high ^{2}D operating pressure of 1150 bar.

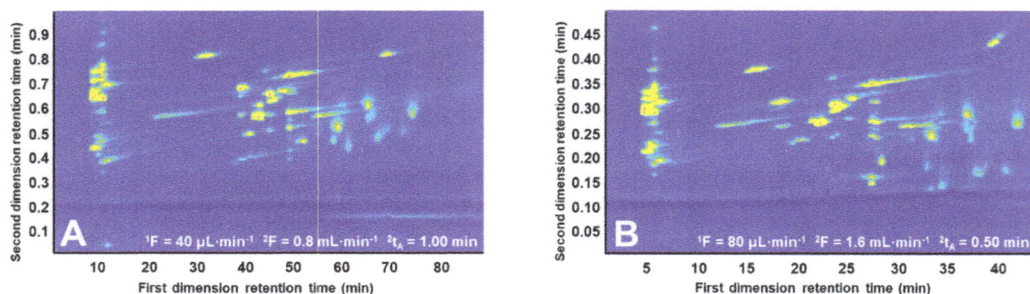

FIGURE 7.9 Comparison of two sets of LC×LC conditions (variables that have been changed are highlighted in the plot body), both using active modulation. All other conditions are the same as those shown in Figure 7.6. Reprinted with permission from Section S-3 of [17] (https://pubs.acs.org/doi/10.1021/acs.analchem.8b05469). Permissions related to further use of this figure should be directed to the American Chemical Society.

With all of these adjustments applied to the original separation (Figure 7.5), the end result is shown in Figure 7.10 with an orthogonality of 0.80 (measured using the Asterisk method), no occurrences of breakthrough, MS compatibility with ^{1}D effluent salts removed, and a significantly reduced analysis time. Of course, our method development does not necessarily end here. Indeed, the tailing observed for some compounds is serious, and resolution and surface coverage could be improved further.

7.4   CASE STUDY #2 – METHOD DEVELOPMENT FOR TRYPTIC PEPTIDES

Analytical scale separations of peptides produced by tryptic digestion of therapeutic proteins are important in the development and quality control of these materials as discussed in Section 10.2.2. In recent work we demonstrated the systematic development of a LC×LC method for separation of such peptides; this work forms the basis of this case study [6].

7.4.1   Phase I – Sample Preparation, Sample Dimensionality, and Separation Modes

Relative to other sample types that can be considerably more complex (e.g., food matrices), mixtures of peptides resulting from tryptic digestion of purified protein material do not require sample preparation prior to LC×LC analysis. The reagents used in the digestion, including the trypsin itself, do not strongly interfere with the separation. The peptides can be characterized as having two primary sample dimensions – charge state (ranging from negatively to positively charged in solution), and hydrophobicity. This is the reason that historically 2D-LC separations of peptides and proteins have relied on IEX separation in one dimension, and RP separation in the other dimension. However, IEX has limitations in this context, including relatively low separation efficiencies, and the requirement of mobile phases with high salt concentrations, which makes coupling to MS detection difficult. A characteristic of peptide chemistry that can be exploited here is that the functional groups of the amino acid side chains that are ionogenic are either weak acids or weak bases. This means not only that they have the potential to be charged in solution, but also that their charge will be highly dependent on the solution pH. As was demonstrated in the seminal paper by Gilar *et al.* on LC separation orthogonality, this characteristic of peptide chemistry can lead to quite different selectivities when separated on RP columns at different pH values (see Fig. 1 of ref. [18]). This realization has since led several groups to explore the potential of online LC×LC separations involving RP columns in both dimensions, but at high (first dimension; pH ~ 10) and low (second dimension; pH ~ 3) pH [6, 19–21]. Although other possible pairings are attractive from the point of view of the

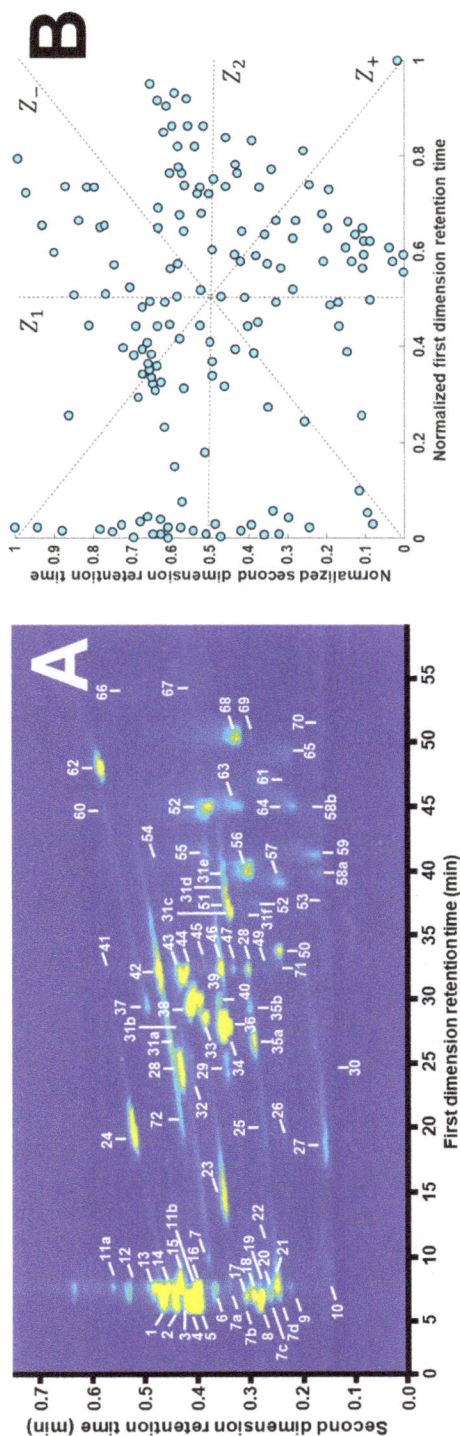

FIGURE 7.10 Final SAX×IPRP separation of a mixture of reference standard dyes. For all peak assignments see Fig. 4 of [17]. There are many peaks that are unretained by the ¹D column, but they are well spread out on the ²D column. Reprinted with permission from [17] (https://pubs.acs.org/doi/10.1021/acs. analchem.8b05469). Permissions related to further use of this figure should be directed to the American Chemical Society.

complementarity of the two selectivities, such as HILIC×RP [18, 22], it seems clear at this point that RP(high pH)×RP(low pH) provides the best compromise of ease of use and separation performance as measured by use of the 2D separation space, peak capacity, and detection sensitivity [6, 19].

Finally, in this work we prioritized the use of MS detection to enable identification of the separated peptides, and focused on 2D-LC analysis times in the range of 2 to 4 hours.

7.4.2   PHASE II – PHYSICAL PARAMETERS

In this work we were actually interested in ultimately striking a balance between the three objectives highlighted in Phase II of Figure 7.1. That is, we were aiming for a LC×LC method with high peak capacity, good sensitivity, and good retention repeatability, all in an analysis time in the range of 2 to 4 hrs.

7.4.2.1 ^{1}D Parameters

To support good repeatability in the ^{1}D retention times, we chose a ^{1}D flow rate of 80 μL/min. This in turn allowed the use of 2.1 mm i.d. ^{1}D columns with lengths of 100 to 200 mm. The dead times of these columns at 80 μL/min. are about 2 to 4 min., which are small compared to the desired 2 to 4 hr analysis time.

7.4.2.2 ^{2}D Parameters

In our work on this method we evaluated the potential of several different combinations of ^{2}D column dimensions (and particle sizes and types) and flow rates:

- Column lengths of 20 to 50 mm, all with 2.1 mm i.d.
- Particle sizes of 1.8 to 2.7 μm fully porous or superficially porous particles
- Flow rates of 1.25 to 2.50 mL/min.
- ^{2}D cycle times of 21 to 90 s

The experimental space covered by combinations of these parameters is expected to contain optimal conditions. The use of 2.1 mm i.d. columns in the second dimension enables fast, highly productive (as measured by peak capacity) ^{2}D separations. This does come at the cost of relatively high operating pressures, however, and precludes the use of certain combinations (e.g., 50 mm x 2.1 mm i.d., 1.9 μm, 2.5 mL/min.). The use of the higher flow rate of 2.5 mL/min. favors high peak capacities, but leads to larger peak volumes, and therefore lower detection sensitivities. These outcomes are nicely summarized in Figure 7.11, which mainly illustrates the tradeoff between 2D peak capacity and ^{2}D operating pressure. The figure also highlights other tradeoffs – for example, higher peak capacity can be obtained at the cost of lower detection sensitivity. Systematic study of these variables yields great quantitative insights into these tradeoffs. From this work we learned that by prioritizing peak capacity we could obtain a 2D peak capacity of 10,000 in an analysis time of 4 hrs; the resulting chromatogram is shown in Figure 10.2. On the other hand, decreasing the emphasis on peak capacity and increasing the priority of a shorter analysis time and more robust operation at a lower ^{2}D operating pressure requires the use of different ^{2}D conditions; these are highlighted by the green box in Figure 7.11, where we see that there are a few different combinations of variables that are viable.

7.4.2.3   Modulation Parameters

From the study of variables highlighted in the preceding section, we found that a ^{2}D cycle time of 30 s was a good compromise of 2D peak capacity and system robustness (i.e., reducing the number of interface valve switches). Given the ^{1}D flow rate of 80 μL/min., we settled on a loop size of 80 μL, which leads to 100% filling of the loop with ^{1}D effluent during each cycle. As is discussed in Section 4.4.13, this is not the best practice for quantitative studies. For quantitative work with this

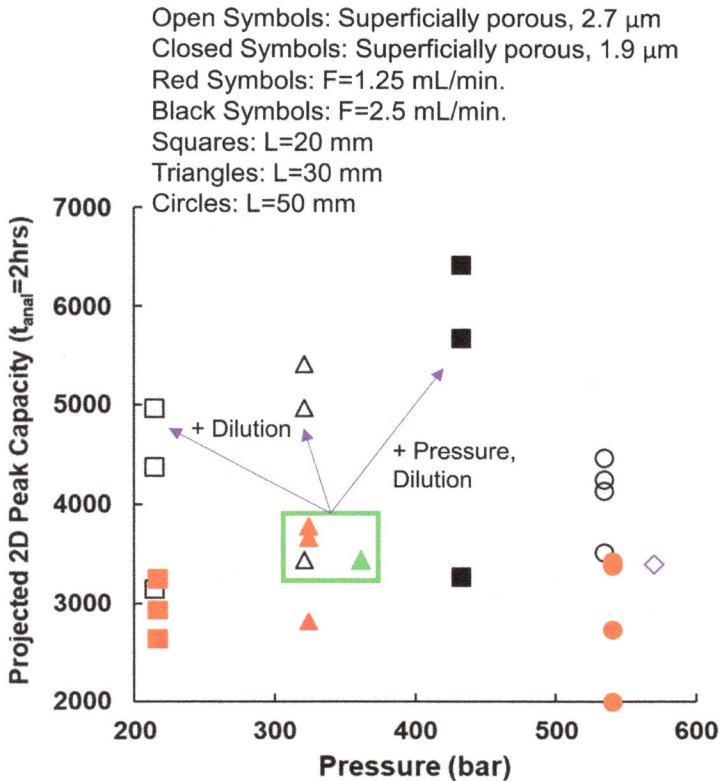

FIGURE 7.11 Relationship between LC×LC peak capacity for tryptic peptides and operating pressure of the ^{2}D column. Chromatographic conditions peak capacity calculation details are described in ref [6]. The single green triangle point corresponds to a 30 mm x 2.1 mm i.d. column packed with 1.8 μm fully porous Zorbax Eclipse Plus C18, operated at 1.25 mL/min. Reprinted from *Journal of Chromatography, B*, 1134–1135, D. Stoll, H. Lhotka, D.C. Harmes, B. Madigan, J. Hsiao, G. Staples, High resolution two-dimensional liquid chromatography coupled with mass spectrometry for robust and sensitive characterization of therapeutic antibodies at the peptide level, In Press, Copyright (2019), with permission from Elsevier.

method we would want to fill the loops to around 80%, which would require a reduction in ^{1}D flow rate, decrease in the ^{2}D cycle time, increase in the loop size, or some combination of these changes. In this case we used active modulation (ASM) and found that an ASM factor of 3 resulted in good ^{2}D peak shapes for a large majority of the ^{2}D peaks observed. The importance of active modulation in this context is shown in Figure 4.22. RP(high pH)×RP(low pH) separations of peptides like this can be run without active modulation, but this results in a serious decrease in detection sensitivity because large volumes of ^{1}D effluent are not tolerated by the ^{2}D column.

7.4.3 PHASE III – OPTIMIZATION

In the case of developing this particular method, most of the optimization effort was dedicated to maximizing the use of the 2D separation space by optimizing shifting gradient elution programs. An example of the impact of this step in the context of a separation like this is shown in Figure 4.34. In our work we did this optimization manually, using experience-guided trial-and-error, because the peak tracking algorithms referred to below (Section 7.5.3) were not available at that time. If we were to repeat this study today, we would certainly do it differently and leverage the recent work on automating this part of the method development process.

7.5 OPTIMIZATION

The experience level of the user significantly impacts the development time for a LC×LC method. Physical parameters can be selected using established routes (e.g., [23]), and a thorough understanding of the sample dimensionality will, in combination with 2D-LC experience, allow the chromatographer to rapidly establish an initial method.

However, once the initial separation conditions have been established, improvements based on further adjustment of method parameters will lead to questions about whether further adjustments are need, or if the method is "fit for purpose". This process is generally referred to as "optimization", a term used rather loosely in the chromatographic literature with divergent intentions. In some cases, "optimization" is used to describe the selection of a more suitable stationary-phase chemistry. In other cases, it is used to describe a process leading to reduction of analysis time, and in others improving separation efficiencies, or reducing solvent consumption. In chemometrics, "optimization" is often reserved for the improvement of an algorithm to maximize its efficiency. This aim is shared by optimization of LC×LC methods in that selection of the best possible methods parameters results in an improvement in metrics of separation performance such as peak capacity and usage of the separation space. Currently, optimization of LC×LC methods is still a bottleneck that impedes implementation of these methods, in part due to the large number of parameters that affect method performance and could be varied as part of an optimization scheme.

7.5.1 DECIDING WHETHER OR NOT TO CONTINUE WITH OPTIMIZATION

Deciding whether to continue with another iteration of method development requires a cost-benefit analysis. Would the additional information gained from improved chromatograms be worth the investment of additional time and effort?

The pair of LC×LC chromatograms shown in Figure 7.12 are useful as concrete examples to illustrate this thought process. The LC×LC chromatogram in the left panel was obtained for a separation of industrial surfactants. With the emphasis of Section 3.3 about maximizing use of the available 2D separation space in mind, one might look at this chromatogram and consider adjusting the ^{2}D elution conditions to make better use of the 2D space. Usage of the space in the regions highlighted by the rectangles A, B, and C obviously could be improved by spreading peaks out in the second dimension. However, if the analytical goal is to separate species already clearly resolved in this chromatogram, then further optimization is not necessary because the goal has already been achieved. On the other hand, the analysis time is long at several hours, and the target of subsequent optimization of the method could be focused on reducing the analysis time instead. This would improve the viability of the method in a product/process control scenario, where rapid production of analytical information is needed to provide feedback for control of the process.

We can roughly divide optimization goals into two classes.

- Sample-dependent or targeted optimization – In this class, efforts to improve the method aim to increase the likelihood of obtaining the desired analytical information that is specific to the sample.
- Untargeted or sample-independent optimization – In this class, the focus is on finding conditions that enable the separation of as many compounds as efficiently as possible.

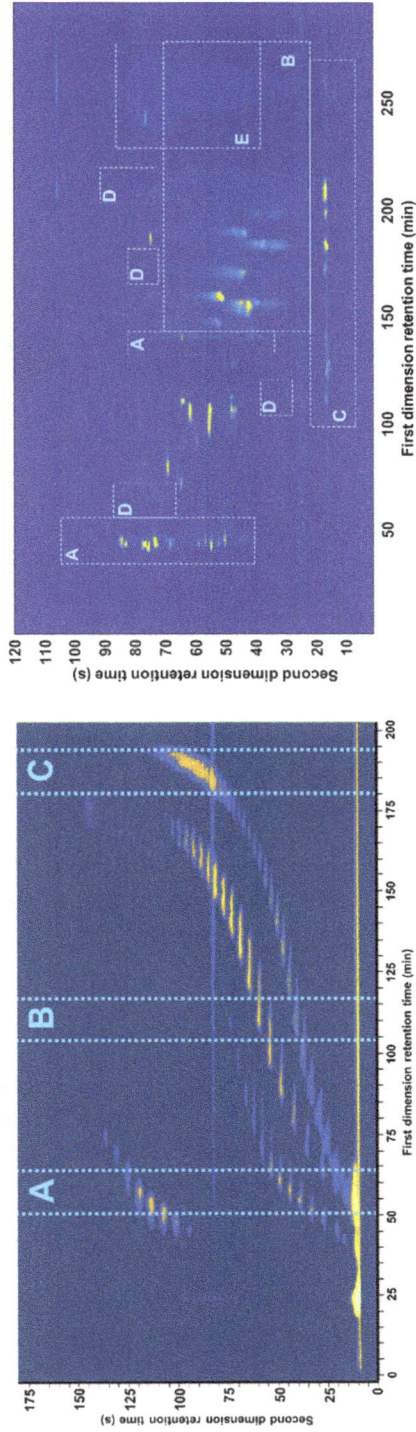

FIGURE 7.12 Left: SAX×RPLC separation of ionic industrial surfactants, adapted from [24]. Right: SAX×IPRP separation of the dye mixture as shown in Figure 7.6, adapted from [24].

The differences between these approaches can be understood by considering the example chromatogram in the right panel of Figure 7.12. Examples of objectives for targeted optimization of this separation include:

1. Improving resolution in specific regions of the chromatogram (A)
2. Reduction of excessive peak tailing to allow quantitation of more species (B)
3. Elimination of analyte breakthrough in the second dimension (C)
4. Increasing the sensitivity for analytes present at trace concentrations (D)
5. Reduction of analysis time to obtain a higher throughput (E).

In contrast, an untargeted optimization approach would prioritize more fully using the available 2D separation space and increasing the 2D peak capacity, which may come at the cost of not meeting several of the objectives mentioned in the preceding list.

Given the fact that targeted optimization requires a detailed and specific analytical goal to establish the objectives for optimization, the remainder of this chapter is focused on strategies for untargeted optimization.

7.5.2 Optimizing Stationary Phase Selection

Once an initial LC×LC chromatogram is in hand, it is common to think about further optimization involving steps to change elution conditions, flow rates, modulation parameters, and so on. But we emphasize here that it is never too late to change the selectivities used in the two dimensions. If a strong correlation in ^{1}D and ^{2}D retention times is observed in the first result, it might be a good idea to consider changing the selectivities used before going any further. In the particular case of choosing RP selectivities, some strategies for choosing from the hundreds of available options are discussed in Section 5.3. These strategies rely heavily on the use of the Hydrophobic Subtraction Model of RP selectivity, and the large, freely available database of column selectivity data (www. hplccolumns.org).

Other approaches have also been used for classification of columns by selectivity, including principal component analysis (PCA) [25–27] and the PRISMA model [28, 29]. For the latter a combined workflow with Drylab [30] was recently developed to select the appropriate column and gradient program [31]. The quest for unique selectivity has even sparked the development of stationary-phase gradients [32–34], although optimization of mobile phase programs is more complex when using such stationary phase gradients [35].

7.5.3 Optimization of Elution Conditions Using Retention Modeling

Historically, the elution conditions for many LC×LC methods have been optimized through trial-and-error adjustment, informed by user experience. By using retention modeling we can use a more systematic approach to this part of optimizing a method. Retention models are based on relationships between analyte retention factors and mobile phase composition.

Using such retention models requires that the retention factor – mobile phase composition relationship is known, either from experimental measurements or some other data source (e.g., databases or literature articles). These data are then fit to a model of the relationship, typically using simple regression. Commonly-used models include the simple log-linear model (often referred to as the linear solvent strength, LSS model, or the Snyder model) [36] for RPLC, the so-called adsorption model for NPLC, HILIC, or IEX [37, 38], the "mixed-mode" model [39, 40], the quadratic model [41], and the non-linear, empirical Neue-Kuss model [42, 43]. Methods for evaluating the suitability of any individual model for a particular dataset include statistical approaches such as the Akaike information criterion (AIC) [44, 45], the F-test of regression [46] and the adjusted coefficient of determination [47], although often the predictive accuracy of a model is used to assess its viability.

Analyte retention factors can be extracted from multiple LCxLC chromatograms collected under conditions where the elution programs are deliberately changed such that the dependence of k on mobile phase conditions can be established. Typically, the most accurate data can be obtained under isocratic elution conditions, however this can be a time-consuming, and highly impractical if the sample at hand is complex [48]. Thus, gradient elution is typically used for this purpose, and the series of methods is referred to as "scanning" or "training" gradients. Users should understand that models resulting from these data are only strictly applicable to methods where the gradient slopes are similar to the slopes used during the scanning/training experiments. If a method uses conditions that involve extrapolation outside of this range, then the predicted chromatograms will not be accurate [49, 50].

Figure 7.13 illustrates how the retention times obtained from scanning LCxLC experiments are used to build retention models for each analyte. Using these models, retention times can be predicted for other elution conditions, leading to prediction of entire LCxLC chromatograms. The goal of the optimization process is to simulate separations for a large array of methods and to identify the optimal separation using separation quality descriptors (see Chapter 3) or chromatographic response functions (CRF). Several different CRFs have been developed and compared [51, 52]. Recently, Alvarez-Segura and co-workers introduced a CRF based on peak prominences and applied it to several complex separations [53]. CRFs specifically applicable to 2D-LC have also been developed, including one combining the number of observed peaks, analysis time, and peak overlap [54], as well as one employing orthogonality metrics as part of the computation [55].

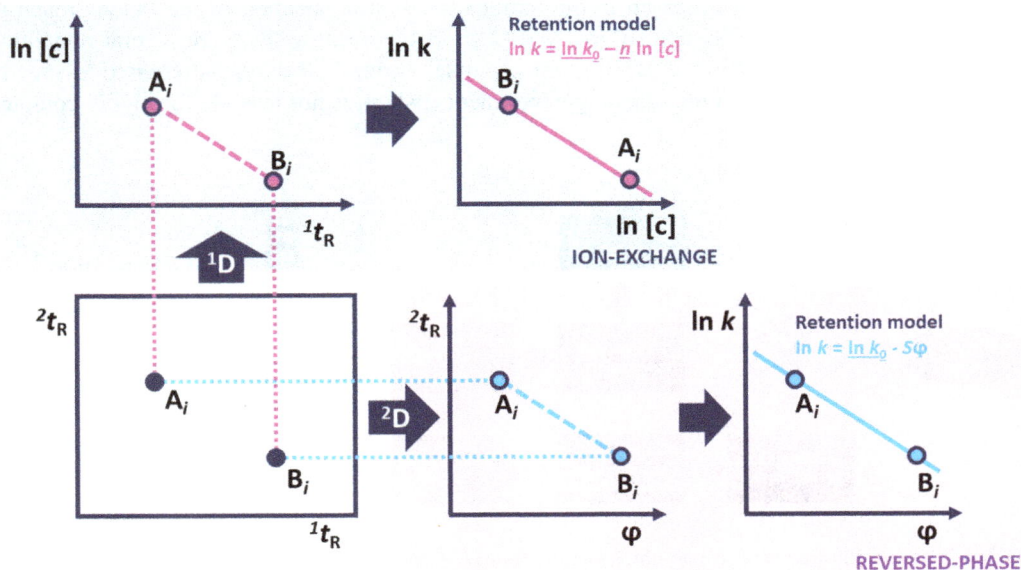

FIGURE 7.13 Illustration of the workflow for building retention models using scanning LCxLC experiments. In each LCxLC experiment, a different retention time is obtained for each compound. Regressing the retention factors against a mobile phase property (i.e., counterion concentration for IEX, and organic solvent fraction for RP) produces characteristic retention parameters (ln k_0 and n for ion exchange and ln k_0 and S for reversed phase) for each compound.

These concepts have been applied to the separation challenge discussed in Case Study #1 (Section 7.3) as an untargeted optimization workflow. The entire process is illustrated in Figure 7.14A. First, LC×LC chromatograms are obtained using gradient elution in both dimensions. For a determination of retention parameters to be useful, the gradient slopes used should be near the expected slopes. As this is unclear with the optimal method not known yet, it is thus useful to sample a wide range of gradient slopes. The retention times obtained are used to build a retention model for each compound. Next, a large array of experiments is simulated for which elution conditions are varied. Next, the quality descriptors (e.g., peak capacity, usage of the separation space, etc.) of the resulting simulations are computed. Using Pareto analysis, the quality descriptors can be plotted against each other for all simulated separations. An example for the dye separation is shown in Figure 7.14B, where a 2D resolution score is plotted against analysis time for 10,368 different conditions [57]. The points connected by the line represent the so-called Pareto-optimal front, which are the optimal methods found within the ranges of parameters used in the simulations. The user can then carry out experimental separations using the conditions for one or more of the Pareto-optimal (PO) points, and compare the experimental data with the prediction. If the differences between the experimental and predicted results are too large, other retention models can be used. In some cases, the increased separation power realized through the optimization process results in the resolution of more peaks than were observed in the initial scanning experiments, and thus a new iteration of scanning experiments may be warranted. This approach has been applied to the LC×LC separation of dyes (Section 7.3) to optimize the shifting gradient programs as shown in Figure 7.7 [17].

While this workflow was first demonstrated for LC×LC by Pirok *et al.* [56], it is not new in chromatography. Indeed, many approaches have been developed for chromatographic optimization in 1D-LC, including Drylab [30], Chromsword [58], PEWS [59] and other commercial packages. Muller *et al.* recently combined the above workflow with an optimization of kinetic and other parameters, including column length, temperature, dilution factor, and flow rate [60].

Despite promising results observed in the initial studies of application of the PO approach to LC×LC, it has not been widely applied in LC×LC, probably because there are several remaining barriers to robust implementation [61, 62]. For example, currently the user is required to specify the location of each analyte in the scanning experiments, which is not feasible for highly complex

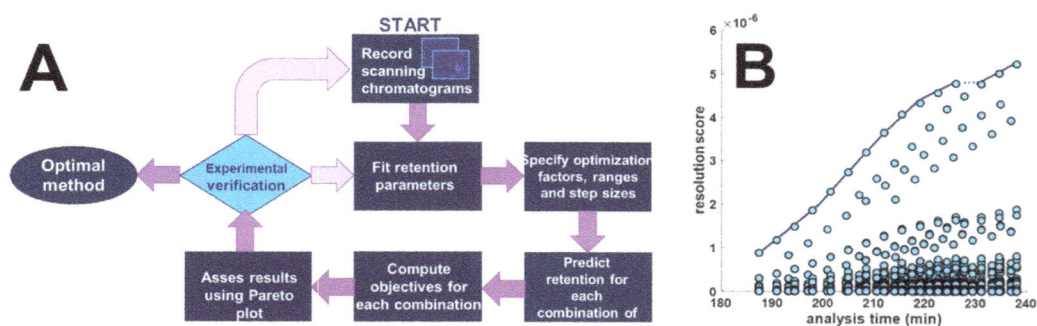

FIGURE 7.14   A: Illustration of an algorithmic optimization procedure [56]. B: Example of a Pareto-optimality plot with each point representing one of the 10,368 simulated LC×LC separations evaluated for separation of a mixture of aged, synthetic dyes. The Pareto-optimal (PO) front (line) reflects all Pareto-optimal points in the plot. These points correspond to conditions where one of the two plotted criteria cannot be improved without compromising the other. Panel A is adapted from *Journal of Chromatography, A,* 1450, B. Pirok, S. Pous-Torres, C. Ortiz-Bolsico, G. Vivo-Truyols, P. Schoenmakers, Program for the interpretive optimization of two-dimensional resolution, 29–37, Copyright (2016), with permission from Elsevier. Plot in Panel B created using MOREPEAKS [56].

samples such as those encountered in LC×LC(-MS/MS) experiments. Recent work on the development of automated peak-tracking algorithms should lower this barrier [63, 64]. Furthermore, improved modeling of retention as well as elution shapes of chromatographic peaks will allow algorithms to better simulate realistic chromatograms. Finally, algorithms capable of handling the multi-gigabyte datasets produced from LC×LC separation coupled with multi-channel detection (i.e., DAD, MS), and improved retention models are needed to move this work forward. Fortunately, several groups have already made tremendous progress, which is discussed in Chapter 8.

7.6 CLOSING REMARKS

The generic approach to LC×LC method development proposed in Section 7.2 and illustrated in Figure 7.1 is an attempt to leverage our own method development experiences, and the knowledge built up in the literature in recent years, to establish a workflow that is approachable by LC×LC users of all experience levels. One of the greatest strengths of LC×LC (its flexibility for application to a variety of analytical problems) also presents challenges in method development because there are so many decisions to make, many of which are interdependent. In Phase II of our approach we simplify the decision-making process by prioritizing different performance metrics. Moving forward, we expect to see the development of software tools that facilitate the optimization of LC×LC methods once an initial result has been obtained. This will be especially helpful in cases where the LC×LC separation is coupled to multi-channel detection, such as mass spectrometry.

REFERENCES

[1] Z.D. Dunn, J. Desai, G.M. Leme, D.R. Stoll, D.D. Richardson, Rapid two-dimensional Protein-A size exclusion chromatography of monoclonal antibodies for titer and aggregation measurements from harvested cell culture fluid samples, MAbs. 12 (2020). doi:10.1080/19420862.2019.1702263

[2] J. Camperi, A. Goyon, D. Guillarme, K. Zhang, C. Stella, Multi-dimensional LC-MS: The next generation characterization of antibody-based therapeutics by unified online bottom-up, middle-up and intact approaches, Analyst. (2021). doi:10.1039/D0AN01963A

[3] D.R. Stoll, H.R. Lhotka, D.C. Harmes, B. Madigan, J.J. Hsiao, G.O. Staples, High resolution two-dimensional liquid chromatography coupled with mass spectrometry for robust and sensitive characterization of therapeutic antibodies at the peptide level, J. Chromatogr. B. 1134–1135 (2019). doi:10.1016/j.jchromb.2019.121832

[4] M. Sarrut, A. D'Attoma, S. Heinisch, Optimization of conditions in on-line comprehensive two-dimensional reversed phase liquid chromatography: Experimental comparison with one-dimensional reversed phase liquid chromatography for the separation of peptides, J. Chromatogr. A. 1421 (2015) 48–59. doi:10.1016/j.chroma.2015.08.052

[5] G. Vanhoenacker, I. Vandenheede, F. David, P. Sandra, K. Sandra, Comprehensive two-dimensional liquid chromatography of therapeutic monoclonal antibody digests, Anal. Bioanal. Chem. 407 (2015) 355–366. doi:10.1007/s00216-014-8299-1

[6] D.R. Stoll, H.R. Lhotka, D.C. Harmes, B. Madigan, J.J. Hsiao, G.O. Staples, High resolution two-dimensional liquid chromatography coupled with mass spectrometry for robust and sensitive characterization of therapeutic antibodies at the peptide level, J. Chromatogr. B. 1134–1135 (2019). doi:10.1016/j.jchromb.2019.121832

[7] V. Elsner, V. Wulf, M. Wirtz, O.J. Schmitz, Reproducibility of retention time and peak area in comprehensive two-dimensional liquid chromatography, Anal. Bioanal. Chem. 407 (2015) 279–284. doi:10.1007/s00216-014-8090-3

[8] D.R. Stoll, X. Wang, P.W. Carr, Comparison of the practical resolving power of one- and two-dimensional high-performance liquid chromatography analysis of metabolomic samples, Anal. Chem. 80 (2008) 268–278. doi:10.1021/ac701676b

[9] P. Petersson, Fluctuating peptide retention in 2D-LC: How to address a moving target, LC-GC North Am. 37 (2019) 740–746.

[10] A.P. Schellinger, D.R. Stoll, P.W. Carr, High speed gradient elution reversed-phase liquid chromatography, J. Chromatogr. A. 1064 (2005) 143–156. doi:10.1016/j.chroma.2004.12.017

[11] C. Seidl, D.S. Bell, D.R. Stoll, A study of the re-equilibration of hydrophilic interaction columns with a focus on viability for use in two-dimensional liquid chromatography, J. Chromatogr. A. 1604 (2019). doi:10.1016/j.chroma.2019.460484

[12] K. Wicht, M. Baert, M. Muller, E. Bandini, S. Schipperges, N. von Doehren, et al., Comprehensive two-dimensional temperature-responsive×reversed phase liquid chromatography for the analysis of wine phenolics, Talanta. 236 (2022). doi:10.1016/j.talanta.2021.122889

[13] B.W.J. Pirok, Making Analytical Incompatible Approaches Compatible, University of Amsterdam, Amsterdam, 2019. https://hdl.handle.net/11245.1/4b32e4f5-bc71-41b0-8fd4-c694b410d425.

[14] A. Moussa, T. Lauer, D. Stoll, G. Desmet, K. Broeckhoven, Numerical and experimental investigation of analyte breakthrough from sampling loops used for multi-dimensional liquid chromatography, J. Chromatogr. A. 1626 (2020). doi:10.1016/j.chroma.2020.461283

[15] B.W.J. Pirok, A.F.G. Gargano, P.J. Schoenmakers, Optimizing separations in online comprehensive two-dimensional liquid chromatography, J. Sep. Sci. 41 (2018) 68–98. doi:10.1002/jssc.201700863

[16] M. Camenzuli, P.J. Schoenmakers, A new measure of orthogonality for multi-dimensional chromatography, Anal. Chim. Acta. 838 (2014) 93–101. doi:10.1016/j.aca.2014.05.048

[17] B.W.J. Pirok, M.J. den Uijl, G. Moro, S.V.J. Berbers, C.J.M. Croes, M.R. van Bommel et al., Characterization of dye extracts from historical cultural-heritage objects using state-of-the-art comprehensive two-dimensional liquid chromatography and mass spectrometry with active modulation and optimized shifting gradients, Anal. Chem. 91 (2019) 3062–3069. doi:10.1021/acs.analchem.8b05469

[18] M. Gilar, P. Olivova, A.E. Daly, J.C. Gebler, Orthogonality of separation in two-dimensional liquid chromatography, Anal. Chem. 77 (2005) 6426–6434. doi:10.1021/ac050923i

[19] M. Sarrut, F. Rouvière, S. Heinisch, Theoretical and experimental comparison of one dimensional versus on-line comprehensive two dimensional liquid chromatography for optimized sub-hour separations of complex peptide samples, J. Chromatogr. A. 1498 (2017) 183–195. doi:10.1016/j.chroma.2017.01.054

[20] I. François, D. Cabooter, K. Sandra, F. Lynen, G. Desmet, P. Sandra, Tryptic digest analysis by comprehensive reversed phase×two reversed phase liquid chromatography (RP-LC×2RP-LC) at different pH's, J. Sep. Sci. 32 (2009) 1137–1144. doi:10.1002/jssc.200800578

[21] M. Gilar, P. Olivova, A.E. Daly, J.C. Gebler, Two-dimensional separation of peptides using RP-RP-HPLC system with different pH in first and second separation dimensions, J. Sep. Sci. 28 (2005) 1694–1703. doi:10.1002/jssc.200500116

[22] A. D'Attoma, S. Heinisch, On-line comprehensive two dimensional separations of charged compounds using reversed-phase high performance liquid chromatography and hydrophilic interaction chromatography. Part II: Application to the separation of peptides, J. Chromatogr. A. 1306 (2013) 27–36. doi:10.1016/j.chroma.2013.07.048

[23] G. Vivó-Truyols, S. Van Der Wal, P.J. Schoenmakers, Comprehensive study on the optimization of online two-dimensional liquid chromatographic systems considering losses in theoretical peak capacity in first- and second-dimensions: A pareto-optimality approach, Anal. Chem. 82 (2010). doi:10.1021/ac101420f

[24] B.W.J. Pirok, A.F.G. Gargano, P.J. Schoenmakers, Optimizing separations in on-line comprehensive two-dimensional liquid chromatography, J. Sep. Sci. 41 (2018) 68–98. doi:10.1002/jssc.201700863

[25] M.R. Euerby, P. Petersson, Chromatographic classification and comparison of commercially available reversed-phase liquid chromatographic columns using principal component analysis, J. Chromatogr. A. 994 (2003) 13–36. doi:10.1016/S0021-9673(03)00393-5

[26] M.R. Euerby, P. Petersson, W. Campbell, W. Roe, Chromatographic classification and comparison of commercially available reversed-phase liquid chromatographic columns containing phenyl moieties using principal component analysis, J. Chromatogr. A. 1154 (2007) 138–151. doi:10.1016/j.chroma.2007.03.119

[27] P. Petersson, M.R. Euerby, Characterisation of RPLC columns packed with porous sub-2 µm particles, J. Sep. Sci. 30 (2007) 2012–2024. doi:10.1002/jssc.200700086

[28] Bischoff chromatography, POPLC, www.bischoff-chrom.de/poplc.html. Accessed May 30, 2022.

[29] S. Nyiredy, C.A.J. Erdelmeier, B. Meier, O. Sticher, The "PRISMA" mobile phase optimization model in thin-layer chromatography – Separation of natural compounds, Planta Med. 51 (1985) 241–246. doi:10.1055/s-2007-969468

[30] J.W. Dolan, D.C. Lommen, L.R. Snyder, Drylab® computer simulation for high-performance liquid chromatographic method development, J. Chromatogr. A. 485 (1989) 91–112. doi:10.1016/S0021-9673(01)89134-2

[31] R.M. Krisko, K. McLaughlin, M.J. Koenigbauer, C.E. Lunte, Application of a column selection system and DryLab software for high-performance liquid chromatography method development, J. Chromatogr. A. 1122 (2006) 186–193. doi:10.1016/j.chroma.2006.04.065

[32] S. Currivan, D. Connolly, B. Paull, Stepped gradients on polymeric monolithic columns by photoinitiated grafting, J. Sep. Sci. 38 (2015) 3795–3802. doi:10.1002/jssc.201500776

[33] V.C. Dewoolkar, L.N. Jeong, D.W. Cook, K.M. Ashraf, S.C. Rutan, M.M. Collinson, Amine gradient stationary phases on in-house built monolithic columns for liquid chromatography, Anal. Chem. 88 (2016) 5941–5949. doi:10.1021/acs.analchem.6b00895

[34] S.L. Stegall, K.M. Ashraf, J.R. Moye, D.A. Higgins, M.M. Collinson, Separation of transition and heavy metals using stationary phase gradients and thin layer chromatography, J. Chromatogr. A. 1446 (2016) 141–148. doi:10.1016/j.chroma.2016.04.005

[35] L.N. Jeong, S.C. Rutan, Simulation of elution profiles in liquid chromatography – III. Stationary phase gradients, J. Chromatogr. A. 1564 (2018) 128–136. doi:10.1016/j.chroma.2018.06.007

[36] P.J. Schoenmakers, H.A.H. Billiet, R. Tijssen, L. De Galan, Gradient selection in reversed-phase liquid chromatography, J. Chromatogr. A. 149 (1978) 519–537. doi:10.1016/S0021-9673(00)81008-0

[37] P. Jandera, M. Holčapek, L. Kolářová, Retention mechanism, isocratic and gradient-elution separation and characterization of (co)polymers in normal-phase and reversed-phase high-performance liquid chromatography, J. Chromatogr. A. 869 (2000) 65–84. doi:10.1016/S0021-9673(99)01216-9

[38] C.M. Roth, K.K. Unger, A.M. Lenhoff, Mechanistic model of retention in protein ion-exchange chromatography, J. Chromatogr. A. 726 (1996) 45–56. doi:10.1016/0021-9673(95)01043-2

[39] A.E. Karatapanis, Y.C. Fiamegos, C.D. Stalikas, A revisit to the retention mechanism of hydrophilic interaction liquid chromatography using model organic compounds, J. Chromatogr. A. 1218 (2011) 2871–2879. doi:10.1016/j.chroma.2011.02.069

[40] G. Jin, Z. Guo, F. Zhang, X. Xue, Y. Jin, X. Liang, Study on the retention equation in hydrophilic interaction liquid chromatography, Talanta. 76 (2008) 522–527. doi:10.1016/j.talanta.2008.03.042

[41] P.J. Schoenmakers, H.A.H. Billiet, R. Tijssen, L. De Galan, Gradient selection in reversed-phase liquid chromatography, J. Chromatogr. A. 149 (1978) 519–537. doi:10.1016/S0021-9673(00)81008-0

[42] U.D. Neue, Nonlinear Retention Relationships in reversed-phase chromatography, Chromatographia. 63 (2006) S45–S53. doi:10.1365/s10337-006-0718-9

[43] U.D. Neue, H.-J. Kuss, Improved reversed-phase gradient retention modeling, J. Chromatogr. A. 1217 (2010) 3794–3803. doi:10.1016/j.chroma.2010.04.023

[44] H. Akaike, A new look at the statistical model identification, IEEE Trans. Automat. Contr. 19 (1974) 716–723. doi:10.1109/TAC.1974.1100705

[45] B.W.J. Pirok, S.R.A. Molenaar, R.E. van Outersterp, P.J. Schoenmakers, Applicability of retention modelling in hydrophilic-interaction liquid chromatography for algorithmic optimization programs with gradient-scanning techniques, J. Chromatogr. A. 1530 (2017) 104–111. doi:10.1016/j.chroma.2017.11.017

[46] L.S. Roca, S.E. Schoemaker, B.W.J. Pirok, A.F.G. Gargano, P.J. Schoenmakers, Accurate modelling of the retention behaviour of peptides in gradient-elution hydrophilic interaction liquid chromatography, J. Chromatogr. A. (2019). doi:10.1016/j.chroma.2019.460650

[47] R.G.D. Steel, J.H. Torrie, Principles and Procedures of Statistics, 1960, McGraw-Hill, New York.

[48] G. Vivó-Truyols, J.R. Torres-Lapasió, M.C. García-Alvarez-Coque, Error analysis and performance of different retention models in the transference of data from/to isocratic/gradient elution, J. Chromatogr. A. 1018 (2003) 169–181. doi:10.1016/j.chroma.2003.08.044

[49] G. Vivó-Truyols, J.. Torres-Lapasió, M.. García-Alvarez-Coque, Error analysis and performance of different retention models in the transference of data from/to isocratic/gradient elution, J. Chromatogr. A. 1018 (2003) 169–181. doi:10.1016/j.chroma.2003.08.044

[50] M.J. den Uijl, P.J. Schoenmakers, G.K. Schulte, D.R. Stoll, M.R. van Bommel, B.W.J. Pirok, Measuring and using scanning-gradient data for use in method optimization for liquid chromatography, J. Chromatogr. A. 1636 (2021). doi:10.1016/j.chroma.2020.461780

[51] E. Tyteca, G. Desmet, A universal comparison study of chromatographic response functions, J. Chromatogr. A. 1361 (2014) 178–190. doi:10.1016/j.chroma.2014.08.014

[52] J.T.V. Matos, R.M.B.O. Duarte, A.C. Duarte, Chromatographic response functions in 1D and 2D chromatography as tools for assessing chemical complexity, Trends Analyt. Chem. 45 (2013) 14–23. doi:10.1016/j.trac.2012.12.013

[53] J.A. Navarro-Huerta, T. Alvarez-Segura, J.R. Torres-Lapasió, M.C. García-Alvarez-Coque, Study of the performance of a resolution criterion to characterise complex chromatograms with unknowns or without standards, Anal. Methods. 9 (2017) 4293–4303. doi:10.1039/C7AY00399D

[54] R.M.B.O. Duarte, J.T. V Matos, A.C. Duarte, A new chromatographic response function for assessing the separation quality in comprehensive two-dimensional liquid chromatography, J. Chromatogr. A. 1225 (2012) 121–131. doi:10.1016/j.chroma.2011.12.082

[55] W. Nowik, M. Bonose, S. Héron, M. Nowik, A. Tchapla, Assessment of two-dimensional separative systems using the nearest neighbor distances approach. Part 2: Separation quality aspects, Anal. Chem. 85 (2013) 9459–9468. doi:10.1021/ac4012717

[56] B.W.J. Pirok, S. Pous-Torres, C. Ortiz-Bolsico, G. Vivó-Truyols, P.J. Schoenmakers, Program for the interpretive optimization of two-dimensional resolution, J. Chromatogr. A. 1450 (2016) 29–37. doi:10.1016/j.chroma.2016.04.061

[57] M.R. Schure, Quantification of resolution for two-dimensional separations, J. Microcolumn Sep. 9 (1997) 169–176. doi:10.1002/(SICI)1520-667X(1997)9:3<169::AID-MCS5>3.0.CO;2-#

[58] ChromSword Off-line (2015).

[59] E. Tyteca, A. Périat, S. Rudaz, G. Desmet, D. Guillarme, Retention modeling and method development in hydrophilic interaction chromatography, J. Chromatogr. A. 1337 (2014) 116–127. doi:10.1016/j.chroma.2014.02.032

[60] M. Muller, A.G.J. Tredoux, A. de Villiers, Predictive kinetic optimisation of hydrophilic interaction chromatography×reversed phase liquid chromatography separations: Experimental verification and application to phenolic analysis, J. Chromatogr. A. 1571 (2018) 107–120. doi:10.1016/j.chroma.2018.08.004

[61] T.S. Bos, W.C. Knol, S.R.A. Molenaar, L.E. Niezen, P.J. Schoenmakers, G.W. Somsen et al., Recent applications of chemometrics in two-dimensional chromatography, J. Sep. Sci. (2020) 1678–1727. doi: 10.1002/jssc.202000011

[62] G. Groeneveld, B.W.J.W.J. Pirok, P.J. Schoenmakers, Perspectives on the future of multi-dimensional platforms, Faraday Discuss. 218 (2019) 72–100. doi:10.1039/C8FD00233A

[63] B.W.J. Pirok, S.R.A. Molenaar, L.S. Roca, P.J. Schoenmakers, Peak-tracking algorithm for use in automated interpretive method-development tools in liquid chromatography, Anal. Chem. 90 (2018) 14011–14019. doi:10.1021/acs.analchem.8b03929

[64] S.R.A. Molenaar, T.A. Dahlseid, G.M. Leme, D.R. Stoll, P.J. Schoenmakers, B.W.J. Pirok, Peak-tracking algorithm for use in comprehensive two-dimensional liquid chromatography – Application to monoclonal-antibody peptides, J. Chromatogr. A. 1639 (2021) . doi:10.1016/j.chroma.2021.461922

8 Data Analysis for Multi-Dimensional Liquid Chromatography

Bob W.J. Pirok, Sarah C. Rutan, and Dwight R. Stoll

CONTENTS

DOI: 10.1201/9781003090557-8

8.1 INTRODUCTION

Upon maximizing the separation power of 2D-LC, particularly in the case of comprehensive 2D-LC, after data acquisition we are often left with a large and rich set of data. It is at this moment that we realize that our gain in peak capacity comes at the cost of much greater data complexity. Indeed, some have referred to the sheer size of our LC×LC-MS/MS datasets as a "tsunami of data" [1]. Making the step from 1D-LC to 2D-LC, and perhaps even adding an extremely powerful high-resolution mass spectrometer, may offer powerful analytical resolution, but all our work would be in vain unless we find a way to find meaning in the resulting data and use it to answer our analytical questions. New analytical tools typically generate more and more complex data, from which it is increasingly difficult to deduce clear answers and useful information. In this chapter, we will address all the relevant steps related to data analysis for 2D-LC. We will learn about the format of the raw data and see how this can be visualized into a two-dimensional plot. Next, post-acquisition corrections such as baseline correction, peak alignment, and removal of undesired background signals will be discussed. We will then focus on those methods that allow for quantification, with an emphasis on techniques for peak detection and for curve resolution.

As we proceed through these steps, we will note that the datasets obtained from (comprehensive) two-dimensional liquid chromatography separations are really a collection of one-dimensional (second dimension) chromatograms. Thus, we will sometimes employ chemometric approaches that focus on analysis of one-dimensional signals. We will also see that the field of chemometrics and data analysis is highly dynamic and active, with a large number of new methods proposed in recent years. While such innovations are advancing the field, making the most of them will require a critical comparison of performance across a variety of datasets to help users understand the strengths and weaknesses of different approaches.

8.2 DATA REFORMATTING AND VISUALIZATION

To rapidly gain insight from the data, one of the first priorities is to visualize the data. As with 1D-LC, this will allow us to manually inspect the chromatogram. In addition, we will see in later sections that the choice of preprocessing technique often depends on the type of data acquired. For heartcut 2D-LC approaches, additional 1D chromatograms for each heartcut can be found in the datafile. These may be visualized by simply plotting the detected signal vs. the time vector as is done with any 1D chromatogram, and analyzed using tools already available for 1D-LC. Thus, most of the following discussion pertains to sLC×LC and LC×LC data.

8.2.1 DATA REFORMATTING

For sLC×LC and LC×LC data, producing the 2D chromatogram requires an additional step. We have learned in Chapter 4 that LC×LC often employs a single detector that monitors the effluent of the ^{2}D separation. Consequently, raw LC×LC datafiles essentially contain single, long, one-dimensional data sequence of the detector signal, where individual ^{2}D chromatograms are serially connected. An example of such a concatenation of the ^{2}D chromatograms is shown in Figure 8.1A. The repeating pattern shows the presence of system peaks in the ^{2}D. The most prominent example of this is the dead-volume marker. Indeed, transferred fractions of ^{1}D effluent contain large quantities of solvents and buffers in the ^{1}D mobile phase. Consequently, these components may each systematically give rise to a peak in the second dimension. The length of each ^{2}D section of the long string of data is, of course, equal to the modulation time.

In order to plot the 2D chromatogram, the data must first be reformatted by dividing the long, one-dimensional data sequence into sections (i.e., vectors) using the modulation time that was used to acquire the data. These vectors are then rearranged into a $M \times N$ matrix, which is plotted as a 2D chromatogram (Figure 8.1B). This process is also sometimes referred to as folding the data. Here, M equals the number of modulations and N the number of datapoints within each modulation (i.e., the product of the ²D detector sampling frequency and the modulation time).

8.2.2 VISUALIZATION

Visualization of the matrix is generally done by plotting the two-dimensional space as a color map or contour plot. An example of the first is shown in Figure 8.1B. As is customary, the ¹D time is plotted on the x-axis and the ²D time on the y-axis. The detector signal level is represented by the color.

At this stage the representation in Figure 8.1B is accurate, but sometimes interpreting a chromatogram in this format can be confusing. Because the sampling frequency of the ¹D data points (i.e., Figure 8.1D) is dictated by the modulation frequency (as discussed in Section 3.4), it can be beneficial to interpolate the ¹D data points to achieve a smoother profile [2]. Thus, LC×LC chromatograms are often smoothed by interpolating the datapoints, resulting in chromatograms like that in Figure 8.1C. The chromatogram shown was obtained using bilinear interpolation [3] (using interp2 function in Matlab) where the linear interpolation is first applied in one dimension and then in the second dimension. Cubic spline interpolation methods are frequently used for either one- or two-dimensional interpolation.

While visualization of these datafiles is not very complicated, it is useful to realize that there are a number of different options in the way we can visualize the data. First of all, summing all datapoints vertically (i.e., the sum of all rows in the MxN matrix) returns the full ¹D chromatogram as if it were recorded by a 1D-LC (Figure 8.1D), but at a data acquisition rate much lower than we would normally use in 1D-LC. Similarly, summing all datapoints horizontally (i.e., the sum of all columns in the matrix) yields the 1D-LC version of the ²D separation (Figure 8.1E). However, we see immediately that the quality of description is rather poor compared to original 1D chromatograms obtained using either the ¹D or the ²D retention mechanisms, carried out under conventional 1D-LC conditions (Figures 8.1F and 8.1G, respectively). This brings us to another realization: if there is no dedicated ¹D detector, all our information about the ¹D separation is obtained indirectly through the ²D detector. Consequently, the ²D separation can be regarded as the detector of the ¹D [4], and, as stated above, the modulation time essentially establishes the ¹D sampling frequency. This underlines the importance of minimizing undersampling, which results in the loss of ¹D resolution by sampling the ¹D effluent at a frequency that is too low (see Section 3.4 for more detail on this topic).

In the event that a multi-channel detector is used (e.g., MS or DAD), the raw data also contains a full spectrum for each time point allowing the assembly of a 3D data matrix.

8.2.3 CORRECTION FOR WRAP-AROUND

When analytes do not completely elute within the modulation period set by the injection of fraction of ¹D effluent that contains them, they may elute entirely or partly in subsequent modulations. This is referred to as "wrap-around". While it is not very common in LC×LC (because gradient elution is frequently used in the second dimension), it is rather common in GC×GC because ²D separations are nominally isothermal.

One approach to correct for this issue is to treat the 2D chromatogram as a continuous 3D cylinder, essentially connecting one modulation to another. Weusten *et al.* developed an algorithm exploiting this principle to correct for wrap-around in the analysis of urine samples using GC×GC-MS [5]. Another approach was developed by Micyus *et al.*, which specifically detects occurrences of wrap-around using an integer fraction of the original modulation period [6]. Using this approach, the algorithm determines the absolute retention times.

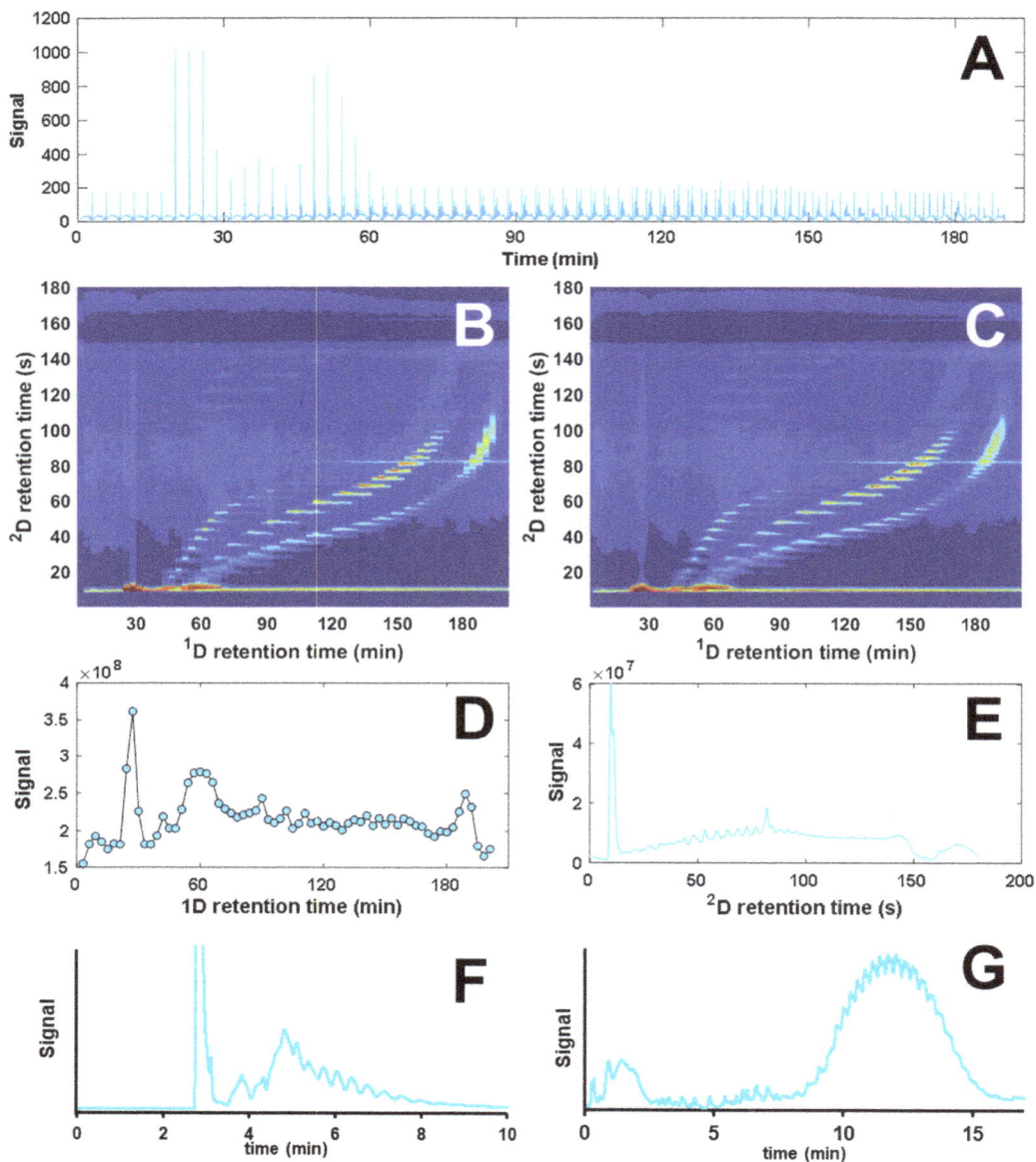

FIGURE 8.1   A) Raw LC×LC detector signal from the separation of some ionic industrial surfactants as acquired from the instrument. B) Two-dimensional LC×LC chromatogram produced from the data shown in A. C) Bilinear interpolation of LC×LC data from 2D plot shown in B. D) Reconstructed ^{1}D chromatogram (i.e., the signal at each time point is the sum of all signals within the modulation period corresponding to that time point). E) Reconstructed chromatogram that would be obtained if the sample were analyzed by 1D-LC using the conditions of the ^{2}D separation (i.e., the signal at each time point is the sum of all signals at that ^{2}D time across the entire 2D separation). F) 1D-LC chromatogram of the same sample using the ^{1}D separation conditions. G) 1D-LC chromatogram of the identical sample using a similar gradient program.

8.3 DATA PREPROCESSING – GOALS AND TECHNIQUES

The need for data preprocessing can perhaps best be understood from the quote that is popular in the field of computer science – "Garbage in, garbage out" – which is often abbreviated simply as GIGO. This phrase was coined to indicate that no matter how good data-processing methods are, they all rely on the quality of the original data. In our case, this means that our ability to distill useful information from the data depends on the quality of the original chromatogram. Here, quality is a nebulous, difficult-to-quantify term that relies heavily on the perspective of the scientist.

For example, if we look at the 1D and 2D chromatographic signals shown in Figure 8.2 from the perspective of a chromatographer, we may see well separated peaks and deem the quality of separation as good. However, from the perspective of a chemometrician, the signal contains undesired distortions (e.g., high-frequency noise and baseline drift) that often complicate data analysis, as we will see as we progress through this chapter. These distortions may impede our ability to obtain accurate, quantitative information, and even prevent us from locating all relevant regions of interest (ROI) in the chromatogram (i.e., those that contain chromatographic peaks).

While it is clear that data preprocessing is crucial to derive accurate conclusions from the data, it must also be noted that preprocessing steps do manipulate the data. Great care must be taken to prevent improper removal of useful information. Often, data preprocessing techniques rely on premises with respect to characteristics of the data and selection of the appropriate technique must be tailored to the dataset; this will also be addressed in this section.

8.3.1 NOISE REDUCTION

As is the case with any analytical signal, a chromatogram comprises several components, each with a different frequency: (low-frequency) baseline drift, the signal of relevance, and (high-frequency) noise. This is shown in Figure 8.2. In LC, drift mainly arises from the gradient programs used, whereas noise is mainly induced by small fluctuations in flow rate, mobile-phase temperature, mobile phase composition fluctuations due to pump imprecision, but also random fluctuations in the signal induced by disturbances in the detector (e.g., shot noise, thermal noise, etc.). Removing these undesired low and high frequency components from the signal will significantly improve quality of the chromatogram, most notably the signal-to-noise ratio (S/N). In addition, the successful performance of derivative-based peak detection methods depends critically

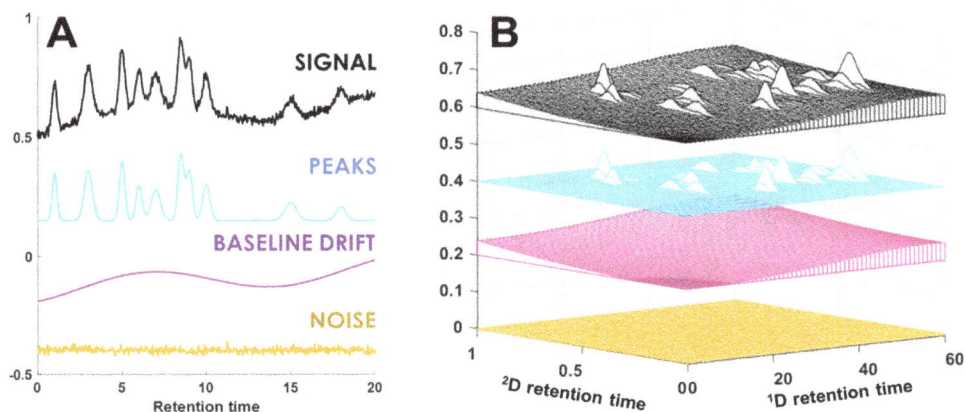

FIGURE 8.2 A chromatographic signal consists of several frequency components, including the actual chromatographic peaks plus low-frequency baseline drift and high-frequency noise. This is shown for A) 1D-LC and B) LCxLC.

on noise reduction. Unsurprisingly, research into removal of these background components goes back to the 1960s [7, 8].

Depending on the origin, noise typically features normally distributed (in terms of signal) distortions occurring at higher frequencies. Since noise may significantly impact both the precision and accuracy of quantitation and the performance of various subsequent data-processing techniques (e.g., derivative-based peak detection, and the determination of peak start and stop times), it is imperative that it must be properly reduced. However, it must be realized that noise-filtering essentially treats a symptom and not the cause. Thus, the GIGO principle applies here and a proper first approach focuses on analysis of the source of any excessive noise.

Noise filtering can be conducted both in the time domain, and the frequency domain, although filtering noise directly in the time domain is equivalent to indirect filtering in the frequency domain. Additionally, noise filters can be classified as low-pass or high-pass filters. Low-pass filters cut off frequencies above a set threshold, whilst allowing lower frequencies through. This renders them ideal for filtering baseline noise. Conversely, high-pass filters cut-off frequencies below a set threshold, making them more suitable for removal of baseline drift.

Some of the most popular filtering approaches in analytical chemistry are the moving average and polynomial filters, also known as Savitzky-Golay filters [9–14]. These filters are based on replacing the center point of a window that moves across the chromatogram with either the average, or the result of a *local* polynomial fit to the data within each window. This is achieved practically by using a weighted average of the windows points, with well-established weighting factors. This type of filter is a low-pass filter. First and higher order of the signal are easily calculated using this method.

The selection of the window width is key in the application of these filters. A filter with a window width of 3 will not be as effective in removing noise as a much wider filter that consults more neighboring points. However, filters with wider windows risk averaging out actual chromatographic peaks, potentially destroying information and thus rendering the method less accurate. This is particularly problematic for modern chromatographic separations executed at UHPLC conditions including very fast ^{2}D separations employed in comprehensive 2D-LC. An example is shown in Figure 8.3, where, regardless of the width of the filter, an increasing fraction of the area belonging to the Gaussian peaks is lost as the filter window width is increased. This example illustrates the importance of carefully selecting preprocessing methods and parameters. Another example is tuned filters, which relate to the concept of matched filtration [15]. Such filters are sometimes used in commercial packages.

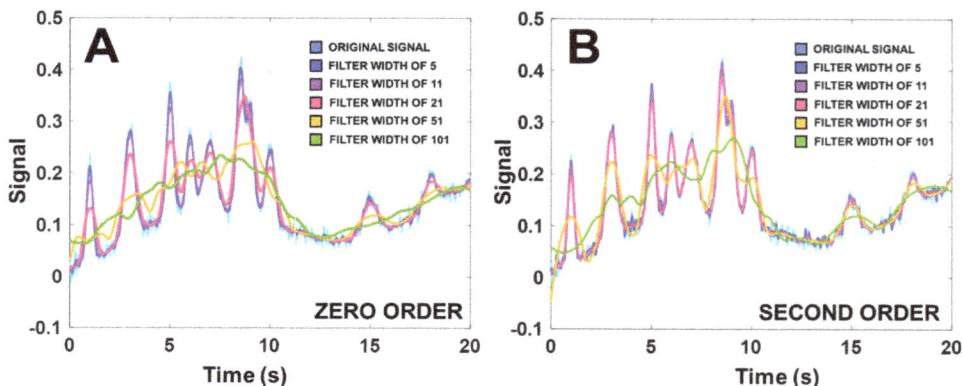

FIGURE 8.3 Two common signal filters applied to a section of a chromatographic signal. A: Zero-order moving average filter, B: Quadratic Savitzky-Golay filter.

The filters discussed above clearly improve the S/N ratio in the time domain. A common approach to filter in the frequency domain utilizes transformation functions. While many transformation functions can be used, the Hadamard and Fourier transformation are two well-known examples. Fourier transformation (FT) exploits the fact that true chromatographic peaks and noise differ in frequency [16, 17]. The largest practical difficulty with this approach is that it is difficult to use on chromatograms with widely varying peak widths.

For chromatographic signals containing vastly different frequency components that must be retained through the filtering process, wavelets can be used which adapt to the signal characteristics using both high-pass and low-pass filtering components. Wavelets automatically apply a narrow window to find high-frequency components and a wide window for low-frequency components [18]. The orthogonal and local functions employed by wavelets render them efficient and effective for processing signals that exhibit a wide range of frequency components. Daszykowski and coworkers have used wavelets for smoothing 2D-electropherograms [19].

Of most interest to users of comprehensive 2D-LC methods, both Fourier transform [14] and wavelet transforms [14, 20] can be applied to two-dimensional data sets. However, to our knowledge neither of these methods have yet been applied to LC×LC datasets.

8.3.2 Baseline Drift and Comprehensive Background Correction Approaches

Removal of baseline drift often involves the use of a curve-fitting approach, and is often combined with noise filtering. Both methods utilize a loss function to fit a curve through the presumed background signal. The combined approach is typically referred to as "background correction".

Relative to background correction for 1D chromatographic data, background correction for LC×LC chromatograms is challenging. Some of the challenges are illustrated in Figure 8.4, which displays 3D plots from the separation of a mixture of industrial surfactants (these are the same data shown in Figure 8.1C, plotted in different ways). As we will see in the remainder of this section, the effectiveness of background correction approaches relies significantly on regional characteristics and sparsity of the data. In the case of Figure 8.4, the chromatogram exhibits very different features in different regions, at different magnitudes, due to different chromatographic phenomena. For example, the "ridge" highlighted in Figure 8.4B is due to elution of components of the ^{1}D effluent that are unretained by the ^{2}D column, and also may arise from artifacts from rapidly changing refractive indices of the mobile phase in UV-visible detection when fast gradient separations are employed [21]. With large volumes of ^{1}D effluent injected into the second dimension, the intensity of this signal is often rather large and thus the ridge feature is prominent. Correction of such a ridge requires a different approach than correction of the rather normal peaks behind it as is clearly visible from Figure 8.4C, which is the same as Figure 8.4B but rotated by 180°. If the ^{1}D column bleeds (i.e., loses stationary phase) under the conditions of the ^{1}D separation, the components of the bleed may be visible in the 2D plot as the systematic occurrence of a ridge in the middle of the chromatogram as shown in Figure 8.4D. In 1D-LC column bleed or mobile phase impurities typically appear as a single (but sometimes broad) peak or feature, in 2D-LC this turns into a significant disturbance spanning the entire 2D chromatogram. Another profound feature frequently encountered in LC×LC are the distortions typically occurring during the equilibration step at the end of each ^{2}D separation when gradient elution is used (Figure 8.4E).

The detection of analytes at trace concentrations is normally a challenge even in 1D-LC. In LC×LC this can be even more challenging because the background correction method used must perform well in both dimensions to facilitate reliable detection of real but small chromatographic features on top of the background (i.e., recognition of peaks in adjacent ^{2}D chromatograms as one 2D peak).

Background correction approaches can be classified as parametric or non-parametric. Parametric methods assume a shape of the baseline defined by a number of parameters. Examples of these types

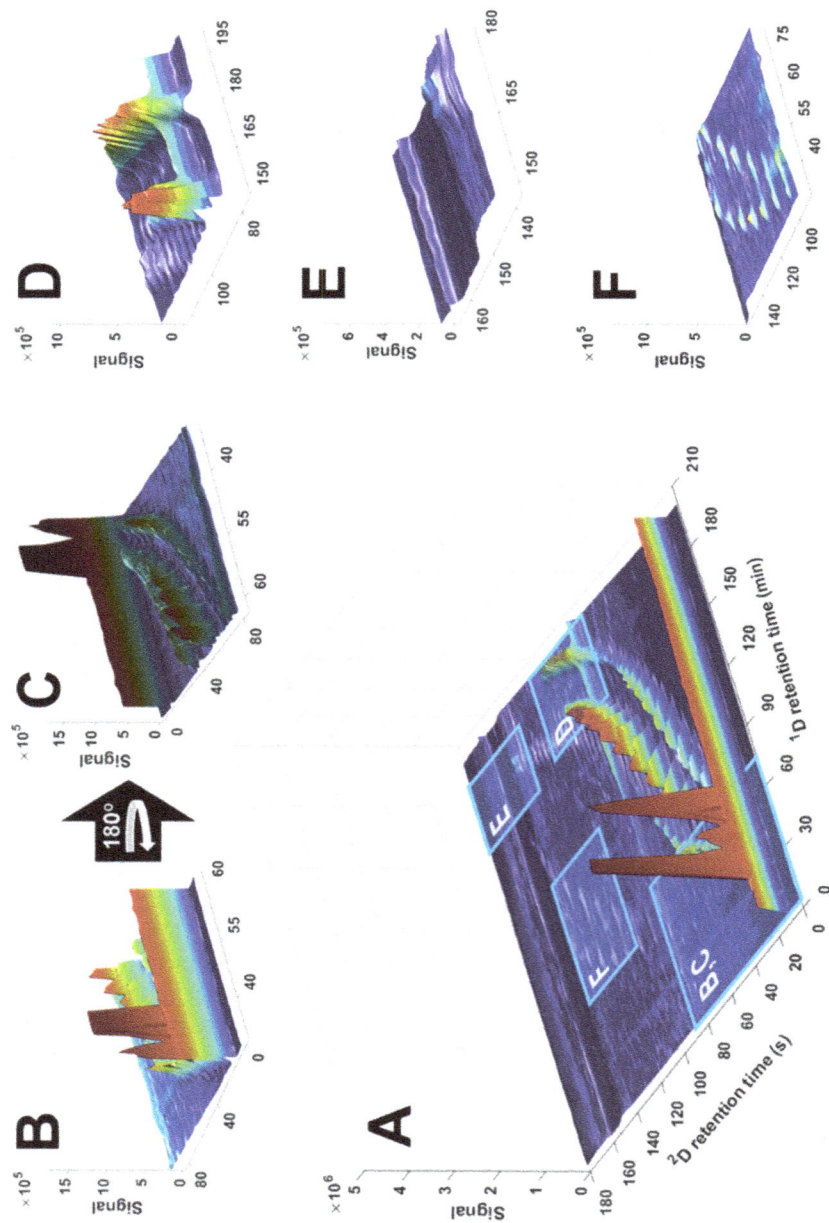

FIGURE 8.4 A) 3D plot of the LCxLC separation shown as a 2D plot in Figure. 8.2C. Various sections of the chromatogram have been selected to create a number of insets. B) The typical "ridge" resulting from the systematic elution of unretained compounds in the second dimension. C) This "ridge" requires a vastly different background correction approach than regular eluting peaks on the 2D plane. D) ^{1}D column bleed results in systematic elution of species in the 2D plots, significantly hindering data analysis. E) Distortions commonly encountered due to the equilibration step of a ^{2}D gradient. F) Separation of two homologous series of compounds.

Source: Adapted with permission from [22].

of approaches are the polynomial regression methods. Conversely, non-parametric approaches do not assume a shape of the background and the number of parameters used for the models depends solely on the data. When a large number of peaks are clustered together, fewer data points are available that describe the background. Consequently, background correction becomes increasingly more challenging with the presence of more clusters of peaks.

8.3.2.1 Penalized Least-Squares Approaches

Penalized least-squares is a smoothing method based on the Whittaker smoothing function [23], and is frequently applied for background correction. The fit of a model to the data, F, expressed as the sum of squares (SSQ), is balanced against its roughness (R) through a smoothing parameter λ. This relationship is also given by Eq. 8.1.

$$Q = F + \lambda R = \sum_{i=1}^{N} \left(x_i - z_i \right)^2 + \lambda \sum_{i=2}^{N} \left(\Delta z_i \right)^2 = \left\| \mathbf{x} - \mathbf{z} \right\|^2 + \lambda \left\| \mathbf{D} \mathbf{z} \right\|^2 \tag{8.1}$$

Here, x_i and z_i represent the data points in the signal and the model, respectively. $\mathbf{D}$ is an N x N-1 difference matrix containing values of 1, -1 and 0 such that $\mathbf{Dz} = \Delta \mathbf{z}$, where $\Delta \mathbf{z}$ is the difference vector for $\mathbf{z}$. Minimization of the cost function (Q) represented in Eq. (8.1) is an example of least-squares minimization with a regularization term to introduce an appropriate penalty (in this case against too much roughness). Solving for $\frac{\partial Q}{\partial z} = 0$ (minimization of the Q function) then returns

$$\left(\mathbf{I} + \lambda \mathbf{D}^{\mathrm{T}} \mathbf{D} \right) \mathbf{z} = \mathbf{x} \tag{8.2}$$

where the superscript T indicates a matrix transpose. For correction of the baseline, a binary matrix, $\mathbf{W}$, can be created that labels whether a datapoint belongs to a detected peak or not [24,25], as given by Eq. 8.3:

$$\left(\mathbf{W} + \lambda \mathbf{D}^{\mathrm{T}} \mathbf{D} \right) \mathbf{z} = \mathbf{W} \mathbf{x} \tag{8.3}$$

The disadvantage of this weighting method immediately becomes clear as it induces the premise that the location of peaks must be known, thus requiring a peak-detection algorithm to be executed *a priori*, which in turn may require baseline correction in order to function properly. In their asymmetrical Least Squares (asLS) method, Eilers introduced an asymmetry parameter to resolve this limitation [26]. This parameter allocates for an increase or reduction of weights imposed on positive and negative deviations of the signal from the baseline. One limitation of asLS, however, was that the asymmetry factor was determined for the entire baseline equivalently. This led to the introduction of adaptive iteratively reweighted penalized least squares (airPLS) [27]. In airPLS, a more-accurate weight vector can be obtained by iteratively solving a weighted PLS case until the difference between the signal and model is at least three orders of magnitude smaller than the signal value. The consequence is that regions of the baseline can be penalized differently.

Nevertheless, the performance of asLS and airPLS is significantly impacted by the presence of noise. Various methods have since then been introduced to improve these strategies, including asymmetrically reweighted PLS (arPLS) [28], modified adaptive iteratively reweighted penalized least squares (MairPLS) [49], and morphologically weighted penalized least squares (MPLS) [29, 30]. The latter method employs morphological analysis to determine the weighting vector more accurately. In MairPLS, the chromatogram is pre-treated before airPLS is performed.

Ultimately, all asymmetric least squares approaches rely on an accurate determination of the λ parameter to describe the baseline. Once a good value for λ has been found, it can be used for the entire dataset.

8.3.2.2 Local Minimum Values (LMV)

Baseline correction can also be carried out using local minimum values (LMVs) [31]. First, the signal is scanned for local minima, by searching all datapoints x_i which are smaller than their neighbors x_{i-1} and x_{i+1}, as is also reflected by the two conditions in Eq. 8.4. An example of these values is shown in Figure 8.5A, where each minimum value is presented as a red dot.

$$x_{i-1} > x_i \tag{8.4a}$$

$$x_i < x_{i+1} \tag{8.4b}$$

Next, the resulting vector of minima, or minimum vector, is stored. At this stage, peaks may still be present as is illustrated by Figures 8.5A and 8.5B. To remove data points corresponding to peaks, an iterative moving-window strategy is employed. All data points corresponding to S/N > 2.5 are treated as outliers and replaced by the median value within that window. This process is repeated until a convergence point is reached (Figure 8.5C).

The resulting vector of datapoints should now exclusively contain datapoints that describe the baseline. Through linear interpolation the vector is subtracted from the original data (Figure 8.5D). One disadvantage of this approach may be the required *a priori* estimation of the width of the moving window.

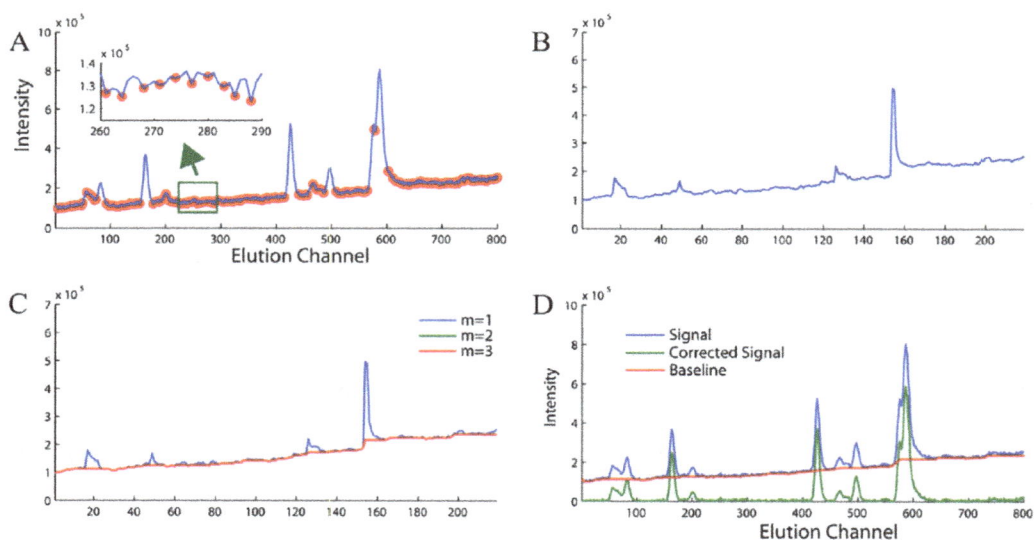

FIGURE 8.5 Example of background correction using local minimum values (LMV). A) Assignment of LMV points. B) Resulting vector of LMV, which still contains peaks. C) Removal of peak points using a moving-window strategy after *m* iterations. D) Overlay of the original signal, LMV vector and the corrected signal. Reprinted from *Journal of Chromatography, A*, 1449, H. Fu, H. Li, Y. Yu, B. Wang, P. Lu, H. Cui, P. Liu, Y. She, Simple automatic strategy for background drift correction in chromatographic data analysis, 89–99, Copyright (2016), with permission from Elsevier.

The LMV method has been compared to other approaches [32] including the moving-window-minimum-value (MWMV) [33], morphological penalized least-squares (MPLS) [34], and the orthogonal subspace projection (BD-OSP) methods [35].

For the comparison with MPLS and MWMV, simulated data were used which comprised singular peaks as well as peaks that overlapped with two to four other peaks [32]. MPLS and MWMV were used for background correction and the resulting peak areas and standard deviations were compared to values obtained after LMW combined with robust statistical analysis (LMW-RSA). The comparison was carried out for different degrees of noise present in the data. The LMV-based approach was found to yield the best accuracy for determination of peak features in all cases except for the highest noise level. MWMV was moderately less accurate, whereas MPLS yielded significantly deviating values with peak area recoveries between 53% and 74%, as opposed to the near 100% recovery by LMW-RSA. The influence of the moving-window width was found to be negligible. Because these methods use neighborhood minima in establishing the baseline, these types of approaches may be particularly useful when generalized to comprehensive 2D-LC data.

Comparison of LMW-RSA with BD-OSP was also carried out using LC-QToF-MS data [32], however the differences in performance were only evaluated on a qualitative level. The BD-OSP approach was unsuccessful in completely removing the background drift; the LMW-RSA method did remove the background drift, but also a portion of the information contained in the total ion-current chromatogram (TIC).

It should be noted that while the study found the influence of the width of the window to be negligible, chromatograms with large domains of overlapping peaks (i.e., few sections containing information regarding the baseline) will be difficult to be processed by the LMW-RSA method.

8.3.2.3 Baseline Estimation and Denoising Using Sparsity (BEADS)

In some cases, chromatograms are not necessarily expected to be completely filled with peaks. Sparsely populated chromatograms feature a large number of datapoints that describe the baseline relative to the number of datapoints describing a peak, something that is exploited by the baseline estimation and denoising using sparsity (BEADS) algorithm as developed by Ning *et al.* [36, 37].

Conceptually, BEADS does not rely on highly restrictive models to describe the frequency components of a signal. Instead, the approach is based on breaking down the chromatographic signal into its basic components:

$$\mathbf{x} = \mathbf{s} + \mathbf{w} = \mathbf{y} + \mathbf{f} + \mathbf{w} \qquad (8.5)$$

Here, $\mathbf{x}$ is the original input chromatogram, comprising a trace of peaks ($\mathbf{y}$), a baseline signal ($\mathbf{f}$), and white Gaussian noise ($\mathbf{w}$). Consequently, Eq. 8.5 implies that $\mathbf{s}$ describes the noise-free input chromatogram (i.e., $\mathbf{y} + \mathbf{f}$). Thus, in the BEADS approach, a peak vector is estimated ($\hat{\mathbf{y}}$) using a regularization approach analogous to the cost function shown in Eq. 8.1. In this case, the sum of squares is subjected to a high-pass filter (low frequency residuals are rejected), and the penalties are such that the pure chromatogram signal ($\mathbf{y}$) and its derivatives are sparse (meaning there is a lot a zero baseline between the peaks), and that the peaks are positive and not negative. The baseline estimate, $\hat{\mathbf{f}}$, is then expressed by Eq. 8.6.

$$\hat{\mathbf{f}} = \mathbf{L}(\mathbf{x} - \hat{\mathbf{y}}) \qquad (8.6)$$

where $\mathbf{L}$ represents a low-pass filter.

The authors developed an optimization algorithm to efficiently search for the minimum. Readers interested in details associated with this concept are referred elsewhere [36]. In their concept paper, the authors compared BEADS to airPLS and backcor approaches applied to both real and simulated

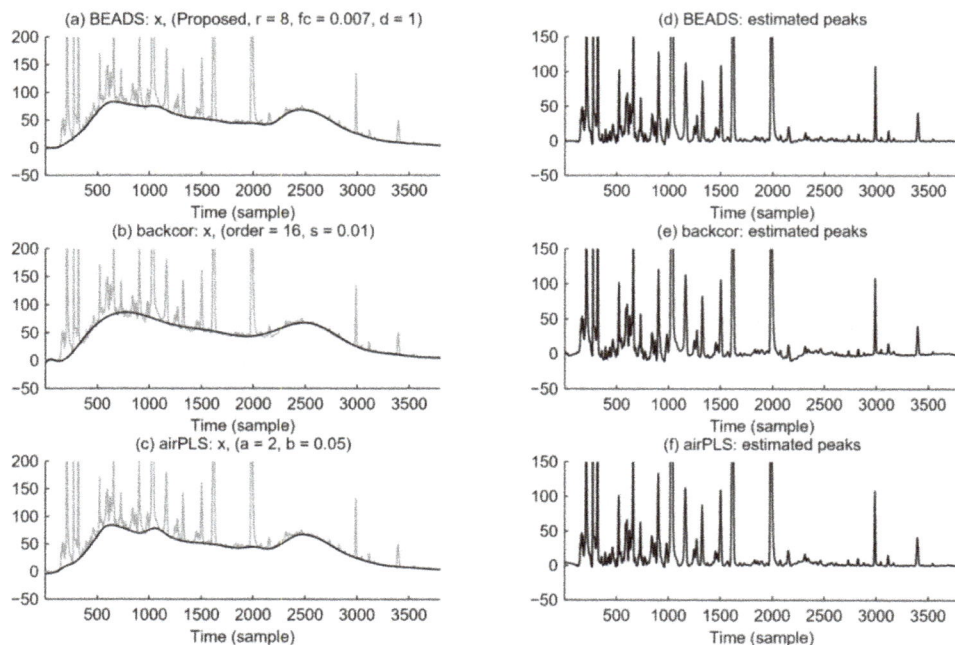

FIGURE 8.6 Comparison of background correction by BEADS (top), backcor (middle) and airPLS (bottom). The left-hand set of traces show the original signal (grey) and the modeled baseline (black), whereas the right-hand panels reflect the resulting background-corrected signal. Reprinted from *Chemometrics and Intelligent Laboratory Systems*, 139, X. Ning, I. Selesnick, L. Duval, Chromatogram baseline estimation and denoising using sparsity (BEADS), 156–167, Copyright (2014), with permission from Elsevier.

data as is also shown in Figure 8.6. Where airPLS and backcor were found to over- and underestimate the baseline, respectively, BEADS was found to be the most accurate method.

However, one disadvantage is that BEADS requires the baseline characteristics at the start to be similar to the baseline at the end. Indeed, in Figures 8.6A and D, the baseline returns to a value similar to that found at the start of the measurement. If this premise was not met, the modeled baseline was found to deviate in that its slope towards the final datapoint would be directed to the starting value of the first datapoint. While the baseline modeling does not require complicated parameters to be set, the parameters for the employed filters must be tailored (e.g., cut-off frequency and cost function parameters).

Small changes in these parameters were found to significantly impact the accuracy of the algorithm. However, while the authors admitted these flaws of the approach, they also noted that this made the algorithm conceptually flexible, as – with correct tweaking of the parameters – the algorithm could work for all types of data. Moreover, Navarro-Huerta *et al.* addressed most of these limitations in their assisted-BEADS algorithm [37], and Selesnick proposed solutions for the endpoint artifacts resulting from non-periodic signals [38].

8.3.2.4 Bayesian Approaches

We have seen for earlier methods that crowded chromatograms (i.e., large regions with co-eluting peaks) complicate background correction using most methods [39]. This is particularly true if the S/N is poor for a given peak of interest. To solve this, Lopatka *et al.* developed a method based on Bayesian statistics using a probabilistic peak-detection algorithm. Their peak-weighted (PW) approach fits a number of different models through a set domain of datapoints using least squares. For each model, the probability of the datapoint belonging to a peak is then computed and expressed as weight vectors. The authors compared their algorithm to several approaches and demonstrated that the PW method performed particularly well for crowded chromatograms [39] as is shown by Figure 8.7.

The authors also applied their method to a comprehensive two-dimensional GC-FID chromatogram obtained from a separation of volatile components from fire debris, as shown in Figures 8.7 C–E, yet were unable to compare the performance of the PW method to other background correction methods due to the absence of accepted benchmark methods.

8.3.2.5   Background Correction Using Profile Spectra from Multi-Channel Detectors

When separations are coupled with mass spectrometric (MS) detection, the recorded spectra may be exploited to enable background correction. Erny and coworkers developed such an approach and applied this to CE-ToF-MS and UHPLC-QToF-MS data [40]. Their approach utilized full spectra rather than centroided spectra, the latter of which are known to merge overlapping peaks [41]. The authors favored this approach over other approaches as it facilitated improved acquisition of base-peak ions.

Prior to background correction, the authors first reduced the number of profiles as their typical dataset contained 141,000 profiles each containing 3,581 points. For selection, the authors removed all profiles with a certain number of non-zero values, as zero values indicate that no ion was detected at a given m/z interval. Consequently, when larger fractions of the mass spectrum contain non-zero values this is a strong indicator of background ions, and the authors used this criterion to reduce the number of profiles used for background correction to 37,000. For the actual background correction arPLS was used and no significant deviations in the total-ion chromatogram were observed, suggesting that no important information was removed.

For an elaborate approach such as this, the computational time was approximately 20 minutes for a 2.9 GB dataset. This makes clear the need for data reduction as a preprocessing step when working with MS data.

8.3.2.6   Dedicated Approaches Relevant to Comprehensive Two-Dimensional Chromatography

Several studies have focused specifically on development of background correction tools for comprehensive 2D chromatography. One approach utilized trilinear decomposition to remove the background drift from LC×LC-DAD data [42]. Using alternating trilinear decomposition (ATLD) to treat the raw dataset, the analytical signal component was separated from the background signal component. Parallel factor analysis (PARAFAC, see Section 8.6.2) and self-weighted alternating trilinear decomposition (SWATLD) have also been used for this purpose [42].

Given the fact that 2D chromatographic data are generally visualized on a plane, image-treatment software has also been applied. Reichenbach *et al.* applied such an approach to GC×GC data using various statistical and structural characteristics of the background from 2D chromatograms, including the white noise properties of noise in chromatographic signals [43]. Both the GC Image and LC Image commercial software packages employ this algorithm [44, 45]. A method by Zeng *et al.* utilizes linear least-squares curve fitting in combination with moving average smoothing to correct all one-dimensional peaks of the ^{2}D chromatograms [46].

A number of algorithms for background correction have been developed in recent years. However, unfortunately these algorithms are rarely compared and limited quantitative information about their performance has been published. As a consequence it is often difficult to discern which algorithm would be best for a particular application. A recent study thus focused on generating objective data for such numerical comparisons [47], and found that the performance of various background-correction algorithms depends largely on specific signal characteristics and are not as generally

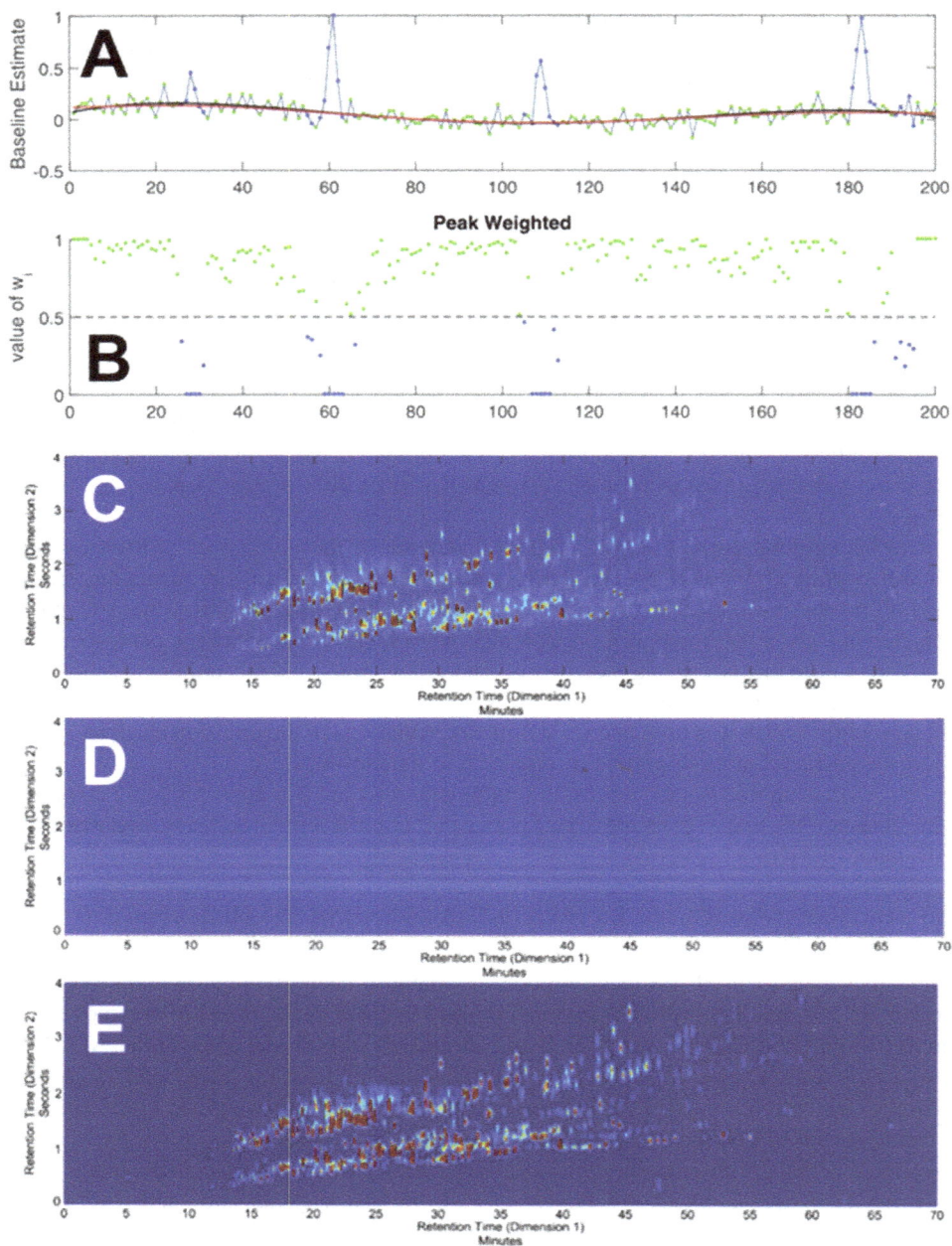

FIGURE 8.7   A) Application of the Bayesian peak-weighted approach for background correction demonstrated using a simulated chromatogram. B) Weight points indicating the magnitude with which each point contributed to the shape of the baseline; green points indicate a strong influence on baseline shape, C) Uncorrected GC×GC-FID separation of fire debris material, D) Peak-weighted estimate of the background, E) Corrected GC×GC-FID chromatogram after removing the background. Adapted from *Journal of Chromatography, A*, 1431, M. Lopatka, A. Barcaru, M. Sjerps, G. Vivó-Truyols, Leveraging probabilistic peak detection to estimate baseline drift in complex chromatographic samples, 122–130, Copyright (2016), with permission from Elsevier.

applicable as we would like; further study is needed in this area. It is thus not surprising that some approaches focus on combining the strengths of a number of tools. A case in point is the orthogonal background correction (OBGC) method developed by Filguiera *et al.* for use in 2D-LC [48]. The OBGC method exploits the fact that the ^{1}D chromatogram features a lower frequency of baseline fluctuations relative to those found in the ^{2}D chromatograms.

8.4 RETENTION-TIME ALIGNMENT

8.4.1 INTRODUCTION

When multiple chromatograms need to be compared, the next step after background correction is the alignment of retention axes. This is particularly important for two-dimensional LC where shifts in retention time are rather common. The actual alignment is generally carried out using either peak tables or the chromatograms themselves, often employing integrated peak-detection and peak-tracking algorithms.

There are two ways that retention time shifts can affect the analysis of comprehensive 2D-LC data. First, there can be shifts between the sequential ^{2}D chromatograms, such that it is difficult to determine whether two or more peaks that appear in adjacent ^{2}D separations are associated with the same compound *within* a single 2D-LC chromatogram. This can be addressed with either alignment algorithms, or directly within the peak detection method, as discussed in Section 8.5. Second, alignment can be used to address retention time shifts *between* chromatograms in order to confirm that peaks in multiple 2D chromatograms are associated with the same compound.

The complexity of algorithms that have been used for alignment varies from relatively simple local approaches including scalar-shift alignment and alignment of a selection of peaks, to global alignment, where multiple regions of the chromatogram are comprehensively aligned. In this section, we will mainly focus on the latter category, which also have found their application in forensics [49] and metabolomics [50].

In addition to the recent developments addressed below, a large number of other 2D approaches have been developed. One example is the algorithm using windowed rank minimization with interpolative stretching by Johnson *et al.* [51], which was applied to GC×GC data obtained for the analysis of naphthalene in jet fuel. Another approach by Pierce *et al.* employs indexing schemes for warping in both dimensions and was applied to GC×GC data [52]. Alignment based on images has also been extensively investigated and applied to comprehensive two-dimensional data for various applications [46, 53, 54]. While most approaches are suitable for three-way analysis, Allen and Rutan developed an algorithm for LC×LC-DAD with four-way data structures [2].

Attention has also been devoted to within-analysis retention shifts from modulation to modulation using PARAFAC in combination with PARAFAC2 [55] (see Section 8.6.2).

8.4.2 CORRELATION-OPTIMIZED WARPING

One well-known approach for alignment is correlation-optimized warping (COW), where the chromatogram is divided into a number of local regions. Next, each section of chromatogram is compressed or stretched and compared to a reference until the correlation is maximized. The approached employs the Pearson correlation coefficient (PCC) defined as

$$PCC = \frac{(r - \bar{r})^T (x - \bar{x})}{\sqrt{(r - \bar{r})^T (r - \bar{r})(x - \bar{x})^T (x - \bar{x})}} \qquad (8.7)$$

with r representing the reference, x the sample chromatogram, and $\bar{r}$ and $\bar{x}$ their mean values (i.e., average chromatograms), respectively. Interestingly, COW has also been expanded to support two-dimensional separations by Zhang et al. [56] and Gros et al. [57]. The latter method was recently successfully applied for alignment of GC×GC chromatograms based on high resolution mass spectrometry (HRMS) data [58].

Using a 2D chromatogram as reference for aligning regions of interests, van Mispelaar *et al.* proposed a correlation-optimized shifting through inner-product correlation for all selected regions in GC×GC chromatograms [59]. Paraster *et al.* introduced a bilinear peak-alignment method based on multivariate-curve resolution (MCR, see Section 8.6.1) and applied it for retention alignment of GC×GC data [60].

8.4.3 AUTOMATIC TIME-SHIFT ALIGNMENT

Zheng *et al.* developed the automatic time-shift alignment (ATSA) protocol, which employs a two-stage alignment protocol [61]. After baseline correction by LMV-RSA (Section 8.3.2.2) and peak detection using multi-scale Gaussian smoothing (Section 8.5.1), the chromatogram is divided into a distinct number of segments. In the first stage of alignment, the authors opted to use the total peak correlation coefficient (TPC) as defined by

$$\text{TPC} = \left(\frac{\sum_{i=1}^{I} w_i \text{PCC}_i}{\sum_{i=1}^{I} w_i} \right) \frac{I}{n_{\text{tot}}} \tag{8.8}$$

where w_i is the ratio between area and width (number of datapoints) of peak i, and I and n_{tot} are the total number of peaks in the sample and reference chromatogram, respectively. Segments that could not be aligned were labeled as outliers and subsequently realigned using PCC instead. Any convoluted or severed segments were corrected using a warping strategy to adjust the boundaries between segments.

The second stage of the protocol focused on precise alignment by again segmenting the preliminarily aligned chromatogram using the number of peaks. Here, boundaries between segments were located precisely in the middle between two chromatographic peaks. Next, each segment was aligned to the reference chromatogram. When no peak is available on the reference chromatogram, the algorithm was programmed to use an average time shift based on the other segments. The segments were reconnected using the principle of warping. In their study, the authors showed that ATSA was able to improve the correlation coefficient from 0.72 to 0.96 and eventually to 0.99 after the first and second stages of alignment, respectively.

ATSA and similar alignment tools require two parameters to be specified before use: the segment size and the initial time shift. While the authors showed that varying the segment size between 1 and 10 minutes had little effect, it was observed that larger sizes (particularly those above 10 minutes) would reduce consumption of computational capacity, yet resulted in erroneous time shifts. The size of the initial time shift, varied in the study between 0.1 and 1 min, was found to have little influence on the outcome.

Nevertheless, utilizing such an alignment tool is not without risk. The warping strategy applied by ATSA may influence peak areas, which may influence quantitation. While the authors did not find any evidence of changes in peak areas, application of ATSA on data obtained from a study focused on degradation of oils yielded a different conclusion [61]. Without ATSA, the data suggested that oil components were degrading, whereas after ATSA correction, the data indicated the opposite.

8.4.4 ALIGNMENT USING MASS SPECTRA

Similar to background correction, consulting mass spectra may also be fruitful for the purpose of retention alignment. One approach by Fu *et al.* is illustrated in Figure 8.8. After background correction by LMV, the algorithm calculates the PCC for each sample and reference peak which falls within a pre-specified time window (Figure 8.8A). In their study, the authors used 0.5 minutes as the time-shift window. Next, a correlation matrix was compiled using the resulting PCC values to establish a maximum-correlation path (Figure 8.8C). The green cells depict unaligned values, whereas the orange boxes represent corrected values. The correction is also apparent from the schematic representation shown in Figure 8.8D.

An underlying assumption that is made with most alignment algorithms is that the elution order is similar between samples. To account for dissimilar elution orders, the authors programmed the algorithm to specify landmark peaks, which are peaks with a correlation coefficient larger than 0.99. The time shifts of the found landmark peaks were then collected in a vector and outliers removed. Based on this vector, time shifts between two landmark peaks are linearly interpolated to calculate an expected time shift. The resulting value is compared to the earlier calculated time shift, and the peak is realigned in the event the difference is significantly larger.

For validation, the alignment tool was applied to GC-MS data comprising the characterization of plant samples [62]. The algorithm was assessed for its performance to correct the 15 most co-eluting peaks across a series of 30 samples. Figure 8.9A displays an overlay of the resulting 30 chromatograms. The extent of misalignment without correction is shown for the peaks within the highlighted box in Figure 8.9C. Here, all 30 chromatograms are displayed in series (y-axis) with the intensity represented by color. Retention-time alignment using the developed approach yielded aligned peaks as shown in Figure 8.9D (fully corrected chromatogram shown in Figure 8.9B).

When the elution order is not expected to change, another method of interest establishes the chromatogram with the largest number of peaks as the reference chromatogram. After background correction and peak detection using automated peak detection and baseline correction (ACPD-BCP, see Section 8.5.3), a rough alignment was carried out in a way similar to the COW approach [63]. In this case, however, a cosine correlation was computed instead of the PCC. The cosine correlation is a measure of the similarity between two vectors of a product space, by calculating the cosine between the two vectors. Such a metric is often used in pattern recognition algorithms [64].

After preliminary alignment, the next stage utilized the relative distances between a particular peak found in a sample chromatogram compared to the reference chromatogram, as well as the cosine values and absolute distances. The resulting differences were collected in an alignment table. However, no actual information on robustness of the algorithm was shown, nor was the algorithm compared to alternative approaches.

8.5 PEAK DETECTION

After completing initial data preprocessing of 2D-LC data, we can proceed to locate all true chromatographic peaks. In signal processing, this process is referred to as peak detection. Similar to data preprocessing, peak detection of higher-order data, such as our two-dimensional chromatograms, relies on lower-order data processing techniques. While direct two-dimensional peak detection is possible, the limited number of data points in the first dimension typically seriously limit this possibility, leaving chemometricians with no choice but to use one-dimensional peak detection for all ^{2}D separations, after which the resulting peaks are clustered into 2D peaks. Thus, again, most of the following paragraphs will concern one-dimensional approaches to peak detection.

Once a peak has been located, determination of a number of elementary characteristics of the peak may be useful. Examples include the peak area, retention time and asymmetry. These and

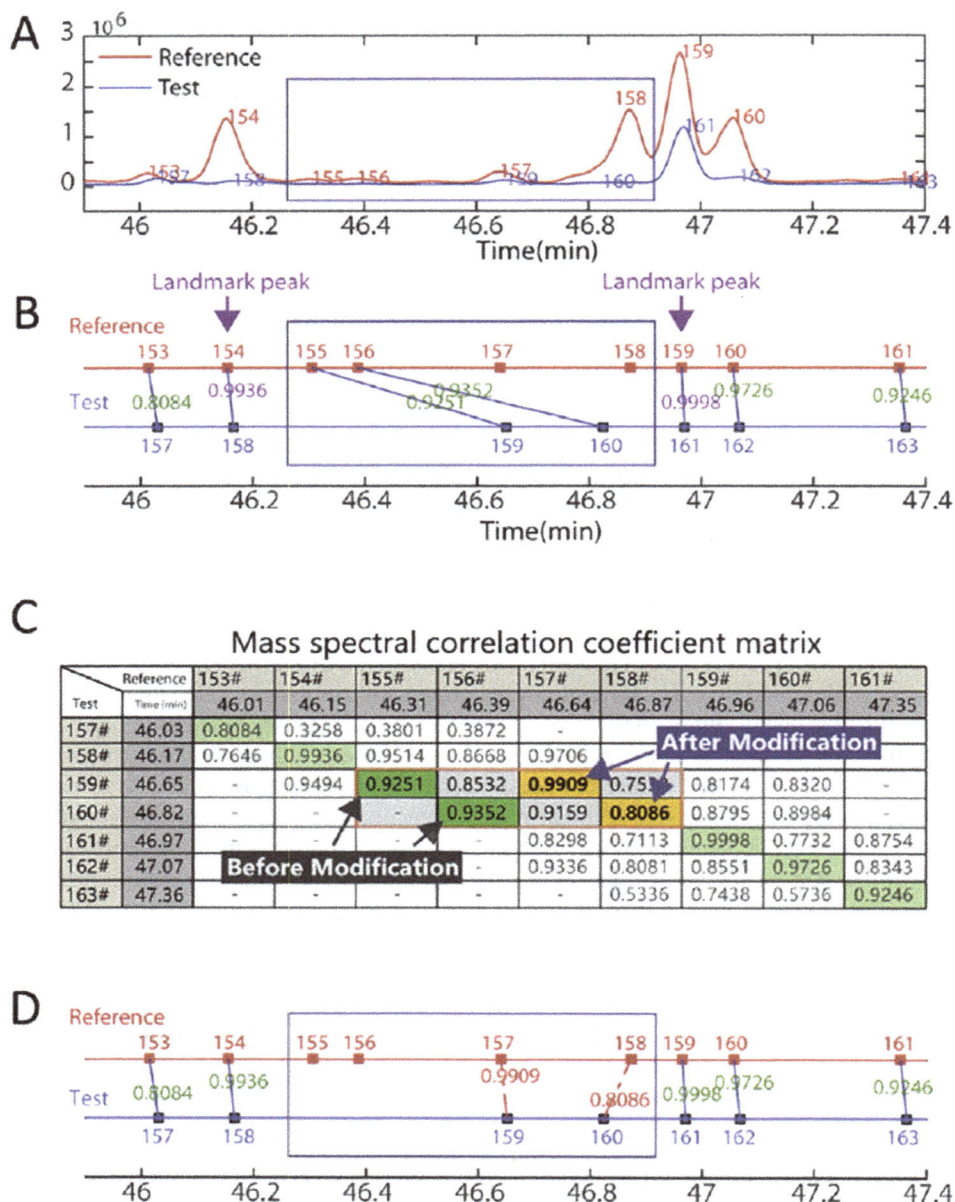

FIGURE 8.8 Peak alignment using the maximum-correlation path and landmark peaks. A) Selected segment from chromatogram, B) misalignment resulting from exclusively consulting mass spectra, C) locations of misaligned peaks in the maximum correlation coefficient path. D) Schematic representation of reference chromatogram and aligned peaks of a test chromatogram after correction, where the x-axis is time in minutes, and the points are peaks associated with peak numbers. Reprinted from *Journal of Chromatography, A*, 1513, H. Fu, Y. Zhang, L. Zhang, J. Song, P. Lu, Q. Zheng, P. Liu, Q. Chen, B. Wang, X. Wang, L. Han, Y. Yu, Mass-spectra-based peak alignment for automatic nontargeted metabolic profiling analysis for biomarker screening in plant samples, 201–209, Copyright (2017), with permission from Elsevier.

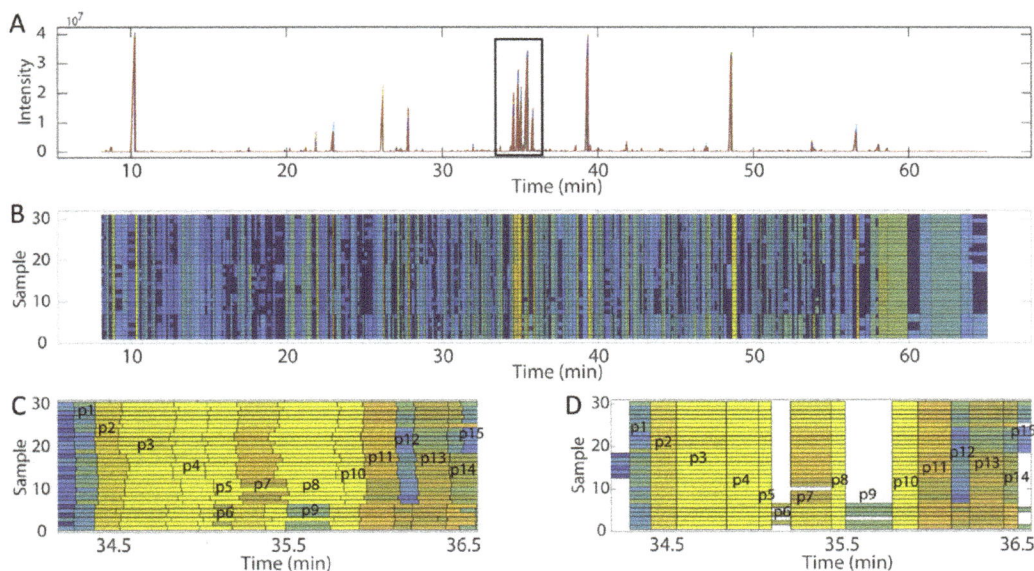

FIGURE 8.9 Illustration of peak alignment using mass spectra. A) Original chromatogram – highlighted box depicts dense region of peaks. B) Alignment of all chromatograms (time, x-axis) for all 30 samples (y-axis). Color depicts signal intensity. Panels C and D show the peak structure for the dense region with C) original chromatogram, D) corrected chromatograms. Adapted from *Journal of Chromatography, A*, 1513, H. Fu, Y. Zhang, L. Zhang, J. Song, P. Lu, Q. Zheng, P. Liu, Q. Chen, B. Wang, X. Wang, L. Han, Y. Yu, Mass-spectra-based peak alignment for automatic nontargeted metabolic profiling analysis for biomarker screening in plant samples, 201–209, Copyright (2017), with permission from Elsevier.

other characteristics are often automatically calculated by the software supplied with the instrument. Generally, the algorithms define the peak start and end points as boundaries, although the user can typically adjust these graphically. A default approach in many instrument software packages is the perpendicular drop method for integrating overlapped peaks. In the case of even moderate coelution, this rather simplistic approach leads to erroneous results, as is illustrated schematically in Figure 8.10.

For a more accurate determination of the various properties of a well-separated peak, the statistical moments may be used [66] (Table 8.1). In Eqs. 8.9–8.13 x_i are time points in the chromatogram, $f(x_i)$ is the signal value at time x_i, and the index i runs from the start to the stop points of the peak. When using curve fitting for peak detection, the function $f(t)$ can be replaced by the model used for fitting. When no model is available, the solution can be numerically determined. The accuracy of the statistical moments relies heavily on the preprocessing and sampling frequency of the detector [67–69].

8.5.1 CLASSICAL PEAK DETECTION

Peak detection is generally performed using either a derivative-based approach or a curve-fitting approach, although other methods such as tuned or matched filters exist [15]. When peaks are well-resolved, the noise is minimal or adequately removed by preprocessing, and the background is reasonably constant, derivative methods, which are usually incorporated into chromatographic data system software, work well. However, when overlapped peaks are present, derivative detection methods are often inadequate. Figure 8.11 illustrates some of these challenges [65]. In Figure 8.11

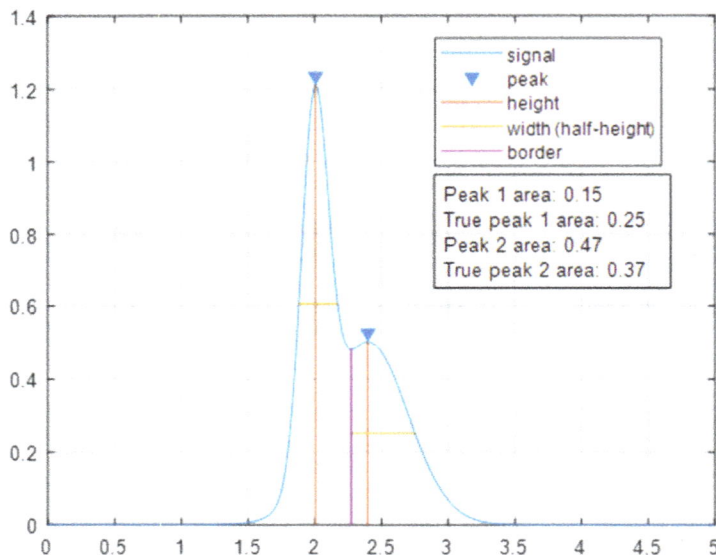

FIGURE 8.10 Illustration of an outcome of simple, automated peak integration as encountered with standard data-analysis software. Inset provides a comparison of the results with the true values for a case with overlapping peaks 1 (left) and 2 (right).

Source: Reproduced from [65].

TABLE 8.1

Overview of Statistical Moments

Moment ordinal	Property	Formula	Eq.
$0\ (m_0)$	Area	$\displaystyle\sum_{i=1}^{N} f(x_i)$	(8.9)
$1\ (m_1)$	Retention time	$\dfrac{\displaystyle\sum_{i=1}^{N} x_i f(x_i)}{m_0}$	(8.10)
$2\ (\mu_2)$	Variance (σ^2)	$\displaystyle\sum_{i=1}^{N} (x_i - m_1)^2 f(x_i)$	(8.11)
$3\ (\widetilde{\mu}_3)$	Skewness	$\dfrac{\displaystyle\sum_{i=1}^{N} (x_i - m_1)^3 f(x_i)}{\mu_2^{\ 3}}$	(8.12)
$4\ (\widetilde{\mu}_4)$	Kurtosis	$\dfrac{\displaystyle\sum_{i=1}^{N} (x_i - m_1)^4 f(x_i)}{\mu_2^{\ 2}}$	(8.13)

Source: [70].

A, a signal containing two overlapped peaks with no noise or baseline drift is shown. If we take the second derivative of this signal, we obtain Figure 8.11C. The new signal shows maxima that reflect the presence of inflection points in the chromatogram. More importantly, the minima indicate the presence of peak apexes in the original chromatogram.

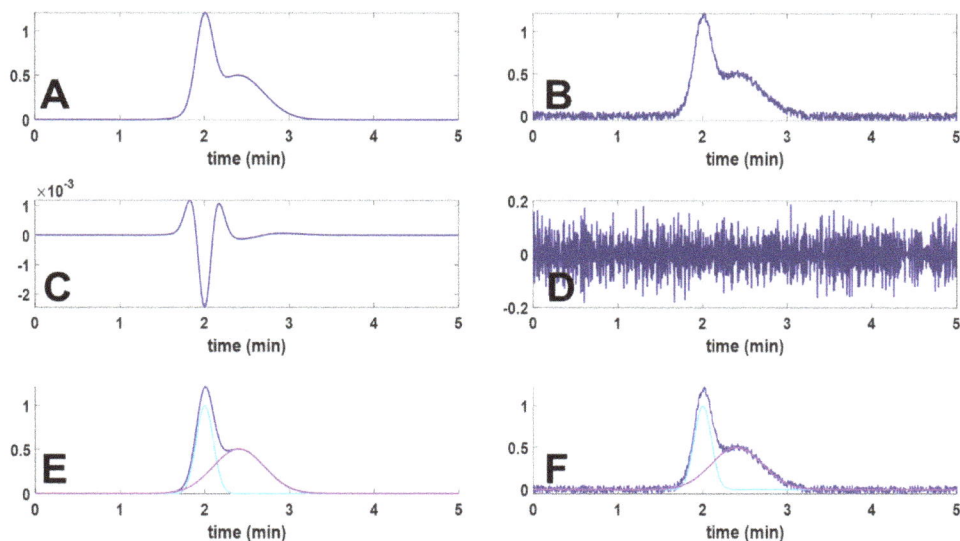

FIGURE 8.11 A) Ideal (noise-free) signal containing two overlapped peaks. B) Equivalent of signal in (A) but superimposed with high frequency noise. C) Second derivative of (A) where a minimum indicates the apex of a peak, D) Second derivative of (B) where the extreme number of peaks induced by the noise, yields an equal number of minima, again yielding noise. E) and F): two curves fitted to (A) and (B), respectively. Source: Reproduced from [71].

Now we consider the noisy chromatogram shown in Figure 8.11B. While the signal to noise ratio and resolution between the two peaks is sufficient to observe two peaks, the second derivative only amplifies the high-frequency noise. To make things worse, our two peaks have now completely disappeared from the derivative signal. From this simple comparison, we can immediately draw a number of conclusions. First, derivative-based peak detection approaches require significant data preprocessing to remove most of the noise. From Section 8.3.1, we also know that too extensive noise filtering is likely to remove useful information. Second, we see that the peaks must be sufficiently resolved, as the resolution in Figure 8.11A, and thus 8.11C, is just sufficient to detect the second peak.

The curve-fitting approach to peak detection suffers much less from the challenges discussed above. Here, a section of the chromatogram, such as the ones in Figures 8.11A/B, is taken and x number of curves as defined by the selected model are fit to the chromatogram. Classically, a Gaussian function is used for fitting, but that has the disadvantage that the algorithm may completely miss non-Gaussian shaped peaks. Indeed, peaks in LC often comprise a tailing component in the distribution function. Consequently, it is often difficult to find a distribution function which describes the peak accurately.

Arguably the biggest challenge in peak detection for 2D-LC, however, is that the algorithm must determine the number of peaks present in the overlapped section of the chromatogram consistently across all modulations in the LC×LC space. Suppose the signal shown in Figure 8.11A actually contains a third, smaller peak buried underneath the other two. For curve-fitting approach to pick this up, it must know *a priori* that three peaks are present. The exercise of determining the correct number of peaks is paradoxically the whole aim of the curve-fitting approach, rendering curve fitting very challenging for chromatograms containing regions with a high degree of peak overlap. Nevertheless, as is shown in Figures 8.11E and F, once the correct number of peaks is known, curve fitting is very robust, even for moderately noisy data.

Most recent developments in peak detection for 1D chromatography have focused on improving either of these two generic peak-detection strategies. Due to the susceptibility of the algorithms to noise, many of these approaches utilize integrated preprocessing algorithms that remove some of the noise prior to the peak detection step.

One example of an integrated approach is the smoothing-based peak-detection method, such as the multi-scale Gaussian-smoothing algorithm developed by Fu *et al.* [72], which operates in three steps. The first and second steps involve removal of background drift and subsequent detection of all local maxima. The key step involves application of a smoothing filter to the signal with various window sizes of the filter. Working on the assumption that true peaks retain a constant location of their maxima after smoothing, noise peaks are automatically removed. By varying the width of the filter, the intensity of the peak can be assessed. This renders the algorithm more robust against noise and baseline drift than classical derivative-based peak detection which tends to be more sensitive to such factors although this highly depends on the specific applications.

8.5.2 Continuous Wavelet Transformation (CWT)

The critical dependence of curve-fitting approaches on the determination of the number of overlapped components (see Section 8.5.1) has also received attention. One example is the development of wavelet-transform based peak detection, where Peters *et al.* applied cross validation to estimate the number of components [73]. Another challenge for curve-fitting approaches is robustness against the presence of a large variation in characteristics between neighboring peaks, and a number of different wavelet morphologies have been suggested [74], including the continuous-wavelet-transform (CWT) approach. CWT is more sensitive to peak characteristics such as symmetry, yielding fewer false positives than classical derivative-based approaches [75, 76] and the multi-scale Gaussian-smoothing approach [77]. CWT has also been incorporated with ridge-detection algorithms [78], which locate peaks using local maxima [74].

Nevertheless, the CWT method is not without flaws. This method produces chromatograms analogous to second derivative analysis, so the focus is more on peak detection rather than profile determination, although some studies have shown that areas can be extracted and calibration curves can be obtained [79, 80]. The CWT needs to be optimized in terms of the selection of the appropriate wavelet function as well as the scale factor. These weaknesses have been addressed using a heuristic and recursive approach that resulted in improved peak detection, as well as determination of peak characteristics, such as area [81]. Despite developments for both CWT and Gaussian-smoothing, both approaches struggle with cases where coelution is severe [72, 82].

An approach completely different from the above-mentioned methods employs Bayesian statistics. Unlike the previous techniques, which yield a binary answer (true or false) for the detection of a peak at a given datapoint, Bayesian methods employ probabilities. While initial implementations struggled with overlapping peaks [83], statistical-overlap theory [84] was incorporated to improve this [85]. This is a characteristic advantage of Bayesian statistics, in that these methods can incorporate prior knowledge. Woldebriel *et al.* further developed a probabilistic model to allow untargeted peak detection of LC-MS without requiring any preprocessing [86], and thus reducing the risk of accidentally destroying information.

8.5.3 Automatic Peak Detection and Background Drift Correction

Another example of background drift correction combined with peak detection is the algorithm developed by Yu *et al.* [76]. This automatic peak detection and background drift correction (ACPD-BDC) approach focuses on start and end points of peaks. A datapoint x_i was considered a starting point if its value was lower than the next three points x_{i+1} to x_{i+3}. Similarly, the end point x_j must be larger than the next three points, as is also reflected by Eqs. 8.14a/b.

$$x_i < x_{i+1} < x_{i+2} < x_{i+3} \qquad (8.14a)$$

$$x_j > x_{j+1} > x_{j+2} > x_{j+3} \qquad (8.14b)$$

The list of start and end points was stored in two vectors; the linear combination of these vectors represents the list of peak elution ranges. Next, all detected peak regions were subtracted from the chromatogram ($\mathbf{x}$), yielding the background ($\mathbf{x}_{filtered}$) as a result. By taking the derivative of this background signal ($d\mathbf{x}_{filtered}$), outliers can be detected and removed using Eq. 8.15.

$$\frac{\left| dx_{filtered,i} - \overline{d\mathbf{x}_{filtered}} \right|}{\sigma} > 3 \qquad (8.15)$$

Here, σ is the standard deviation of the $d\mathbf{x}_{filtered}$ vector. By iteratively removing outliers, the noise level – which is the first-order derivative of $d\mathbf{x}_{filtered}$ – is fine-tuned.

The removed regions that contain peaks in $\mathbf{x}_{filtered}$ are then linearly interpolated, thus constructing $\mathbf{x}_{background}$. This signal is then filtered again using a moving-average filter with a width of 3 points. Meanwhile, the first- and second-order derivatives of the original signal (x) are compared. In this approach, peaks are only considered a true peak if 1) $|dx_i|$ is five times larger than 3σ, and 2) the second-order derivative crosses zero fewer than eight times. Finally, the background, $\mathbf{x}_{background}$, is subtracted from $\mathbf{x}$, yielding a background-corrected chromatogram with peaks detected.

The authors of this elaborate approach compared their ACPD-BDC algorithm with airPLS and MairPLS, and applied the three methods to: 1) simulated data; 2) a GC separation of a plant-based flavor extract; and 3) a LC separation of pharmaceuticals in water. MairPLS and ACPD-BDC were found to perform better than airPLS (Figure 8.12) [76].

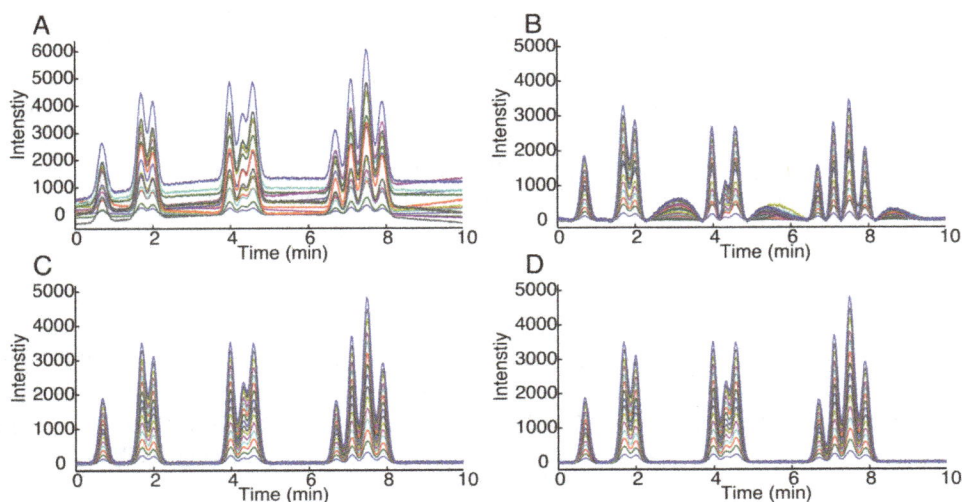

FIGURE 8.12 Comparison of three background correction and peak detection approaches. A) Raw data, B) airPLS, C) MairPLS, and D) ACPD-BDC. Reprinted from *Journal of Chromatography, A*, 1359, Y. Yu, Q. Xia, S. Wang, B. Wang, F. Xie, X. Zhang, Y. Ma, H. Wu, Chemometric strategy for automatic chromatographic peak detection and background drift correction in chromatographic data, 262–270, Copyright (2014), with permission from Elsevier.

8.5.4   Comprehensive Two-Dimensional Approaches

For single-channel detectors, two categories of approaches have been used for peak detection in comprehensive 2D chromatography. The first category detects peaks on the 1D signal and subsequently clusters them into 2D peaks [87]. The second category employs the image-based watershed algorithm [45].

Difficulties in peak detection mainly arise from the variability in the second dimension times from one modulation to another in LCxLC, data but most importantly undersampling in the first dimension (Chapter 3). A robust algorithm must be able to handle situations where the degree of undersampling varies within a single LCxLC chromatogram. Shifts in the retention of individual compounds in adjacent ^{2}D separations are also challenging. These shifts can result from either natural variation in retention (i.e., retention precision is not perfect), or deliberate changes in ^{2}D conditions, as in the case when using shifting gradient programs (see Section 4.5).

In addition, multi-way methodologies have been developed for peak detection, specifically for datasets recorded using multi-channel detectors. These are discussed in Section 8.6.

8.5.4.1   Two-Step Peak Detection Using Peak Clustering

One method of peak detection was introduced by Peters *et al.* and utilizes two steps [87]. First, 1D peak detection algorithms, such as described in the previous sections, are used to for peak detection across the entire string of concatenated ^{2}D chromatograms. In their study, Peters *et al.* used the Savitzky-Golay method to detect the 1D peaks based on their derivatives (see Section 8.3.1). Indeed, this strategy allows the determination of various peak properties, such as the peak height, as well as the start and end points.

In the second step, an algorithm is applied to merge all 1D peaks across different modulations belonging to the same compound [87]. This process is generally referred to as peak clustering or peak merging, and remains a difficult challenge in data analysis for two-dimensional LC. The clustering algorithm associates the peaks belonging to the same compound together using retention-time alignment (see Section 8.4), but since shifting gradients deliberately induce substantial retention time shifts, merging peaks properly when shifting gradients are used is particularly challenging.

For the actual merging, the algorithm employs overlap and unimodality criteria to determine whether the 1D peaks belong together. The overlap criterion assesses the degree of overlap between two regions (*a* and *b*) in which the peak elutes in adjacent modulations (Figure 8.13). The ratio of overlap is computed by Eq. 8.16 and a threshold must be set to determine which peaks will be merged. The unimodality criterion is used to determine whether ^{2}D maxima belong to the same peak and investigates the maxima of the peak profile in the first dimension.

$$OV = \frac{b}{a} \cdot 100\% \qquad\qquad (8.16)$$

For determination of the peak area, Peters *et al.* used a trapezoidal method, essentially summing the areas of peaks in adjacent ^{2}D separations that are associated with elution of the same compound [87]. This method has been applied to LCxLC data [88, 89]. In order to allow the method to profit from four-way data, which was demonstrated to be needed by Bailey and Rutan [90], Vivó-Truyols developed a Bayesian two-step approach [83].

An approach that is analogous to the approach described by Peters *et al.* is the well-known msPeak algorithm that utilizes the normal-exponential-Bernoulli (NEB) model to describe peaks [91]. This method includes a means for resolving overlapped peaks in the second dimension. The approach combines preprocessing with a scan for regions with co-eluting peaks. This approach was recently improved with the development of the normal-gamma-Bernoulli (NGB) model, which, unlike the NEB model, does not have an analytical solution [92]. The authors demonstrated, however, that their

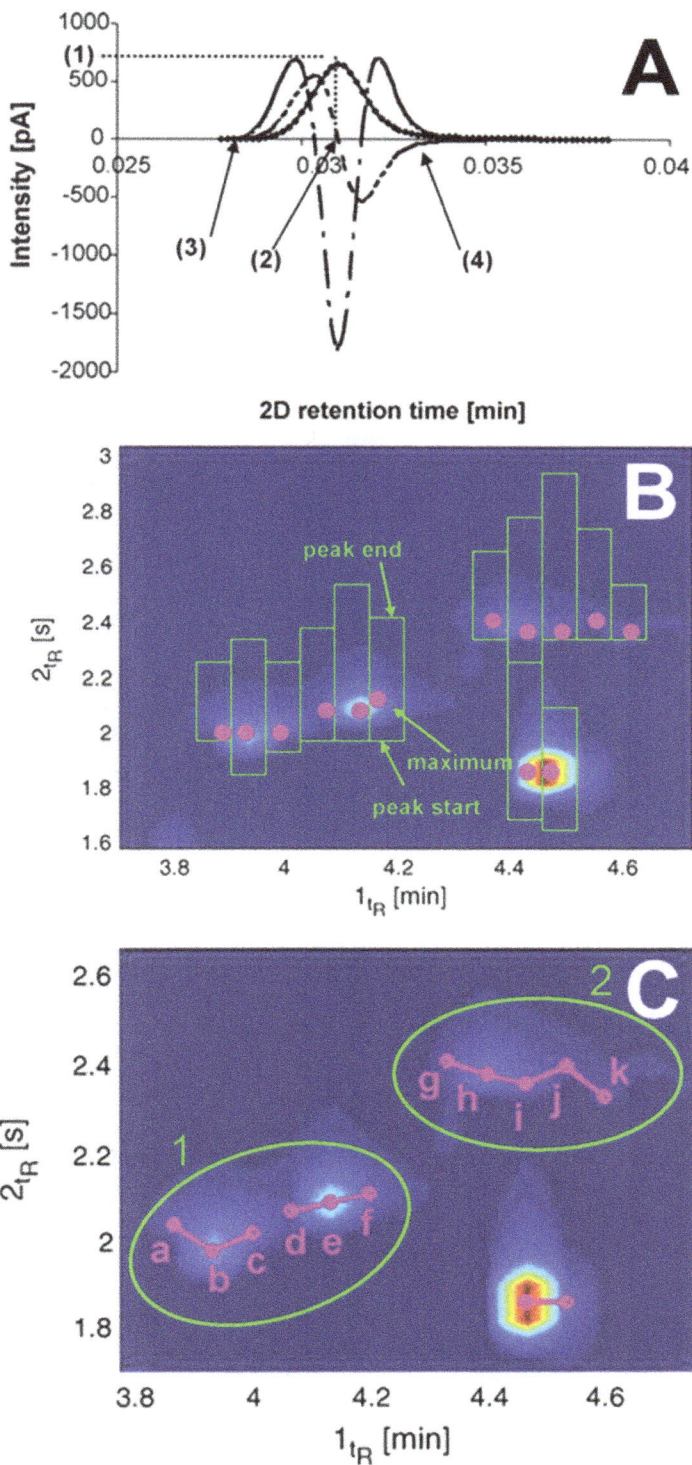

FIGURE 8.13 Illustration of the peak detection in LCxLC by first employing 1D peak detection methods on 1D data, and then clustering related ^{2}D peaks to form 2D peaks. A) Peak detection using derivative signal processing on 1D data – solid line, original signal; dashed line, first derivatives; dashed-dot line, second

FIGURE 8.13 (Continued)

derivatives. Indicated characteristics are (1) peak maximum; (2) ^{2}D retention time; (3) peak start; and (4) peak end. B) Locations of detected peaks in 2D chromatograms. The purple points in series belong to the same peak and peak-clustering algorithms aim to connect these correctly for each analyte. C) Clustered peaks as depicted by the connected purple dots. Adapted from *Journal of Chromatography, A*, 1156, S. Peters, G. Vivó-Truyols, P. Marriott, P. Schoenmakers, Development of an algorithm for peak detection in comprehensive two-dimensional chromatography, 14–24, Copyright (2007), with permission from Elsevier.

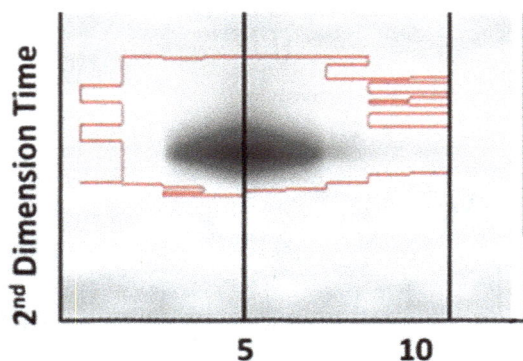

FIGURE 8.14 Contour plot of a peak after background correction using LC Image software [90]. The (red) line is the peak boundary as determined by the LC Image software via the watershed algorithm. Adapted from *Journal of Chromatography, A*, 1218, H. Bailey, S. Rutan, P. Carr, Factors that affect quantification of diode array data in comprehensive two-dimensional liquid chromatography using chemometric data analysis, 8411–8422, Copyright (2011), with permission from Elsevier.

NGB model yielded better fits of the peaks and an improved true positive rate for detection of chromatographic peaks with low total ion currents. Both models were applied to GC×GC-ToF datasets, where the NGB model was shown to find more true positives.

8.5.4.2 Watershed Algorithm

A completely different approach relies on processing the two-dimensional chromatogram as a surface. The inverted watershed algorithm leverages this concept and defines the boundaries of peaks using the topology of the surface [93]. This can be imagined as viewing the chromatogram as a mountain landscape, turning it upside down and filling it with water until the different peak maxima can no longer be distinguished. The algorithm will continue this process until it reaches the background signal, thus rendering it vulnerable to noise or artifacts. Figure 8.14 shows an LC×LC peak, with the boundaries as identified by the watershed algorithm. While the watershed algorithm is routinely applied (e.g., [94]), and also is available in commercial applications (e.g., LC Image software), it has been shown to be prone to erroneous results when the modulations are not correctly aligned and the noise levels are high [95]. Figure 8.14 illustrates this point, where the peaks for the outer modulations tend to appear and disappear in a single ^{2}D chromatogram. This issue was addressed by Latha *et al.* in a study which applied skew correction to improve the watershed algorithm [96]. In a comparison, the authors concluded that the improved watershed algorithm outperformed the two-step algorithm. In any case, both methods are known to be sensitive to large degrees of coelution and noise levels, and improved methods are needed.

Another commercially available program for peak detection in comprehensive 2D-LC, ChromSquare, is available and can be used in conjunction with a Shimadzu data system. However, details related to the algorithm(s) used in this software have not been published to the best of our knowledge.

8.6 MULTI-WAY APPROACHES

Most of the methods discussed above treat single-channel data (e.g., flame ionization detection in GC) or one channel in the case of multi-channel detectors (e.g., total ion chromatogram in MS detection). In contrast, multi-way analysis leverages all of the available data produced by multi-channel detectors. In general, these methods rely on modeling the data as a sum of linearly independent (i.e., not correlated) components, where ideally these components correspond directly to the chromatographic and spectroscopic signatures of real chemical species. These methods often treat the background contributions as additional components, which removes the need for an independent background subtraction algorithm. Examples include multivariate curve resolution-alternating least squares (MCR-ALS) and parallel factor analysis (PARAFAC) and related methods.

8.6.1 MULTIVARIATE CURVE RESOLUTION-ALTERNATING LEAST SQUARES

Multivariate curve resolution-alternating least squares (MCR-ALS) is a tool for resolving overlapped signals (i.e., overlapped peaks, or peaks overlapped with background signals) resulting from a wide range of analytical measurements, and numerous chromatographic applications have been reported [97,98]. In this section we introduce the basics of MCR-ALS for analysis of single 1D chromatograms with full spectrum detection (i.e., diode array or mass spectrometry), and then extend the concept to show how single or multiple 2D-LC chromatograms can be analyzed. Some specific advantages of this approach are that often extensive preprocessing of the data is not required, and that when multiple chromatograms are analyzed simultaneously, quantitative results can be obtained as a direct outcome of the algorithm.

MCR decomposes a data matrix ($\mathbf{X}$) into a set of chromatographic components ($\mathbf{C}$), a corresponding set of spectral components ($\mathbf{S}$) and the error ($\mathbf{E}$) which ideally only contains the noise.

$$\mathbf{X} = \mathbf{C} \cdot \mathbf{S}^T + \mathbf{E} \tag{8.19}$$

The matrix $\mathbf{X}$ (R x S) consists of R rows, corresponding to R chromatographic time points and S columns, corresponding to S wavelengths or mass channels and is an LC-DAD or LC-MS chromatogram generally containing contributions from multiple chemical species. Upon application of the algorithm, the matrix $\mathbf{C}$ (R x N) contains estimates for the "pure" component chromatograms corresponding to N components. Each of these N components may be directly associated with a chemical compound or may be associated with instrumental contributions to the signal such as background (e.g., solvent impurities). The recovered matrix $\mathbf{S}$ (S x N) consists of N spectra of these same components.

The implementation of the algorithm requires that initial estimates for each component (either the chromatograms or spectra; here the illustration is for initial estimates for the spectra, $\mathbf{S}$) are made. These can be obtained from the data itself using a method such as principal components analysis (PCA) [99, 100], and/or *a priori* knowledge of the component spectra (e.g., from a DAD or MS spectral library). Several other approaches have been developed to obtain the initial estimates, including key-set factor analysis [101], self-modeling [102], and orthogonal projection [103].

Using ALS, equation (8.19) is then iteratively optimized and solved as

$$\mathbf{C} = \mathbf{XS} \cdot \left(\mathbf{S}^T \mathbf{S} \right)^{-1} \tag{8.20a}$$

$$\mathbf{S}^T = \left(\mathbf{C}^T \mathbf{C} \right)^{-1} \cdot \mathbf{C}^T \mathbf{X} \tag{8.20b}$$

The -1 superscript indicates a matrix inverse; Eqs. 8.20a/b are least squares solutions for $\mathbf{C}$ and $\mathbf{S}$, respectively.

A key aspect of MCR-ALS that makes it powerful is the possibility to apply chemically meaningful constraints during the optimization process. These constraints can include non-negativity (i.e., analyte concentrations should not be negative), unimodality (i.e., well-behaved chromatographic peaks are singlets), and/or predefined elution profiles and spectra [104]. These constraints can be applied to the chromatograms and/or spectral profiles, and to one or more of the individual components, and make it more likely that the algorithm converges to a chemically reasonable solution.

The MCR-ALS method can be extended to resolve overlapped peaks in comprehensive 2D-LC as well. Figure 8.15A shows the format of the matrix $\mathbf{X}$ for a single LC-DAD (or LC-MS) chromatogram. Figure 8.15B shows the format of the array when the multiple modulations of subsequent chromatograms, i.e., ^{2}D are stacked together, and subsequently unfolded, as discussed above. This unfolded form of the $\mathbf{X}$ matrix is analyzed as described above, recognizing that the resulting

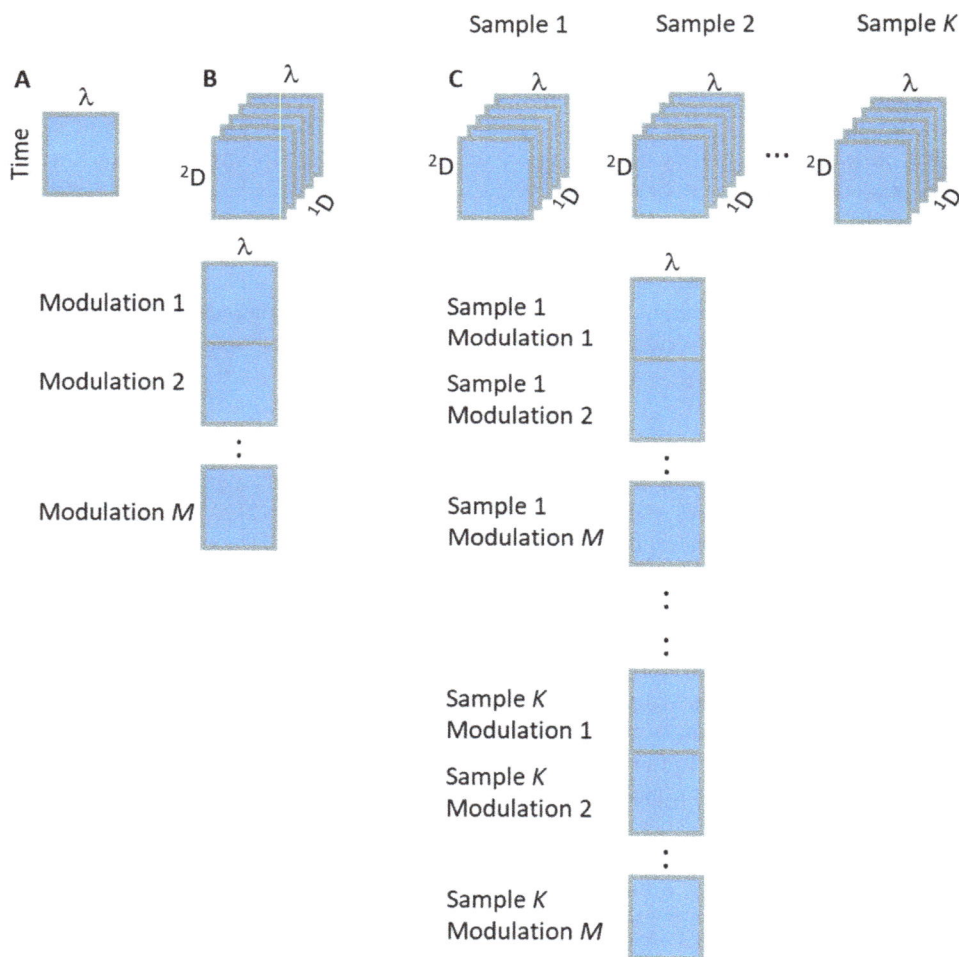

FIGURE 8.15 Schematic illustrating extension of the MCR-ALS approach to LC×LC separations and multiple samples. A) Data structure for LC-DAD or LC-MS data; B) Data structure for LC×LC-DAD or LC ×LC-MS data; C) Data structure for LC×LC-DAD or LC×LC-MS data for multiple samples.

resolved chromatograms contained in **C** will also be in this unfolded format. Finally, it is advantageous to analyze multiple samples simultaneously, as shown in Figure 8.15C. In this case the **X** matrix now contains all the ^{2}D chromatograms for all samples appended end to end; again, the resulting **C** matrix will have this same structure. A schematic of the curve resolution results from the data structure shown in Figure 8.15C is provided in Figure 8.16.

Inherent in the MCR-ALS algorithm is the requirement that the pure component spectra present in the matrix **S** must be consistent across all modulations and samples; this requirement is referred to in mathematical terms as bilinearity. This offers a degree of smoothing because there are multiple instances of the spectra across the data set. Additionally, if the K samples treated simultaneously consist of both standards and unknown samples, then areas under the resolved chromatograms can be used to construct "pseudo-univariate" calibration curves and provide quantitative results for the unknown samples, as discussed by Olivieri [105].

One challenge with LC×LC-DAD data is that the sample is continuously diluted during the analysis procedure, such that by the time the analytes enter the detector upon exiting the ^{2}D column, significant dilution has occurred. Cook *et al.* proposed incorporating a DAD column at the exit of the ^{1}D, when the sample has not been as diluted, and, hence, the S/N is much larger, and combining these data with the data from the ^{2}D detector [106]. MCR-ALS was then used to resolve the overlapped peaks.

A few comments on the implementation of MCR-ALS are in order here. First, as commonly implemented, the chromatogram is typically analyzed in small segments containing less than 10 components, making this approach somewhat tedious, although some progress has been made in automating this procedure [107, 108]. Second, the correct number of components must be chosen to achieve successful resolution; this is not always straightforward, and has proven to be difficult to automate in a way that is, according to the authors, robust. Additionally, the manual intervention required has so far prevented widespread implementation in chromatographic data system software, which presents a large barrier to use of these approaches by researchers who are not expert chemometricians. However, there are available toolboxes that function within the Matlab computing environment from Tauler *et al.* [109, 110] and Olivieri *et al.* [111], as well as a commercially available toolbox (PLS Toolbox) from Eigenvector [112].

While the above discussion focused primarily on DAD detector data as the basis for the curve resolution, MS detection can be advantageous because of the improved selectivity and structural information available from mass spectral data. The emergence of hyphenated LC-MS has been accompanied by a significant increase in dataset size for a single experiment. Depending on whether MS is carried out in tandem (i.e., MS/MS), at high resolution (e.g., TOF, ICR), and the chromatographic analysis time, datasets can easily reach up to 80 GB per analysis. One aim in data analysis has been the compression of the data to a manageable size. A conventional data reduction approach is binning, where the m/z axis is separated into segments, usually to unit resolution. Numerous applications of MCR-ALS to these types of LC-MS data have been reported [113–115]. Of course, this reduction is also accompanied by a loss of mass resolution, and this can result in multiple compounds being represented within a single mass data point.

Another method that has been proposed to address the large volumes of data resulting from high resolution LC-MS experiments is the region-of-interest (ROI) strategy [116]. Here, data regions with a high information density are selected using criteria such as signal intensity and the fact that peaks are typically adjacent to "data voids". This strategy allowed for peak detection without loss of mass resolution [117].

The ROI approach has been combined with MCR-ALS to provide for resolution of overlapped peaks [118]. A protocol has been reported to carry out this procedure, which includes all steps from the export of the data from the chromatographic data system to complete resolution of a set of metabolomic data [119]. The ROI-MCR-ALS method has been applied to LC×LC-HRMS data of the rice metabolome by Navarro-Reigh *et al.* [120]. The authors used ROIs that were selected using S/N, mass accuracy, and the minimum number of subsequent occurrences of the same m/z. In addition, the authors applied wavelet compression [121, 122] to further compress the data up to

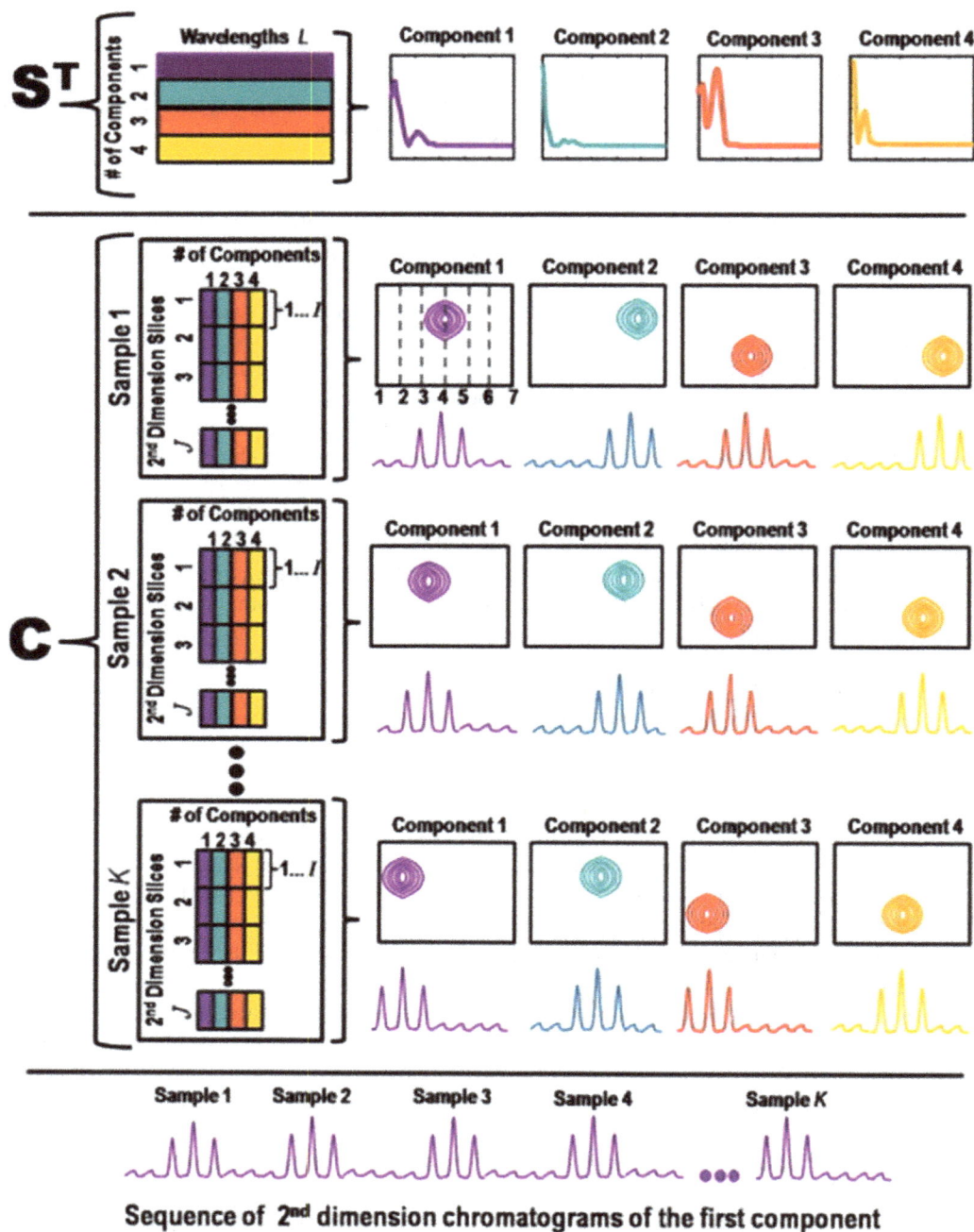

FIGURE 8.16 Example of curve resolution results from MCR-ALS for the analysis of multiple LCxLC-DAD chromatograms. The top panel shows the resolved spectra ($\mathbf{S}^T$) for the hypothetical four peaks found in the chromatogram. The next panel shows the resolved chromatograms (C) for these four peaks. On the left side, the concatenated ^{2}D chromatograms are shown in matrix format appended end to end, while on the right, the contour plots for each of the four resolved peaks are shown, with the sequence of individual ^{2}D peaks shown below the contour plot. The third panel shows the corresponding sequence of ^{2}D chromatograms for component 1 for all samples (1–K) as an example. Reprinted from *Chemometrics and Intelligent Laboratory Systems*, 106, H. Bailey, S. Rutan, Chemometric resolution and quantification of four-way data arising from comprehensive 2D-LC-DAD analysis of human urine, 131–141, Copyright (2011), with permission from Elsevier.

a total reduction of the data by a factor of 50. The reduced data were processed by MCR-ALS and 154 metabolites were detected. Although the protocol for this strategy is fairly complex, a detailed step-by-step procedure has been provided to aid in its implementation [119].

8.6.2 PARAFAC AND PARAFAC2

Another chemometric method that can also be used for quantification is parallel factor analysis (PARAFAC) [123, 124]. The PARAFAC method relies on an extension of the bilinearity concept described above. In the case of the analysis of a single LCxLC-DAD or LCxLC-MS experiment, as depicted in Figure 8.15 B, it is assumed that the data are trilinear, meaning that the ^{2}D chromatograms for individual pure components are perfectly reproducible across all modulations (as well as the spectra, as above), with no retention time shifts. The difference between trilinear and non-trilinear data is shown in Figure 8.17; note the differences in retention times. When multiple samples are treated simultaneously, as depicted in Figure 8.15 C, the data are assumed to have a quadrilinear structure, meaning that both the ^{2}D and ^{1}D chromatograms of the pure components are reproducible from sample to sample. In the case of either trilinear or quadrilinear data structure, the PARAFAC solution is achieved using an alternating least squares method. PARAFAC algorithms have also been developed which allow for constraints [125] such as non-negativity to be applied, but generally the implementation of constraints is much less flexible than for MCR-ALS. Algorithms for PARAFAC are available from Bro [125–127] and in Eigenvector's PLS Toolbox [112].

As LCxLC data is acquired as a series of 1D chromatograms, which are combined into a two-dimensional plane, the occurrence of retention-time shifts from one modulation to another and/or one sample to another is not uncommon. This means that relatively few applications of PARAFAC

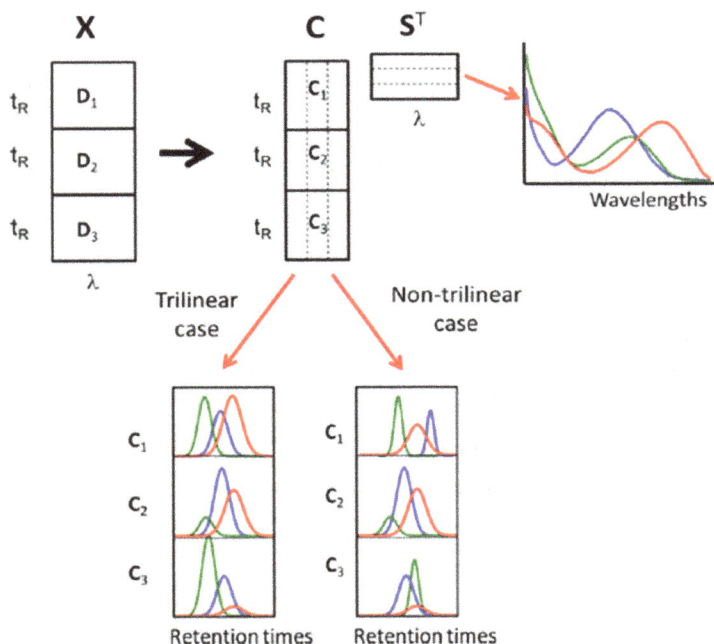

FIGURE 8.17 Schematic showing the difference between trilinear and non-trilinear data. Reprinted with permission from M. Bauza, G. Ibanez, R. Tauler, A. Olivieri, Sensitivity equation for quantitative analysis with multivariate curve resolution-Alternating Least-Squares: Theoretical and experimental approach, *Analytical Chemistry* 84 (2012), 8697–8706. Copyright 2012 American Chemical Society.

to raw LC×LC data have been reported [128]. Allen and Rutan have described a semi-automated method that incorporates initialization with MCR-ALS, along with appropriate alignment to correct for retention time shifts in both the ^{1}D and the ^{2}D to enable the use of PARAFAC to analyze both simulated data and an application for quantification of phenytoin in waste water by LC×LC-DAD with quadrilinear PARAFAC [129]. Recoveries were adequate, and in the case of the phenytoin analysis, quantitation accuracy and precision matched that of the reference LC-LC-MS/MS method.

PARAFAC2 is a variant of PARAFAC that relaxes the rigid multilinearity constraint in one of the dimensions [127, 130]. This is achieved by using a mathematical constraint that the matrix cross product of the matrix containing the shifting dimension is a constant. Effectively, this constraint allows for retention time shifts, as long as the shape of the profile is not significantly changed. PARAFAC2 has been applied to GC×GC datasets [55].

Navarro-Reig *et al.* compared MCR-ALS, PARAFAC and PARAFAC2 for the analysis of corn oil samples by LC×LC-MS [131]. As others have found, the data did not rigorously follow the trilinearity assumption, so that PARAFAC2 and MCR-ALS yielded improved results as compared to PARAFAC. The authors particularly commented on the flexibility of the MCR-ALS in this regard. These authors also noted that it is possible to implement a flexible trilinearity constraint within the MCR-ALS algorithm, which can also accommodate retention time shifts.

8.7   CLASSIFICATION

Often the results from chromatographic analysis are used for pattern recognition; either supervised or unsupervised methods can be used. Once peak tables are obtained from the analysis or compounds identified and quantified, numerous, well-established methods can be used [132–134], and their application is not different from the case of data obtained from 1D chromatographic methods. In this section, we will only discuss selected methods that have been applied within the framework of the 2D-LC data analysis workflow.

Pierce *et al.* have developed a method for detecting components in GC×GC-MS chromatograms that help to classify samples based on differences in concentration using a Fisher ratio method. This method carries out the comparison on a point-by-point basis, requiring that there is no misalignment between chromatograms. Subsequently, Marney *et al.* [135] and Parsons *et al.* [136] developed a tile-based Fisher ratio method that was less sensitive to misalignment, and resulted in fewer false positives (i.e., identification regions showing significant differences in concentrations, when such differences were not really present).

Bailey *et al.* have compared the Fisher ratio method [137] to the similarity index method of Windig [133] for the screening and classification of wine samples using LC×LC-DAD data [137]. The similarity index method was originally developed for 1D LC-MS data and is based on correlation coefficients. Both simulated data and wine analysis data were evaluated. Both methods were successfully able to identify peaks that showed significant concentration differences, which could be subsequently targeted for a more thorough analysis.

Reichenbach *et al.* developed an application to distinguish two classes of patients using LC×LC-DAD data of urine samples [138]. This method was based on creating a template based on the peaks or regions of the chromatograms that were common across multiple samples. Like the tile-based Fisher ratio method described above, this method does not require precise peak alignment to successfully identify those regions of the chromatogram showing significant concentration differences. Subsequently, two conventional pattern recognition techniques, K-nearest neighbor and support vector machines, were used for classification.

8.8   SUMMARY

Alongside advances in instrumentation for 2D-LC in recent years, there have been many advances in data analysis approaches to address the "data tsunami" from 2D-LC experiments. We have summarized many of them in this chapter. While many significant and important advances have been made, there is a desperate need for more improvements in some key areas. First, the background contributions to signals in LC×LC data are generally larger and more variable than in conventional 1D-LC experiments. Additionally, many creative methods have been applied to 1D-LC experimental data, but many of these could be better tailored to LC×LC data, for example, the very short rapid gradients in ^{2}D chromatograms lead to pronounced background features (especially for DAD detection).

There is also room for significant improvement in peak detection. On the one hand, the watershed algorithm treats the 2D data holistically, but it requires very high quality data with clean preprocessing. On the other hand, two-step methods with peak clustering can better handle less ideal data, but require a complex series of supervised steps as part of the process.

Multi-way data methods show great promise, especially when full spectrum DAD or MS detectors are employed with LC×LC separations. These methods can typically resolve background contributions to the signal quite well, and can lead directly to quantitative results. However, the limitations of these methods include the fact that the chromatogram usually needs to be segmented to process small portions of the dataset at one time, and full automation of the process across the entire chromatogram has yet to be achieved.

Because of the numerous data analysis strategies already developed, and the increasing number of new algorithms being reported, the field would benefit from a dataset repository where reference datasets could be used to compare different data analysis methods, whether background subtraction, peak alignment, peak detection, and/or quantitation. Thus, data analysis in comprehensive 2D-LC continues to be an area ripe for creative and novel developments.

REFERENCES

[1] S.J. Qin, Process data analytics in the era of big data, AIChE J. 60 (2014) 3092–3100. doi:10.1002/aic.14523

[2] R.C. Allen, S.C. Rutan, Investigation of interpolation techniques for the reconstruction of the first dimension of comprehensive two-dimensional liquid chromatography – Diode array detector data, Anal. Chim. Acta. 705 (2011) 253–260. doi:10.1016/j.aca.2011.06.022

[3] W.H. Press, S.A. Teukolsky, W.T. Vetterling, B.P. Flannery, Numerical Recipes in C: The Art of scientific computing, 2nd ed., Cambridge University Press, New York, 1992.

[4] D.R. Stoll, X. Li, X. Wang, P.W. Carr, S.E.G. Porter, S.C. Rutan, Fast, comprehensive two-dimensional liquid chromatography, J. Chromatogr. A. 1168 (2007) 3–43. doi:10.1016/j.chroma.2007.08.054

[5] J.J.A.M. Weusten, E.P.P.A. Derks, J.H.M. Mommers, S. van der Wal, Alignment and clustering strategies for GC×GC–MS features using a cylindrical mapping, Anal. Chim. Acta. 726 (2012) 9–21. doi:10.1016/j.aca.2012.03.009

[6] N.J. Micyus, S.K. Seeley, J. V. Seeley, Method for reducing the ambiguity of comprehensive two-dimensional chromatography retention times, J. Chromatogr. A. 1086 (2005) 171–174. doi:10.1016/j.chroma.2005.06.016

[7] J.D. Wilson, C.A.J. McInnes, The elimination of errors due to baseline drift in the measurement of peak areas in gas chromatography, J. Chromatogr. A. 19 (1965) 486–494. doi:10.1016/s0021-9673(01)99489-0

[8] G.A. Pearson, A general baseline-recognition and baseline-flattening algorithm, J. Magn. Reson. 27 (1977) 265–272. doi:10.1016/0022-2364(77)90076-2

[9] A. Savitzky, M.J.E. Golay, Smoothing and differentiation of data by simplified least squares procedures, Anal. Chem. 36 (1964) 1627–1639. doi:10.1021/ac60214a047

[10] M.U.A. Bromba, H. Ziegler, Application hints for Savitzky-Golay digital smoothing filters, Anal. Chem. 53 (1981) 1583–1586. doi:10.1021/ac00234a011

[11] C.G. Enke, T.A. Nieman, Signal-to-noise ratio enhancement by least-squares polynomial smoothing, Anal. Chem. 48 (1976) 705A–712A. doi:10.1021/ac50002a769

[12] D.F. Thekkudan, S.C. Rutan, Denoising and Signal-to-Noise Ratio Enhancement: Classical Filtering, in: Compr. Chemom ., Elsevier, Amsterdam, 2009: pp. 9–24. doi:10.1016/B978-044452701-1.00098-3

[13] P.D. Wentzell, C.D. Brown, Signal processing in analytical chemistry, in: Encycl. Anal. Chem., John Wiley & Sons, Ltd, Chichester, UK, 2000: pp. 9764–9800. doi:10.1002/9780470027318.a5207

[14] F. Vogt, Data filtering in instrumental analyses with applications to optical spectroscopy and chemical imaging, J. Chem. Educ. 88 (2011) 1672–1683. doi:10.1021/ed100984c

[15] A. Felinger, Peak Detection, in: Data Anal. Signal Process. Chromatogr., Elsevier, Amsterdam, 1998: pp. 183–190.

[16] R. Bracewell, The Fourier Transform and Its Applications, 3rd edition, McGraw-Hill Science, New York, 1999.

[17] A. Felinger, T.L. Pap, J. Inczédy, Improvement of the signal-to-noise ratio of chromatographic peaks by Fourier transform, Anal. Chim. Acta. 248 (1991) 441–446. doi:10.1016/S0003-2670(00)84661-9

[18] B. Walczak, Wavelets in Chemistry, 1st edition, Elsevier, Amsterdam, 2000.

[19] M. Daszykowski, I. Stanimirova, A. Bodzon-Kulakowska, J. Silberring, G. Lubec, B. Walczak, Start-to-end processing of two-dimensional gel electrophoretic images, J. Chromatogr. A. 1158 (2007) 306–317. doi:10.1016/j.chroma.2007.02.009

[20] M. Li Vigni, J.M. Prats-Montalban, A. Ferrer, M. Cocchi, Coupling 2D-wavelet decomposition and multivariate image analysis (2D WT-MIA), J. Chemom. 32 (2018) e2970. doi:10.1002/cem.2970

[21] R.C. Allen, M.G. John, S.C. Rutan, M.R. Filgueira, P.W. Carr, Effect of background correction on peak detection and quantification in online comprehensive two-dimensional liquid chromatography using diode array detection, J. Chromatogr. A. 1254 (2012) 51–61. doi:10.1016/j.chroma.2012.07.034

[22] B.W.J. Pirok, J.A. Westerhuis, Challenges in obtaining relevant information from one- and two-dimensional LC experiments, LC-GC North Am. 6 (2020) 8–14. www.chromatographyonline.com/view/challenges-obtaining-relevant-information-one-and-two-dimensional-lc-experiments.

[23] E.T. Whittaker, On a new method of graduation, Proc. Edinburgh Math. Soc. 41 (1922) 63–75. doi:10.1017/S001309150000359X

[24] J. Carlos Cobas, M.A. Bernstein, M. Martín-Pastor, P.G. Tahoces, A new general-purpose fully automatic baseline-correction procedure for 1D and 2D NMR data, J. Magn. Reson. 183 (2006) 145–151. doi:10.1016/j.jmr.2006.07.013

[25] Z.-M. Zhang, S. Chen, Y.-Z. Liang, Z.-X. Liu, Q.-M. Zhang, L.-X. Ding et al., An intelligent background-correction algorithm for highly fluorescent samples in Raman spectroscopy, J. Raman Spectrosc. 41 (2009) 659–669. doi:10.1002/jrs.2500

[26] P.H.C.C. Eilers, A perfect smoother, Anal. Chem. 75 (2003) 3631–3636. doi:10.1021/ac034173t

[27] Z.-M.M. Zhang, S. Chen, Y.-Z.Z. Liang, Baseline correction using adaptive iteratively reweighted penalized least squares, Analyst. 135 (2010) 1138. doi:10.1039/b922045c

[28] S.-J. Baek, A. Park, Y.-J. Ahn, J. Choo, Baseline correction using asymmetrically reweighted penalized least squares smoothing, Analyst. 140 (2015) 250–257. doi:10.1039/C4AN01061B

[29] R. Perez-Pueyo, M.J. Soneira, S. Ruiz-Moreno, Morphology-based automated baseline removal for raman spectra of artistic pigments, Appl. Spectrosc. 64 (2010) 595–600. doi:10.1366/000370210791414281

[30] Z. Li, D.-J. Zhan, J.-J. Wang, J. Huang, Q.-S. Xu, Z.-M. Zhang et al., Morphological weighted penalized least squares for background correction, Analyst. 138 (2013) 4483. doi:10.1039/c3an00743j

[31] H.-Y. Fu, H.-D. Li, Y.-J. Yu, B. Wang, P. Lu, H.-P. Cui, et al. Simple automatic strategy for background drift correction in chromatographic data analysis, J. Chromatogr. A. 1449 (2016) 89–99. doi:10.1016/j.chroma.2016.04.054

[32] H.-Y.Y. Fu, H.-D.D. Li, Y.-J.J. Yu, B. Wang, P. Lu, H.-P.P. Cui et al., Simple automatic strategy for background drift correction in chromatographic data analysis, J. Chromatogr. A. 1449 (2016) 89–99. doi:10.1016/j.chroma.2016.04.054

[33] P. Yaroshchyk, J.E. Eberhardt, Automatic correction of continuum background in Laser-induced Breakdown Spectroscopy using a model-free algorithm, Spectrochim. Acta – Part B At. Spectrosc. 99 (2014) 138–149. doi:10.1016/j.sab.2014.06.020

[34] Z. Li, D.-J. Zhan, J.-J. Wang, J. Huang, Q.-S. Xu, Z.-M. Zhang et al., Morphological weighted penalized least squares for background correction, Analyst. 138 (2013) 4483. doi:10.1039/c3an00743j

[35] Y.-J. Yu, H.-L. Wu, H.-Y. Fu, J. Zhao, Y.-N. Li, S.-F. Li et al., Chromatographic background drift correction coupled with parallel factor analysis to resolve coelution problems in three-dimensional chromatographic data: Quantification of eleven antibiotics in tap water samples by high-performance liquid chromatography, J. Chromatogr. A. 1302 (2013) 72–80. doi:10.1016/j.chroma.2013.06.009

[36] X. Ning, I.W. Selesnick, L. Duval, Chromatogram baseline estimation and denoising using sparsity (BEADS), Chemom. Intell. Lab. Syst. 139 (2014) 156–167. doi:10.1016/j.chemolab.2014.09.014

[37] J.A. Navarro-Huerta, J.R. Torres-Lapasió, S. López-Ureña, M.C. García-Alvarez-Coque, Assisted baseline subtraction in complex chromatograms using the BEADS algorithm, J. Chromatogr. A. 1507 (2017) 1–10. doi:10.1016/j.chroma.2017.05.057

[38] I. Selesnick, Sparsity-assisted signal smoothing (revisited), in: ICASSP, IEEE Int. Conf. Acoust. Speech Signal Process. – Proc., 2017: pp. 4546–4550. doi:10.1109/ICASSP.2017.7953017

[39] M. Lopatka, A. Barcaru, M.J. Sjerps, G. Vivó-Truyols, Leveraging probabilistic peak detection to estimate baseline drift in complex chromatographic samples, J. Chromatogr. A. 1431 (2016) 122–130. doi:10.1016/j.chroma.2015.12.063

[40] G.L. Erny, T. Acunha, C. Simó, A. Cifuentes, A. Alves, Background correction in separation techniques hyphenated to high-resolution mass spectrometry – Thorough correction with mass spectrometry scans recorded as profile spectra, J. Chromatogr. A. 1492 (2017) 98–105. doi:10.1016/j.chroma.2017.02.052

[41] A. Kaufmann, P. Butcher, Strategies to avoid false negative findings in residue analysis using liquid chromatography coupled to time-of-flight mass spectrometry, Rapid Commun. Mass Spectrom. 20 (2006) 3566–3572. doi:10.1002/rcm.2762

[42] Y. Zhang, H.-L. Wu, A.-L. Xia, L.-H. Hu, H.-F. Zou, R.-Q. Yu, Trilinear decomposition method applied to removal of three-dimensional background drift in comprehensive two-dimensional separation data, J. Chromatogr. A. 1167 (2007) 178–183. doi:10.1016/j.chroma.2007.08.055

[43] S.E. Reichenbach, M. Ni, D. Zhang, E.B. Ledford, Image background removal in comprehensive two-dimensional gas chromatography, J. Chromatogr. A. 985 (2003) 47–56. doi:10.1016/S0021-9673(02)01498-X

[44] S.E. Reichenbach, P.W. Carr, D.R. Stoll, Q. Tao, Smart templates for peak pattern matching with comprehensive two-dimensional liquid chromatography, J. Chromatogr. A. 1216 (2009) 3458–3466. doi:10.1016/j.chroma.2008.09.058

[45] S.E. Reichenbach, V. Kottapalli, M. Ni, A. Visvanathan, Computer language for identifying chemicals with comprehensive two-dimensional gas chromatography and mass spectrometry, J. Chromatogr. A. 1071 (2005) 263–269. doi:10.1016/j.chroma.2004.08.125

[46] Z.-D. Zeng, S.-T. Chin, H.M. Hugel, P.J. Marriott, Simultaneous deconvolution and re-construction of primary and secondary overlapping peak clusters in comprehensive two-dimensional gas chromatography, J. Chromatogr. A. 1218 (2011) 2301–2310. doi:10.1016/j.chroma.2011.02.028

[47] L.E. Niezen, P.J. Schoenmakers, B.W.J. Pirok, Critical comparison of background correction methodology used in chromatography and spectroscopy, Anal. Chim. Acta. (2022) submitted.

[48] M.R. Filgueira, C.B. Castells, P.W. Carr, A simple, robust orthogonal background correction method for two-dimensional liquid chromatography, Anal. Chem. 84 (2012) 6747–6752. doi:10.1021/ac301248

[49] T. Gröger, M. Schäffer, M. Pütz, B. Ahrens, K. Drew, M. Eschner, et al., Application of two-dimensional gas chromatography combined with pixel-based chemometric processing for the chemical profiling of illicit drug samples, J. Chromatogr. A. 1200 (2008) 8–16. doi:10.1016/j.chroma.2008.05.028

[50] T. Gröger, R. Zimmermann, Application of parallel computing to speed up chemometrics for GCxGC-TOFMS based metabolic fingerprinting, Talanta. 83 (2011) 1289–1294. doi:10.1016/j.talanta.2010.09.015

[51] K.J. Johnson, B.J. Prazen, D.C. Young, R.E. Synovec, Quantification of naphthalenes in jet fuel with GCxGC/Tri-PLS and windowed rank minimization retention time alignment, J. Sep. Sci. 27 (2004) 410–416. doi:10.1002/jssc.200301640

[52] K.M. Pierce, L.F. Wood, B.W. Wright, R.E. Synovec, A comprehensive two-dimensional retention time alignment algorithm to enhance chemometric analysis of comprehensive two-dimensional separation data, Anal. Chem. 77 (2005) 7735–7743. doi:10.1021/ac0511142

[53] R.K. Nelson, B.M. Kile, D.L. Plata, S.P. Sylva, L. Xu, C.M. Reddy, et al., Tracking the weathering of an oil spill with comprehensive two-dimensional gas chromatography, Environ. Forensics. 7 (2006) 33–44. doi:10.1080/15275920500506758

[54] C. Cordero, E. Liberto, C. Bicchi, P. Rubiolo, S.E. Reichenbach, X. Tian et al., Targeted and non-targeted approaches for complex natural sample profiling by GCxGC-qMS, J. Chromatogr. Sci. 48 (2010) 251–261. doi:10.1093/chromsci/48.4.251

[55] T. Skov, J.C. Hoggard, R. Bro, R.E. Synovec, Handling within run retention time shifts in two-dimensional chromatography data using shift correction and modeling, J. Chromatogr. A. 1216 (2009) 4020–4029. doi:10.1016/j.chroma.2009.02.049

[56] D. Zhang, X. Huang, F.E. Regnier, M. Zhang, Two-dimensional correlation optimized warping algorithm for aligning GC×GC–MS data, Anal. Chem. 80 (2008) 2664–2671. doi:10.1021/ac7024317

[57] J. Gros, D. Nabi, P. Dimitriou-Christidis, R. Rutler, J.S. Arey, Robust algorithm for aligning two-dimensional chromatograms, Anal. Chem. 84 (2012) 9033–9040. doi:10.1021/ac301367s

[58] Y. Zushi, J. Gros, Q. Tao, S.E. Reichenbach, S. Hashimoto, J.S. Arey, Pixel-by-pixel correction of retention time shifts in chromatograms from comprehensive two-dimensional gas chromatography coupled to high resolution time-of-flight mass spectrometry, J. Chromatogr. A. 1508 (2017) 121–129. doi:10.1016/j.chroma.2017.05.065

[59] V.G. van Mispelaar, A.C. Tas, A.K. Smilde, P.J. Schoenmakers, A.C. van Asten, Quantitative analysis of target components by comprehensive two-dimensional gas chromatography, J. Chromatogr. A. 1019 (2003) 15–29. doi:10.1016/j.chroma.2003.08.101

[60] H. Parastar, M. Jalali-Heravi, R. Tauler, Comprehensive two-dimensional gas chromatography (GC×GC) retention time shift correction and modeling using bilinear peak alignment, correlation optimized shifting and multivariate curve resolution, Chemom. Intell. Lab. Syst. 117 (2012) 80–91. doi:10.1016/j.chemolab.2012.02.003

[61] Q.-X. Zheng, H.-Y. Fu, H.-D. Li, B. Wang, C.-H. Peng, S. Wang, et al., Automatic time-shift alignment method for chromatographic data analysis, Sci. Rep. 7 (2017) 256. doi:10.1038/s41598-017-00390-7

[62] H.Y. Fu, O. Hu, Y.M. Zhang, L. Zhang, J.J. Song, P. Lu, et al., Mass-spectra-based peak alignment for automatic nontargeted metabolic profiling analysis for biomarker screening in plant samples, J. Chromatogr. A. 1513 (2017) 201–209. doi:10.1016/j.chroma.2017.07.044

[63] Y.-J. Yu, H.-Y. Fu, L. Zhang, X.-Y. Wang, P.-J. Sun, X.-B. Zhang, et al., A chemometric-assisted method based on gas chromatography – Mass spectrometry for metabolic profiling analysis, J. Chromatogr. A. 1399 (2015) 65–73. doi:10.1016/j.chroma.2015.04.029

[64] P. Xia, L. Zhang, F. Li, Learning similarity with cosine similarity ensemble, Inf. Sci. (Ny). 307 (2015) 39–52. doi:10.1016/j.ins.2015.02.024

[65] T.S. Bos, W.C. Knol, S.R.A. Molenaar, L.E. Niezen, P.J. Schoenmakers, G.W. Somsen et al., Recent applications of chemometrics in two-dimensional chromatography, J. Sep. Sci. (2020) submitted.

[66] E. Grushka, M.N. Myers, P.D. Schettler, J.C. Giddings, Computer characterization of chromatographic peaks by plate height and higher central moments, Anal. Chem. 41 (1969) 889–892. doi:10.1021/ac60276a014

[67] S.B. Howerton, C. Lee, V.L. McGuffin, Additivity of statistical moments in the exponentially modified Gaussian model of chromatography, Anal. Chim. Acta. 478 (2003) 99–110. doi:10.1016/S0003-2670(02)01472-1

[68] Y. Vanderheyden, K. Broeckhoven, G. Desmet, Comparison and optimization of different peak integration methods to determine the variance of unretained and extra-column peaks, J. Chromatogr. A. 1364 (2014) 140–150. doi:10.1016/j.chroma.2014.08.066

[69] P.G. Stevenson, X.A. Conlan, N.W. Barnett, Evaluation of the asymmetric least squares baseline algorithm through the accuracy of statistical peak moments, J. Chromatogr. A. 1284 (2013) 107–111. doi:10.1016/j.chroma.2013.02.012

[70] D.W. Morton, C.L. Young, Analysis of peak profiles using statistical moments, J. Chromatogr. Sci. 33 (1995) 514–524. doi:10.1093/chromsci/33.9.514

[71] T.S. Bos, W.C. Knol, S.R.A. Molenaar, L.E. Niezen, P.J. Schoenmakers, G.W. Somsen et al., Recent applications of chemometrics in one- and two-dimensional chromatography, J. Sep. Sci. 43 (2020) 1678–1727. doi:10.1002/jssc.202000011

[72] H.-Y. Fu, J.-W. Guo, Y.-J. Yu, H.-D. Li, H.-P. Cui, P.-P. Liu, et al., A simple multi-scale Gaussian smoothing-based strategy for automatic chromatographic peak extraction, J. Chromatogr. A. 1452 (2016) 1–9. doi:10.1016/j.chroma.2016.05.018

[73] S. Peters, H.-G. Janssen, G. Vivó-Truyols, A new method for the automated selection of the number of components for deconvolving overlapping chromatographic peaks, Anal. Chim. Acta. 799 (2013) 29–35. doi:10.1016/j.aca.2013.08.041

[74] Z.-M. Zhang, X. Tong, Y. Peng, P. Ma, M.-J. Zhang, H.-M. Lu et al., Multiscale peak detection in wavelet space, Analyst. 140 (2015) 7955–7964. doi:10.1039/C5AN01816A

[75] J. Lu, M.J. Trnka, S.-H. Roh, P.J.J. Robinson, C. Shiau, D.G. Fujimori et al., Improved peak detection and deconvolution of native electrospray mass spectra from large protein complexes, J. Am. Soc. Mass Spectrom. 26 (2015) 2141–2151. doi:10.1007/s13361-015-1235-6

[76] Y.-J. Yu, Q.-L. Xia, S. Wang, B. Wang, F.-W. Xie, X.-B. Zhang, et al., Chemometric strategy for automatic chromatographic peak detection and background drift correction in chromatographic data, J. Chromatogr. A. 1359 (2014) 262–270. doi:10.1016/j.chroma.2014.07.053

[77] V.P. Andreev, T. Rejtar, H.-S. Chen, E. V. Moskovets, A.R. Ivanov, B.L. Karger, A universal denoising and peak picking algorithm for LC–MS based on matched filtration in the chromatographic time domain, Anal. Chem. 75 (2003) 6314–6326. doi:10.1021/ac0301806

[78] R.A.R.A. Carmona, W.L.W.L. Hwang, B. Torresani, B. Torrésani, Multiridge detection and time-frequency reconstruction, IEEE Trans. Signal Process. 47 (1999) 480–492. doi:10.1109/78.740131

[79] E. Dinç, E. Büker, A new application of continuous wavelet transform to overlapping chromatograms for the quantitative analysis of amiloride hydrochloride and hydrochlorothiazide in tablets by ultra-performance liquid chromatography, J. AOAC Int. 95 (2012) 751–756. doi:10.5740/jaoacint.SGE_Dinc

[80] X. Shao, L. Sun, An application of the continuous wavelet transform to resolution of multicomponent overlapping analytical signals, Anal. Lett. 34 (2001) 267–280. doi:10.1081/AL-100001578

[81] X. Tong, Z. Zhang, F. Zeng, C. Fu, P. Ma, Y. Peng, et al., Recursive wavelet peak detection of analytical signals, Chromatographia. 79 (2016) 1247–1255. doi:10.1007/s10337-016-3155-4

[82] Y.-J. Yu, Q.-X. Zheng, Y.-M. Zhang, Q. Zhang, Y.-Y. Zhang, P.-P. Liu et al., Automatic data analysis workflow for ultra-high performance liquid chromatography-high resolution mass spectrometry-based metabolomics, J. Chromatogr. A. 1585 (2019) 172–181. doi:10.1016/j.chroma.2018.11.070

[83] G. Vivó-Truyols, Bayesian approach for peak detection in two-dimensional chromatography, Anal. Chem. 84 (2012) 2622–2630. doi:10.1021/ac202124t

[84] J.M. Davis, J.C. Giddings, Statistical theory of component overlap in multicomponent chromatograms, Anal. Chem. 55 (1983) 418–424. doi:10.1021/ac00254a003

[85] M. Lopatka, G. Vivó-Truyols, M.J. Sjerps, Probabilistic peak detection for first-order chromatographic data, Anal. Chim. Acta. 817 (2014) 9–16. doi:10.1016/j.aca.2014.02.015

[86] M. Woldegebriel, G. Vivó-Truyols, Probabilistic model for untargeted peak detection in lc–ms using bayesian statistics, Anal. Chem. 87 (2015) 7345–7355. doi:10.1021/acs.analchem.5b01521

[87] S. Peters, G. Vivó-Truyols, P.J. Marriott, P.J. Schoenmakers, Development of an algorithm for peak detection in comprehensive two-dimensional chromatography, J. Chromatogr. A. 1156 (2007) 14–24. doi:10.1016/j.chroma.2006.10.066

[88] J. Pól, B. Hohnová, M. Jussila, T. Hyötyläinen, Comprehensive two-dimensional liquid chromatography–time-of-flight mass spectrometry in the analysis of acidic compounds in atmospheric aerosols, J. Chromatogr. A. 1130 (2006) 64–71. doi:10.1016/j.chroma.2006.04.050

[89] M. Kivilompolo, T. Hyötyläinen, Comprehensive two-dimensional liquid chromatography in analysis of Lamiaceae herbs: Characterisation and quantification of antioxidant phenolic acids, J. Chromatogr. A. 1145 (2007) 155–164. doi:10.1016/j.chroma.2007.01.090

[90] H.P. Bailey, S.C. Rutan, Chemometric resolution and quantification of four-way data arising from comprehensive 2D-LC-DAD analysis of human urine, Chemom. Intell. Lab. Syst. 106 (2011) 131–141. doi:10.1016/j.chemolab.2010.07.008

[91] S. Kim, M. Ouyang, J. Jeong, C. Shen, X. Zhang, A new method of peak detection for analysis of comprehensive two-dimensional gas chromatography mass spectrometry data, Ann. Appl. Stat. 8 (2014) 1209–1231. doi:10.1214/14-AOAS731

[92] S. Kim, H. Jang, I. Koo, J. Lee, X. Zhang, Normal–Gamma–Bernoulli peak detection for analysis of comprehensive two-dimensional gas chromatography mass spectrometry data, Comput. Stat. Data Anal. 105 (2017) 96–111. doi:10.1016/j.csda.2016.07.015

[93] S.E. Reichenbach, M. Ni, V. Kottapalli, A. Visvanathan, Information technologies for comprehensive two-dimensional gas chromatography, Chemom. Intell. Lab. Syst. 71 (2004) 107–120. doi:10.1016/j.chemolab.2003.12.009

[94] B. Li, S.E. Reichenbach, Q. Tao, R. Zhu, A streak detection approach for comprehensive two-dimensional gas chromatography based on image analysis, Neural Comput. Appl. (2018). doi:10.1007/s00521-018-3917-z

[95]   G. Vivó-Truyols, H.-G. Janssen, Probability of failure of the watershed algorithm for peak detection in comprehensive two-dimensional chromatography, J. Chromatogr. A. 1217 (2010) 1375–1385. doi:10.1016/j.chroma.2009.12.063

[96]   I. Latha, S.E. Reichenbach, Q. Tao, Comparative analysis of peak-detection techniques for comprehensive two-dimensional chromatography, J. Chromatogr. A. 1218 (2011) 6792–6798. doi:10.1016/j.chroma.2011.07.052

[97]   R. Tauler, A. Smilde, B. Kowalski, Selectivity, local rank, three-way data analysis and ambiguity in multivariate curve resolution, J. Chemom. 9 (1995) 31–58. doi:10.1002/cem.1180090105

[98]   A. de Juan, R. Tauler, Multivariate Curve Resolution (MCR) from 2000: Progress in concepts and applications, Crit. Rev. Anal. Chem. 36 (2006) 163–176. doi:10.1080/10408340600970005

[99]   R.W. Hendler, R.I. Shrager, Deconvolutions based on singular value decomposition and the pseudoinverse: a guide for beginners, J. Biochem. Biophys. Methods. 28 (1994) 1–33. doi:10.1016/0165-022X(94)90061-2

[100] Y. Nagai, W.Y. Sohn, K. Katayama, An initial estimation method using cosine similarity for multivariate curve resolution: application to NMR spectra of chemical mixtures, Analyst. 144 (2019) 5986–5995. doi:10.1039/C9AN01416K

[101] E.R. Malinowski, Obtaining the key set of typical vectors by factor analysis and subsequent isolation of component spectra, Anal. Chim. Acta. 134 (1982) 129–137. doi:10.1016/S0003-2670(01)84184-2

[102] W. Windig, J. Guilment, Interactive self-modeling mixture analysis, Anal. Chem. 63 (1991) 1425–1432. doi:10.1021/ac00014a016

[103] F. Cuesta Sánchez, B. Van Den Bogaert, S.C. Rutan, D.L. Massart, Multivariate peak purity approaches, Chemom. Intell. Lab. Syst. 34 (1996) 139–171. doi:10.1016/0169-7439(96)00020-2

[104] A. De Juan, J. Jaumot, R. Tauler, Multivariate Curve Resolution (MCR). Solving the mixture analysis problem, Anal. Methods. 6 (2014) 4964–4976. doi:10.1039/C4AY00571F

[105] M.C. Bauza, G.A. Ibañez, R. Tauler, A.C. Olivieri, Sensitivity equation for quantitative analysis with multivariate curve resolution-alternating least-squares: Theoretical and experimental approach, Anal. Chem. 84 (2012) 8697–8706. doi:10.1021/ac3019284

[106] D.W. Cook, S.C. Rutan, D.R. Stoll, P.W. Carr, Two dimensional assisted liquid chromatography – A chemometric approach to improve accuracy and precision of quantitation in liquid chromatography using 2D separation, dual detectors, and multivariate curve resolution, Anal. Chim. Acta. 859 (2015) 87–95. doi:10.1016/j.aca.2014.12.009

[107] X. Domingo-Almenara, A. Perera, J. Brezmes, Avoiding hard chromatographic segmentation: A moving window approach for the automated resolution of gas chromatography – Mass spectrometry-based metabolomics signals by multivariate methods, J. Chromatogr. A. 1474 (2016) 145–151. doi:10.1016/j.chroma.2016.10.066

[108] R. Wehrens, E. Carvalho, D. Masuero, A. de Juan, S. Martens, High-throughput carotenoid profiling using multivariate curve resolution, Anal. Bioanal. Chem. 405 (2013) 5075–5086. doi:10.1007/s00216-012-6555-9

[109] R. Tauler, A. de Juan, J. Jaumot, MCR-ALS toolbox, www.mcrals.info/. Accessed May 30, 2022.

[110] J. Jaumot, R. Gargallo, A. de Juan, R. Tauler, A graphical user-friendly interface for MCR-ALS: A new tool for multivariate curve resolution in MATLAB, Chemom. Intell. Lab. Syst. 76 (2005) 101–110. doi:10.1016/j.chemolab.2004.12.007

[111] A.C. Olivieri, H.-L. Wu, R.-Q. Yu, MVC2: A Matlab graphical interface toolbox for second-order multivariate calibration, Chemom. Intell. Lab. Syst. 96 (2009) 246–251. doi:10.1016/j.chemolab.2009.02.005

[112] Eigenvector PLS Toolbox, https://eigenvector.com/software/pls-toolbox/. Accessed May 30, 2022.

[113] G. Vivó-Truyols, J.R. Torres-Lapasió, M.C. García-Alvarez-Coque, P.J. Schoenmakers, Towards unsupervised analysis of second-order chromatographic data: Automated selection of number of components in multivariate curve-resolution methods, J. Chromatogr. A. 1158 (2007) 258–272. doi:10.1016/j.chroma.2007.03.005

[114] M. Navarro-Reig, J. Jaumot, A. García-Reiriz, R. Tauler, Evaluation of changes induced in rice metabolome by Cd and Cu exposure using LC-MS with XCMS and MCR-ALS data analysis strategies, Anal. Bioanal. Chem. 407 (2015) 8835–8847. doi:10.1007/s00216-015-9042-2

[115] M. Farrés, B. Piña, R. Tauler, Chemometric evaluation of Saccharomyces cerevisiae metabolic profiles using LC–MS, Metabolomics. 11 (2015) 210–224. doi:10.1007/s11306-014-0689-z

[116] R. Stolt, R.J.O. Torgrip, J. Lindberg, L. Csenki, J. Kolmert, I. Schuppe-Koistinen et al., Second-order peak detection for multicomponent high-resolution LC/MS data, Anal. Chem. 78 (2006) 975–983. doi:10.1021/ac050980b

[117] M. Pérez-Cova, C. Bedia, D.R. Stoll, R. Tauler, J. Jaumot, MSroi: A pre-processing tool for mass spectrometry-based studies, Chemom. Intell. Lab. Syst. 215 (2021) 104333. doi:10.1016/j.chemolab.2021.104333

[118] E. Gorrochategui, J. Jaumot, R. Tauler, ROIMCR: A powerful analysis strategy for LC-MS metabolomic datasets, BMC Bioinformatics. 20 (2019). doi:10.1186/s12859-019-2848-8

[119] R. Tauler, E. Gorrochategui, J. Jaumot, R. Tauler, A protocol for LC-MS metabolomic data processing using chemometric tools, Protoc. Exch. (2015) 1–46. doi:10.1038/protex.2015.102

[120] M. Navarro-Reig, J. Jaumot, A. Baglai, G. Vivó-Truyols, P.J. Schoenmakers, R. Tauler, Untargeted comprehensive two-dimensional liquid chromatography coupled with high-resolution mass spectrometry analysis of rice metabolome using multivariate curve resolution, Anal. Chem. 89 (2017) 7675–7683. doi:10.1021/acs.analchem.7b01648

[121] I. Daubechies, Ten Lectures on Wavelets, Society for Industrial and Applied Mathematics, 1992. doi:10.1137/1.9781611970104

[122] J. Trygg, N. Kettaneh-Wold, L. Wallbäcks, 2D wavelet analysis and compression of on-line industrial process data, J. Chemom. 15 (2001) 299–319. doi:10.1002/cem.681

[123] D.W. Cook, S.C. Rutan, Chemometrics for the analysis of chromatographic data in metabolomics investigations, J. Chemom. 28 (2014) 681–687. doi:10.1002/cem.2624

[124] A. Smilde, R. Bro, P. Geladi, Multi-Way Analysis with Applications in the Chemical Sciences, John Wiley & Sons, Ltd, Chichester, UK, 2004. doi:10.1002/0470012110

[125] R. Bro, PARAFAC. Tutorial and applications, Chemom. Intell. Lab. Syst. 38 (1997) 149–171. doi:10.1016/S0169-7439(97)00032-4

[126] C.A. Andersson, R. Bro, The N-way Toolbox for MATLAB, Chemom. Intell. Lab. Syst. 52 (2000) 1–4. doi:10.1016/S0169-7439(00)00071-X

[127] H.A.L. Kiers, J.M.F. ten Berge, R. Bro, PARAFAC2 – Part I. A direct fitting algorithm for the PARAFAC2 model, J. Chemom. 13 (1999) 275–294. doi:10.1002/(SICI)1099-128X(199905/08)13:3/4<275::AID-CEM543>3.3.CO;2-2

[128] S.E.G. Porter, D.R. Stoll, S.C. Rutan, P.W. Carr, J.D. Cohen, Analysis of Four-Way Two-Dimensional Liquid Chromatography-Diode Array Data: Application to Metabolomics, Anal. Chem. 78 (2006) 5559–5569. doi:10.1021/ac0606195

[129] R.C. Allen, S.C. Rutan, Semi-automated alignment and quantification of peaks using parallel factor analysis for comprehensive two-dimensional liquid chromatography – Diode array detector data sets, Anal. Chim. Acta. 723 (2012) 7–17. doi:10.1016/j.aca.2012.02.019

[130] R. Bro, C.A. Andersson, H.A.L. Kiers, PARAFAC2 – Part II. Modeling chromatographic data with retention time shifts, J. Chemom. 13 (1999) 295–309. doi:10.1002/(SICI)1099-128X(199905/08)13:3/4<295::AID-CEM547>3.0.CO;2-Y

[131] M. Navarro-Reig, J. Jaumot, T.A. van Beek, G. Vivó-Truyols, R. Tauler, Chemometric analysis of comprehensive LC×LC-MS data: Resolution of triacylglycerol structural isomers in corn oil, Talanta. 160 (2016) 624–635. doi:10.1016/j.talanta.2016.08.005

[132] L.C. Lee, C.-Y. Liong, A.A. Jemain, Partial least squares-discriminant analysis (PLS-DA) for classification of high-dimensional (HD) data: A review of contemporary practice strategies and knowledge gaps, Analyst. 143 (2018) 3526–3539. doi:10.1039/C8AN00599K

[133] W. Windig, W.F. Smith, W.F. Nichols, Fast interpretation of complex LC/MS data using chemometrics, Anal. Chim. Acta. 446 (2001) 465–474. doi:10.1016/S0003-2670(01)01276-4

[134] K. Vanden Branden, M. Hubert, Robust classification in high dimensions based on the SIMCA Method, Chemom. Intell. Lab. Syst. 79 (2005) 10–21. doi:10.1016/j.chemolab.2005.03.002

[135] L.C. Marney, W. Christopher Siegler, B.A. Parsons, J.C. Hoggard, B.W. Wright, R.E. Synovec, Tile-based Fisher-ratio software for improved feature selection analysis of comprehensive two-dimensional gas chromatography – Time-of-flight mass spectrometry data, Talanta. 115 (2013) 887–895. doi:10.1016/j.talanta.2013.06.038

[136] B.A. Parsons, L.C. Marney, W.C. Siegler, J.C. Hoggard, B.W. Wright, R.E. Synovec, Tile-based Fisher ratio analysis of comprehensive two-dimensional gas chromatography time-of-flight mass spectrometry (GC×GC–TOFMS) data using a null distribution approach, Anal. Chem. 87 (2015) 3812–3819. doi:10.1021/ac504472s

[137] H.P. Bailey, S.C. Rutan, Comparison of chemometric methods for the screening of comprehensive two-dimensional liquid chromatographic analysis of wine, Anal. Chim. Acta. 770 (2013) 18–28. doi:10.1016/j.aca.2013.01.062

[138] S.E. Reichenbach, X. Tian, Q. Tao, D.R. Stoll, P.W. Carr, Comprehensive feature analysis for sample classification with comprehensive two-dimensional LC, J. Sep. Sci. 33 (2010) 1365–1374. doi:10.1002/jssc.200900859

9 Applications of Two-Dimensional Liquid Chromatography for Analysis of Synthetic Pharmaceutical Materials

Alexandre Goyon, Gregory O. Staples, Dwight R. Stoll, and Kelly Zhang

CONTENTS

DOI: 10.1201/9781003090557-9

9.1   PREFACE

Chapters 9 and 10 address applications of two-dimensional liquid chromatography (2D-LC) in modern pharmaceutical analysis. Given the dramatic growth in the use of 2D-LC in the last decade and the diversity of current 2D-LC methods in use, we felt it would be helpful to address the work in two different chapters. There are multiple ways one could divide the content, no one of which provides a clear dividing line. We have chosen to use the origin of the pharmaceutical material as a rough demarcation. Molecules that are synthetic in origin are addressed in Chapter 9, including short (< 200 bases) oligonucleotides and therapeutic peptides that are synthesized chemically. Chapter 10 then deals with materials (mostly proteins) that are produced using recombinant organisms and undergo extensive purification prior to use as pharmaceutical products. Of course, overlap is unavoidable, for example in the need to determine the concentration of free, un-bound synthetic small molecule drug in antibody-drug conjugate (ADC) drug product. Indeed, these overlaps reflect some of the challenges of contemporary pharmaceutical analysis.

9.2   INTRODUCTION

The applications of 2D-LC in small molecule pharmaceutical analysis have improved significantly in the past decade in terms of their effectiveness, sophistication, and ease-of-use [1–3]. 2D-LC has evolved from being a research tool providing data for complex samples, to one routinely used for sample characterization, and in some areas, 2D-LC has become a mainstream technology. These advances in 2D-LC applications have largely resulted from improved robustness and user-friendliness of commercially available options for the instrument hardware and software. Over this same time period there has also been a significant shift in the perspective of practitioners using 2D-LC in the pharmaceutical industry. Once perceived as primarily useful for characterization of complex samples containing hundreds of compounds, users now also view 2D-LC as a viable option for improving the effectiveness and throughput for analysis of samples of moderate complexity. In some cases, 2D-LC provides the only viable option to solve some particularly challenging problems in the pharmaceutical application space. In this chapter we will discuss these challenges, and provide examples of how they can be addressed using 2D-LC.

9.3   OVERVIEW OF CHALLENGES ENCOUNTERED IN PHARMACEUTICAL ANALYSIS

Due to the limited selectivity and peak capacity of any single 1D-LC method, complete sample profiling is very challenging in the pharmaceutical setting, as illustrated by Figure 9.1. Over the past decade several groups have developed creative solutions to some of these challenges using 2D-LC, and in the following sections we will provide examples that illustrate both the problem and how 2D-LC can be used to solve it. First, though, brief descriptions of some of the challenges illustrated in Figure 9.1 is warranted.

A)   A large majority of LC methods in use in the pharmaceutical industry involve nonvolatile buffer components such as phosphate due to their transparency to UV absorbance detectors, favorable effects on peak shape and selectivity, low cost, and safety. If, in the course of running these methods a new peak emerges that is unknown, rapid identification of the unknown species is challenging because the 1D-LC method cannot be directly coupled to mass spectrometric (MS) detection.

B)   Historically much of the development of chiral chromatography has focused on resolving the enantiomers of molecules with single chiral centers. In recent years the number of drug candidates with multiple chiral centers has increased significantly. Generally speaking, current chiral column technologies do not exhibit the selectivity needed to separate all of

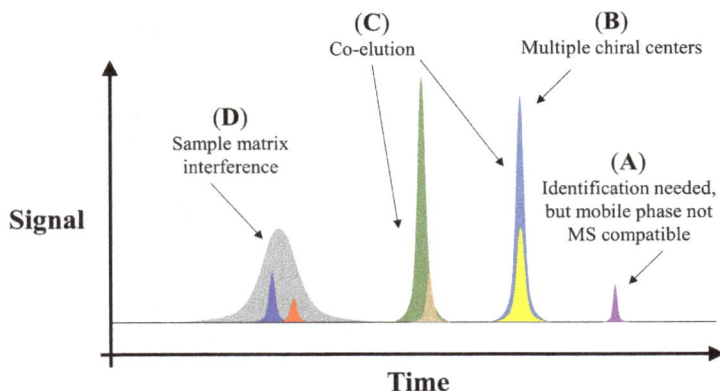

FIGURE 9.1 Challenges commonly encountered in pharmaceutical sample profiling illustrated using a hypothetical chromatogram Adapted from [1].

  the enantiomers of a molecule with multiple chiral centers; this then leads to coelution in separations of these molecules by 1D-LC.

C) Coelution problems are commonplace in pharmaceutical analyses. Even with modern high efficiency columns, compounds with highly similar structures may coelute if the column does not have sufficient selectivity to separate them.

D) Many pharmaceutical materials also contain sample components present at concentrations that span several orders of magnitude. When excipients, which are usually present at high concentration, are complex (e.g., polyethylene glycol), their constituents can show up as major peaks covering a larger portion of a 1D-LC chromatogram. When a target analyte is present at low concentration, it can easily be obscured by the high abundance of sample matrix components, making detection and accurate quantitation extremely difficult.

9.4 COUPLING NON-MS COMPATIBLE LC SEPARATIONS TO MS (FIGURE 9.1A)

Online "desalting" using 2D-LC is widely used to address real-world challenges. LC methods used in the pharmaceutical quality control (QC) environment to establish substance purity commonly employ mobile phase components that are generally considered MS-incompatible; these include nonvolatile salts such as phosphates and chlorides, as well as ion-pairing reagents. These reagents are used for their chromatographic benefits, including UV transparency, and retention and/or selectivity enhancement. However, these reagents prevent direct identification of new peaks that may appear over time, because the LC separation cannot be directly coupled to a mass spectrometer. Instead, identification requires translation of the method to conditions that are MS-compatible, which typically involves a buffer exchange to an ammonium formate or acetate buffer. The buffer exchange is then frequently accompanied by a selectivity change, which in turn requires that peaks are "tracked" from one separation to the other, usually on the basis of peak area. This can be quite challenging because MS-compatible buffers are usually not very UV transparent, and because the mobile phase composition can affect the absorptivity of the analytes of interest and the separation performance (e.g., selectivity, peak shape, and resolution).

  To address these challenges, heartcutting 2D-LC approaches have been used frequently for the direct identification of impurity peaks observed in an existing 1D-LC separation. In the 2D-LC context, the existing separation becomes the first dimension (^{1}D) separation, and fractions of ^{1}D effluent are collected online and injected into the second dimension (^{2}D) column where nonvolatile salts are separated from the impurities of interest, and discarded. Zhang and coworkers demonstrated that implementing this approach was straightforward and enabled unambiguous peak tracking using the ^{2}D separation with MS detection, while also providing excellent quantitative performance using UV absorbance data from the ^{1}D separation running with a UV-transparent phosphate buffer [4]. Wang and coworkers have published several studies that demonstrate the use of optimal ^{1}D reversed-phase liquid chromatography (RPLC) separations using phosphate buffers followed by a ^{2}D RP desalting method and MS detection to identify the impurities present in the azithromycin, cefpiramide, eroxithromycin, erythromycin, and leucomycin drug substances [5, 6]. Long and coworkers have developed a RP-RP-MS method that involves a weak anion exchange (WAX) trap column in the interface between the two dimensions for online removal of the ion-pairing reagent 1-octanesulfonate used in the first dimension, followed by impurity characterization by MS [7]. Luo and coworkers described an application where perchlorate was used as a 1D-LC mobile phase additive to improve the separation of a peptide and its impurities. Using the RP-RP-MS approach, the perchlorate was removed by the ^{2}D column, enabling direct identification of peptide impurities by MS [8]. In another study, the authors reported the need to combine ion-pairing agents and phosphate to provide sufficient retention of the hydrophilic 5-aminolevulinic acid (ALA) on RP columns. Using this as the first dimension of an ion-pairing reversed-phase (IPRP)-RP-MS method, six impurities present in the ALA material following forced degradation were identified [9]. In this case a RP trap column was used in the interface to separate the impurities of interest from the phosphate and IP reagents prior to the ^{2}D separation. This is an interesting approach, however it is not clear how the RP trap was cleaned, nor how many analyses could be carried out before cleaning the trap.

The analysis of hydrophilic nucleotides by RPLC requires the use of alkylamine ion-pairing reagents (IPRs) to provide enough retention for separation. These IPRs can significantly suppress the MS signal for these compounds when using electrospray ionization; they can also lead to undesired formation of adducts between analytes of interest and the IPRs long after they have been introduced into the MS, unless the MS is carefully cleaned. Thus, using the second dimension of a 2D-LC separation to prevent exposure of the MS to IPRs used in the first dimension is attractive. Wang and coworkers developed an IPRP-mixed-mode (MM) chromatography method with the use of volatile salts and an Acclaim Trinity Q1 column in the second dimension to identify unknown impurities in nucleotide analogues [10]. The ^{2}D MM column was coupled to MS and the structure elucidation of seven impurities of vidarabine monophosphate was achieved by MS and MS/MS [10]. 2D-LC has also been used to facilitate coupling of MM oligonucleotide separations to MS detection; these are discussed in detail in Section 9.6.1.

9.5   CHIRAL SEPARATION (FIGURE 9.1B)

Chiral separation is one of the most successful applications of 2D-LC in small molecule pharmaceutical analysis. We note here that Chapter 13 deals with enantioselective 2D-LC in general. In this chapter we have focused on those aspects of chiral 2D-LC that are most relevant to small molecule pharmaceutical analysis. We refer interested readers to Chapter 13 for more detailed discussion on other aspects of chiral 2D-LC. Applications in small molecule pharmaceutical analysis can be summarized in three main categories that are related to each other:

9.5.1   ACHIRAL-CHIRAL 2D-LC TO SEPARATE ACHIRAL INTERFERENCES FROM CHIRAL COMPOUNDS

Due to the limited peak capacity of chiral separations in general, enantiomers frequently coelute with impurities and other components in a sample. Chiral separations with UV detection lack specificity for the separation and quantification of stereoisomers present in complex mixtures. Mass spectrometry can be used to improve detection specificity, however, the possible coelution of impurities may compromise quantification due to ion suppression or enhancement phenomena [11, 12]. The lifetimes of chiral columns are also known to be reduced when injecting crude reaction mixtures [13]. A common application of 2D-LC in this context is to first separate chiral components from achiral components using an achiral column in the first dimension, and then resolve chiral isomers that coelute from the ^{1}D column using a chiral column in the second dimension. A plethora of achiral-chiral 2D-LC applications have been reported [13–20]. In this case, 2D-LC instruments are commonly operated in heartcut modes. Most achiral-chiral separations have been performed using RPLC separations in the first dimension, and chiral separations in the second dimension. 2D-LC separation using achiral normal phase liquid chromatography (NPLC) in the first dimension and chiral NP separation in the second dimension has also been reported [21]. Two-dimensional separation involving a RP ^{1}D separation and chiral supercritical fluid chromatography (SFC) in the second dimension has been demonstrated [22], but challenges related to the incompatibility of the mobile phases used in these two separations, as well as concerns about the robustness of this type of setup, have prevented use of this solution for routine analysis. The use of chiral-chiral 2D-LC separations beyond the scope of conventional small pharmaceutical molecules has included the analysis of hydrophilic amino acids (log P < 1) [14, 17]. Readers interested in this topic are referred to Chapter 13.

9.5.2 Compounds with Multiple Chiral Centers

An important current trend in the pharmaceutical industry is that more and more drug candidates contain multiple chiral centers. Given that the number of enantiomers increases exponentially with the number of chiral centers, it quickly becomes challenging to separate all chiral isomers using a single 1D-LC method. In the case of compounds with three or more chiral centers, it can be very difficult to separate all of the isomers, even when using multiple 1D-LC methods, each focused on separating a subset of species in the sample. Achiral-chiral 2D-LC separation as discussed above with multiple heartcutting is one effective way to address this challenge [17, 19, 20]. Figure 9.2 illustrates a commonly used concept; diastereomers and other achiral impurities are first separated by an achiral column (typically RP), and then each diastereomer is transferred to the second dimension for enantiomer separation using a chiral column.

Regalado *et al.* reported a significant reduction in analysis time for a mixture containing multiple enantiomers using offline 2D-LC and combining ^{1}D achiral RP separation with UV-triggered fraction collection, and subsequent ^{2}D chiral RP separation. Complete resolution of 12 components of a mixture of warfarin and hydroxywarfarin isomers was achieved [23].

Chiral-chiral separations have been performed in the multiple heartcutting and comprehensive modes for a pharmaceutical drug containing multiple chiral centers. Twelve stereoisomers of posaconazole were separated in the multiple heartcutting mode using a ^{1}D Chiralpak IB column and the combination of Chiralpak IC and IF3 ^{2}D columns [24]. In another study, the use of ultrafast ^{2}D chiral separation (48 s) enabled the use of the comprehensive mode and the complete resolution of the enantiomers of a synthetic intermediate as well as warfarin and hydroxywarfarin stereoisomers [19].

9.5.3 Integrated Method Development and Sample Analysis

Although significant research effort has been invested to improve the speed of 2D-LC separations [25–27], historically not much attention has been given to method development workflows for 2D-LC. In the case of chiral separations and analytes with multiple chiral centers, it could take an extraordinarily long time to develop a 2D-LC method. When it comes to fast-paced pharmaceutical labs that deal with many different drug candidates, the time-consuming method development step is more prominent, and a platform approach that enables rapid development of 2D-LC methods for different molecules is essential.

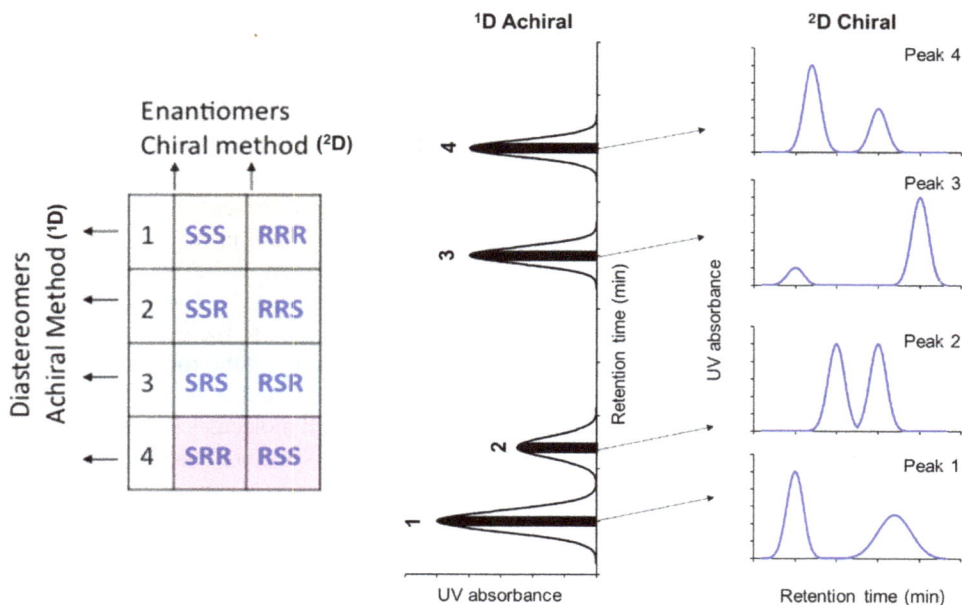

FIGURE 9.2 Illustration of achiral-chiral multiple heartcut 2D-LC separation for a compound with three chiral centers.

To address these issues, Lin *et al.* developed a LC-*m*LC strategy that employed multiple heartcutting (MHC) and multiple chiral columns with different chiral selectors in the second dimension [20], as shown in Figure 9.3A. Peaks with potential coeluting-compounds in the first dimension are cut and screened by different columns in the second dimension [20]. The best separation result from the different column screenings in the second dimension is selected for reporting purity, and combining the purity information obtained from ^{1}D, the total impurity profile of the sample is obtained from method screening. Using this strategy, very little adjustment of separation conditions is needed for each new molecule or sample encountered. The strategy was also applied for characterization of both chiral and achiral components present in an in-process sample "on the fly". Eight stereoisomers of a pharmaceutical compound with three chiral centers were fully separated with a total method development time of less than two hours, resulting in a 24-min 2D-LC method including column equilibration time, as shown in Figure 9.3B.

Makey *et al.* reported a different strategy to tackle the arduous 2D-LC method development issue by using computer-assisted modeling [28]. The authors used multifactorial modeling (gradient slope, column temperature, and different column and mobile phase combinations) supported by ACD/Labs LC simulator software to map the retention landscape for each analyte in both the first and second dimensions, which enables facile optimization of offline and online 2D-LC methods for both analytical and purification methods. Please note that a complementary discussion of this topic of integrated method development strategies for 2D-LC can be found in Chapter 13 of this book.

9.6 PEAK COELUTION, PEAK PURITY, AND STABILITY INDICATING METHOD ASSESSMENT (FIGURE 9.1C)

In pharmaceutical HPLC analysis, both "targeted" and "non-targeted" impurities can be encountered during method development. The term "targeted" here means that the impurities are known, and standards are available. In this case it is easy to check for coelution with other compounds based on retention time, or with sample spiking experiments. The term "non-targeted" here means that there is

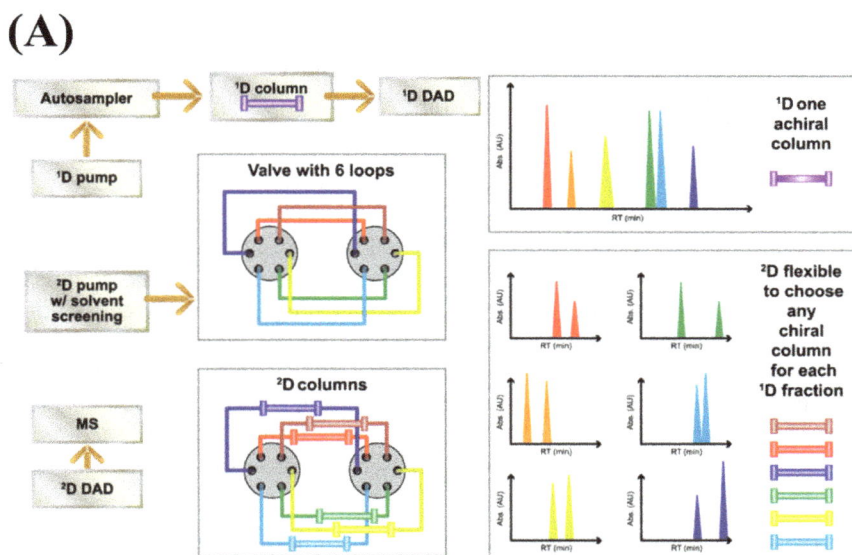

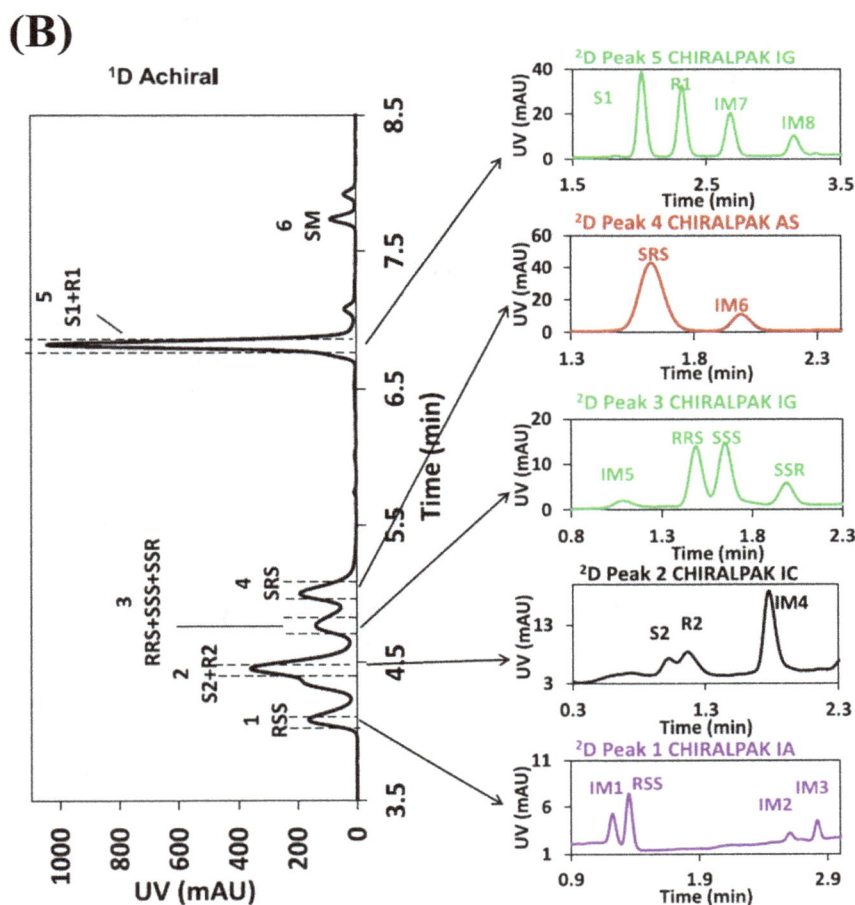

FIGURE 9.3 A) Illustration of an instrument setup for multiple heartcut LC-*m*LC that enables automated use of different columns in the second dimension. B) LC-*m*LC chromatograms obtained from the separation of an in-process sample involving a target molecule with three chiral centers, and multiple structurally similar achiral

FIGURE 9.3 (Continued)

impurities. Adapted from *Journal of Chromatography, A*, 1620, J. Lin, C. Tsang, R. Lieu, K. Zhang, Fast chiral and achiral profiling of compounds with multiple chiral centers by a versatile two-dimensional multicolumn liquid chromatography (LC-*m*LC) approach, In Press, Copyright (2020), with permission from Elsevier.

no expectation to observe a particular impurity *a priori* – for example, based on known reactivity of particular chemical structures. In this case standards will not be available, and detection of coelution is more difficult. Here, orthogonal method screening is essential to increase confidence that no coelution is occurring. Although orthogonal 1D-LC methods (e.g., with different stationary phases and/or mobile phases) can be used, tracking peaks across chromatograms with different retention patterns is challenging, especially for low-level impurities and complex chromatograms with many peaks. In this context 2D-LC can provide unambiguous peak tracking in addition to the orthogonality provided by complementary ^{1}D and ^{2}D selectivities. In the following sections we discuss several examples of how 2D-LC methods have been used to address these challenges in practice.

9.6.1 USE OF 2D-LC FOR CHARACTERIZATION OF SYNTHETIC THERAPEUTIC PEPTIDES

Synthetic peptide therapeutics are being developed to address a widening field of clinical targets including those important in the fields of oncology, infectious disease, and neurology. A recent review article by Lian *et al.* has summarized emerging trends in the field, quality control and analytical challenges that arise from the manufacturing, handling, and storage of the peptides, and current trends in the development and application of various analytical methods for characterization of these important materials [29]. Whereas 2D-LC has been an important and routinely applied tool for peptide identification and quantification in the proteomics field for more than two decades, there has been surprisingly little work published on the use of 2D-LC for characterization of therapeutic peptides. To the best of our knowledge the only published work describing online 2D-LC applied to synthetic therapeutic peptides is a recent paper by Karongo *et al.* [30]. In this work they evaluated the potential for the ^{2}D separation of a sLC×LC method to resolve impurities or degradants that co-purify with the peptide active pharmaceutical ingredient (API) under RPLC conditions with an acidic mobile phase (e.g., 0.1% TFA). The major factors they studied were the effects of the ^{2}D column stationary phase chemistry (12 different phases were evaluated), the ^{2}D mobile phase buffer type and concentration, and the interface conditions on the overall ability of the method to separate and identify impurities and degradants. They focused on the use of RP separations in both dimensions, and found that a combination of two different C18 phases provided the best overall performance as measured by resolution and peak shape. Since the ^{1}D mobile phase was fixed at low pH to mimic the conditions used in quality control labs, the authors explored different buffers for use in the second dimension that both provided complementary selectivity at high pH (around 10), and were amenable to parallel detection using UV-Vis absorbance, charged aerosol detection (CAD), and mass spectrometry (MS). Within these constraints they found that 10 mM ammonium hydroxide buffers were highly useful in the second dimension. They then went on to demonstrate the utility of this sLC×LC-UV-CAD-MS system for identification of multiple impurities present in four different synthetic therapeutic peptides as purified by standard preparative RPLC methods. It seems this type of methodology is highly flexible and has tremendous potential to streamline method development and improve QC procedures for these peptide therapeutics.

9.6.2 USE OF 2D-LC FOR CHARACTERIZATION OF THERAPEUTIC OLIGONUCLEOTIDES

Oligonucleotide therapeutics are a rapidly growing class of treatments for diseases such as cancer, metabolic disorders, and neurological disorders. These drugs can be DNA- or RNA-based, and are designed to specifically bind to genes, mRNA, or proteins in order to modulate gene expression, protein translation, or protein function, respectively [31]. There is a growing list of drugs within the oligonucleotide class including aptamers, small-interfering RNAs (siRNAs), microRNAs, and anti-sense oligonucleotides (ASOs). Developments in the gene editing tool CRISPR-Cas have given rise to an additional group of oligonucleotide therapeutics in the form of guide RNAs that are necessary to provide specificity in directing the Cas protein to the proper genomic target for editing [32].

Characterization of oligonucleotide therapeutics typically involves separation using liquid chromatography, and here IPRP and AEX are the most common modes. A review on this subject was recently published by Goyon et al., which the reader is encouraged to consult for a detailed background [33]. A primary goal of oligonucleotide separations is to quantify impurities in the preparation, and here MS is an invaluable tool. For especially complex samples, one-dimensional separation methods can be insufficient for resolving impurities. With the increasing popularity of oligonucleotide therapeutics, and recognizing the need for additional peak capacity, investigations of offline and online 2D separation approaches have been explored. These include offline fractionation using IPRP or IEX followed by capillary gel electrophoresis [34], and online approaches, such as the direct coupling of HILIC with IPRP in a 2D-LC format involving intermediate trapping [35].

Several options for potential ^{1}D separations were explored recently by Roussis et al. for use in 2D-LC separations of oligonucleotides where IPRP is used in the second dimension, coupled with MS detection [36]. These included SEC, RP, and SAX. IPRP was used in the second dimension due to the acceptable MS compatibility and general acceptance of this method. Multiple heartcut RP-IPRP, making use of an intermediate trapping column, was employed for stability studies of monomethoxytrityl (MMT) modified oligonucleotides exposed to long-term storage at different pH values. The MMT moeity is used during oligonucleotide synthesis to enable purification using RP, and analysis of these preparations using 2D-LC avoids production of MMT-off artifacts that would otherwise be created using offline sample preparation. In separate SAX-RP experiments, isobaric impurities ("n+O" vs "n+S-O") that would be difficult to resolve using conventional 1D IPRP-MS approaches were characterized. While IPRP is MS compatible as an option for detection in 2D-LC, ion suppression is still a concern. Hence, desalting approaches are also being developed as alternatives [37].

Koshel et al. leveraged the flexibility of 2D-LC in order to develop a method to characterize impurities in preparations of fluorescently labeled oligonucleotides [38]. Unlike unmodified oligonucleotides, those with a fluorophore do not separate according to their length during IPRP separations using traditional TEA/HFIP mobile phases [39]. Stronger ion pairing reagents, such as hexylammonium acetate (HAA), are needed for length-driven separations, but this reagent is not nearly as MS friendly as TEA/HFIP. To address these challenges, the authors designed an IPRP-IPRP-MS system utilizing at-column dilution and with the same RP chemistries in the first and second dimensions. The mobile phases in the first and second dimensions used different ion pairing reagents, with the stronger HAA in the ^{1}D separation and the weaker, more MS-friendly TEA in the ^{2}D separation. Failed (e.g. n-1) sequences of Cy3 conjugated 25-mer oligonucleotides were separated based on size in the ^{1}D separation, followed by heartcutting of these impurities, further separation in the second dimension, and identification using MS detection. The workflow enables low nanogram level impurity detection and identification that would not have been possible using a 1D separation.

The first oligonucleotide separation using HILIC in the second dimension of a multiple heartcut 2D-LC platform was recently reported. The work utilized AEX or IPRP in the first dimension [40]. The platform was applied to impurity analysis of an ASO with a complete backbone of phosphorothioate bonds. For these samples, impurities can include shortmers, longmers, phosphodiester substitutions, or incomplete deprotection. Careful optimization of the HILIC separation using mobile phase pH and additives and column temperature was performed, which permitted detection of very low-level impurities, improving upon the sensitivity of existing IPRP-MS methods.

While the body of published 2D-LC work focused on oligonucleotides is small today, our expectation is that new methods will continue to be developed and implemented, especially in pharmaceutical labs. The use of 2D-LC in this context continues to increase the method development design space [41], which was already large considering the combination of available separation modes and unique oligonucleotide modalities that must be characterized. These emerging methods will be especially useful for determining the presence of low levels of impurities in complex samples, and as well for providing separations that are compatible with MS for mass determination. Going forward, the versatility of 2D-LC will likely make the technology a routine tool to support the development of oligonucleotide drugs.

9.6.3 OTHER EXAMPLES OF DIFFERENT 2D-LC MODES USED TO RESOLVE COELUTION FOR PHARMACEUTICAL MATERIALS

In 2008, Huidobro *et al.* performed stability studies of alprazolam tablets by comprehensive offline RP×RP using SB-CN and BEH-C18 stationary phases [42]. An SB-CN stationary phase proved to be highly complementary to the BEH-C18 stationary phase and significantly increased the number of observed peaks to 147, whereas the largest number of peaks observed in a 1D-LC separation involving either column was just 21 [42].

Zhang *et al.* assessed peak coelution in a stability sample using a 2D-LC method and a new degradant "hidden" in the tail of the API peak was found by the ^{2}D separation. This finding helped to guide improvement of the original 1D-LC method intended for use in quality control as a stability indicating method [4].

The stability of the unconjugated drug and related small molecule species in a complex ADC drug matrix at different pH levels (5, 6, and 7) was studied by SEC-RPLC [43]. A slight increase in the concentration of unconjugated drug was observed at higher pH and the formation of two drug-related impurities was observed [43]. In 2017, a 2D-LC method was validated for quality control testing purposes and it demonstrated suitable specificity [44]. In particular, a regioisomer could only be separated from the API on the ^{2}D CSH Phenyl-Hexyl column while other impurities were only separated on the ^{1}D Zorbax Eclipse XDB C-18 column [44].

2D-LC setups are commonly operated in the selective comprehensive mode (sLC×LC) to collect multiple fractions across the API main peak as it elutes from the ^{1}D column [45, 46]. Venkatramani *et al.* separated a diastereomer impurity coeluting with the API on the ^{1}D RP column using sLC×LC [47]. In a different study, the same author resolved 15 impurities coeluting with a linker drug intermediate using sLC×LC, and achieved ultra-low detection limits of 0.01% [48]. An impurity coeluting with indole species was also revealed during purity assessment by selective RP×RP separation [49]. In another study by Shackman and Kleintop, a product containing a dibasic API, a zwitterionic API, and an acidic API, was studied by online 2D-LC [50]. A 1D-LC method was developed for the drug product (DP) in order to separate 16 known impurities previously observed for the three APIs. However, a new degradant was generated during a stability study of the zwitterionic API drug substance (DS) that could not be separated using the 1D-LC method for the DP [50]. The DP method was then implemented in the first dimension of a 2D-LC method, while the 1D-LC method for the zwitterionic API DS was implemented in the second dimension. One impurity coeluting with the API in the first dimension was separated in the second dimension and found in six fractions collected across the zwitterionic API main peak [50]. Finally, in a study by Li and Lämmerhofer, the purity of the main peak for the antisense strand of a siRNA molecule was evaluated by sLC×LC [41]. An IPRP×RP method was developed and impurities were found in all four fractions collected across the main peak.

Using a unique approach to offline 2D-LC, Lee *et al.* reported the determination of peak purity and the sensitive detection of a low-level impurity that was not detected using a reference analytical method for the DS [51]. Of particular interest, the ^{1}D separation was run multiple times, each time collecting the same fraction of the ^{1}D peak and diverting it to a vial using an autosampler. After five such ^{1}D separations, the entire volume collected in the vial was injected at once into the ^{2}D column to improve the detection sensitivity of the method. Using this approach, three different regions of the main peak were targeted and an additional impurity (0.8% relative peak area) was detected and identified as a diastereomer.

The use of single heartcut and multiple heartcut 2D-LC separation modes is preferred in the pharmaceutical industry whereas comprehensive 2D-LC separation modes have been more frequently employed in academic laboratories. The need for more sophisticated instrumentation, time constraints upon the ^{2}D analysis time, system robustness, and time-consuming data treatment in the case of LC×LC separations may partially explain why the pharmaceutical industry prefers the use of heartcut approaches.

9.6.4 COLUMN SCREENING IN THE SECOND DIMENSION TO IDENTIFY COMPLEMENTARY STATIONARY PHASES

Ideally, comprehensive analysis of complex and unknown mixtures would use orthogonal separations and universal detection to ensure that no coelutions are missed, and all analytes are detected, respectively. In 2013, Zhang *et al.* developed a multiple heartcut 2D-LC setup similar to that illustrated in Figure 9.3 that enabled automated screening of selectivities for use in the second dimension. The system was equipped with six interface loops, and six columns with different chemistries in the second dimension, coupled with UV-CAD-MS for comprehensive sample profiling [4]. In this work the authors demonstrated solutions to several of the challenges illustrated in Figure 9.1:

1) Detection of a coeluting impurity using a RP-RP separation with complementary stationary phases in the two dimensions;
2) Detection of an impurity that was not retained on the ^{1}D RP column using a HILIC separation in the second dimension; and
3) Use of phosphate buffer for optimal performance in the first dimension (e.g., peak shape and UV transparency) while using MS compatible buffer in the second dimension for impurity identification by MS.

Through these examples, the authors demonstrated that both the complementary separations used in the two dimensions and coupling the separation with different detectors (i.e., CAD, UV, MS) are important for complete impurity profiling. Multiple heartcut approaches involving the use of a column selector in the second dimension to choose between different ^{2}D stationary phases have been found by other groups to be "fit for purpose" in a variety of applications [16, 20, 52]. Please note that a complementary discussion of this topic of automated screening of ^{2}D chiral columns can be found in Chapter 13 of this book.

9.6.5 USE OF MULTIPLE COMBINATIONS OF SEPARATION MECHANISMS TO SOLVE DIFFICULT COELUTION PROBLEMS

In a very different application, Yang *et al.* developed SEC-RP-MS and RP-SEC-MS methods to characterize the impurities of a high molecular weight polyethylene (PEG) intermediate (40 kDa) used for protein conjugation [53]. The heterogeneity of the intermediate follows from the polydispersity of the PEG, as well as impurities resulting from the different functionalization steps used to synthesize a PEG-maleimide from the PEG starting material [53]. The coupling of a ^{1}D RPLC separation with a ^{2}D SEC separation helped to identify a smaller impurity (11 kDa). Conversely, a SEC-RP method was developed to separate additional size variants in the first dimension with a focus on collecting all 40 kDa species together in a single fraction, and then separating species according to their different functionalities using the ^{2}D RP column [53].

When analyzing highly complex mixtures, comprehensive 2D-LC has the potential to deliver higher resolution of compounds on average compared to 1D-LC. This has been demonstrated in the context of pharmaceutical analysis by Iguiniz et al., who characterized two mixtures containing an API and related synthesis intermediates [2]. A thorough optimization of the RP×RP conditions was performed, including stationary phase chemistry and mobile phase composition and an effective peak capacity of 1,000 was achieved within a 50-min analysis time [2]. Overall, the number of compounds separated was increased about 40% by LC×LC in comparison to 1D-LC. In another study, a LC×LC method with temperature responsive LC (TRLC) in the first dimension and RPLC in the second dimension was developed for the analysis of complex drug formulations using a commercial system [54]. TRLC is based on the use of polymers that exhibit significant changes in their properties due to an environmental stimulus. In this case the hydrophobicity of poly (N-isopropylacrylamide) changed in response to a change in the column temperature. This makes the coupling of TRLC (when used in the first dimension) with conventional RPLC in the second dimension easier because TRLC can be used with a fully aqueous eluent. Successful separation of a mixture containing steroids as well as impurities present at 0.05% level from beclomethasone prepared at 2 mg/mL were presented [54]. In other work, Alexander and Ma presented three pharmaceutical applications of RPLC×RPLC separations including the analysis of a drug degradation mixture [55]. The orthogonality between the two RPLC separations was achieved by using a different mobile phase pH in order to separate an oxidation product (1%, w/w) and a photo-degradant (0.5%, w/w) that were not separated by the reference 1D-LC method. Sensitivity was suitable as demonstrated by the separation and detection of 0.05% buspirone (w/w) in 1 mg/ml caffeine [55].

9.7   RESOLVING SAMPLE MATRIX EFFECTS USING 2D-LC (FIGURE 9.1D)

9.7.1   DRUG PRODUCT MATRIX INTERFERENCE REMOVAL

The determination and characterization of unconjugated small molecule payloads in ADC products, or excipients in monoclonal antibody (mAb) formulations, is often hampered by the presence of a high concentration of proteins in the analytical sample. Removal of the protein can be achieved via solvent precipitation, but the manual approach is time consuming and may result in incomplete recovery of unconjugated payloads. Automated online 2D-LC approaches have been developed to overcome these problems, and provide accurate quantitative information to help understand drug product stability and formulation development.

Single heartcut 2D-LC methods have been developed in order to separate the small molecule species (unconjugated payload, linker and linker-payload) present in ADC samples using a ^{1}D SEC column [43, 56–58]. The small molecule fraction of the sample eluting from the SEC column is then transferred to a ^{2}D RP column where the different small molecule species can be separated and identified using MS detection.

Polysorbates are important excipients in protein drug formulations. Li et al. investigated the use of MM and cation exchange (CEX) separations in the first dimension of a heartcut 2D-LC method to separate polysorbate 20 from proteins and other excipients, and then separate the different polysorbate 20 species on a ^{2}D RP column based on their aliphatic chain length and polar head type [59]. The ^{2}D RP column was coupled to an MS instrument and the different polysorbate ester subspecies were identified. Different degradation rates were observed for different polysorbate ester species. Later, He et al. coupled hydrophobic interaction chromatography (HIC) to RPLC to characterize polysorbate 80 impurities using MS detection [60]. He et al. separated and quantified several excipients (sucrose, histidine and succinate) using a single heartcut SEC-MM separation [61]. In this case, the excipients isolated by the ^{1}D SEC separation were subsequently separated using a ^{2}D Trinity P1 MM column for various samples including mAbs, ADCs, and vaccines. In another study, histidine degradation was studied using a single heartcut SEC-HILIC separation [62]. Particular care was taken to limit solvent mismatch between the aqueous ^{1}D SEC and ^{2}D HILIC separations (i.e., use of small loop volume). Trans-urocanic acid was identified as a major histidine deamination product and quantified via stable-isotope dilution.

9.7.2 TRACE ANALYSIS

Detection sensitivity can be an issue in 2D-LC due to strong dilution of analyte zones as they proceed from the sample injector to the outlet of the ^{2}D column [63]. Detection sensitivity in comprehensive 2D-LC separation of pharmaceutical drug impurities was evaluated by Stoll *et al.* [64]. Notably, impurities could be detected at 0.05% level (relative peak area to the total area) under appropriate interface conditions. The results demonstrated the possibility of obtaining information about low-concentration compounds in complex matrices and paved the way for the quantification of low-level impurities by 2D-LC. Later, Iguiniz *et al.* studied the quantification of pharmaceutical drug impurities by comprehensive 2D-LC analysis and achieved a limit of quantification (LOQ) at 0.1% [65]. In other work, Yamamoto *et al.* reported the determination of two intermediates in the 0.25–250 ppm range using a 2D-LC method coupled with an online solid phase extraction (SPE) pre-concentration step [66].

Sometimes, a simple 2D-LC method can provide critical information about a sample that cannot be obtained using any other technology. Dai *et al.* reported the study of a trace-level degradation product from an API present at just 0.6 ppm that was enabled by 2D-LC-MS [67]. Due to the high concentration of PEG in the DP, and the microdose nature of the drug, 1D-LC-MS, nuclear magnetic resonance (NMR), and other technologies were unable to detect the degradant due to strong interference from the high concentration of PEG in the sample. By using a simple single heartcut RP-RP method, the interference was separated not only by the first dimension, but also further separated by the second dimension, as shown in Figure 9.4. The resolved degradant peak was identified by MS, which in turn revealed the cause of the degradation.

9.8 HIGH THROUGHPUT ANALYSIS

As suggested by Liu *et al.* in 2013, achiral-chiral 2D-LC separations hold the potential to be implemented in high throughput environments [15]. The blossoming of fast achiral-chiral 2D-LC methods has been recently implemented in a high throughput experimentation environment. For example, Goyon and coworkers used 2D-LC to study the yield and stereoselectivity of asymmetric reactions performed in the 96-well plate format [13]. The total analysis time per sample was 10 min, and a preferred chiral ligand family was identified by the high throughput analysis platform [13]. In other work, Barhate and coworkers demonstrated 2D-LC separations involving isocratic chiral separations with analysis times of less than 1 min in the second dimension [19]. This area is rich with opportunity for further innovation and creative applications of 2D-LC in the near future.

9.9 CONCLUDING REMARKS AND OUTLOOK

2D-LC now plays a critical role in addressing the challenges encountered in contemporary small molecule pharmaceutical analysis. Although the applications of 2D-LC in this area have increased significantly in the past decade, the broader implementation of 2D-LC in the pharmaceutical industry has not yet reached its full potential.

More widespread application of 2D-LC for everyday use will benefit from the development of "generic" methods as starting points for method development. Without such starting points, method development often takes a long time, especially for inexperienced users. In the context of the fast-paced environment of pharmaceutical laboratories, the time-consuming method development step

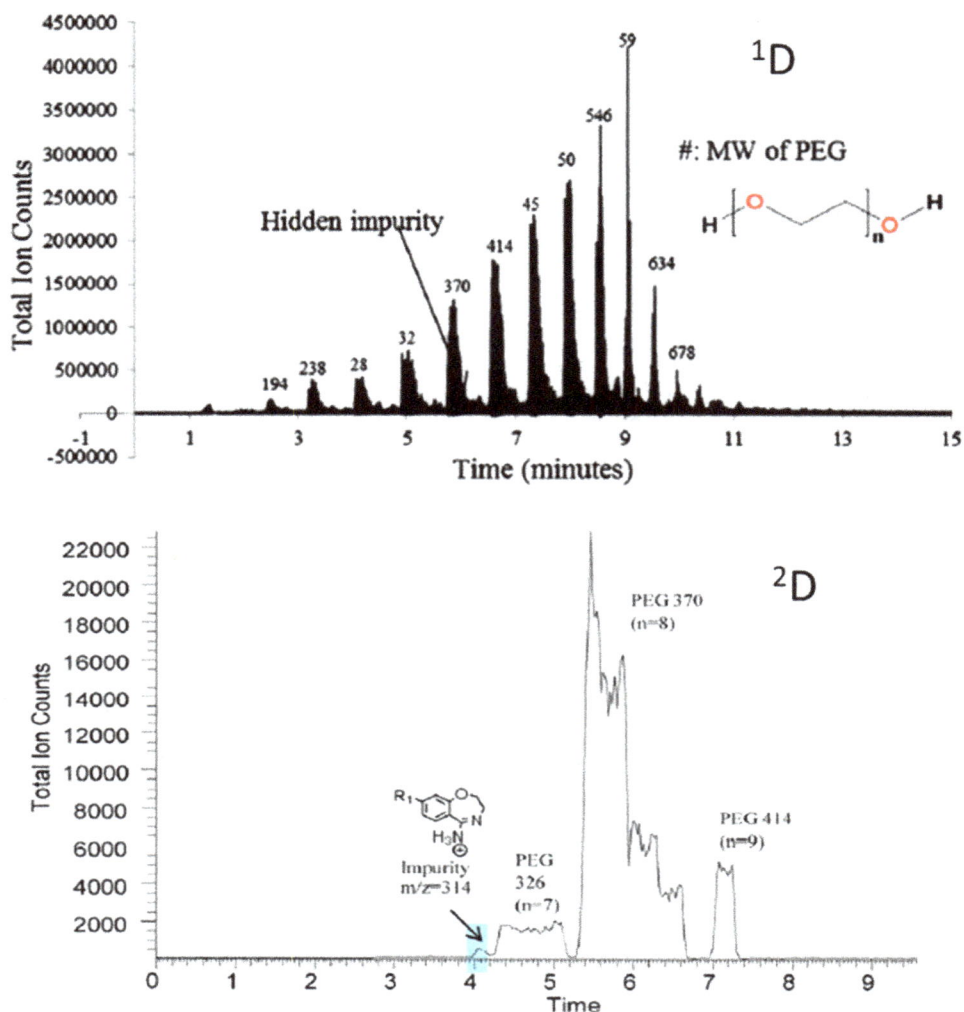

FIGURE 9.4 Identification of a degradation impurity resulting from a low concentration API (0.6 ppm) in the presence of 4% PEG using RP-RP-MS Adapted from *Journal of Pharmaceutical and Biomedical Analysis*, 137, L. Dai, G. Yeh, Y. Ran, P. Yehl, K. Zhang, Compatibility study of a parenteral microdose polyethylene glycol formulation in medical devices and identification of degradation impurity by 2D-LC/MS, 182–188, Copyright (2017), with permission from Elsevier.

frequently becomes a serious obstacle to implementation of 2D-LC to solve real problems. A platform approach that enables rapid development of 2D-LC methods for different types of molecules is essential. This also requires the instrument setup to be versatile and robust. Integrating decision-making tools into commercial software for 2D-LC would also be beneficial. Visualization of the sample flow path and simplification of data treatment would provide a solid foundation for more routine use of 2D-LC.

As a result of the rapid evolution of the pharmaceutical landscape in recent years, small molecule drugs now exhibit properties that are beyond the traditional Lipinski's rule of five [68]. New drug modalities are becoming more complex and diverse, and together with innovations in drug delivery technologies, the resulting drug products are more sophisticated and complex than they were in the past. The analytical challenges these new modalities present provide new opportunities for 2D-LC applications, and are also pushing the limits of current 2D-LC methodologies.

Finally, active collaborations between academic labs, pharmaceutical companies, and instrument vendors will certainly accelerate the broader implementation of 2D-LC.

REFERENCES

[1] K. Zhang, J. Wang, M. Tsang, L. Wigman, N. Chetwyn, Two-dimensional hplc in pharmaceutical analysis, Am. Pharm. Rev. 16 (2013) 39–44.

[2] M. Iguiniz, F. Rouvière, E. Corbel, N. Roques, S. Heinisch, Comprehensive two dimensional liquid chromatography as analytical strategy for pharmaceutical analysis, J. Chromatogr. A. 1536 (2018) 195–204. https://doi.org/10.1016/j.chroma.2017.08.070.

[3] D.R. Stoll, T.D. Maloney, Recent advances in two-dimensional liquid chromatography for pharmaceutical and biopharmaceutical analysis, LCGC N. Am. 35 (2017) 680–687.

[4] K. Zhang, Y. Li, M. Tsang, N.P. Chetwyn, Analysis of pharmaceutical impurities using multiheartcutting 2D LC coupled with UV-charged aerosol MS detection, J. Sep. Sci. 36 (2013) 2986–2992. https://doi.org/10.1002/jssc.201300493.

[5] J. Wang, G. Liu, Y. Xu, B. Zhu, Z. Wang, Separation and characterization of new components and impurities in leucomycin by multiple heart-cutting two-dimensional liquid chromatography combined with ion trap/time-of-flight mass spectrometry, Chromatographia. 82 (2019) 1333–1344. https://doi.org/10.1007/s10337-019-03754-5.

[6] J. Wang, Y. Xu, C. Wen, Z. Wang, Application of a trap-free two-dimensional liquid chromatography combined with ion trap/time-of-flight mass spectrometry for separation and characterization of impurities and isomers in cefpiramide, Anal. Chim. Acta. 992 (2017) 42–54. https://doi.org/10.1016/j.aca.2017.08.028.

[7] Z. Long, Z. Zhan, Z. Guo, Y. Li, J. Yao, F. Ji, C. Li, X. Zheng, B. Ren, T. Huang, A novel two-dimensional liquid chromatography – Mass spectrometry method for direct drug impurity identification from HPLC eluent containing ion-pairing reagent in mobile phases, Anal. Chim. Acta. 1049 (2019) 105–114. https://doi.org/10.1016/j.aca.2018.10.031.

[8] H. Luo, W. Zhong, J. Yang, P. Zhuang, F. Meng, J. Caldwell, B. Mao, C.J. Welch, 2D-LC as an on-line desalting tool allowing peptide identification directly from MS unfriendly HPLC methods, J. Pharm. Biomed. Anal. 137 (2017) 139–145. https://doi.org/10.1016/j.jpba.2016.11.012.

[9] H. Wang, S. Xie, Identification of impurities in 5-aminolevulinic acid by two-dimensional column-switching liquid chromatography coupled with linear ion trap mass spectrometry, Chromatographia. 79 (2016) 1469–1478. https://doi.org/10.1007/s10337-016-3165-2.

[10] H. Wang, T. Xu, J. Yuan, The use of online heart-cutting high-performance liquid chromatography coupled with linear ion trap mass spectrometry in the identification of impurities in vidarabine monophosphate, J. Sep. Sci. 40 (2017) 1674–1685. https://doi.org/10.1002/jssc.201601320.

[11] D. Remane, M.R. Meyer, D.K. Wissenbach, H.H. Maurer, Ion suppression and enhancement effects of co-eluting analytes in multi-analyte approaches: Systematic investigation using ultra-high-performance liquid chromatography/mass spectrometry with atmospheric-pressure chemical ionization or electrospray ionization, Rapid Commun. Mass Spectrom. RCM. 24 (2010) 3103–3108. https://doi.org/10.1002/rcm.4736.

[12] J. Lin, C. Masui, R. Lieu, K. Zhang, High-throughput experimentation: Where does mass spectrometry fit?, Current Trends in Mass Spectrometry. 17 (2019) 8–15.

[13] A. Goyon, C. Masui, L.E. Sirois, C. Han, P. Yehl, F. Gosselin, K. Zhang, Achiral–chiral two-dimensional liquid chromatography platform to support automated high-throughput experimentation in the field of drug development, Anal. Chem. 92 (2020) 15187–15193. https://doi.org/10.1021/acs.analchem.0c03754.

[14] C.J. Venkatramani, L. Wigman, K. Mistry, N. Chetwyn, Simultaneous, sequential quantitative achiral-chiral analysis by two-dimensional liquid chromatography, J. Sep. Sci. 35 (2012) 1748–1754. https://doi.org/10.1002/jssc.201200005.

[15] Q. Liu, X. Jiang, H. Zheng, W. Su, X. Chen, H. Yang, On-line two-dimensional LC: A rapid and efficient method for the determination of enantiomeric excess in reaction mixtures, J. Sep. Sci. 36 (2013)
3158–3164. https://doi.org/10.1002/jssc.201300412.

[16] R.S. Hegade, K. Chen, J.-P. Boon, M. Hellings, K. Wicht, F. Lynen, Development of an achiral-chiral
2-dimensional heart-cutting platform for enhanced pharmaceutical impurity analysis, J. Chromatogr.
A. 1628 (2020) 461425. https://doi.org/10.1016/j.chroma.2020.461425.

[17] U. Woiwode, S. Neubauer, W. Lindner, S. Buckenmaier, M. Lämmerhofer, Enantioselective multiple
heartcut two-dimensional ultra-high-performance liquid chromatography method with a Coreshell
chiral stationary phase in the second dimension for analysis of all proteinogenic amino acids in a
single run, J. Chromatogr. A. 1562 (2018) 69–77. https://doi.org/10.1016/j.chroma.2018.05.062.

[18] J.C. Barreiro, K.L. Vanzolini, Q.B. Cass, Direct injection of native aqueous matrices by achiral-
chiral chromatography ion trap mass spectrometry for simultaneous quantification of pantoprazole
and lansoprazole enantiomers fractions, J. Chromatogr. A. 1218 (2011) 2865–2870. https://doi.org/
10.1016/j.chroma.2011.02.064.

[19] C.L. Barhate, E.L. Regalado, N.D. Contrella, J. Lee, J. Jo, A.A. Makarov, D.W. Armstrong, C.J. Welch,
Ultrafast chiral chromatography as the second dimension in two-dimensional liquid chromatography
experiments, Anal. Chem. 89 (2017) 3545–3553. https://doi.org/10.1021/acs.analchem.6b04834.

[20] J. Lin, C. Tsang, R. Lieu, K. Zhang, Fast chiral and achiral profiling of compounds with multiple chiral
centers by a versatile two-dimensional multicolumn liquid chromatography (LC–mLC) approach, J.
Chromatogr. A. (2020) 460987. https://doi.org/10.1016/j.chroma.2020.460987.

[21] K.H. Kim, H.W. Yun, H.J. Kim, H.J. Park, P.W. Choi, Coupled column chromatography in chiral separation of salmeterol, Arch. Pharm. Res. 21 (1998) 212–216. https://doi.org/10.1007/BF02974030.

[22] C.J. Venkatramani, M. Al-Sayah, G. Li, M. Goel, J. Girotti, L. Zang, L. Wigman, P. Yehl, N. Chetwyn,
Simultaneous achiral-chiral analysis of pharmaceutical compounds using two-dimensional reversed
phase liquid chromatography-supercritical fluid chromatography, Talanta. 148 (2016) 548–555.
https://doi.org/10.1016/j.talanta.2015.10.054.

[23] E.L. Regalado, J.A. Schariter, C.J. Welch, Investigation of two-dimensional high performance liquid
chromatography approaches for reversed phase resolution of warfarin and hydroxywarfarin isomers,
J. Chromatogr. A. 1363 (2014) 200–206. https://doi.org/10.1016/j.chroma.2014.08.025.

[24] F. Xu, Y. Xu, G. Liu, M. Zhang, S. Qiang, J. Kang, Separation of twelve posaconazole related
stereoisomers by multiple heart-cutting chiral-chiral two-dimensional liquid chromatography, J.
Chromatogr. A. 1618 (2020) 460845. https://doi.org/10.1016/j.chroma.2019.460845.

[25] D.R. Stoll, X. Li, X. Wang, P.W. Carr, S.E.G. Porter, S.C. Rutan, Fast, comprehensive two-
dimensional liquid chromatography, J. Chromatogr. A. 1168 (2007) 3–43. https://doi.org/10.1016/
j.chroma.2007.08.054.

[26] D.R. Stoll, J.D. Cohen, P.W. Carr, Fast, comprehensive online two-dimensional high performance
liquid chromatography through the use of high temperature ultra-fast gradient elution reversed-
phase liquid chromatography, J. Chromatogr. A. 1122 (2006) 123–137. https://doi.org/10.1016/j.chr
oma.2006.04.058.

[27] A.P. Schellinger, D.R. Stoll, P.W. Carr, High speed gradient elution reversed-phase liquid chromatography, J. Chromatogr. A. 1064 (2005) 143–156. https://doi.org/10.1016/j.chroma.2004.12.017.

[28] D.M. Makey, V. Shchurik, H. Wang, H.R. Lhotka, D.R. Stoll, A. Vazhentsev, I. Mangion, E.L.
Regalado, I.A.H. Ahmad, Mapping the separation landscape in two-dimensional liquid chromatography: Blueprints for efficient analysis and purification of pharmaceuticals enabled by computer-
assisted modeling, Anal. Chem. 93 (2021) 964–972. https://doi.org/10.1021/acs.analchem.0c03680.

[29] Z. Lian, N. Wang, Y. Tian, L. Huang, Characterization of synthetic peptide therapeutics using liquid
chromatography – mass spectrometry: Challenges, solutions, pitfalls, and future perspectives, J. Am.
Soc. Mass Spectrom. 32 (2021) 1852–1860. https://doi.org/10.1021/jasms.0c00479.

[30] R. Karongo, T. Ikegami, D.R. Stoll, M. Lämmerhofer, A selective comprehensive reversed-
phasexreversed-phase 2D-liquid chromatography approach with multiple complementary detectors as
advanced generic method for the quality control of synthetic and therapeutic peptides, J. Chromatogr.
A. 1627 (2020) 461430. https://doi.org/10.1016/j.chroma.2020.461430.

[31] K. Takakura, A. Kawamura, Y. Torisu, S. Koido, N. Yahagi, M. Saruta, Clinical Potential of
Oligonucleotide Therapeutics against Pancreatic Cancer (2019). https://doi.org/10.20944/preprint
s201905.0239.v1.

[32] H. Babačić, A. Mehta, O. Merkel, B. Schoser, CRISPR-cas gene-editing as plausible treatment of neuromuscular and nucleotide-repeat-expansion diseases: A systematic review, PloS One. 14 (2019). https://doi.org/10.1371/journal.pone.0212198.

[33] A. Goyon, P. Yehl, K. Zhang, Characterization of therapeutic oligonucleotides by liquid chromatography, J. Pharm. Biomed. Anal. 182 (2020). https://doi.org/10.1016/j.jpba.2020.113105.

[34] P. Álvarez Porebski, F. Lynen, Combining liquid chromatography with multiplexed capillary gel electrophoresis for offline comprehensive analysis of complex oligonucleotide samples, J. Chromatogr. A. 1336 (2014) 87–93. https://doi.org/10.1016/j.chroma.2014.02.007.

[35] Q. Li, F. Lynen, J. Wang, H. Li, G. Xu, P. Sandra, Comprehensive hydrophilic interaction and ion-pair reversed-phase liquid chromatography for analysis of di- to deca-oligonucleotides, J. Chromatogr. A. 1255 (2012) 237–243. https://doi.org/10.1016/j.chroma.2011.11.062.

[36] S.G. Roussis, I. Cedillo, C. Rentel, Two-dimensional liquid chromatography-mass spectrometry for the characterization of modified oligonucleotide impurities, Anal. Biochem. 556 (2018) 45–52. https://doi.org/10.1016/j.ab.2018.06.019.

[37] F. Li, X. Su, S. Bäurer, M. Lämmerhofer, Multiple heart-cutting mixed-mode chromatography-reversed-phase 2D-liquid chromatography method for separation and mass spectrometric characterization of synthetic oligonucleotides, J. Chromatogr. A. 1625 (2020). https://doi.org/10.1016/j.chroma.2020.461338.

[38] B. Koshel, R. Birdsall, W. Chen, Two-dimensional liquid chromatography coupled to mass spectrometry for impurity analysis of dye-conjugated oligonucleotides, J. Chromatogr. B. 1137 (2020). https://doi.org/10.1016/j.jchromb.2019.121906.

[39] A. Apffel, J.A. Chakel, S. Fischer, K. Lichtenwalter, W.S. Hancock, Analysis of oligonucleotides by hplc-electrospray ionization mass spectrometry, Anal. Chem. 69 (1997) 1320–1325. https://doi.org/10.1021/ac960916h.

[40] A. Goyon, K. Zhang, Characterization of antisense oligonucleotide impurities by ion-pairing reversed-phase and anion exchange chromatography coupled to hydrophilic interaction liquid chromatography/mass spectrometry using a versatile two-dimensional liquid chromatography setup, Anal. Chem. 92 (2020) 5944–5951. https://doi.org/10.1021/acs.analchem.0c00114.

[41] F. Li, M. Lämmerhofer, Impurity profiling of siRNA by two-dimensional liquid chromatography-mass spectrometry with quinine carbamate anion-exchanger and ion-pair reversed-phase chromatography, J. Chromatogr. A. 1643 (2021). https://doi.org/10.1016/j.chroma.2021.462065.

[42] A.L. Huidobro, P. Pruim, P. Schoenmakers, C. Barbas, Ultra rapid liquid chromatography as second dimension in a comprehensive two-dimensional method for the screening of pharmaceutical samples in stability and stress studies, J. Chromatogr. A. 1190 (2008) 182–190. https://doi.org/10.1016/j.chroma.2008.02.114.

[43] Y. Li, C. Gu, J. Gruenhagen, K. Zhang, P. Yehl, N.P. Chetwyn, C.D. Medley, A size exclusion-reversed phase two dimensional-liquid chromatography methodology for stability and small molecule related species in antibody drug conjugates, J. Chromatogr. A. 1393 (2015) 81–88. https://doi.org/10.1016/j.chroma.2015.03.027.

[44] S.H. Yang, J. Wang, K. Zhang, Validation of a two-dimensional liquid chromatography method for quality control testing of pharmaceutical materials, J. Chromatogr. A. 1492 (2017) 89–97. https://doi.org/10.1016/j.chroma.2017.02.074.

[45] S.R. Groskreutz, M.M. Swenson, L.B. Secor, D.R. Stoll, Selective comprehensive multi-dimensional separation for resolution enhancement in high performance liquid chromatography. Part I: Principles and instrumentation, J. Chromatogr. A. 1228 (2012) 31–40. https://doi.org/10.1016/j.chroma.2011.06.035.

[46] S.R. Groskreutz, M.M. Swenson, L.B. Secor, D.R. Stoll, Selective comprehensive multidimensional separation for resolution enhancement in high performance liquid chromatography. Part II: Applications, J. Chromatogr. A. 1228 (2012) 41–50. https://doi.org/10.1016/j.chroma.2011.06.038.

[47] C.J. Venkatramani, J. Girotti, L. Wigman, N. Chetwyn, Assessing stability-indicating methods for coelution by two-dimensional liquid chromatography with mass spectrometric detection, J. Sep. Sci. 37 (2014) 3214–3225. https://doi.org/10.1002/jssc.201400590.

[48] C.J. Venkatramani, S.R. Huang, M. Al-Sayah, I. Patel, L. Wigman, High-resolution two-dimensional liquid chromatography analysis of key linker drug intermediate used in antibody drug conjugates, J. Chromatogr. A. 1521 (2017) 63–72. https://doi.org/10.1016/j.chroma.2017.09.022.

[49] I.A.H. Ahmad, A. Blasko, A. Clarke, S. Fakih, Two-dimensional liquid chromatography (2D-LC) in pharmaceutical analysis: applications beyond increasing peak capacity, Chromatographia. 81 (2018) 401–418. https://doi.org/10.1007/s10337-018-3474-8.

[50] J.G. Shackman, B.L. Kleintop, Peak purity assessment in a triple-active fixed-dose combination drug product related substances method using a commercial two-dimensional liquid chromatography system, J. Sep. Sci. 37 (2014) 2688–2695. https://doi.org/10.1002/jssc.201400515.

[51] C. Lee, J. Zang, J. Cuff, N. McGachy, T.K. Natishan, C.J. Welch, R. Helmy , F. Bernardoni, Application of heart-cutting 2D-LC for the determination of peak purity for a chiral pharmaceutical compound by HPLC, Chromatographia. 76 (2013) 5–11. https://doi.org/10.1007/s10337-012-2367-5.

[52] H. Wang, H.R. Lhotka, R. Bennett, M. Potapenko, C.J. Pickens, B.F. Mann, I.A.H. Ahmad, E.L. Regalado, Introducing online multicolumn two-dimensional liquid chromatography screening for facile selection of stationary and mobile phase conditions in both dimensions, J. Chromatogr. A. (2020). https://doi.org/10.1016/j.chroma.2020.460895.

[53] S.H. Yang, B. Chen, J. Wang, K. Zhang, Characterization of high molecular weight multi-arm functionalized peg–maleimide for protein conjugation by charge-reduction mass spectrometry coupled to two-dimensional liquid chromatography, Anal. Chem. 92 (2020) 8584–8590. https://doi.org/10.1021/acs.analchem.0c01567.

[54] K. Wicht, M. Baert, A. Kajtazi, S. Schipperges, N. von Doehren, G. Desmet, A. de Villiers, F. Lynen, Pharmaceutical impurity analysis by comprehensive two-dimensional temperature responsive×reversed phase liquid chromatography, J. Chromatogr. A. 1630 (2020). https://doi.org/10.1016/j.chroma.2020.461561.

[55] A.J. Alexander, L. Ma, Comprehensive two-dimensional liquid chromatography separations of pharmaceutical samples using dual Fused-Core columns in the 2nd dimension, J. Chromatogr. A. 1216 (2009) 1338–1345. https://doi.org/10.1016/j.chroma.2008.12.063.

[56] Y. Li, C. Stella, L. Zheng, C. Bechtel, J. Gruenhagen, F. Jacobson, C.D. Medley, Investigation of low recovery in the free drug assay for antibody drug conjugates by size exclusion-reversed phase two dimensional-liquid chromatography, J. Chromatogr. B Analyt. Technol. Biomed. Life. Sci. 1032 (2016) 112–118. https://doi.org/10.1016/j.jchromb.2016.05.011.

[57] A. Goyon, L. Sciascera, A. Clarke, D. Guillarme, R. Pell, Extending the limits of size exclusion chromatography: Simultaneous separation of free payloads and related species from antibody drug conjugates and their aggregates, J. Chromatogr. A. 1539 (2018) 19–29. https://doi.org/10.1016/j.chroma.2018.01.039.

[58] R.E. Birdsall, S.M. McCarthy, M.C. Janin-Bussat, M. Perez, J.-F. Haeuw, W. Chen, A. Beck, A sensitive multidimensional method for the detection, characterization, and quantification of trace free drug species in antibody-drug conjugate samples using mass spectral detection, MAbs. 8 (2016) 306–317. https://doi.org/10.1080/19420862.2015.1116659.

[59] Y. Li, D. Hewitt, Y.K. Lentz, J.A. Ji, T.Y. Zhang, K. Zhang, Characterization and stability study of polysorbate 20 in therapeutic monoclonal antibody formulation by multidimensional ultrahigh-performance liquid chromatography–charged aerosol detection–mass spectrometry, Anal. Chem. 86 (2014) 5150–5157. https://doi.org/10.1021/ac5009628.

[60] Y. He, P. Brown, M.R. Bailey Piatchek, J.A. Carroll, M.T. Jones, On-line coupling of hydrophobic interaction column with reverse phase column-charged aerosol detector/mass spectrometer to characterize polysorbates in therapeutic protein formulations, J. Chromatogr. A. 1586 (2019) 72–81. https://doi.org/10.1016/j.chroma.2018.11.080.

[61] Y. He, O.V. Friese, M.R. Schlittler, Q. Wang, X. Yang, L.A. Bass, M.T. Jones, On-line coupling of size exclusion chromatography with mixed-mode liquid chromatography for comprehensive profiling of biopharmaceutical drug product, J. Chromatogr. A. 1262 (2012) 122–129. https://doi.org/10.1016/j.chroma.2012.09.012.

[62] C. Wang, S. Chen, J.A. Brailsford, A.P. Yamniuk, A.A. Tymiak, Y. Zhang, Characterization and quantification of histidine degradation in therapeutic protein formulations by size exclusion-hydrophilic interaction two dimensional-liquid chromatography with stable-isotope labeling mass spectrometry, J. Chromatogr. A. 1426 (2015) 133–139. https://doi.org/10.1016/j.chroma.2015.11.065.

[63] M.R. Schure, Limit of detection, dilution factors, and technique compatibility in multidimensional separations utilizing chromatography, capillary electrophoresis, and field-flow fractionation, Anal. Chem. 71 (1999) 1645–1657. https://doi.org/10.1021/ac981128q.

[64] D.R. Stoll, E.S. Talus, D.C. Harmes, K. Zhang, Evaluation of detection sensitivity in comprehensive two-dimensional liquid chromatography separations of an active pharmaceutical ingredient and its degradants, Anal. Bioanal. Chem. 407 (2015) 265–277. https://doi.org/10.1007/s00216-014-8036-9.

[65] M. Iguiniz, E. Corbel, N. Roques, S. Heinisch, Quantitative aspects in on-line comprehensive two-dimensional liquid chromatography for pharmaceutical applications, Talanta. 195 (2019) 272–280. https://doi.org/10.1016/j.talanta.2018.11.030.

[66] E. Yamamoto, J. Niijima, N. Asakawa, Selective determination of potential impurities in an active pharmaceutical ingredient using HPLC-SPE-HPLC, J. Pharm. Biomed. Anal. 84 (2013) 41–47. https://doi.org/10.1016/j.jpba.2013.05.033.

[67] L. Dai, G.K. Yeh, Y. Ran, P. Yehl, K. Zhang, Compatibility study of a parenteral microdose poly-ethylene glycol formulation in medical devices and identification of degradation impurity by 2D-LC/MS, J. Pharm. Biomed. Anal. 137 (2017) 182–188. https://doi.org/10.1016/j.jpba.2017.01.036.

[68] C.A. Lipinski, F. Lombardo, B.W. Dominy, P.J. Feeney, Experimental and computational approaches to estimate solubility and permeability in drug discovery and development settings, Adv. Drug Deliv. Rev. 23 (1997) 3–25. https://doi.org/10.1016/S0169-409X(96)00423-1.

10 Application of Two-Dimensional Liquid Chromatography to Analysis of Biologics

Zachary Dunn, Douglas D. Richardson, Gregory O. Staples, and Dwight R. Stoll

CONTENTS

10.1 INTRODUCTION TO CHARACTERIZATION OF BIOLOGICS

Biologics, sometimes referred to as biotherapeutics, have continued to grow as primary treatment options across a wide range of disease areas [1]. This growth has been focused on both cancer and non-cancer treatment options, including those for COVID-19 [2]. Many biotherapeutics are complex proteins produced recombinantly from host organisms such as mammalian cells, *E.coli*, or yeast cells that have been genetically modified to express the target protein of interest. Production of biotherapeutics consists of cell line expansion, upstream cell culture or fermentation followed by downstream purification and formulation. Historically, production has occurred in batch or fed-batch processes with separate, disconnected unit operations across the upstream and downstream workflows. These fed-batch processes have typically occurred in stainless-steel bioreactors at the scale of 500–2000 L for clinical development, and ≥15,000 L for commercial manufacturing. Advances in both host organism productivity, chemically defined media, continuous capture purification, and single-use technologies have resulted in a transition toward intensified processes that enable connected continuous manufacturing at much smaller scales compared to batch production. The transition toward novel manufacturing approaches for biotherapeutics has in turn created opportunities for more modern, novel analytical tools and control strategies. Momentum for investments in future manufacturing processes, along with regulatory encouragement to modernize, is enabling growth and investments in process analytical technology (PAT) that will ultimately enable advanced process control for manufacturing of biotherapeutics.

The most popular mammalian host for biotherapeutic production is Chinese hamster ovary (CHO) cells. CHO-produced monoclonal antibodies (mAbs) are heterogenous mixtures of the target protein (i.e., drug substance) with various post translational modifications (PTMs). This mixture of molecules includes a range of charge variants including acidic and basic species, all having affinity for the therapeutic target. The complexity of these protein mixtures poses a challenge during characterization of the structure and function of these biotherapeutics and their multiple critical quality attributes (CQAs). Following final sequence selection during drug development (i.e., the primary amino acid sequence of the protein), opportunities to control the CQAs of biotherapeutics are primarily limited to the upstream cell culture where the target protein is expressed from the host. Purity of the harvested protein can be increased throughout the downstream purification process, but at the cost of reduction in the overall yield of the target protein. Additionally, host- and process-related impurities present in the protein product make characterization of the protein even more challenging. Nevertheless, comprehensive characterization of these impurities is critical for control of the processes that lead to final formulated drug substance and drug product. Identification of levers for CQAs that enable active control and optimization will enable expanded applications of advanced process control during manufacturing.

Figure 10.1 illustrates the chemistry manufacturing and control (CMC) analytical toolkit for biologics, which is constantly expanding to enable better characterization of the structure-function relationship of biotherapeutics. A typical certificate of analysis (CoA) for biotherapeutics includes approximately 25 assays for both the bulk drug substance (DS) as well as the drug product (DP), many of which involve a separation component. This highlights the testing burden on quality control (QC) labs responsible for analytical results needed for product release and evaluation of product stability. Clearly there are opportunities to modernize the analytical toolkit for biotherapeutics. Currently the toolkit addresses five categories associated with biotherapeutic quality attributes, including primary structure, secondary/tertiary structure, quaternary structure, impurities, and biological potency. Common biotherapeutic CQAs measured by these separation methods include titer (concentration), aggregation, fragmentation, charge heterogeneity, glycosylation, and oxidation. These CQAs are part of the primary and quaternary structure categories and monitored with conventional one-dimensional liquid chromatographic separation techniques. These techniques include Protein A affinity (ProA), size exclusion (SEC), ion exchange (IEX), reversed-phase (RP), hydrophobic interaction (HIC), and hydrophilic interaction (HILIC) liquid chromatography, typically coupled with ultraviolet (UV), florescence, and/or mass spectrometric detection. These purity assays can be applied to upstream and downstream intermediates sampled from biotherapeutic production processes to assess product quality during production. Typically, product quality comparisons for mAbs are done following Protein A purification, and in many cases are automated at small scale to enable more rapid process development. Due to the heterogeneous nature of the biotherapeutic, complexity of the sample matrices encountered in intermediate process samples, and the limited resolving power of these 1D-LC methods, each assay is typically only utilized to measure a single CQA per analysis. Progress has been made to leverage 2D-LC to both increase the number of CQAs measured per analysis, and enable direct measurement of titer and purity of the mAb using clarified cell culture samples pulled throughout the production process. Additional investments in automation workflows are being made to expand this concept to three-dimensional analyses and beyond. 2D-LC provides

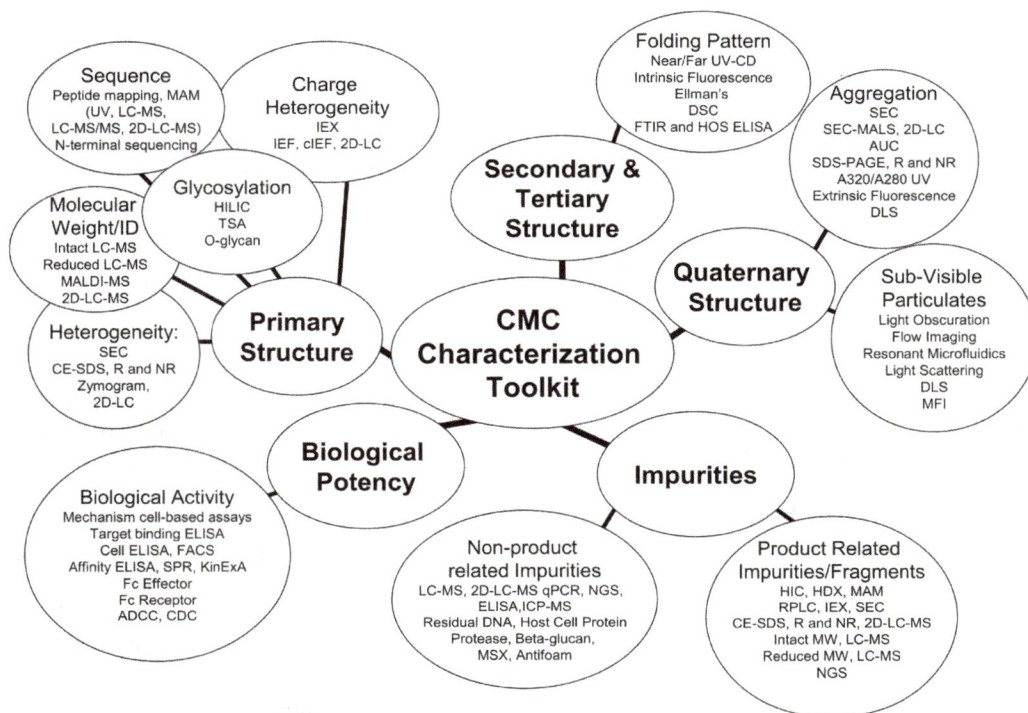

FIGURE 10.1 Assays used for CMC characterization of biologics. Please see the glossary for definitions of acroynms.

opportunities to both reduce the complexity of the sample ahead of analysis, and to increase the number of CQAs measured within a single experiment; these benefits will motivate continued development of multi-dimensional separations for characterization of biotherapeutics.

10.2 CHARACTERIZATION OF THERAPEUTIC MONOCLONAL ANTIBODIES AND RELATED MOLECULES

10.2.1 INTACT AND SUBUNIT LEVEL

10.2.1.1 Determination of Titer and Aggregation

Methods for the determination of titer and aggregation are two of the most commonly used assays for biotherapeutic antibodies. Aggregation is a CQA for mAbs that has been implicated in safety because of its effect on immunogenicity. The most common way to measure aggregation of mAbs is by SEC, which nominally separates molecules by their size. However, typical SEC separations require that the target mAb is first separated from the contaminating culture material found in the bioreactor and harvested cell culture fluid (HCCF) samples. This means that the mAb must be purified in some way before SEC analysis. Protein A affinity chromatography is very effective for this purification due to its ability to selectively bind the target mAb. At the same time, the concentration of the mAb can be determined during elution of the mAb from the ProA column, usually using UV detection.

Multiple attempts have been made to combine the ProA and SEC assays, starting with connecting a ProA and SEC column in series [3, 4]. While these experiments had limited success, it wasn't until more recent advances in 2D-LC technology that online ProA-SEC methods (i.e., truly two-dimensional) were developed. Sandra *et al.* were the first to use heartcut 2D-LC to reliably determine titer and aggregation using a ProA-SEC separation. While the aggregation values obtained from the ^{2}D SEC separations of the 2D-LC method were correlated with 1D SEC results, the absolute values were slightly different. Building on this work, Dunn *et al.* found that some mAb is lost when fractions of ^{1}D effluent are transferred to the ^{2}D SEC, which is the suspected cause of the observed difference between 1D SEC and 2D ProA-SEC data [5]. To prevent this loss, Wang *et al.* used a capillary HIC column between dimensions instead of an open tubular loop, and the resulting ProA-SEC data matched the 1D SEC data [6]. Advances in understanding from these ProA-SEC studies can be used to develop other quantitative methods with a ^{1}D ProA separation that serves as a cleanup step to multiple possible ^{2}D separations [7].

10.2.1.2   Determination of Charge Variants and Aggregation

Since mAbs that are produced in bioreactors are complex heterogeneous mixtures, peaks observed in a 1D separation are often sub-mixtures of the sample, where the coeluting species only differ by small modifications. In these cases, a 2D separation can help elucidate relationships between analytes that coelute in the first dimension. One example of this is the relationship between charge variants separated by IEX and aggregation measured by SEC. Zhi *et al.* developed LC-LC methods for CEX-SEC, AEX-SEC, and SEC-IEX to examine this relationship [8]. By first isolating certain fragments by AEX or CEX, they were able to show that for their particular mAb, species that eluted later in both the AEX and CEX separations exhibited a higher level of aggregation than species that eluted earlier. Since this study used a salt gradient for elution in both the AEX and CEX separations, it was hypothesized that the aggregates eluted later because they are more hydrophobic and would have greater interactions with the stationary phase. This type of interaction has been previously noted in the literature [9]. Additionally, by reversing the order of the separations (i.e., SEC-CEX rather than CEX-SEC), Zhi *et al.* were able to show that higher order oligomers and dimers both eluted later in AEX and CEX than monomers, which confirmed the results of their first experiments [8].

While this study is a great example of using 2D-LC to relate two orthogonal properties of mAbs, there is still considerable research area that remains unexplored. In typical analyses of biopharmaceuticals, AEX and CEX are performed with elution facilitated by gradients in both pH and salt concentration so that proteins elute based on their pI and hydrophobicity. Additional studies are needed to test whether the relationship between sizes and charges of the variants discussed above holds under conditions where simultaneous gradients in pH and salt concentration are used for elution from the AEX and CEX columns.

10.2.1.3   Determination of Charge Variants and Subunits

To further characterize the charge variants separated during ion-exchange chromatography, MS could be used to determine the masses of different species. However, the mobile phases typically used for CEX and AEX contain non-volatile salts, which are not MS compatible. This incompatibility can be overcome by adding a second dimension of separation using a RP column to separate the non-volatile salts from the proteins of interest before elution into the MS. The first applications of this approach using online 2D-LC were the heartcut methods developed by Alvarez *et al.* [10]. However, multiple heartcut methods can only transfer a limited amount of material to the second dimension for further analysis. Sorensen *et al.* expanded upon this by developing a comprehensive CEX×RP method to compare biosimilars of cetuximab, trastuzumab, and infliximab that were digested using the *IdeS* enzyme for subunit analysis [11]. By combining CEX and RP separations the method was able to separate species based on differences in their glycosylation, amino acid composition, and post-translational modifications. Additionally, the RP separation served the desalting role needed to enable direct mass determination by MS and facilitate identification of each peak. Altogether, this method was able to assess the similarity of originator and biosimilar samples that included changes in amino acid sequence, glycosylation, and post-translational modifications with a total analysis time under one hour for each sample [11].

10.2.1.4 Determination of Glycoforms and Subunits

While the CEX×RP method discussed above was able to separate some subunit glycoforms, many species still coeluted. Hydrophilic interaction liquid chromatography (HILIC) is a separation mode that can effectively separate glycoforms of mAb subunits that differ by as little as a single sugar. However, this method by itself has two limitations: the multiple subunits resulting from an *IdeS* digestion will still coelute, and the TFA used to improve the efficiency of these HILIC separations will also significantly reduce MS signal. Both of these drawbacks can be addressed by adding a second dimension RP separation, running with a MS-friendly formic acid mobile phase. One challenge encountered in coupling HILIC and RP separations is that the ACN-rich HILIC mobile phase can lead to significant breakthrough of protein peaks in the ^{2}D RP separation (see Sections 4.4.4 and 5.2.1). Using active solvent modulation (ASM) Stoll *et al.* were able to overcome the mobile phase mismatch problem and develop a comprehensive HILIC×RP-MS method to characterize the different glycoforms of multiple mAbs [12]. The RP separation was able to separate glycosylated Fc/2 and Fd subunits that coeluted during the ^{1}D HILIC separation and the TFA was separated from protein analytes by the ^{2}D RP column prior to elution into the MS. This method is a convincing example of adding a ^{2}D RP separation to an existing separation (HIILIC in this case) to not only improve the chromatographic separation, but also facilitate acquisition of MS data. Compared to the previous example of CEX×RP-MS, this HILIC×RP-MS was able to separate additional peaks and glycoforms, albeit with a doubling of the analysis time.

10.2.1.5 Measuring Bispecific Antibody Mispairing

Bispecific antibodies (bsAbs) that target two antigens add another level of complexity to the characterization of these materials. Unlike a mAb that contains two sets of identical heavy chains (HC) and light chains (LC), bsAbs are composed of two distinct heavy chains and light chains in a single molecule, such that the resulting molecule has two different binding sites. It is important for the function of these molecules that each HC is paired with its corresponding LC, which means analytical methods are needed that are capable of detecting incorrect pairings. As shown by Wang *et al.*, the different forms of pairings can be separated using HIC, but the identity of these peaks must be determined using an orthogonal method, such as MS [13]. However, HIC relies on very high ionic strength mobile phases (e.g., > 1 M sodium sulfate), and thus cannot be directly coupled with MS.

Much like the HILIC×RP-MS example discussed in the preceding section, Wang *et al.* demonstrated the utility of adding a ^{2}D RP separation to a ^{1}D HIC separation, followed by MS detection [13]. Again, the ^{2}D RP separation serves both to desalt fractions of ^{1}D effluent, and to separate protein species that coelute in the first dimension. Using the HIC-RP method for analysis of non-reduced bsAbs, they were able to identify which peak in the ^{1}D HIC chromatogram corresponded to the correct pairing of heavy and light chains. Additionally, a similar experiment was performed that included a disulfide bond reduction step after loading protein into the ^{2}D RP column, but before elution into the MS. The results of this experiment helped identify which subunits were fragmented or degraded. This is a great example of performing a chemical transformation in between the first and second dimensions to obtain the most information possible in a single analysis (see Section 4.4.11).

10.2.2 Peptide level

A common way to identify site-specific modifications and amino acid substitutions in proteins is to use bottom-up proteomics. In this approach the protein sample is first enzymatically digested into peptide fragments before analysis by LC-MS. For proteins as large as mAbs, this leads to dozens

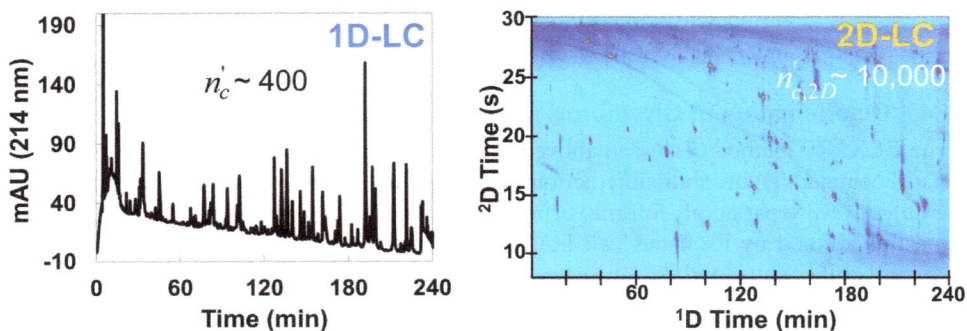

FIGURE 10.2 Comparison of 1D-LC and LC×LC separations of a tryptic digest of a mAb, showing the 25-fold improvement in peak capacity in the same analysis time. The method development process for this separation is discussed in Section 7.4. Other conditions are described in [15]. Reprinted from *Journal of Chromatography, B*, 1134–1135, D. Stoll, H. Lhotka, D.C. Harmes, B. Madigan, J. Hsiao, G. Staples, High resolution two-dimensional liquid chromatography coupled with mass spectrometry for robust and sensitive characterization of therapeutic antibodies at the peptide level, In Press, Copyright (2019), with permission from Elsevier.

of unique peptide sequences, and each sequence can be present in multiple forms due to PTMs, glycosylation, degradation, and so on, leading to extremely complex mixtures. Sarrut *et al.* showed that an optimized comprehensive RP(high pH)×RP(low pH) method has significantly greater peak capacity than a 1D RP method, especially for analysis times that are greater than 60 min [14]. The 2D RP×RP method capitalizes on the fact that amino acids are zwitterionic and change charge states at different pH, thus the combination of high pH RP in the first dimension and low pH RP in the second dimension provides good complementarity even though both dimensions use chemically similar stationary phases and mobile phase organic solvents. Sandra *et al.* used a 75-min RP×RP method to separate and identify a single amino acid difference (substitution) in the primary sequences of tocilizumab produced in different cell lines, highlighting the resolving power of this type of method [7]. Additionally, this method used a shifting gradient in the second dimension, which allowed for better utilization of the entire separation space. Stoll *et al.* continued to optimize method parameters of the RP×RP method, which included using ASM to overcome mobile phase mismatch problems that would otherwise cause poor peak shape and breakthrough in the ^{2}D RP separation [15]. This allowed for the use of longer ^{1}D columns and larger fractions for injection into the second dimension, which lead to higher peak capacities and detection sensitivities, respectively. Altogether the optimized conditions used in this work led to separations that have an effective peak capacity of 10,000 within an analysis time of 4 hours. Figure 10.2 shows a comparison of 1D- and 2D-LC separations of the same tryptic digest of a mAb, which makes clear the dramatic difference between the highly crowded 1D separation and the 2D separation with much better resolution and open expanses of baseline between peaks. The improved resolving power translates to less coelution of peptides entering the MS, increasing the ability to detect modified peptides present at low concentrations.

10.2.3 ONLINE DIGESTION OF MABS

As discussed in Section 10.2.1, analysis of mAbs at the intact and subunit level can be extremely powerful. Recently, groups have shown it is possible to isolate single peaks from a ^{1}D separation and perform the necessary sample preparation online, mid-method, before a final ^{2}D RP separation and MS detection [16–20]. This workflow, illustrated in a generalized way in Figure 10.3, uses a

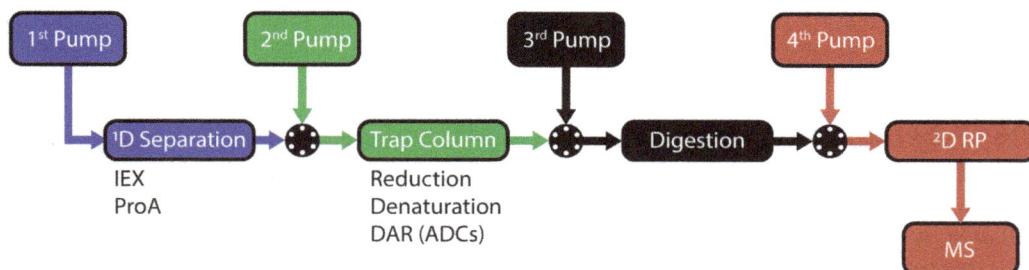

FIGURE 10.3 Generalized illustration of the multi-dimensional approach used for fully automated, online characterization of mAbs. Different separation selectivities can be used in the first stage, depending on the property of the molecule being interogated. The second stage can be used for online reduction and/or denaturation of the protein. The third stage is typically where online digestion of the protein occurs using an immobilized trypsin column. The final stage of separation usually involves a RP separation followed by MS detection.

fully automated system involving four columns and four pumps to replace the traditional workflow where protein reduction, denaturation, and digestion steps are all carried out offline, and often with considerable manual sample manipulation. The first example of this, published by Gstöttner *et al.*, used a first dimension IEX separation to separate mAb species by charge. Fractions of ^{1}D IEX peaks were transferred to a RP column where they were reduced and denatured online by flowing the reducing and denaturing agents through the column. The reduced and denatured protein was then eluted from the RP column and into an immobilized trypsin column for digestion of the protein. The outlet of this reactor was serially connected to a second RP column, which provides a second dimension of separation that is complementary to the ^{1}D IEX separation. Elution from this ^{2}D RP column was performed like a typical 1D peptide map for detection of individual amino acid modifications [16]. Once the entire method is completed for one fraction of ^{1}D effluent, another fraction collected from the ^{1}D separation can be subjected to all of the subsequent steps and repeated until all of the fractions of interest from the ^{1}D separation have been characterized. The traditional workflow to identify these species would require multiple separations to collect enough of the same fraction to use in an independent RP peptide mapping experiment, which typically takes weeks of time and significantly more sample. This online methodology described by Gstöttner produces comparable results, but in significantly less time of about 90 min. for each ^{1}D fraction studied [19]. We note here that this workflow is described as 4D-LC in the literature, but these are not 4D separations as defined by Giddings [21], primarily because the second (reducing and denaturing) and third (digestion) steps in this case do not add any actual separation of the mixture. Only the ^{1}D IEX separation and the ^{2}D RP separation actually separate the constituents of the mixture, thus it is a 2D-LC separation. Nevertheless, this methodology enables fully automated identification of mAb species from the original ^{1}D IEX separation at the analytical scale.

Other groups have expanded on this methodology to develop additional capabilities. Camperi *et al.* used a ^{1}D ProA separation to analyze HCCF samples directly, which allows for quantitation of glycoforms and PTMs for a mAb in 80 min [18]. Also, Goyon *et al.* used this technique to analyze antibody-drug conjugates (ADCs), and were able to use the ^{2}D RP separation to determine the drug-antibody ratio (DAR), which is a critical quality attribute for ADCs [22]. Currently the most common use for this workflow is the identification of impurities observed in the ^{1}D separation. However, if the technique becomes more widespread and optimized it could be used for quantitation of different species in a sample. Work in this direction is already underway, and an inter-lab study has shown that quantitation of modifications between labs is consistent, and comparable to those obtained with traditional 1D reversed peptide mapping methods [20]. Further studies are needed to determine if the 1D or 2D-LC methods produce more accurate results, and if either method introduces artifacts leading to the discrepancies currently observed between 1D and 2D methods.

10.2.4 RELEASED GLYCAN AND GLYCOPEPTIDE LEVELS

One area of mAb characterization where online 2D-LC methods have not been used as yet is in the analysis of released glycans and glycopeptides. While the HILIC×RP method described above by Stoll *et al.* is able to detect different glycosylations on mAb subunits, the subunit approach is only applicable to biotherapeutics from which subunits can easily be generated [12]. Methods capable of specifically targeting glycopeptides in tryptic digests would be more broadly applicable, especially to modalities with multiple glycosylation sites and with more complex glycosylation patterns. To the best of our knowledge, the only 2D-LC studies that have focused on separating and detecting glycopeptides in tryptic digests have been offline methods, thus there is opportunity for development of online 2D-LC methods in this area.

One such example of offline 2D-LC analysis of glycopeptides was performed by Dong *et al.* [23]. Their study utilized a RP(high pH)×RP(low pH) 2D-LC method as described in Section 10.2.2, and they found that the 2D-LC method was able to detect almost twice as many glycoforms in NISTmAb compared to a conventional 1D RPLC method [23]. The study demonstrates that further development of 2D-LC methods for glycopeptides would provide deeper glycoform characterization for therapeutic proteins. On the topic of released glycan analysis, we are unaware of any online 2D-LC method that demonstrates improved detection of low abundance species that go undetected in current gold-standard 1D-LC workflows. The increasing complexity of glycosylation on emerging protein therapeutics is expected to drive the need for online 2D-LC methods capable of separating both glycopeptides and released glycans.

10.3 CHARACTERIZATION OF HOST-CELL PROTEINS

Perhaps one of the most analytically challenging tasks in the characterization and quality control of biologics is the identification and quantification of host-cell proteins (HCPs). These process-related impurities originate from the cell line in which the drug is produced and can directly affect its quality. One notable example is the degradation of polysorbates, which are excipients added to stabilize protein formulations, catalyzed by the host-cell phospholipases [24]. The presence of such HCPs has led to new control strategies, for example genetic manipulation of cell lines in order to reduce lipase expression [25] and implementation of quantitative assays utilizing triple quadrupole LC-MS to monitor lipase clearance [26].

The industry standard platform for monitoring HCPs and releasing drug products is based on the enzyme-linked immunosorbent assay (ELISA). A high throughput assay, ELISA lacks the specificity needed to profile individual HCPs. LC-MS has become increasingly valuable for HCP analysis and can be considered a complement to ELISA for in-process testing. As is the case for proteomics, LC-MS analysis of HCPs involves detecting and quantifying low abundance proteins and requires careful consideration of the ability to detect and quantify proteins across a wide dynamic range. 2D-LC directly addresses the dynamic range challenge, and numerous reports have described the coupling of 2D-LC with MS for HCP analysis. We summarized the body of 2D-LC work published up to 2018 in a recent review [27]. Many of the reports we discussed utilized a 2D-LC-MS^E system where tryptic peptides were eluted using a step-gradient from a [1]D high pH RP separation, captured on an intermediate trapping column following online dilution, and further separated on a [2]D RP column using a low pH mobile phase directly connected to the MS inlet [28].

Since 2018, progress in applying 2D-LC to the determination of HCPs has continued. Yang *et al.* investigated the reliability of detection of HCPs present at < 10 ppm using a system that involved offline high pH fractionation, concatenation of fractions, and subsequent [2]D separation at low pH followed by MS detection [29]. Previous studies documented the benefits of fraction concatenation in proteomics [30]. As applied to HCP analysis, the authors demonstrated that concatenation of fractions from the high pH separation into six pools enabled detection of seven proteins that were spiked into a mAb preparation at 10 ppm levels, whereas only four were detected in nonconcatenated control experiments. The study also addressed issues of repeatability of HCP detection and intermediate precision. These experiments included sample preparation performed by multiple analysts, and executing the 2D-LC method with ^{1}D and ^{2}D columns from different manufacturing lots. Overall, the study presents a detailed view of the capability of the method in terms of repeatability and showed that proteins spiked in at levels between 8–12 ppm could be detected among three replicates about 86% of the time.

Reiter *et al.* applied the previously described 2D-LC-MS^E system in support of improving purification of recombinant ExoProtein A, which is a carrier protein used to enhance vaccine immunogenicity [31]. The 2D-LC-MS workflow detected only one HCP at 730 ppm following an improved purification process, compared to 57 HCPs ranging from 40–5,541 ppm using the previous process. As we pointed out in our 2018 review, a major limitation of the published 2D-LC methods for HCP analysis is the instrument time needed per sample. In the interest of improving throughput, Reiter and colleagues went on to develop a 1D-LC-MS method, which identified 59 HCPs ranging from 21–3,436 ppm on samples from the original process, and six HCPs totaling 1,255 ppm in a sample from the improved process. The authors rightfully point out that it is difficult to compare their 2D and 1D experiments, especially because the MS platforms, acquisition modes, and informatics platforms used were different. Nevertheless, it is clear that 2D-LC methods for HCP analysis need to be significantly shortened in order for them to be more useful in supporting bioprocess development.

The reports published to date on the use of 2D-LC to improve HCP identification leverage the increased peak capacity provided by the technique, especially useful in this case due to the high excess of drug product in the sample. Simply stated, 2D-LC provides a higher chance of separating, and thus detecting, any given low abundance HCP-derived peptide in the presence of abundant protein drug. Common to both proteomics and HCP workflows is the issue of proteins being present across a wide dynamic range, but while proteomics samples are highly complex, HCP samples are typically much less complex. The dynamic range issue could be eliminated for the case of HCP investigations by removing the protein drug product from the sample, but this has generally been avoided due to concerns about the potential for false negatives that could originate from an HCP interacting strongly with the drug protein and thus being pulled out of the sample during sample preparation [32].

Several recent studies describe alternative sample preparation strategies to mitigate the risk of such false negative results. The improvements offered by these new methods could be directly compared to existing reports, since they each made use of the NISTmAb reference material [33]. Huang *et al.* employed a strategy of native, overnight tryptic digestion of NISTmAb followed by reduction and heating. Precipitated NISTmAb was removed using centrifugation. With the mAb removed, 1D-LC-MS was used to identify 60 HCPs, compared to 34 that were previously identified using 2D-LC coupled with ion-mobility MS [34]. Chen went on to report detection of 164 HCPs in the NISTmAb reference material using 1D nano-LC/MS [35]. In this case the sample was treated with surfactants before depletion of the NISTmAb using a 50 kDa MWCO filtration step ahead of tryptic digestion. Finally, Ma went on to show that, utilizing the sample preparation approach of Huang and with a specialized 1D nano-LC system, 55 HCPs could be detected in the NISTmAb material in 21 minutes, representing a large gain in productivity (i.e., HCPs identified per unit time). Offline, high-pH fractionation of the samples followed by separation using the nano-LC-MS system identified 171 HCPs with an acquisition time of 3 hours, representing the deepest identification of NISTmAb HCPs to date.

HCP analysis using LC-MS is now accepted as an assay that is orthogonal and complementary to traditional ELISA approaches. It is difficult to say what proportion of pharmaceutical laboratories are utilizing 2D-LC (chiefly in the form of offline fractionation or step gradient elution with intermediate trapping) as compared to 1D-LC approaches. As illustrated from the work of Ma *et al.*, even with drug protein depletion strategies, sample fractionation using 1D high-pH reversed phase chromatography yields deeper HCP coverage, albeit with a reduction in the number of HCPs identified per unit of acquisition time. The productivity of HCP analysis can be improved further using other available 2D separation modes, such as online LCxLC. Recent studies optimizing LCxLC for the analytical scale separation of peptides provide a starting point for such experiments [15].

10.4   CHARACTERIZATION OF ANTIBODY-DRUG CONJUGATES (ADCS)

ADCs are mAbs that are decorated with cytotoxic small molecules through linker molecules covalently bound to both the mAb and the small molecule (sometimes referred to as the payload). The resulting ADCs possess the cell targeting specificity of the mAb plus the cytotoxicity of the small molecule payload. The linker molecule is most often bound to the mAb through cysteine residues in the protein that are ordinarily involved in inter-chain disulfide linkages. This modification chemistry results in many variants (i.e., different numbers of small molecule drugs attached, typically in pairs – 2, 4, 6, 8) and isomers (e.g., two isomers that have four small molecule drugs in each case, but bound at different cysteine residues). HIC is highly selective for the number of small molecule drugs attached to the protein, as these are typically hydrophobic and thus have a significant impact on the hydrophobicity of the ADC surface that can interact with the HIC stationary phase. HIC has been used extensively to determine the drug-to-antibody ratio (DAR) of specific ADC samples, a method that can be used with UV detection in quality control settings. However, coupling these HIC separations directly with MS detection is not feasible because of the high salt concentrations (e.g., > 1 M sodium sulfate) needed for the HIC mechanism to work properly. This has been a major impetus for several groups to develop 2D-LC involving HIC in one of the dimensions, as well as a number of other combinations of LC separation modes to characterize ADCs [7, 36–41].

In our book chapter from 2018 we reviewed the development of 2D-LC methods for characterization of ADCs [27], and that history is not repeated here. Readers interested in a comprehensive review of this area are referred to the recent contribution by Chapel *et al.* [42]. Strategies involving 2D-LC for characterization of the small molecule payload in ADC samples are discussed in Sections 9.6.3 and 9.7.1. Most of the early work aimed at characterization of the entire ADC (i.e., protein+linker+payload) involved RPLC separation in one of the dimensions, which denatures the protein. Work since 2018 has been focused on developing methods that allow the use of non-denaturing conditions in both dimensions of the 2D-LC separation. Ehkrich *et al.* demonstrated HICxSEC separations of the ADC brentuximab vedotin, which were coupled with ion mobility mass spectrometric (IM-MS) detection [43]. While Gilroy *et al.* had previously demonstrated the utility of the HIC-SEC-MS coupling in a heartcut format for ADC analysis [41], extending this to the comprehensive format makes the analysis more efficient, particularly when multiple peaks are observed in the ^{1}D HIC separation. The primary role of the ^{2}D SEC separation in this case is to separate the proteins of interest from the non-volatile salt present in the ^{1}D effluent prior to MS detection. Most recently, Goyon *et al.* have demonstrated the use of SEC-HIC in single and multiple heartcut formats for characterization of an intact ADC [22]. The ^{1}D SEC separation is used to separate monomeric ADC from dimeric and trimeric species. The ^{2}D HIC separation then yields information about the DAR associated with the monomeric protein compared to the different aggregated forms. Inline dilution with the ^{2}D HIC mobile phase was used to mitigate effects of mismatch between the SEC and HIC mobile phases. In the same paper the authors described the use of inline digestion of proteins separated by the ^{1}D SEC separation prior to separation of the resulting peptides in a ^{2}D RP separation (i.e., see Section 10.2.3).

10.5 CHARACTERIZATION OF EXCIPIENTS

Therapeutic proteins are often formulated with excipients to promote stability of the drug product and prevent degradation during storage. The types of excipients used range from inorganic ions and amino acids, to complex surfactants such as polysorbates. Each of the excipient components in a formulation plays a specific role in protecting the protein from degradation, and it is important to be able to analyze the concentration and stability of these excipients to determine that the drug product has been properly formulated. In samples of drug product, the concentrations of these excipients are often low compared to the therapeutic protein, which complicates the determination of the excipients. Multiple literature reports have demonstrated the use of online 2D-LC to overcoming this difficulty, with the first dimension separating protein from the target excipients, and the second dimension providing further separation prior to detection.

One study by Wang *et al.* monitored the degradation of histidine in samples of biotherapeutics stored at elevated temperatures for five weeks [44]. This study used SEC in the first dimension to separate the protein from the excipients of interest, and HILIC in the second dimension to separate histidine and its most common degradant, trans-urocanic acid (t-UA). During the method development, the authors optimized various aspects of the 2D-LC separation including the flow rates, loop size, and tubing configuration. Because the ^{1}D SEC mobile phase contained only 20–40% organic solvent compared to the 90% organic solvent used as the starting point in the HILIC separation, significant breakthrough was observed for analyses using loop volumes greater than 100 uL [44]. This type of breakthrough could be prevented using active modulation techniques (see Section 4.4.4). Instead, Wang *et al.* overcame the solvent mismatch problem by using a small loop volume, only partially filling the loops, and increasing the ^{2}D flow rate while using a short gradient time. Together, these parameters gave adequate resolution of histidine from t-UA in the second dimension. The authors were also surprised to find that a small amount of cis-urocanic acid (c-UA) was being formed in the mobile phase due to phototransformation occurring in the flow cell of the ^{1}D diode-array detector. By turning off the detector they could prevent any c-UA from forming, but then they would lose the information from the ^{1}D detector. This problem is a good example of the additional elements that must be considered when developing 2D-LC methods. The final method was used to show that the rate of histidine degradation was dependent on the protein present in the sample. One of the five protein samples studied had roughly double the amount of t-UA compared to the other four samples [44].

While histidine is a relatively simple excipient, some can be very complex. Polysorbate is a common surfactant, used in many biotherapeutic formulations, that has a complex mass profile and no UV chromophore. The complex MS profile of polysorbate is due to the synthetic strategy used during manufacturing that creates a mixture of poly-(oxyethylene) (POE), sugar cores, and esters. Moreover, each of these molecules can degrade through pathways that lead to related species. All of this, taken together with the fact that there is a large amount of protein also present in formulated drug product, makes it difficult to measure degradation of polysorbate in samples of biotherapeutics used for stability studies, especially if 1D-LC is used. To address these challenges, Li *et al.* developed two different 2D-LC methods involving CAD and MS to measure changes in polysorbate species concentration [45]. By splitting the ^{2}D effluent flow between the MS and CAD detectors, the group was able to simultaneously identify peaks separated in the second dimension using the MS, and obtain relative abundance information from the CAD. The first 2D separation method – IEX-RP – capitalized on the fact that the POE is hydrophobic relative to the protein, while the protein carries a net charge in solution. This allows POE to flow through the IEX column unretained and can be transferred to the ^{2}D RP for further separation, while the charged species including protein were retained by the IEX column. The ^{2}D RP column then separates the POEs by chain length. The second 2D separation method used a mixed-mode column in the first dimension so that protein flowed through unretained, and the esters in the polysorbate were retained and subsequently transferred in a heartcut to a ²D RP column. This method is complementary to the IEX-RP method, and together they were used to quantify all the major ester and POE subspecies present in the samples. Li *et al.* were able to identify differences in the degradation of polysorbate in the presence or absence of protein, showing that the protein does indeed have an effect on polysorbate degradation [45].

Both studies discussed in this section highlight the utility of 2D-LC for determination of excipient levels in drug formulations containing high concentrations of protein that would not be possible using 1D-LC. These studies also showed the importance of this type of analysis since the degradation of polysorbate and histidine was different in the presence of protein, and even the identity of the protein caused differences in the way that histidine degraded. While only two types of excipients were discussed in this section, many other types of molecules are used in biotherapeutic formulations. Some excipients are difficult to analyze, such as lipid nanoparticles, or are included in complex formulations, such as adjuvants for vaccines, and could benefit from 2D-LC studies in the future.

10.6   CLOSING REMARKS

Currently, the adoption of online 2D-LC methods for analysis of biologics is inconsistent. A few specific implementations have developed rapidly over the past few years and are becoming routine in the laboratories of biopharmaceutical companies around the world. Prominent examples including the use of ProA-SEC to determine titer and purity for mAbs in HCCF, and automated online multi-dimensional methods for protein characterization that replace tedious multi-step off-line methods, which are time-consuming and prone to sample contamination and introduction of artifacts. At the same time, in this chapter we have highlighted several areas where online 2D-LC seems poised to have a big impact, but are relatively unexplored so far. Examples include the use of online LC×LC-MS for HCP analysis, and characterization of modalities with complex glycosylation characteristics. Thus, there are tremendous opportunities for expansion and optimization of the online 2D-LC method portfolio. This continued growth will be driven in part by the continually increasing complexity of new biotherapeutic modalities.

Advances in continuous manufacturing for biotherapeutics along with investments in PAT have positioned the industry for additional growth. Regulatory encouragement for integration of analytical technology with novel manufacturing platforms will define the future of biotherapeutic supply for the world. 2D-LC can contribute to this growth through automation and analytical methods that enable advanced process control to accelerate process development, manufacturing and product release testing. Continued technology development that designs in robustness, automation, and expansion in the scope of applicability for 2D-LC will be critical for further use of 2D-LC in process and product understanding, both for current biotherapeutics, and more complex modalities expected in the near future.

REFERENCES

[1]   Novel Drug Approvals for 2021, United States Food and Drug Administration, n.d. www.fda.gov/drugs/new-drugs-fda-cders-new-molecular-entities-and-new-therapeutic-biological-products/novel-drug-approvals-2021 (accessed November 27, 2021).

[2]   H. Kaplon, J.M. Reichert, Antibodies to watch in 2021, MAbs. 13 (2021). https://doi.org/10.1080/19420862.2020.1860476.

[3]   X. Gjoka, M. Schofield, A. Cvetkovic, R. Gantier, Combined Protein A and size exclusion high performance liquid chromatography for the single-step measurement of mAb, aggregates and host cell proteins, J. Chromatogr. B. 972 (2014) 48–52. https://doi.org/10.1016/j.jchromb.2014.09.017.

[4]   M. Lemmerer, A.S. London, A. Panicucci, C. Gutierrez-Vargas, M. Lihon, P. Dreier, Coupled affinity and sizing chromatography: A novel in-process analytical tool to measure titer and trend Fc-protein aggregation, J. Immunol. Methods. 393 (2013) 81–85. https://doi.org/10.1016/j.jim.2013.04.008.

chromatography x reversed phase liquid chromatography hyphenated to high resolution mass spectrometry. II- Identification of sub-units for the characterization of even and odd load drug species, J. Chromatogr. B. 1032 (2016) 91–102. https://doi.org/10.1016/j.jchromb.2016.06.049.

[38] M. Sarrut, A. Corgier, S. Fekete, D. Guillarme, D. Lascoux, M.-C. Janin-Bussat, A. Beck, S. Heinisch, Analysis of antibody-drug conjugates by comprehensive on-line two-dimensional hydrophobic interaction chromatography x reversed phase liquid chromatography hyphenated to high resolution mass spectrometry. I – Optimization of separation conditions, J. Chromatogr. B. 1032 (2016) 103–111. https://doi.org/10.1016/j.jchromb.2016.06.048.

[39] K. Sandra, G. Vanhoenacker, I. Vandenheede, M. Steenbeke, M. Joseph, P. Sandra, Multiple heart-cutting and comprehensive two-dimensional liquid chromatography hyphenated to mass spectrometry for the characterization of the antibody-drug conjugate ado-trastuzumab emtansine, J. Chromatogr. B. 1032 (2016) 119–130. https://doi.org/10.1016/j.jchromb.2016.04.040.

[40] E. Largy, A. Catrain, G. Van Vyncht, A. Delobel, 2D-LC–MS for the analysis of monoclonal antibodies and antibody–drug conjugates in a regulated environment, Current Trends in Mass Spectrometry. 14 (2016) 29–35.

[41] J.J. Gilroy, C.M. Eakin, Characterization of drug load variants in a thiol linked antibody-drug conjugate using multidimensional chromatography, J. Chromatogr. B. 1060 (2017) 182–189. https://doi.org/10.1016/j.jchromb.2017.06.005.

[42] S. Chapel, F. Rouvière, M. Sarrut, S. Heinisch, Two-dimensional liquid chromatography coupled to high-resolution mass spectrometry for the analysis of ADCs, in: L.N. Tumey (Ed.), Antibody-Drug Conjugates, Springer, NY, 2020: pp. 163–185. https://doi.org/10.1007/978-1-4939-9929-3_11.

[43] A. Ehkirch, A. Goyon, O. Hernandez-Alba, F. Rouviere, V. D'Atri, C. Dreyfus, J.-F. Haeuw, H. Diemer, A. Beck, S. Heinisch, D. Guillarme, S. Cianférani, A novel online four-dimensional SECxSEC-IMxMS methodology for characterization of monoclonal antibody size variants, Anal. Chem. (2018) 13929–13937. https://doi.org/10.1021/acs.analchem.8b03333.

[44] C. Wang, S. Chen, J.A. Brailsford, A.P. Yamniuk, A.A. Tymiak, Y. Zhang, Characterization and quantification of histidine degradation in therapeutic protein formulations by size exclusion-hydrophilic interaction two dimensional-liquid chromatography with stable-isotope labeling mass spectrometry, J. Chromatogr. A. 1426 (2015) 133–139. https://doi.org/10.1016/j.chroma.2015.11.065.

[45] Y. Li, D. Hewitt, Y. Lentz, J. Ji, T. Zhang, K. Zhang, Characterization and stability study of polysorbate 20 in therapeutic monoclonal antibody formulation by multidimensional ultrahigh-performance liquid chromatography–charged aerosol detection–mass spectrometry, Anal. Chem. 86 (2014) 5150–5157. https://doi.org/10.1021/ac5009628.

11 Applications of Two-Dimensional Liquid Chromatography to Chemical Analysis in Academic Research and Industry

Peilin Yang and Matthias Pursch

CONTENTS

11.1 2D-LC FOR CHEMICAL ANALYSIS

The growing desire and demand from consumers for material performance, cost, safety, and sustainability have brought tremendous opportunities as well as challenges to the chemical industry and academic research. In recent years, 2D-LC has emerged as an advanced analytical tool that has the potential to solve some of the toughest measurement and characterization challenges encountered in new material development, production, and formulation. Driven by increasing consumer awareness of material safety, residual small molecule analysis at low levels in complex polymer matrices has become very important. Heartcutting 2D-LC is a suitable technique for quick analysis of target compounds in complex matrices.

At the other end of the molecular weight spectrum, synthetic polymers are used ubiquitously. Deeper understanding of polymer compositions can enable the development of better materials with improved properties. Comprehensive 2D-LC has been shown to be very useful for polymer characterization that nicely complements other analytical techniques. In certain applications, 2D-LC exhibited unique advantages over other techniques for the analysis of polymer samples with high complexity. In this chapter, we will focus on representative research efforts in academia as well as the analytical needs of the chemical industry. We will provide some examples of how 2D-LC is

DOI: 10.1201/9781003090557-11

applied for different types of chemical analysis, focusing on the period from 2014–2021. For earlier work we refer to dedicated review articles [1–6].

11.2 TARGET ANALYSIS – SMALL MOLECULES IN COMPLEX MATRICES

Historically, the large majority of 2D-LC separations have dealt with qualitative analysis, such as structure elucidations or comparisons of main features of complex samples. However, applications in recent years have highlighted the potential of 2D-LC for quantitative analysis of target compounds as well.

There are several examples in which small molecules needed to be determined in polymeric or other complex matrices. Sometimes this is a separation challenge – there are too many different compounds to separate them all by conventional 1D-LC, and changing selectivity and/or increasing column efficiency cannot overcome this problem- that is the peak capacity of 1D-LC is not sufficient. In other cases a major challenge arises from the fact that the sample can only be dissolved in a strong solvent, which in turn compromises the quality of 1D-LC. In this case, pre-separating the sample in the first dimension of a 2D-LC system and cutting out the components of interest can quickly provide a solution to this challenge.

In a study completed in 2014, trace amounts of hexabromocyclododecane (HBCD) flame retardant had to be determined in a polystyrene (PS) matrix. This measurement problem came up when HBCD was receiving increasing attention from environmental agencies and alternative flame retardants had to be found. HBCD had been present in PS typically at 1–2% levels, and RPLC separation using a column packed with superficially porous C18 particles (SPP) yielded satisfactory quantitative data. But when trying to quantify HBCD below 0.1% levels in PS, significant overlapping of HBCD peaks with the peaks of other trace components was observed [7].

To solve the problem, Pursch *et al.* used multiple heartcutting 2D-LC (mLC-LC) methodology, which had just become commercially available at that time. In order to obtain good peak shape and sufficient retention in the second dimension, a 40 µL loop volume was chosen. A phenyl stationary phase was selected for the ¹D column, and solvent gradient elution was used with a water/MeOH/THF mobile phase. The separation used in the first dimension of the 2D-LC system actually looked much worse than the 1D-LC method in use at the time; the main target components were fully buried under the PS matrix peaks.

After adding a C18 column in the second dimension, it was observed that the ²D separation nicely complemented the ¹D one, facilitating the resolution of compounds that were not resolved by the ¹D phenyl column. The method also proved to be highly repeatable with peak area RSDs better than 3% for individual cuts where HBCD was detected, and less than 1% for the sum of ²D peak areas for HBCD across all cuts where it was detected [7]. One downside of this approach at the time was that several cuts had to be made across the ¹D peaks resulting in a relatively slow 2D-LC method (25–30 min analysis time). It should be noted here that this separation would be quite different if it were developed today. Using active modulation technologies (e.g., ASM) one would be able to make just two cuts to capture all of the HBCD from the first dimension rather than seven, and it is also possible that larger sample loops could be used, which would result in a more than two-fold reduction in analysis time. Nevertheless, the method was successfully applied at that time to study the manufacturing processes involving HBCD mixed with PS.

Another small molecule target application by Pursch *et al.* has dealt with determination of bisphenol-A and its diglycidyl ether in polymeric epoxy novolac matrix [8]. Novolacs are condensation products of phenols and aldehydes. They are used in the coatings industry. In this application the ¹D separation was carried out in UHP-SEC mode using a silica-based advanced polymer chromatography (APC) column with THF as the mobile phase. The target compounds were sampled using sLC×LC approach in the region of the ¹D separation where the low molecular weight species elute. The ²D separation was carried out using a C18 SPP column with water/ACN as the mobile phase.

For this case, ASM had to be utilized to avoid poor peak shape in the second dimension due to mobile phase mismatch. Indeed, when the ²D separation was carried out without ASM, significant analyte breakthrough occurred at the dead time of the ²D separation. This was not surprising, as the combination of a 40 µL injection volume and the high solvent strength of THF is too much to overcome when using a 50 mm x 3.0 mm i.d. C18 column with a ²D gradient that starts with a highly aqueous mobile phase. With the use of ASM, however, very good peak shapes were obtained and the target components could be quantified with good precision and accuracy [8]. The sLC×LC approach embodied in this application can eliminate tedious sample preparation

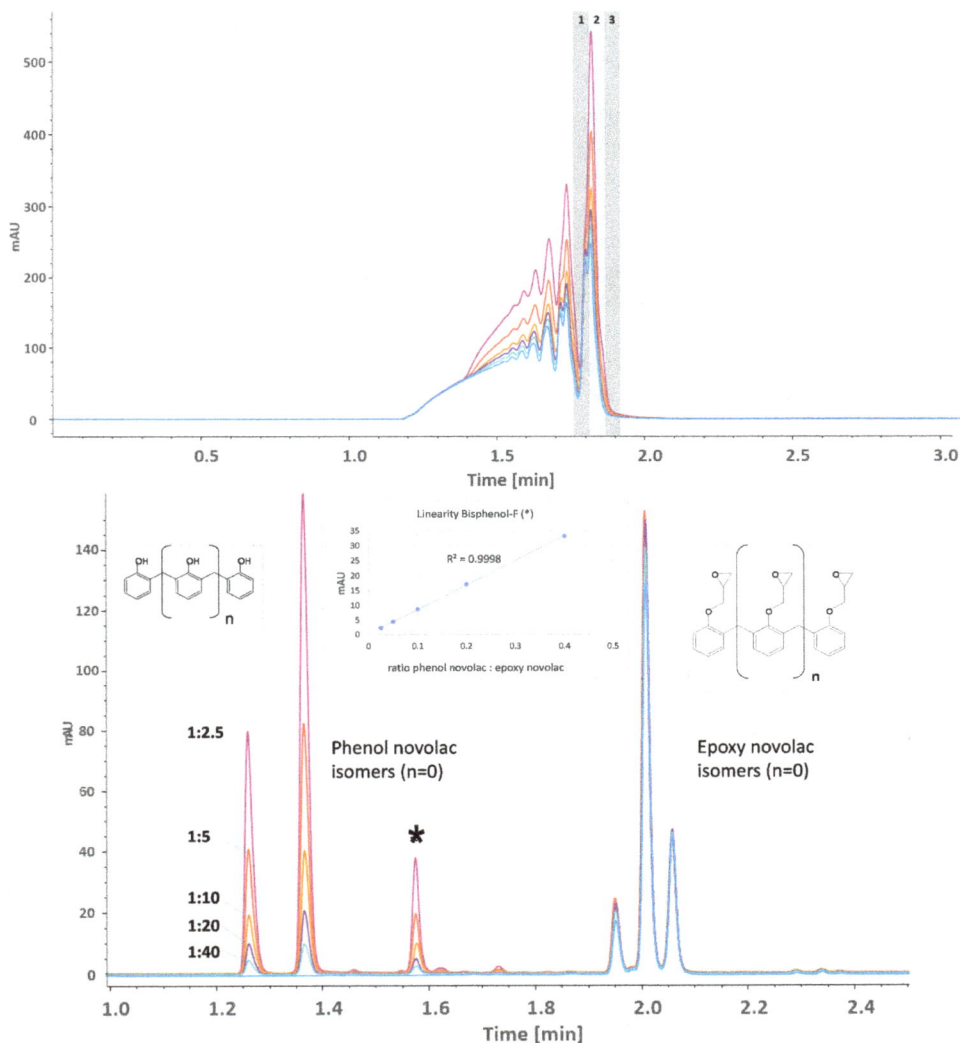

FIGURE 11.1 Determination of bisphenol-F isomer (*) in a synthetic blend of phenol novolac and epoxy novolac. Top chromatogram shows the ¹D UHP-SEC separation in THF; bottom chromatogram shows ²D separation using an Ascentis Express C18 SPP column. Reprinted from *Journal of Chromatography, A*, 1562, Pursch, Wegener, and Buckenmaier, Evaluation of active solvent modulation to enhance two-dimensional liquid chromatography for target analysis in polymeric matrices, 78–86, Copyright (2018), with permission from Elsevier.

efforts, such as removing the polymeric matrices from samples by precipitation. Instead, samples can simply be dissolved in a strong solvent that fully solubilizes the matrix and the analytes of interest. The first dimension is then used for a pre-separation of most of the matrix from the analytes, and then the second dimension completes the resolution of the target analytes, which in turn enables accurate quantitation.

Using the same strategy, bisphenol-F isomers were determined in a synthetic blend of phenol novolac and epoxy novolac oligomers [8]. Chromatograms from the 2D-LC separation and a representative calibration curve are shown in Figure 11.1.

11.3 CHARACTERIZATION OF OLIGOMERIC/ISOMERIC COMPOUNDS: SURFACTANTS, LUBRICANTS, OILS, CHEMICAL PROCESS INTERMEDIATES, LIGNIN, WASTEWATER

Surfactants, polyethers, oils, and chemical process intermediates constitute a class of compounds that is characterized by its unique sample complexity. Due to the manufacturing process, surfactants are heterogeneous in more than one chemical dimension. First, there is a variation of alkyl chain lengths. Second, many surfactants are ethoxylates that exhibit a large degree of variation in oligomer distribution. Third, surfactants often have different end-groups, such as hydroxyl, sulfonate, sulfate, carboxyl, amide, phosphate, and others. Combining all these variables yields samples with tens to hundreds of individual constituents. Therefore, it is not surprising that most of these complex sample characterization schemes are carried out by use of comprehensive 2D-LC separations.

A great example of a detailed surfactant characterization by LC×LC was presented by Elsner et al. [9]. The sample was a mixture of alkyl sulfates, alkyl ethoxylates, alkyl ethoxylate sulfates, and other surfactant types. A Zic-HILIC zwitterionic stationary phase was used in the first dimension, which provided separation according to degree of ethoxylation. The ^{2}D separation employed a C8 stationary phase, which is capable of separating on the basis of analyte hydrophobicity, or alkyl chain length. Both dimensions separate by surfactant type as well. Detection was carried out using a Q-TOF MS system with electrospray ionization (ESI) in the positive mode. The resulting LC×LC chromatogram in Figure 11.2 shows the great power of 2D-LC for the analysis of complex chemical samples. Very high orthogonality is obtained, as peaks are well distributed across the available 2D separation space. The method was also used for the quantitative analysis of individual components. Average peak area/volume RSDs were in the range of 4–5%.

Li et al. investigated the structures of "Polysorbate 20" esters in therapeutic mAb formulations using heartcutting 2D-LC [10]. Polysorbates are fatty acids of varying alkyl chain length attached to sorbitan and isosorbitan ethoxylates. The ^{1}D separation was carried out using a mixed mode column, which has both anion-exchange (AEX) and RP characteristics. The separation conditions were chosen such that the proteins were fully separated from the polysorbates. The second dimension used the RP mode. In a second 2D-LC method, cation-exchange (CEX) separation was selected for ^{1}D separation prior to analysis by RP in the second dimension. Detection was carried out using charged aerosol detection (CAD) and MS. It was found that polysorbates undergo degradation in the protein formulations over three years of storage. It was also discovered that degradation occurs at different rates for the sorbate and isosorbitan containing species.

Another heartcutting 2D application for surfactants was presented by Escrig-Domenech et al. [11]. Interestingly, just one pump was used for both the ^{1}D and ^{2}D separations. Fatty alcohol ethoxylates (FAEs) were derivatized with phthalic anhydride and subject to ^{1}D separation using a C8 column and gradient elution with MeOH/water. The separation was based on differences in the alkyl chain lengths of the ethoxylates. Peaks of interest were transferred to the second dimension, which also used a C8 column and gradient elution, but with an ACN/water mobile phase. The ^{2}D separation was governed by difference in the degree of ethoxylation. Several industrial FAEs from various producers were compared and differences in alkyl distribution and degree of ethoxylation were observed.

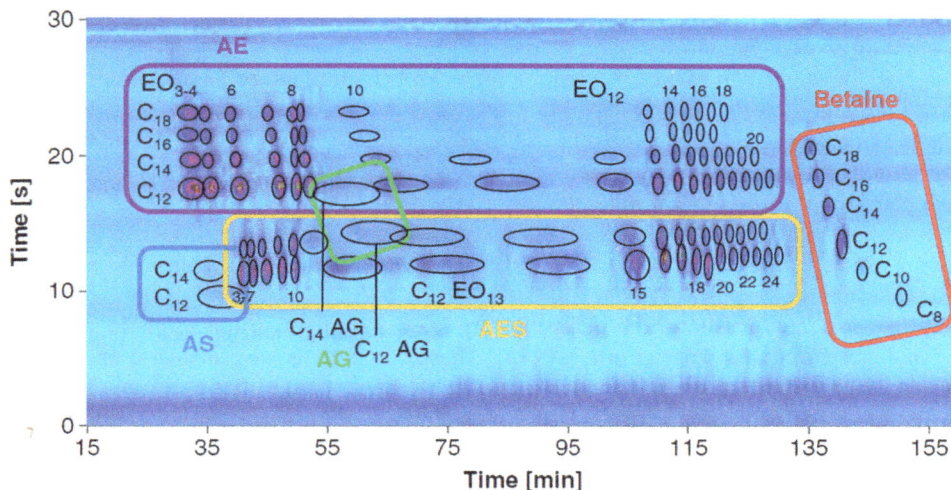

FIGURE 11.2 LC×LC analysis of a surfactant consisting of alkyl sulfates (AS), alkyl ethoxylates (AE), alkyl ethoxylates sulfates (AES), alkyl glucosides (AG) and betaines. Reprinted by permission from: Springer Nature, *Analytical and Bioanalytical Chemistry*, Reproducibility of retention time and peak area in comprehensive two-dimensional liquid chromatography, Elsner, Wulf, Wirtz, and Schmitz, Copyright (2014).

Gargano *et al.* compared different LC×LC instrumental conditions for separation of tristyryl phenol ethoxylates (TSP) and corresponding ethoxylate phosphates [12]. Like the work of Elsner *et al.*, the separations were carried out using HILIC with bare silica columns as the first dimension. A second dimension RPLC mode used C18 columns in a variety of formats mode and UV absorbance (diode array) for detection. A comparison of three different approaches was made: 1) "standard" LC×LC mode using 400 bar instrumentation; 2) UHPLC mode (1200 bar); and 3) UHPLC with active modulation (AM). Analysis times were improved from 200 to 80, and then 40 min, respectively. The peak capacities estimated for the separations varied slightly depending on the calculation used, however, on average they were roughly the same for all three approaches (n_c ~ 700–1000). However, peak capacity per unit of analysis, time (i.e., n_c/min.) was significantly higher when UHPLC conditions were used, and even more so when AM was used. The optimal separation was used to compare TSPs and TSP phosphates from four different producers. Different peak patterns were observed for each sample showing the ability of LC×LC to track production processes.

One current trend in chemical industry is to use renewable resources to produce "greener" products. Natural oils are a predominant source of feedstock; the selection includes soybean, sunflower, canola, palm, and castor oils. These oils are inherently very complex, as they contain fatty acids with different alkyl chain lengths and different degrees of unsaturation. During further chemical processing, such as ethoxylation, additional complexity is introduced. Recent work by Groeneveld *et al.* on the use of 2D-LC-MS to characterize castor oil ethoxylates demonstrates how complex these materials are. As with many prior 2D-LC separations of ethoxylate mixtures, the ¹D separation was also carried out in HILIC mode using a silica column, followed by a RPLC separation with a phenyl column in the second dimension. Very high separation orthogonality was observed, and more than 400 individual components were identified using Q-TOF MS detection [13]. In this sample fatty acid ethoxylates as well as ethoxylated fatty acid dimers, trimers and tetramers were observed, which led to such high sample complexity.

The same concept was also used for separation of a synthetic mixture of glycerin ethoxylates (EO), glycerin propoxylates (PO), and glycerin EO/PO oligomers. The EO/PO oligomers could be separated according to the number of EO units in HILIC mode and according to PO units in RPLC mode, yielding a very symmetric separation pattern [13].

Pirok *et al.* developed an LC×LC separation for a synthetic mixture of 54 dyes commonly found in cultural heritage objects. The [1]D separation used a strong anion exchange (SAX) column. This separation mode was selected because many of the dyes contain anionic functional groups, thus the SAX column provides the selectivity needed to separate the analytes by charge. The second dimension was carried out using ion-pairing RPLC. An orthogonality of 63% was achieved, and the peak capacity was estimated at 2200, with a total analysis time of about three hours. UV absorbance detection was used, and UV/vis spectra were recorded, which facilitated identification of the dyes based on their unique spectra [14].

Zhu *et al.* presented an ultra-high peak capacity LC×LC separation for aromatic amines taken as intermediates from an industrial process [15]. These compounds have isomeric and oligomeric structures that contribute to a very complex mixture. The separation utilized six pentafluorophenyl (PFP) columns coupled in series for the first dimension, and a Zorbax PAH column in the second dimension. Even though both dimensions were operated in RP mode, satisfactory orthogonality was obtained. This complementarity of the two columns used was attributed to dipolar interactions with the PFP column and shape selectivity/hydrophobicity selectivity with the PAH column. The use of shifted gradients in the second dimension helped to enhance the surface coverage in the 2D plot. Within 20 h of analysis time a peak capacity of greater than 11,000 was obtained. About 900 individual peaks were observed.

Heavy oil is another class of samples that are complex in composition. Van Beek *et al.* investigated the "maltenes" structures of heavy oil using LC×LC with UV detection [16]. For this separation challenge, a cyanopropyl stationary phase was used in the first dimension, while the second dimension used a C18 column. The resulting LC×LC separation is capable of resolving various groups such as saturates, aromatics, and resins and can be used for oil manufacturing process studies.

Bio-oils produced from pyrolysis of biomass (e.g. Eucalyptus sawdust, spent coffee grounds) are also very complex samples. Lazzari *et al.* used an LC×LC scheme for quantitative analysis of up to 28 components in these two matrices [17]. The aqueous phases resulting from pyrolysis were studied in detail. Reversed-phase separations were used in both dimensions; an Xbridge amide column was used in the first dimension, and a Poroshell C18 column was used in the second dimension. Detection was carried out by UV absorbance and ESI-MS (ion trap). The resulting contour plots are shown in Figure 11.3. Calibration curves were made using various phenolic, aldehyde, or hydroxyl ketone standards covering two orders of magnitude in concentration. In the literature several approaches for determination of orthogonality and actual peak capacities exist [18, 19]. In this work, effective peak capacities of up to 1,000 were achieved in about 1 h of analysis time. Orthogonality is ranging from 33–50%, which means that a significant part of the 2D separation space is not used. Recoveries ranged from 92–113%, and intraday precision (n=9) ranged from to 0.3 to 14.5%. Many of the identified phenolics and ketones are of potential interest as feedstocks in the chemical industry.

Sun and colleagues developed LC×SFC methodology for analysis of depolymerized lignin [20]. Their setup included a phenyl hexyl trapping column in the modulation interface. The [1]D separation was carried out in RP mode using a Zorbax Eclipse Plus C18 column. Several [2]D columns were screened and a Torus UPC^2 diol stationary phase was found to provide the best overall separation and good orthogonality. The method was developed using a set of 40 phenolic reference compounds and applied to processed lignin.

Leonhard *et al.* studied wastewater samples with a micro-LC×LC setup [21]. The first dimension was run with nano-flow conditions (200 nL/min) and a Hypercarb capillary column, while the [2]D separation was executed using a Sunshell C18 column. MS/MS was used for detection. A synthetic mixture composed of about 100 components was used as reference standard. 2D-LC/MS

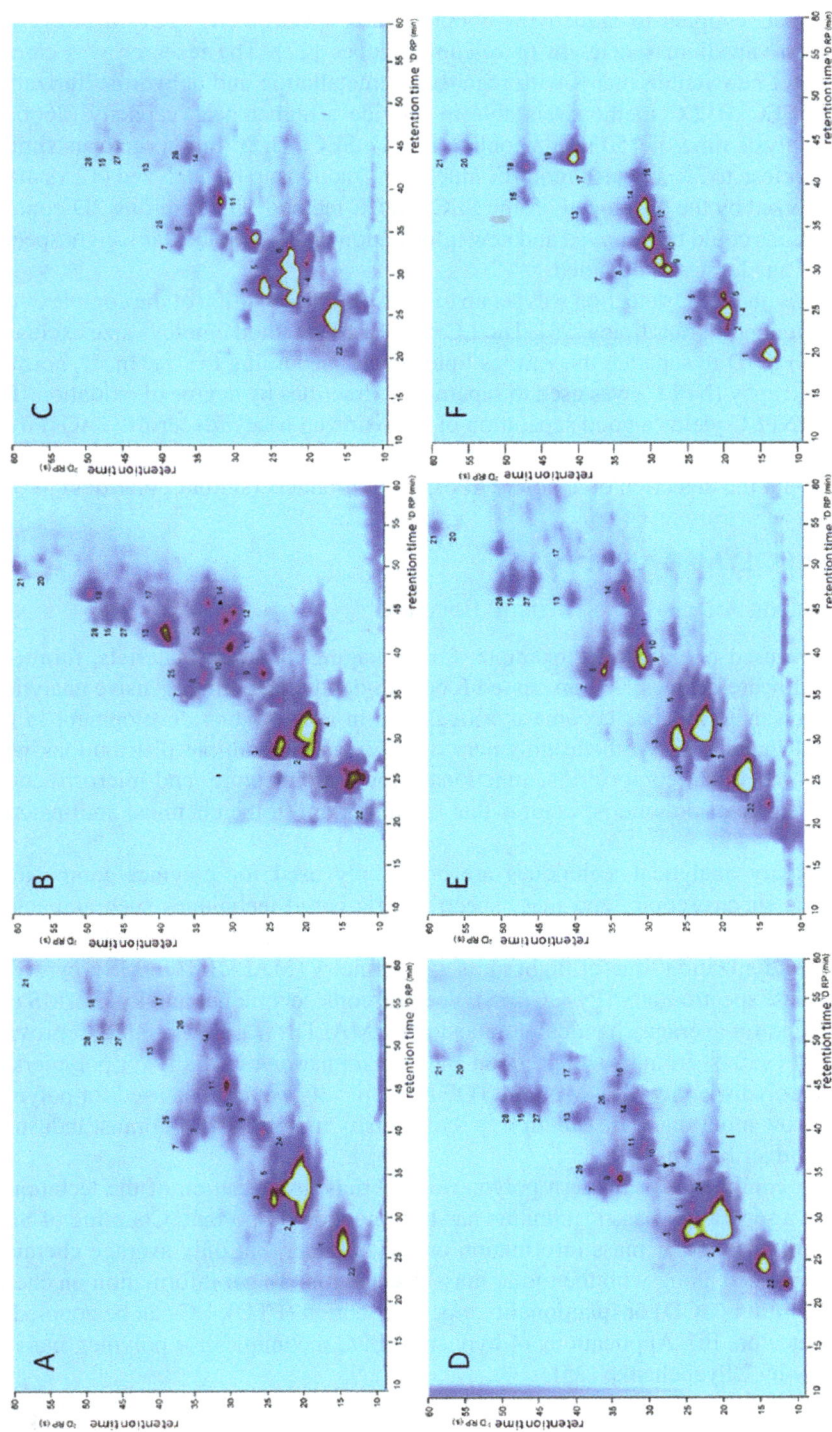

FIGURE 11.3 LCxLC-UV contour plots for analysis of different bio-oil aqueous phases produced by pyrolysis from: rice husk (A), peanut shell (B), spent coffee grounds (C), peach core (D) and eucalyptus sawdust (E). Panel F shows the separation obtained for the mixture of commercial standards analyzed (50 µg/mL each). Reprinted from *Journal of Chromatography, A*, Quantitative analysis of aqueous phases of bio-oils resulting from pyrolysis of different biomasses by two-dimensional comprehensive liquid chromatography, 1602, Lazzari, Arena, Caramao, and Herrero, 359–367, Copyright (2019), with permission from Elsevier.

measurements were compared with 1D-LC/MS, and the authors found that 2D-LC/MS improved the detectability of some of the target components.

A study using LC×LC coupled to inductively coupled mass spectrometry (ICP-MS/MS) was explored for sulfur and vanadium species in petroleum residues [22]. The research was carried out with a focus to meet new requirements with regard to demetallation and dehydrosulfurization processes. An offline SEC×RPLC method was able to provide a higher peak capacity (2600 vs 1700) for the same analysis time of 150 min as online RPLC×SEC. The dilution factor is similar with both approaches (close to 30) but also requires much less fractions to be analyzed (12 vs 400). Asphaltenes were analyzed by the developed offline SEC×RPLC method. The resulting 2D-contour plots show that co-elutions could be removed and new information on high molecular weight species containing sulfur and vanadium was obtained.

An LC×LC method with ELSD detection was set up to study the composition of the complex mixture of oxidized polar and non-polar lipids [23]. The LC×LC-ELSD method employs size exclusion chromatography (SEC) in ^{1}D to separate the various lipid species according to size. In ^{2}D, normal-phase liquid chromatography (NPLC) was used to separate the fractions by degree of oxidation. The coupling of SEC with NPLC yields a good separation of the oxidized triacylglycerols (TAGs) from the large excess of non-oxidized TAGs. In addition, it allows the isolation of non-oxidized species that usually interfere with the detection of a variety of oxidized products (similar polarities).

11.4   2D-LC FOR POLYMER ANALYSIS

11.4.1   TECHNIQUES FOR ANALYSIS OF SYNTHETIC POLYMERS

Synthetic polymers are used broadly – for example, for packaging, building materials, furniture, paints, detergents, healthcare devices, and processed foods. In-depth and comprehensive analytical characterization of such materials has become a critical step in new product development in the chemical and material industries. Synthetic polymers usually exhibit multiple distributions with respect to molar mass, chemical composition, functional group, architecture, and microstructure. Figure 11.4 shows a variety of polymer structures that are important in the chemical and material industries.

Several complementary analytical techniques are commonly used for polymer composition analysis. These include spectroscopic- and mass spectrometric-based techniques such as nuclear magnetic resonance spectroscopy (NMR), Fourier-transform infrared spectroscopy (FTIR), matrix-assisted laser desorption/ionization time-of flight mass spectrometry (MALDI-TOF MS), pyrolysis gas chromatography/mass spectrometry (Py-GC/MS). Spectroscopic techniques and Py-GC/MS can only provide information on average chemical compositions. MALDI-TOF MS is able to provide molecular speciation but suffers from ion suppression and discrimination especially for polymers or copolymers with high polydispersity [24]. MALDI-TOF MS typically can only be used for polymer samples with very narrow molar mass distribution (polydispersity index < 1.2), but most industrial polymers have broader distributions [3].

With the increasing complexity of modern polymeric materials, hyphenation of the techniques mentioned above with a separation-based technique has become more important. Coupling of SEC with NMR or FTIR provides molar mass information on the polymer, but only average chemical compositions can be obtained along with the molar mass distribution. To get information on chemical composition distribution (CCD) or functionality type distribution (FTD), LC can be coupled to these spectroscopic detectors [6]. Applications of hyphenated LC techniques for polymer analysis were reviewed recently by Uliyanchenko [25].

Even when combined with advanced detectors, 1D separations are often found to be insufficient for resolution of complex polymer samples. LC×LC is a powerful technique for separations of polymers as the two separation dimensions can provide information on the distribution of species along two different chemical axes (e.g., chain length and degree of ethoxylation), and also reveal a

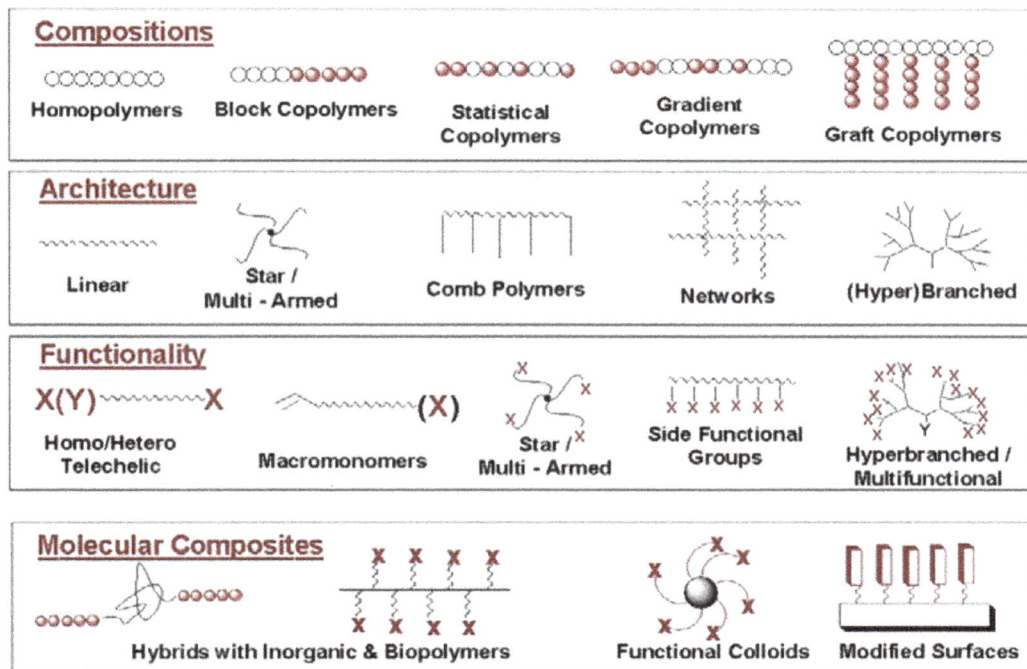

FIGURE 11.4 Examples of polymer molecular structures. Reprinted from *Materials Today*, 8, K. Matyjaszewski, J. Spanswick, Controlled/living radical polymerization, 26–33, Copyright (2005), with permission from Elsevier.

dependence of one of these two distributions on the other. Many different combinations of separation modes are possible, and typically the preferred combination is sample-dependent. Coupling of two-dimensional separations with selective or information-rich detectors can further extend the utility of LCxLC in polymer characterization [26].

LC at critical conditions (LCCC) is one of the important chromatographic techniques used for polymer separation and often used in the first dimension of 2D-LC separations [5]. In LCCC the separation according to molar mass is suppressed, hence it is considered orthogonal to SEC. Despite many reported LCCCxSEC separations of polymers, some major drawbacks of LCCC have prevented this technique from being broadly adopted by industrial labs. Polymer separations under truly critical conditions are very sensitive to chain architectures and end-group variations. Critical conditions are also very difficult to maintain over weeks or months. Subtle changes in experimental conditions can lead to shifts in the critical point causing irreproducible separations. LCCC can suffer from low analyte recovery and has limited use in the separation of copolymers since critical conditions do not exist for random copolymers and many block copolymers [4, 5]. SEC has been used frequently as a ²D separation because it involves isocratic conditions that are easier to implement in practice.

Besides LCCC and SEC, liquid chromatography in adsorption mode (LAC) can also be applied to a broad range of polymer systems to provide valuable information on chemical composition [27]. In LAC, the retention time increases with polymer molecular weight. LAC can be performed in both normal and reversed phase mode. For simplicity the term "LC" is used. Gradient polymer elution chromatography (GPEC) has been used as a term to describe LC separation of polymers with gradient elution [28]. In contrast to LCCC, GPEC is essentially a quick way to separate a broad range of polymers and copolymers without the extensive experimental effort needed to find the critical point of adsorption, which is required for LCCC. Modern instrumentation and software make it possible to use SEC in the first dimension followed by very fast gradient LC separation in the second dimension. Pros and cons of LC×SEC and SEC×LC have been discussed in a recent review article by Schoenmakers *et al.* [29].

In the following sections, recent examples of comprehensive 2D-LC separations of synthetic polymers will be discussed. Although some of these examples are still considered fundamental studies, the capability of 2D-LC for solving a wide range of polymer analysis problems has been clearly demonstrated.

Several analytical techniques have been well established for simple polymeric materials such as homopolymers with linear architectures and narrow molar mass distributions. These materials will not be the focus of this section. Advanced polymeric materials often require incorporation of multiple monomers with distinct properties or blends of different polymer constituents. These more complex materials present new challenges for analytical characterization and require more advanced analytical tools such as 2D-LC.

11.4.2 Copolymer Analysis by 2D-LC

Copolymers can incorporate more molecular design principles to achieve unique properties that cannot be achieved using pure homopolymers for specific applications. Statistical (random) copolymers, block copolymers, and grafted copolymers are just a few examples of the many possible copolymer structures.

Segmented copolymers are an important type of advanced polymer material. They can be further classified into linear block copolymers, graft copolymers, star block copolymers, and so on, as illustrated in Figure 11.4. Segmented copolymers have some inherent distributions such as molar mass, chemical composition, block length, tacticity, and topology. Homopolymers can also be present in segmented copolymers which makes characterization of these materials quite challenging and often requires many different techniques [30, 31]. In 2014 Malik *et al.* provided a comprehensive review of approaches used for characterization of segment copolymers using various separation techniques [32]. Multi-dimensional separations of different copolymer systems achieved by coupling different separation modes are summarized in this review article.

One important class of block copolymers that is broadly used in drug delivery and other applications is poloxamers, which are nonionic triblock copolymers of poly(ethylene oxide) (PEO) and poly(propylene oxide) (PPO) with PPO as the central chain. Malik *et al.* reported 2D-LC separations of triblock copolymer samples according to the PEO block in one dimension and PPO block in the other dimension. RP gradient LCCC in the first dimension enabled separation based on the number of propylene oxide (PO) units at the critical adsorption point (CAP) of PEO, while the ^{2}D NPLC separation at the CAP of PPO enabled separation on the basis of the number of EO units [33]. Al Samman *et al.* reported 2D separation of polymers according to topology and molar mass using solvent gradient interaction chromatography in the first dimension, and SEC in the second dimension. Linear and hyperbranched polyesters were separated which made it possible to estimate ratios of linear and branched polymers in blend materials [34].

Statistical copolymers are another very important type of polymer used in many industrial applications. The chemical composition distributions of these materials are determined by monomer ratios, monomer reactivity parameters and synthesis process conditions, which can strongly affect the properties of the resulting material. For example, if a less soluble homopolymer is formed, the dissolution profile can be adversely impacted in the case of tablet coating [35]. Gradient LC in combination with SEC as the second separation mode in 2D-LC can be used as a quick way to separate a wide range of homo- or co-polymeric species. The LC×SEC combination is the single most commonly used configuration for 2D-LC separations of polymers and is broadly applied to copolymer analysis [1, 29].

When SEC is used in the second dimension, refractive index (RI) and multi-angle light scattering (MALS) detectors can be used for quantitation. Challenges associated with use of SEC in the second dimension include a limited number of commercially available SEC columns suitable for fast separation, undesired molecular-weight dependence of the first dimension LC separation, and undersampling of first dimension peaks due to relatively long second dimension run time and relatively narrow first dimension peaks. On the other hand, when SEC is used in the first dimension in combination with gradient LC in the second dimension, polymer fractions that exhibit the same size can be separated according to chemical composition [36, 37]. Yang *et al.* reported a few practical industrial applications of SEC×RPLC for separation of polymer dispersants and rheology modifiers [37–39].

In gradient LC, polymer breakthrough can occur when the sample solvent is stronger than the composition of the initial eluent composition. Breakthrough is observed when part of the polymer sample travels with the injection solvent band through the column without being retained by the stationary phase [40, 41]. This phenomenon can be problematic for the particular configuration of SEC×LC when a strong organic solvent is used as the mobile phase for the SEC separation.

Yang *et al.* recently reported the use of ASM in the SEC×RPLC separation of vinyl acetate/acrylic acid copolymers and vinyl acetate/itaconic acid/acrylic acid terpolymers. Breakthrough of the more polar copolymer components in the second dimension was effectively avoided by applying optimized ASM parameters. Using this approach, heterogeneous copolymer composition was revealed in random copolymer and terpolymer samples produced under certain process conditions [38]. This information can provide important insights to industrial polymer process development and optimization [38]. In a LC×SEC setup, online dilution can also help minimize the shift of ^{2}D peaks due to adsorption of weak solvent from the first dimension on the ^{2}D SEC column.

Molecular topology is one of the microstructure characteristics that has been shown to have great influence on the thermal and mechanical behaviors of polymers. Maiko *et al.* reported separation of polybutadiene based on molar mass and microstructure [42]. Polybutadiene samples containing different mol% of syndiotactic 1,2-polybutadiene, isotactic 1,2-polybutadiene, trans-1,4-polybutadiene, and cis-1,4-polybutadiene were successfully separated using LC×LC with gradient LC and SEC in the first and second dimensions, respectively.

Separations according to tacticity are very uncommon and difficult for synthetic polymers. Rode *et al.* investigated tacticity distribution of two amorphous poly-1-octene samples synthesized using different catalysts by online LC×SEC and offline SEC×LC with NMR detection [43]. The offline approach was more successful than the online approach because the LC separation using a Hypercarb™ column showed a strong dependence on molar mass which can potentially obscure a separation according to tacticity. Bivariate distribution with regard to tacticity and molar mass was observed from the offline 2D separation.

11.4.3 COMPLEX POLYMER MIXTURES

Given the ever-growing demand for better performance, lower cost and more environmentally friendly products, a single polymer can seldom meet all requirements in demanding applications. Blending several polymers to provide optimally balanced properties has become common practice in the development of new materials. Polymer additives are also very frequently used in formulated products to allow fine tuning of product properties in many end-use consumer products. These complex polymer systems containing a mixture of macromolecular substances with distinct properties are very difficult to analyze by spectroscopic techniques and 1D separations. It is even more challenging when there are great variations in the amounts of each polymeric constituent present in the complex mixture or formulated product.

LC×LC separation of commercial detergent samples to identify polymer dispersants was recently reported by Yang *et al.* [39]. By coupling SEC and RPLC separations in the comprehensive mode,

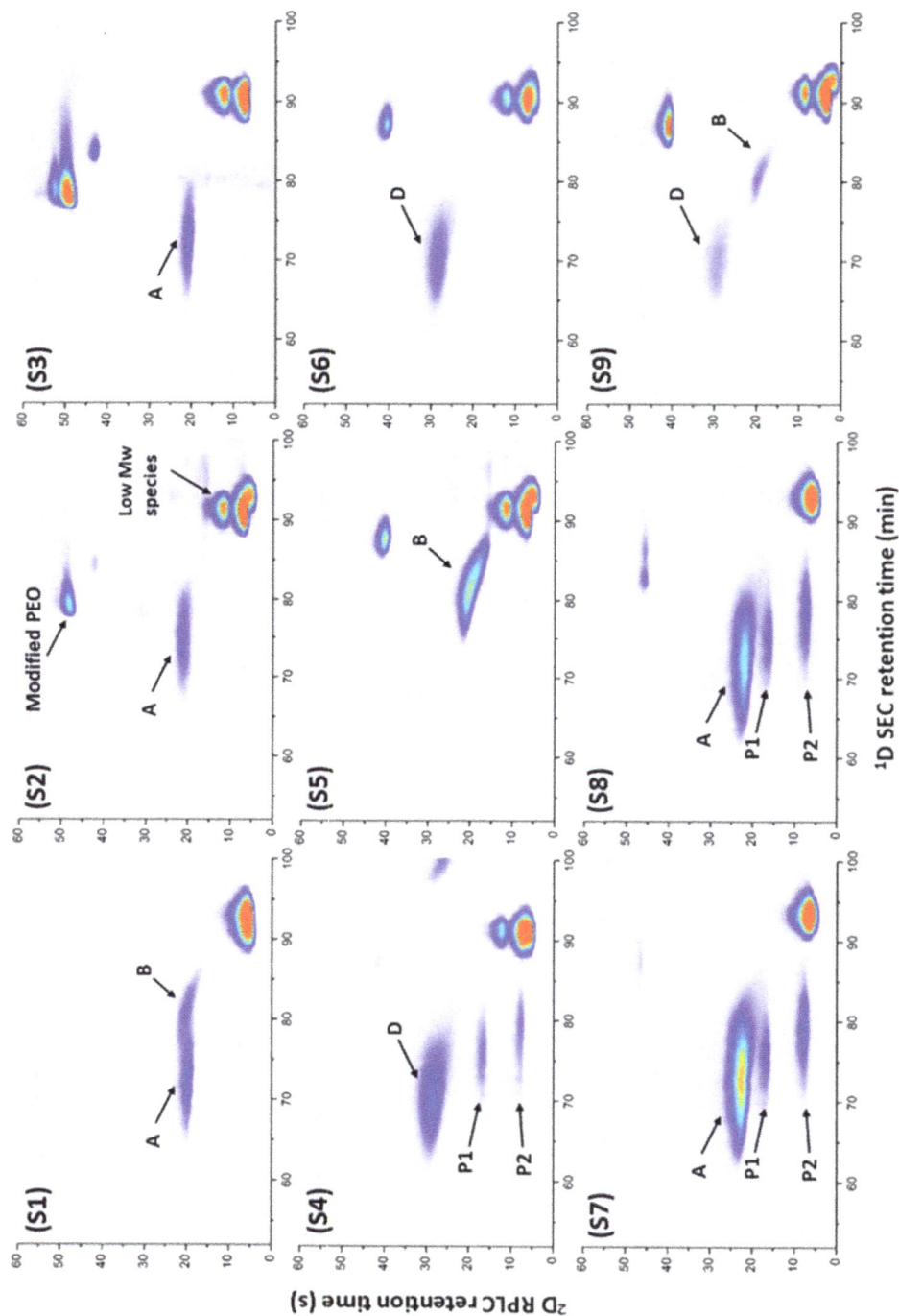

FIGURE 11.5 SECxRPLC analysis of commercial detergent samples for identification of polymeric dispersants. A, B, D refer to known polymeric dispersant ingredient with different chemical composition and molecular weight distributions. P1 and P2 are unknown polymeric ingredients in the detergent. Reprinted from *Journal of Chromatography, A,* 1566, P. Yang, W. Gao, J. Shulman, Y. Chen, Separation and identification of polymeric dispersants in detergents by two-dimensional liquid chromatography, 111–117, Copyright (2018), with permission from Elsevier.

a screening method was developed. By comparing unknown samples with available reference materials, polymer dispersants used in different detergent products were quickly identified in nine commercial detergent samples as shown in Figure 11.5, demonstrating the potential of comprehensive 2D-LC for routine industrial polymer analysis.

Complex polymer mixtures are often encountered in polymer synthesis when starting materials, desired products, and undesired byproducts coexist in process samples and final materials. Complete separation of all of the polymeric constituents in a polymer reaction mixture is complicated due to the presence of multiple distributions for each constituent in the mixture. A 1D separation can seldom provide sufficient information on polymer process samples. 2D-LC has not been routinely used in the analysis of complex polymer process samples, but its separation power makes it a very promising technique to enable deeper understanding of polymer synthesis and process control.

Jeong *et al.* reported characterization of polystyrene samples with unique bicyclic architecture synthesized by click chemistry [44]. Using NPLC×SEC separation, bicyclic polymer, monocyclic polymer, and branched polymer byproduct were successfully separated. Similar NPLC×SEC separation of figure-eight-shaped and cage-shaped cyclic polystyrenes was reported by Lee *et al.* [45]. These applications demonstrated the potential of 2D-LC for characterization of complex reaction mixtures.

11.4.4 HIGH-TEMPERATURE AND TEMPERATURE GRADIENT 2D-LC METHODS

High-temperature methods cannot be ignored when discussing polymer separations. Many polymeric materials such as polyolefins are insoluble in organic solvents at room temperature. High-temperature SEC (HT-SEC) has been routinely applied for determination of molar mass distributions. For LC separations based on chemical composition distribution (CCD), crystallization-based separation techniques such as temperature rising elution fractionation (TREF) and crystallization elution fractionation (CEF) have been used for samples with high crystallinity [46, 47]. For amorphous polymers or polymers with low crystallinity, high-temperature solvent gradient interaction chromatography (HT-SGIC) using a porous graphitic carbon (PGC) column has become a very useful approach [48].

Mekap *et al.* characterized polyolefins using HT 2D-LC with HT-SGIC and HT-SEC in the first and second dimensions, respectively [49]. Using stacked injections, they were able to improve the detector response for ethylene/1-octene copolymer blends. Cheruthazhekatt *et al.* reported characterization of impact polypropylene copolymer (IPC) using a combination of preparative scale TREF and HT-SGIC×HT-SEC separation. Ethylene-propylene (blocked and segmented) copolymers, low molar mass isotactic polypropylene (iPP) homopolymer, as well as high molar mass iPP and polyethylene (PE) homopolymer were observed in the IPC sample [50]. Prabhu *et al.* studied bimodal high-density polyethylene (BiHDPE) containing HDPE and linear low density PE (LLDPE). Using HT 2D-LC with LC separation in the first dimension followed by SEC in the second dimension with IR and ELSD detection, separation and detection of the two constituents were achieved [51].

HT 2D-LC has become a promising technique for the characterization of polymer materials that are insoluble at room temperature. It is important to note that conducting 2D-LC separations at high temperature (typically > 150°C) with organic solvent mobile phase like trichlorobenzene (TCB) requires special instrumentation that can mitigate potential safety hazards and maintain the high temperature for the entire separation and detection flow path. Commercial instruments designed for this purpose have become available in recent years making the routine practice of HT 2D-LC possible in industrial settings.

In the temperature gradient interaction chromatography (TGIC) approach developed by Chang *et al.* column temperature is varied to control retention of polymeric species during isocratic elution [52]. The advantage of TGIC over SGIC is the improved compatibility with RI and light scattering detectors due to the use of isocratic elution. Lee *et al.* reported LC×LC separation of polystyrene (PS) star polymers using reversed-phase temperature gradient interaction chromatography (RP-TGIC) in the first dimension and SEC in the second dimension followed by triple detection (light scattering, concentration, and viscosity). Unreacted arm (linear) and star-shape polymers with different numbers of arms were separated on the basis of branch number [53]. In a more recent study, Murima *et al.* demonstrated comprehensive branching analysis of polydisperse star-shaped polystyrenes using RP-TGIC×SEC [54]. Ndiripo *et al.* studied oxidized waxes having different levels of oxidation by SGIC or TGIC coupled with HT-SEC in a 2D-LC system [55]. Other polymer applications using TGIC in combination with SEC or LCCC in comprehensive 2D-LC configurations have also been reported recently for the separation of various polymer systems such as polystyrene/polybutadiene block copolymers and polystyrene-graft-polyisoprene [56, 57].

11.4.5   OTHER POLYMER APPLICATIONS INVOLVING UNCONVENTIONAL COMBINATIONS OF SEPARATION MECHANISMS

Some unconventional couplings of separation mechanisms have also been reported in recent years for polymer applications. Pirok *et al.* exploited a unique combination of hydrodynamic chromatography (HDC) and SEC for the separation of hydrophobic polystyrene (PS) and poly(-methyl methacrylate) (PMMA) nanoparticles for simultaneous determination of particle size distribution (PSD) and molecular-weight distribution [58]. Nanoparticles eluting from the ^{1}D HDC column were dissolved in tetrahydrofuran (THF) in an Agilent Jet Weaver mixer and then injected onto the ^{2}D SEC column. Stationary-phase-assisted modulation (SPAM) with C18 silica trap cartridges were used in the modulator loop to bridge the very different mobile phases used in the first and second dimensions, which would be otherwise incompatible in this particular case. The same group later reported separation of hydrophilic charged acrylic particles with a modified setup and automatic correction for HDC band broadening. The modified instrument setup is shown in Figure 11.6.

Another unconventional combination of separation mechanisms for polymer analysis was reported by Edam *et al.* for the separation of linear and long-chain branched polystyrenes with the same hydrodynamic size [59]. Molecular-topology fractionation (MTF) was applied in the first dimension to achieve branching selectivity for linear, Y-shaped and X-shaped molecules, and SEC was used in the second dimension to separate based on hydrodynamic size. In the ^{1}D MTF separation a monolithic column with flow-through channels only slightly larger than the polymer led to shear deformation of the polymer chains based on their topology which resulted in a change in migration rates. As shown in Figure 11.7, separation based on number of arms was achieved for a star-polymer sample.

11.5   OUTLOOK

Just less than a decade ago, 2D-LC had been mostly viewed as an academic research tool with limited applications in chemical industry. This was mainly due to the high complexity of hardware, lack of easy-to-use software for LC×LC data analysis, and large number of parameters to optimize with conflicting objectives. Advances in 2D-LC instrumentation and software in recent years led to an emergence of practical applications that are now routinely deployed in the chemical industry. Collaborations between research institutes, instrument manufacturers, and industrial laboratories have also accelerated the development of 2D-LC methods for industrial applications. We are optimistic that 2D-LC will continue to proliferate as an advanced analytical tool for the chemical industry and in academia. Using 2D-LC routinely in more laboratories may become more common in the future.

While 2D-LC has great potential for solving real-world problems, several challenges are still faced by industrial analytical chemists regarding 2D-LC method development and implementation. Instrument and method robustness and reliability are essential for implementation and routine use

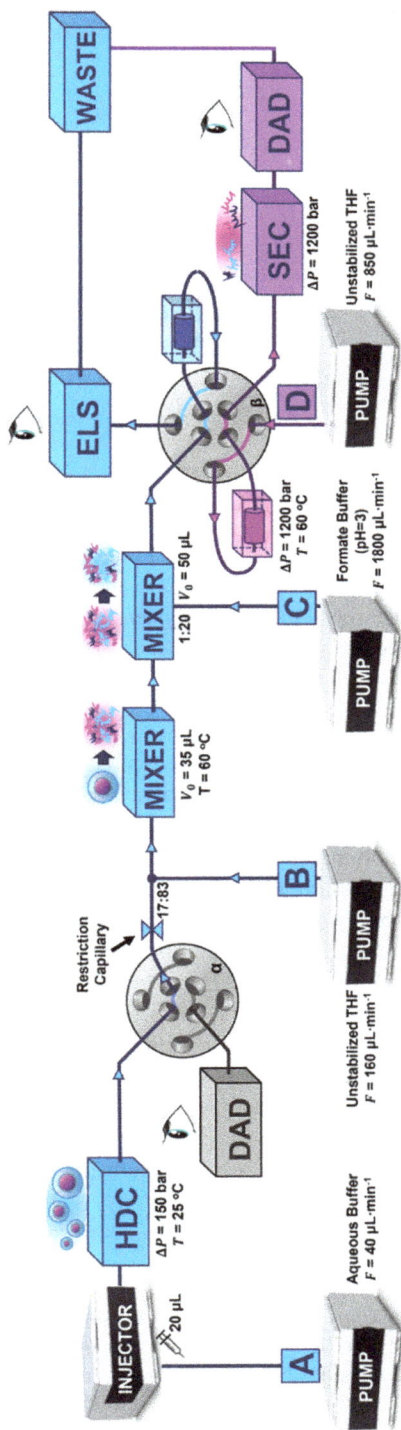

FIGURE 11.6 2D-LC setup for nanoparticle analysis. Reprinted from *Analytica Chimica Acta*, 1054, B. Pirok, N. Abdulhussain, T. Broojmans, T. Nabuurs, J. de Bont, M. Schellekens, R. Peters, P. Schoenmakers, Analysis of charged acrylic particles by on-line comprehensive two-dimensional liquid chromatography and automated data-processing, 184–192, Copyright (2019), with permission from Elsevier.

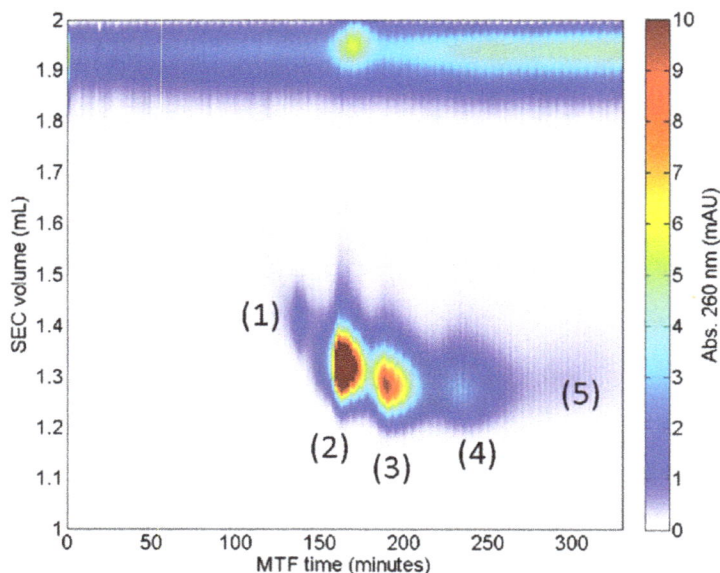

FIGURE 11.7 MTF×SEC separation of single arm (1), two-arm (2), three-arm (3), four-arm (4) star polymers, and higher coupling products (5). Reprinted from *Journal of Chromatography, A*, 1366, R. Edam, E. Mes, D. Meunier, F. Van Damme, P. Schoenmakers, Branched polymers characterized by comprehensive two-dimensional separations with fully orthogonal mechanisms: Molecular-topology fractionation×size-exclusion chromatography, 54–64, Copyright (2014), with permission from Elsevier.

in laboratories that have limited resources and expertise. Understanding of 2D-LC concepts and gaining proficiency in the use of hardware and software operations does take time to develop.

For polymer analysis, quantification within LC×LC separations remains very challenging. UV detection has limited use since most polymers do not contain strong UV absorbing constituents. RI detection is commonly used in SEC separation. However, it is less sensitive, is only compatible with isocratic separation, and can suffer from baseline problems when used as the ^{2}D detector in 2D-LC separations. In the case of universal aerosol-based detectors such as the evaporative light scattering detector (ELSD) and charged aerosol detector (CAD), non-linear detector response is a major problem; the non-linear impact of molecular weight, composition, and mobile phase conditions on detector response compromises performance for quantitative LC×LC analysis of polymers. For more information on the challenges in quantitative polymer analysis, readers can refer to a recent review article by Knol *et al.* [60]. Universal detectors with a more linear response are badly needed for polymer analysis in general, and especially for LC×LC. Development and commercialization of more SEC columns for fast separation (e.g., in the second dimension of LC×LC separations) is also highly desirable for polymer analysis by 2D-LC.

REFERENCES

[1] A. van der Horst, P.J. Schoenmakers, Comprehensive two-dimensional liquid chromatography of polymers, J. Chromatogr. A 1000(1–2) (2003) 693–709. https://doi.org/10.1016/S0021-9673(03)00495-3.

[2] H. Pasch, M. Adler, F. Rittig, S. Becker, New developments in multidimensional chromatography of complex polymers, Macromol. Rapid. Comm. 26(6) (2005) 438–444. https://doi.org/10.1002/marc.200400610.

[3] S. Weidner, J. Falkenhagen, R.-P. Krueger, U. Just, Principle of two-dimensional characterization of copolymers, Anal. Chem. 79(13) (2007) 4814–4819. https://doi.org/10.1021/ac062145f.

[4] D. Berek, Two-dimensional liquid chromatography of synthetic polymers, Anal Bioanal Chem 396(1) (2010) 421–441. https://doi.org/10.1007/s00216-009-3172-3.

[5] A. Baumgaertel, E. Altuntas, U.S. Schubert, Recent developments in the detailed characterization of polymers by multidimensional chromatography, J. Chromatogr. A 1240 (2012) 1–20. https://doi.org/10.1016/j.chroma.2012.03.038.

[6] H. Pasch, Hyphenated separation techniques for complex polymers, Polym Chem-Uk 4(9) (2013) 2628–2650. https://doi.org/10.1039/C3PY21095B.

[7] M. Pursch, S. Buckenmaier, Loop-based multiple heart-cutting two-dimensional liquid chromatography for target analysis in complex matrices, Anal. Chem. 87(10) (2015) 5310–5317. https://doi.org/10.1021/acs.analchem.5b00492.

[8] M. Pursch, A. Wegener, S. Buckenmaier, Evaluation of active solvent modulation to enhance two-dimensional liquid chromatography for target analysis in polymeric matrices, J. Chromatogr. A 1562 (2018) 78–86. https://doi.org/10.1016/j.chroma.2018.05.059.

[9] V. Elsner, V. Wulf, M. Wirtz, O.J. Schmitz, Reproducibility of retention time and peak area in comprehensive two-dimensional liquid chromatography, Anal. Bioanal. Chem. 407(1) (2015) 279–284. https://doi.org/10.1007/s00216-014-8090-3.

[10] Y. Li, D. Hewitt, Y.K. Lentz, J.A. Ji, T.Y. Zhang, K. Zhang, Characterization and stability study of polysorbate 20 in therapeutic monoclonal antibody formulation by multidimensional ultrahigh-performance liquid chromatography–charged aerosol detection–mass spectrometry, Anal. Chem. 86(10) (2014) 5150–5157. https://doi.org/10.1021/ac5009628.

[11] A. Escrig-Doménech, M. Beneito-Cambra, E. Simó-Alfonso, G. Ramis-Ramos, Single-pump heart-cutting two-dimensional liquid chromatography applied to the determination of fatty alcohol ethoxylates, J. Chromatogr. A 1361 (2014) 108–116. https://doi.org/10.1016/j.chroma.2014.07.092.

[12] A.F. Gargano, M. Duffin, P. Navarro, P.J. Schoenmakers, Reducing dilution and analysis time in online comprehensive two-dimensional liquid chromatography by active modulation, Anal. Chem. 88(3) (2016) 1785–1793. https://doi.org/10.1016/j.chroma.2018.07.054.

[13] G. Groeneveld, M.N. Dunkle, M. Rinken, A.F. Gargano, A. de Niet, M. Pursch, E.P. Mes, P.J. Schoenmakers, Characterization of complex polyether polyols using comprehensive two-dimensional liquid chromatography hyphenated to high-resolution mass spectrometry, J. Chromatogr. A 1569 (2018) 128–138.

[14] B.W. Pirok, J. Knip, M.R. van Bommel, P.J. Schoenmakers, Characterization of synthetic dyes by comprehensive two-dimensional liquid chromatography combining ion-exchange chromatography and fast ion-pair reversed-phase chromatography, J. Chromatogr. A 1436 (2016) 141–146. https://doi.org/10.1016/j.chroma.2016.01.070.

[15] K. Zhu, M. Pursch, S. Eeltink, G. Desmet, Maximizing two-dimensional liquid chromatography peak capacity for the separation of complex industrial samples, J. Chromatogr. A (2019) 460–457.

[16] F.T. van Beek, R. Edam, B.W.J. Pirok, W.J.L. Genuit, P.J. Schoenmakers, Comprehensive two-dimensional liquid chromatography of heavy oil, J. Chromatogr. A 1564 (2018) 110–119. https://doi.org/10.1016/j.chroma.2018.06.001.

[17] E. Lazzari, K. Arena, E.B. Caramao, M. Herrero, Quantitative analysis of aqueous phases of bio-oils resulting from pyrolysis of different biomasses by two-dimensional comprehensive liquid chromatography, J. Chromatogr. A 1602 (2019) 359–367. https://doi.org/10.1016/j.chroma.2019.06.016.

[18] M. Camenzuli, P.J. Schoenmakers, A new measure of orthogonality for multi-dimensional chromatography, Anal. Chim. Acta 838 (2014) 93–101. https://doi.org/10.1016/j.aca.2014.05.048.

[19] B.W.J. Pirok, A.F.G. Gargano, P.J. Schoenmakers, Optimizing separations in online comprehensive two-dimensional liquid chromatography, J. Sep. Sci. 41(1) (2018) 68–98. https://doi.org/10.1002/jssc.201700863.

[20] M. Sun, M. Sandahl, C. Turner, Comprehensive on-line two-dimensional liquid chromatography×supercritical fluid chromatography with trapping column-assisted modulation for depolymerised lignin analysis, J. Chromatogr. A 1541 (2018) 21–30. https://doi.org/10.1016/j.chroma.2018.02.008.

[21] J. Leonhardt, T. Teutenberg, J. Tuerk, M.P. Schlüsener, T.A. Ternes, T.C. Schmidt, A comparison of one-dimensional and microscale two-dimensional liquid chromatographic approaches coupled to high resolution mass spectrometry for the analysis of complex samples, Anal. Methods 7(18) (2015) 7697–7706. https://doi.org/10.1039/C5AY01143D.

[22] M. Bernardin, A.L. Masle, F. Bessueille-Barbier, C.-P. Lienemann, S. Heinisch, Comprehensive two-dimensional liquid chromatography with inductively coupled plasma mass spectrometry detection for the characterization of sulfur, vanadium and nickel compounds in petroleum products, J. Chromatogr. A 1611 (2020) 460–605. https://doi.org/10.1016/j.chroma.2019.460605.

[23] E. Lazaridi, H.-G. Janssen, J.-P. Vincken, B. Pirok, M. Hennebelle, A comprehensive two-dimensional liquid chromatography method for the simultaneous separation of lipid species and their oxidation products, J. Chromatogr. A 1644 (2021). https://doi.org/10.1016/j.chroma.2021.462106.

[24] R. Murgasova, D.M. Hercules, MALDI of synthetic polymers – An update, Int. J. Mass Spectrom. 226(1) (2003) 151–162. https://doi.org/10.1016/S1387-3806(02)00971-5.

[25] E. Uliyanchenko, Applications of hyphenated liquid chromatography techniques for polymer analysis, Chromatographia 80(5) (2017) 731–750. https://doi.org/10.1007/s10337-016-3193-y.

[26] J.M. Davis, S.C. Rutan, P.W. Carr, Relationship between selectivity and average resolution in comprehensive two-dimensional separations with spectroscopic detection, J. Chromatogr. A 1218(34) (2011) 5819–5828. https://doi.org/10.1016/j.chroma.2011.06.086.

[27] J. Engelke, J. Brandt, C. Barner-Kowollik, A. Lederer, Strengths and limitations of size exclusion chromatography for investigating single chain folding – current status and future perspectives, Polym Chem-Uk 10(25) (2019) 3410–3425. https://doi.org/10.1039/c9py00336c.

[28] P.J. Cools, A.M. Van Herk, A.L. German, W. Staal, Critical retention behavior of homopolymers, J. Liq. Chromatogr. Relat. Technol. 17(14-15) (1994) 3133–3143. https://doi.org/10.1080/10826079408013195.

[29] P. Schoenmakers, P. Aarnoutse, Multi-Dimensional separations of polymers, Anal. Chem. 86(13) (2014) 6172–6179. https://doi.org/10.1021/ac301162b.

[30] J. Dawkins, Statistical, gradient, block and graft copolymers by controlled/living radical polymerizations. KA Davis and K Matyjaszewski. Springer-Verlag, Berlin, Heidelberg, Germany, 2002. pp 191,. Polym. Int. 52(9) (2003) 1553–1554. doi: 10.1002/pi.1258

[31] K. Matyjaszewski, J. Spanswick, Controlled/living radical polymerization, Mater. Today 8(3) (2005) 26–33. https://doi.org/10.1016/S1369-7021(05)00745-5.

[32] M.I. Malik, H. Pasch, Novel developments in the multidimensional characterization of segmented copolymers, Prog. Polym. Sci. 39(1) (2014) 87–123. https://doi.org/10.1016/j.progpolymsci.2013.10.005.

[33] M.I. Malik, S. Lee, T. Chang, Comprehensive two-dimensional liquid chromatographic analysis of poloxamers, J. Chromatogr. A 1442 (2016) 33–41. https://doi.org/10.1016/j.chroma.2016.03.008.

[34] M. Al Samman, W. Radke, Two-dimensional chromatographic separation of branched polyesters according to degree of branching and molar mass, Polymer 99 (2016) 734–740. https://doi.org/10.1016/j.polymer.2016.07.077.

[35] W. Radke, Polymer separations by liquid interaction chromatography: principles – prospects – limitations, J. Chromatogr. A 1335 (2014) 62–79. https://doi.org/10.1016/j.chroma.2013.12.010.

[36] J.-A. Raust, A. Brüll, C. Moire, C. Farcet, H. Pasch, Two-dimensional chromatography of complex polymers: 6. Method development for (meth) acrylate-based copolymers, J. Chromatogr. A 1203(2) (2008) 207–216. https://doi.org/10.1016/j.chroma.2008.07.067.

[37] P. Yang, L. Bai, W.Q. Wang, J. Rabasco, Analysis of hydrophobically modified ethylene oxide urethane rheology modifiers by comprehensive two dimensional liquid chromatography, J. Chromatogr. A 1560 (2018) 55–62. https://doi.org/10.1016/j.chroma.2018.05.033.

[38] P. Yang, W. Gao, T. Zhang, M. Pursch, J. Luong, W. Sattler, A. Singh, S. Backer, Two-dimensional liquid chromatography with active solvent modulation for studying monomer incorporation in copolymer dispersants, J. Sep. Sci. (2019). https://doi.org/10.1002/jssc.201900283.

[39] P. Yang, W. Gao, J.E. Shulman, Y.S. Chen, Separation and identification of polymeric dispersants in detergents by two-dimensional liquid chromatography, J. Chromatogr. A 1566 (2018) 111–117. https://doi.org/10.1016/j.chroma.2018.06.063.

[40] E. Reingruber, F. Bedani, W. Buchberger, P. Schoenmakers, Alternative sample-introduction technique to avoid breakthrough in gradient-elution liquid chromatography of polymers, J. Chromatogr. A 1217(42) (2010) 6595–6598. https://doi.org/10.1016/j.chroma.2010.07.073.

[41] V. Pepermans, S. Chapel, S. Heinisch, G. Desmet, Detailed numerical study of the peak shapes of neutral analytes injected at high solvent strength in short reversed-phase liquid chromatography

columns and comparison with experimental observations, J. Chromatogr. A 1643 (2021). https://doi.org/10.1016/j.chroma.2021.462078.

[42] K. Maiko, H. Pasch, Comprehensive microstructure and molar mass analysis of polybutadiene by multidimensional liquid chromatography, Macromol. Rapid Comm. 36(24) (2015) 2137–2142. https://doi.org/10.1002/marc.201500381.

[43] K. Rode, F. Malz, J.-H. Arndt, T. Macko, G. Horchler, Y. Yu, R. Brüll, Studying the bivariate tacticity distribution of poly-1-octene using two-dimensional liquid chromatography coupled with NMR, Polymer 174 (2019) 77–85. https://doi.org/10.1016/j.polymer.2019.04.036.

[44] J. Jeong, K. Kim, R. Lee, S. Lee, H. Kim, H. Jung, M.A. Kadir, Y. Jang, H.B. Jeon, K. Matyjaszewski, Preparation and analysis of bicyclic polystyrene, Macromolecules 47(12) (2014) 3791–3796. https://doi.org/10.1021/ma500391z.

[45] T. Lee, J. Oh, J. Jeong, H. Jung, J. Huh, T. Chang, H.-j. Paik, Figure-eight-shaped and cage-shaped cyclic polystyrenes, Macromolecules 49(10) (2016) 3672–3680. https://doi.org/10.1021/acs.macromol.6b00093.

[46] S. Nakano, Y. Goto, Development of automatic cross fractionation: combination of crystallizability fractionation and molecular weight fractionation, J. Appl. Polym. Sci. 26(12) (1981) 4217–4231. https://doi.org/10.1002/app.1981.070261222.

[47] B. Monrabal, J. Sancho-Tello, N. Mayo, L. Romero, Crystallization Elution Fractionation: A New Separation Process for Polyolefin Resins, Macromolecular Symposia, Wiley Online Library, 2007, pp. 71–79.

[48] T. Macko, H. Pasch, Separation of linear polyethylene from isotactic, atactic, and syndiotactic polypropylene by high-temperature adsorption liquid chromatography, Macromolecules 42(16) (2009) 6063–6067. https://doi.org/10.1021/ma900979n.

[49] D. Mekap, T. Macko, R. Brüll, R. Cong, W. deGroot, A. Parrott, W. Yau, Multiple-injection method in high-temperature two-dimensional liquid chromatography (2D HT-LC), macromol. Chem. Phys. 215(4) (2014) 314–319. https://doi.org/10.1002/macp.201300649.

[50] S. Cheruthazhekatt, H. Pasch, Fractionation and Characterization of Impact Poly (propylene) Copolymers by High Temperature Two-Dimensional Liquid Chromatography, Macromolecular Symposia, Wiley Online Library, 2014, pp. 51–57 https://doi.org/10.1002/masy.201450306.

[51] K. Prabhu, R. Brüll, T. Macko, K. Remerie, J. Tacx, P. Garg, A. Ginzburg, Separation of bimodal high density polyethylene using multidimensional high temperature liquid chromatography, J. Chromatogr. A 1419 (2015) 67–80. https://doi.org/10.1016/j.chroma.2015.09.078.

[52] T. Chang, H.C. Lee, W. Lee, S. Park, C. Ko, Polymer characterization by temperature gradient interaction chromatography, Macromol. Chem. Phys. 200(10) (1999) 2188–2204. https://doi.org/10.1002/(SICI)1521-3935(19991001)200:10<2188::AID-MACP2188>3.0.CO;2-F.

[53] S. Ahn, H. Lee, S. Lee, T. Chang, Characterization of branched polymers by comprehensive two-dimensional liquid chromatography with triple detection, Macromolecules 45(8) (2012) 3550–3556. https://doi.org/10.1021/ma2021985.

[54] D. Murima, H. Pasch, Comprehensive branching analysis of star-shaped polystyrenes using a liquid chromatography-based approach, Analytical and bioAnal. Chem. (2019) 1–16. https://doi.org/10.1007/s00216-019-01846-7.

[55] A. Ndiripo, H. Pasch, Comprehensive analysis of oxidized waxes by solvent and thermal gradient interaction chromatography and two-dimensional liquid chromatography, Anal. Chem. 90(12) (2018) 7626–7634. https://doi.org/10.1021/acs.analchem.8b01480.

[56] S. Lee, H. Choi, T. Chang, B. Staal, Two-dimensional liquid chromatography analysis of polystyrene/polybutadiene block copolymers, Anal. Chem. 90(10) (2018) 6259–6266. https://doi.org/10.1021/acs.analchem.8b00913.

[57] S. Lee, H. Lee, T. Chang, A. Hirao, Synthesis and characterization of an exact polystyrene-graft-polyisoprene: A failure of size exclusion chromatography analysis, Macromolecules 50(7) (2017) 2768–2776. https://doi.org/10.1021/acs.macromol.6b02811.

[58] B.W.J. Pirok, N. Abdulhussain, T. Aalbers, B. Wouters, R.A.H. Peters, P.J. Schoenmakers, Nanoparticle analysis by online comprehensive two-dimensional liquid chromatography combining hydrodynamic chromatography and size-exclusion chromatography with intermediate sample transformation, Anal. Chem. (2017). https://doi.org/10.1021/acs.analchem.7b01906.

[59] R. Edam, E.P. Mes, D.M. Meunier, F.A. Van Damme, P.J. Schoenmakers, Branched polymers characterized by comprehensive two-dimensional separations with fully orthogonal mechanisms: Molecular-topology fractionation×size-exclusion chromatography, J. Chromatogr. A 1366 (2014) 54–64 https://doi.org/10.1016/j.chroma.2014.09.011.

[60] W.C. Knol, B.W.J. Pirok, R.A.H. Peters, Detection challenges in quantitative polymer analysis by liquid chromatography, J. Sep. Sci. 44(1) (2021) 63–87. https://doi.org/10.1002/jssc.202000768.

Applications of Two-Dimensional Liquid Chromatography to Natural Product and Food Analysis

Magriet Muller and André de Villiers

CONTENTS

12.1 INTRODUCTION

Food and natural products are complex biological materials containing nutrients and bioactive molecules that are vital for human health. Along with both major (lipids, proteins and carbohydrates) and minor (vitamins and polyphenols) nutrients, these products may also contain non-nutritional additives and toxic contaminants. Advanced analytical methods are essential tools in the investigation of the chemical composition of food and natural products, and play a vital role in ensuring their safety, assessing quality, and detecting adulterants. In addition, these methods are extensively used in the detection and identification of influential and/or novel bioactive molecules in natural products and derived food commodities.

The analysis of natural products and food requires analytical methods that are capable of determining both major and minor constituents with vastly different properties and concentrations in highly complex mixtures. Due to the non-volatile nature of most important nutrient and metabolite classes, HPLC is widely applied for this purpose. Hyphenation of LC to mass spectrometry (MS) significantly extends the utility of the method by providing structural information as well as a highly selective and sensitive method of detection [1–4]. Sample preparation is often essential to remove interfering matrix compounds and to concentrate analytes of interest [5–7]. However,

DOI: 10.1201/9781003090557-12

sample clean-up can be laborious and may introduce errors, while insufficiently separated matrix compounds can interfere with MS detection. Thus further improvements are needed in the chromatographic separation of complex food and natural product mixtures. A range of options are available to improve the performance of one-dimensional HPLC (1D-LC; as outlined in Chapter 2), and indeed these have been used in food and natural product analysis. While the performance of 1D-LC has improved drastically in recent years, it still remains ultimately limited by instrumental and selectivity constraints.

Multi-dimensional LC (MD-LC) is a promising alternative approach, where two (or more) independent separations are coupled to significantly increase the resolving power. In two-dimensional LC (2D-LC), this involves transferring fractions of effluent from one column to a second column, which provides complementary selectivity to further separate analytes that co-eluted from the first column. Heartcutting 2D-LC (LC-LC)[1] methods are ideal for the targeted analysis of specific food components, where the first dimension is often used as a sample fractionation step to separate the analytes of interest from interfering matrix components, thereby reducing the need for sample preparation. A further application of LC-LC is the preparative scale isolation of natural compounds for further characterization. Comprehensive 2D-LC (LC×LC) on the other hand provides exceptional performance for analysis of the complete sample. Untargeted LC×LC screening of natural products is used for fingerprint profiling and the characterization of complex samples, whereas targeted LC×LC methods aim to increase selectivity and sensitivity for quantitative analysis of specific components. Both online and offline MD-LC methods have found application in the analysis of foods and natural products. Because time constraints on the ^{2}D separation are not as serious in the offline 2D-LC approach as compared to those in online 2D-LC, offline 2D-LC can provide exceptionally high peak capacities and is mainly used for the detailed characterization of highly complex natural products. While online 2D-LC is technically more challenging to implement in practice, it offers the advantage of automation and reduced risk of sample alteration, two aspects of importance especially in routine analysis, and is the most common mode of 2D-LC used in food and natural product analysis today [8].

This chapter provides an overview of the use of 2D-LC in the analysis of food and natural products. The goal is not to provide an exhaustive review of all applications in the field[2] – for a more comprehensive list the reader is referred to several recent reviews [1,9–17] – but rather to illustrate typical applications of 2D-LC for particular classes of food and natural product constituents, with an emphasis on particularly interesting applications of 2D-LC to illustrate the potential and versatility of the technique.

12.2   MAJOR FOOD AND NATURAL PRODUCT COMPONENTS

12.2.1   CARBOHYDRATES

Carbohydrates are one of the most important energy sources in food, and comprise monosaccharides that can be joined by glycosidic linkages to form oligosaccharides and polysaccharides. From a nutritional perspective, the four essential classes of carbohydrates are sugars, starch, prebiotic oligosaccharides, and dietary fiber. Analysis of these compounds is essential to determine the quality, authenticity, and nutritional content of food and natural products, and GC, CE, and HPLC have all been used for this purpose. In the case of HPLC, hydrophilic interaction LC (HILIC) [18] and anion-exchange chromatography (AEX) [19] are commonly applied [20], while the use of MD-LC in this area has been very limited to date.

In order to overcome the incompatibility of AEX mobile phases and electrospray ionization (ESI)-MS, Klavins *et al.* [21] developed an online multiple heartcutting (MHC) LC method for the quantification of sugar phosphates, where the ^{2}D separation was used as a desalting step prior to MS detection. AEX peaks were transferred via a sample loop to a porous graphitized carbon column where the sugar phosphates were separated from the unretained Na^+ ions.

A single heartcut LC-LC method coupled with refractive index detection was applied for the quantification of mono- and disaccharides in milk powder [22]. The method eliminated the need for complicated sample preparation procedures by using the first dimension as a fractionation step. Sugars were separated from the other milk constituents (mainly proteins) using a reversed-phase C4 column, and transferred as a single fraction to an amino HILIC column for further separation. LC-LC has also been used in combination with isotope ratio mass spectrometry to determine the stable carbon isotope ratios of gluconic acid in honey [23].

Comprehensive 2D-LC using HILIC and RP separations (HILIC×RP) has been used for the separation of prebiotic oligosaccharides [24]. An amide column (2.1 mm i.d.) was used in the first dimension, and a C18 column (4.6 mm i.d.) was used in the second dimension. Two C18 trapping columns along with an aqueous make-up flow were used in the interface (i.e., see Figure) 4.22 to overcome mobile phase mismatch problems.

12.2.2 LIPIDS

Lipids are a structurally and functionally diverse group of hydrophobic compounds that contribute to the taste, flavor and nutritional value of food. In nature, lipids generally occur as triacylglycerols (TAGs) and phospholipids (PL). The configuration (*cis* or *trans*) and number of double bonds have a major influence on the health-related properties of TAGs, and their analysis is important in the assessment of food quality as well as authenticity [25]. Reversed-phase LC, specifically non-aqueous RP (NARP), is commonly applied for the analysis of TAGs [26]. Elution follows increasing partition numbers (PNs), defined as the number of carbon atoms minus twice the number of double bonds. Silver-ion LC (Ag-LC, or "argentation" chromatography) can be used to separate TAGs that have the same PN, but differ in the number, position, and configuration of double bonds through the formation of weak charge transfer complexes between silver ions and TAG double bonds [27].

The complementary separation mechanisms of NARP and Ag-LC have been exploited for the analysis of TAGs in various food samples [28–40]. In online Ag-LC×NARP applications using passive modulation [28, 30, 31, 33, 34, 36, 37], ^{2}D injection band broadening is typically minimized by coupling narrow-bore (1.0 or 2.1 mm i.d.) ^{1}D columns operated at low flows (7–20 µL/min) to wide-bore (4.6 mm i.d.) ^{2}D columns operated at high flows (3–4 mL/min). Beccaria et al. [39] reported stop-flow and offline Ag-LC×NARP methods, where the relaxed time restrictions were exploited to use longer (150 mm) ^{2}D columns. By exchanging the ^{1}D solvent in the offline method, the approach improved ^{2}D peak focussing and allowed the identification of more than 250 TAGs in fish oil. Byrdwell swapped the two separations in an online NARP×Ag-LC method for the analysis of seed oil TAGs as shown in Figure 12.1 [40].

Supercritical fluid chromatography (SFC) has also been used in the multi-dimensional analysis of TAGs [41, 42] and their derived fatty acids (FAs) [43, 44]. SFC in the first dimension offers the advantage of avoiding solvent incompatibility issues (since the CO_2 mobile phase evaporates once depressurized), but the modulation process is more complicated. François *et al.* [45] designed an interface using a ten-port switching valve with C18 guard columns and aqueous make-up flow to trap analytes during CO_2 expansion. This design was used for the online hyphenation of silver-ion SFC to both RP [44] and NARP [42] for the analysis of derivatized FAs and TAGs, respectively, in fish oil.

Normal-phase LC (NP-LC) separates lipids based on differences in their polar head groups, with the fatty acid composition having little influence on retention. It is therefore useful for fractionating total lipid extracts into different lipid classes. This property was utilized in an online

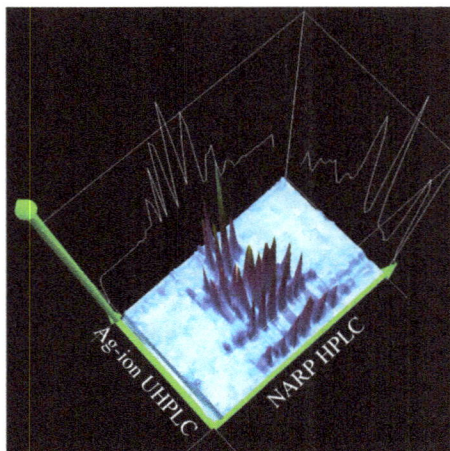

FIGURE 12.1 3D surface plot for the online NARP×Ag-LC-ESI-MS analysis of seed oil TAGs. Conditions: ^{1}D – C18 column (2 x 250 x 4.6 mm i.d., 5 μm); Gradient time, 118 min.; Flow rate, 0.2 mL/min.; Modulation, passive modulation; Sampling time, 1.91 min.; Loop volume, 100 μL (^{1}D flow split); ^{2}D – Ag-ion column (100 mm×2.1 mm i.d., 3 μm); Gradient time, 1.86 min. Reprinted with permission from W. Byrdwell, Comprehensive dual liquid chromatography with quadruple mass spectrometry (LC1MS2×LC1MS2=LC2MS4) for Analysis of *Parinari Curatellifolia* and other seed oil triacylglycerols, *Analytical Chemistry* 89 (2017), 10537–10546. Copyright 2017 American Chemical Society.

LC-LC approach that employed NP in the first dimension to separate phosphatidylethanolamines and phosphatidylcholines from other food lipid classes [46]. The NP separation used a silica-based polyvinyl alcohol column, and the fractions containing the target phospholipid classes were trapped on C18 traps with a make-up flow of weak eluent, prior to backflushing to a RP column where molecular species were separated.

HILIC also separates lipid classes on the basis of polarity, but offers the advantage of using aqueous-organic mobile phases, which are miscible with traditional RP solvents and compatible with ESI-MS. The combination of HILIC and RP has been used in heartcutting [47] and comprehensive [48–50] 2D-LC analyses of food lipids. A stop-flow HILIC×RP method was developed for the analysis of phospholipids in cow's milk [48]. In the first dimension, a bare silica HILIC column separated the phospholipids into six main classes, which were further resolved into 50 molecular species on a C18 column operated isocratically. Stop-flow operation proved critical to extend the analysis time in the second dimension and improve separation according to PN. Both HILIC×RP and RP×HILIC have been used in lipidomic studies of food samples. Donato *et al.* [50] coupled a 2.1 mm i.d. ^{1}D HILIC column online via passive modulation to a 4.6 mm i.d. ^{2}D C18 column to profile the lipid content of Mediterranean mussels. Navarro-Reig *et al.* [49] reported a chemometric approach to analyze the complex data generated from the untargeted RP×HILIC-MS/MS lipidomic analysis of Japanese rice exposed to arsenic.

12.2.3 PROTEINS

Proteins are essential for the existence of all organisms and fulfill many roles in cells, including enzymatic catalysis of metabolic reactions. Proteomics, the study of the entire set of proteins in an organism or system, is used to identify protein composition, determine potential sites for drug interaction, study metabolic pathways, and monitor gene expression as a response to stress factors. Given the complexity of the proteome, it is not surprising that MD-LC-MS has been used extensively in proteomics research [51–55]. Studies dealing with the entire proteome of natural products will not be covered in this chapter; rather, we will limit the discussion to a few applications directly related to the quality or nutritional value of foods and consumable natural products.

Guillén-Casla *et al.* [56] studied the effect of electron-beam irradiation, used to sterilize food products, on the enantiomeric composition of amino acids in meat and dairy products. Racemization of three marker amino acids was evaluated through online heartcutting combining an achiral racemic [1]D separation on a C18 column followed by enantiomeric separation on a chiral column coupled to UV detection. The combination of RP and chiral LC using different chiral stationary phases has also been used to determine the ratios of D- and L-amino acids in rice vinegar [57] and tea [58] following derivatization to enable fluorescence (FL) or UV detection.

Soybeans are a rich source of plant-based proteins and are often added to processed food to increase their total protein content. An offline LC-LC method, using perfusion LC in the first dimension followed by tryptic digestion and RP-MS/MS analysis, was used to identify and monitor marker proteins in heat-treated meat products to detect adulteration [59]. Soybean is however also a common food allergen, and online LC-LC has also been used to quantify Gly m 4, one of the major soybean allergen proteins [60]. A strong anion-exchange (SAX) column was coupled via a ten-port valve equipped with two C4 trapping columns to a [2]D C4 RP column hyphenated to UV and high resolution (HR)-MS detection. In further work on soybean anti-nutritional factors, Kunitz trypsin inhibitor was isolated and quantified by preparative 2D-LC using weak anion-exchange (WAX) separation followed by size-exclusion chromatography (SEC) [61].

In an interesting offline 2D-LC application, Liu *et al.* [62] studied whey protein hydrolysates (WPH) to identify peptides responsible for the bitter taste introduced during protein hydrolysis. Commercial WPH powder was separated into high and low molecular weight fractions through ultrafiltration, and the latter was fractionated using a preparative C18 column with an acidified mobile phase (pH 2); 26 collected fractions were freeze-dried and re-dissolved in water for sensory analysis. Three fractions rated as the highest on the bitterness scale were further separated on the same C18 column operated at pH 5 to obtain about 58 [2]D fractions. These fractions were again sensorially evaluated, and bitter sub-fractions were subsequently analyzed by RP-HR-MS, leading to the identification of four bitter peptides.

12.3 MINOR FOOD AND NATURAL PRODUCT COMPONENTS

12.3.1 POLYPHENOLS

Polyphenols are a diverse class of secondary plant metabolites that are important due to their purported health benefits and the role they play in the quality of many food products [63, 64]. They can be broadly classified as flavonoids, based on a C6-C3-C6 backbone, and non-flavonoids [65], and often occur in foods and plants as O- and/or C-glycosides.

Polyphenols are most often analyzed using RPLC hyphenated to UV and MS detection [66–68]. Due to the large number of structurally related and isomeric polyphenols, however, 1D-LC often provides inadequate separation, while their low levels of occurrence and the lack of commercial standards present further analytical challenges. Coupled with the importance of polyphenols, this has led to numerous 2D-LC methods being developed for their improved analytical performance; the following discussion is limited to publications in the last 10 years (readers are referred to dedicated reviews for further information [9, 10, 68, 69]). RP×RP and HILIC×RP separations are by far the most popular methods for the analysis of phenolic compounds in food and natural products and account for about 90% of LC×LC applications. The speed and automation of online systems make this the preferred approach, although 23% of published LC×LC methods are offline 2D-LC that provide more flexibility with respect to [2]D analysis time and addressing mobile phase mismatch problems.

12.3.1.1 RP×RP Separations

RP×RP methods have the advantage of straightforward hyphenation, but can suffer from poor orthogonality for polyphenols. The performance of RP×RP methods involving three types of ^{2}D gradients – namely full in fraction (FIF), segmented in fraction (SIF), and continuous shifting (CS) gradients – were compared for the separation of red wine [70] and sugarcane [71] polyphenols (see Section 4.5 for more detail about these elution schemes). In both studies, SIF gradients showed the best utilization of the 2D separation space for partially correlated systems. With the introduction of dedicated LC×LC software, the use of shifting ^{2}D gradients (SGs) has become popular. For example, Donato *et al.* [72] reported a 20% increase in the usage of the 2D separation space for red wine phenolics through the introduction of segmented gradients. Multi-step [73] and multi-segmented SG [74] variations have been applied for the analysis of phenolics and saponins in licorice extracts. The combination of cyanopropyl (CN) and C18 columns has been popular for the RP×RP analysis of phenolics in licorice [74], pistachio [75], and pomegranate [76]. Other column combinations used for RP×RP analysis of polyphenols include CN×RP-amide [77, 78], pentafluorophenyl (PFP)×C18 [79], RP-amide×C18 [80], and C18×C18 [81]. To improve orthogonality, different organic modifiers are often used, with methanol (MeOH) and ACN being the most common [73, 79, 81–83].

12.3.1.2 HILIC×RP Separations

HILIC×RP separation offers an orthogonal combination especially useful for the analysis of polyphenols [84]. The high elution strength of HILIC mobile phases in RP can however result in severe peak distortion in the ^{2}D RP separation, which complicates method development [85, 86].

For ^{1}D HILIC separations, diol [85, 87–96] and amide [97–103] columns are commonly employed, although bare silica [104], amino [105, 106], and zwitterionic [107–109] stationary phases have also been used.

HILIC×RP separations have been applied to most classes of phenolic compounds (see Figure 12.2). For example, green tea proanthocyanidins, flavonols, and flavones were separated by offline HILIC×RP-MS using a combination of diol and C18 columns with UV, FL, and HR-MS detection [85]. Second dimension injection band broadening was minimized by re-injecting only 4% of the collected 50 µL fractions of ^{1}D effluent. For glycosylated flavonoids, retention in HILIC increased with the number of attached sugar moieties, while RP offered separation based on the nature of the aglycone and sugar moieties, thus providing group-type 2D separation to facilitate compound identification. A similar method was applied for the separations of dihydrochalcones, flavanones, flavones, and flavonols in rooibos tea [87].

Anthocyanins, a diverse group of flavonoid pigments, exhibit unique chromatographic behavior [110], and dedicated off- and online HILIC×RP methods using highly acidic mobile phases have been developed for their analysis in various fruits and vegetables [98, 106] and red wine [97].

Procyanidins-oligomeric and polymeric flavanol species- are notoriously difficult to analyze by 1D-LC, and great emphasis has been placed on their separation by 2D-LC [89–94, 111–113]. HILIC×RP is extensively used here, since HILIC provides exquisite selectivity for procyanidins, which elute in order of increasing degree of polymerization, whereas isomers with the same degree of polymerization can be separated in RP. A practical approach for the optimization of offline, stop-flow and online HILIC×RP separations of procyanidins was used to evaluate the advantages and limitations of each methodology [91, 113]. Montero *et al.* [93] found a monolithic C18 column to be more suitable than a superficially porous (SPP) C18 column in the ^{2}D for the online HILIC×RP-MS analysis of grape seed procyanidins, and used a similar method for the analysis of apple polyphenols [90] and pholorotannins in brown algae [95, 96]. The same group investigated diol, zwitterionic and polyethylene glycol HILIC phases for the HILIC×RP-MS analysis of proanthocyanidins and stilbenoids in grapevine canes, and found that the diol phase provided the highest orthogonality (78% according to the asterisk method; 72% when recalculated using the convex hull method) when used in combination with a C18 column in the second dimension [89]. In an interesting application

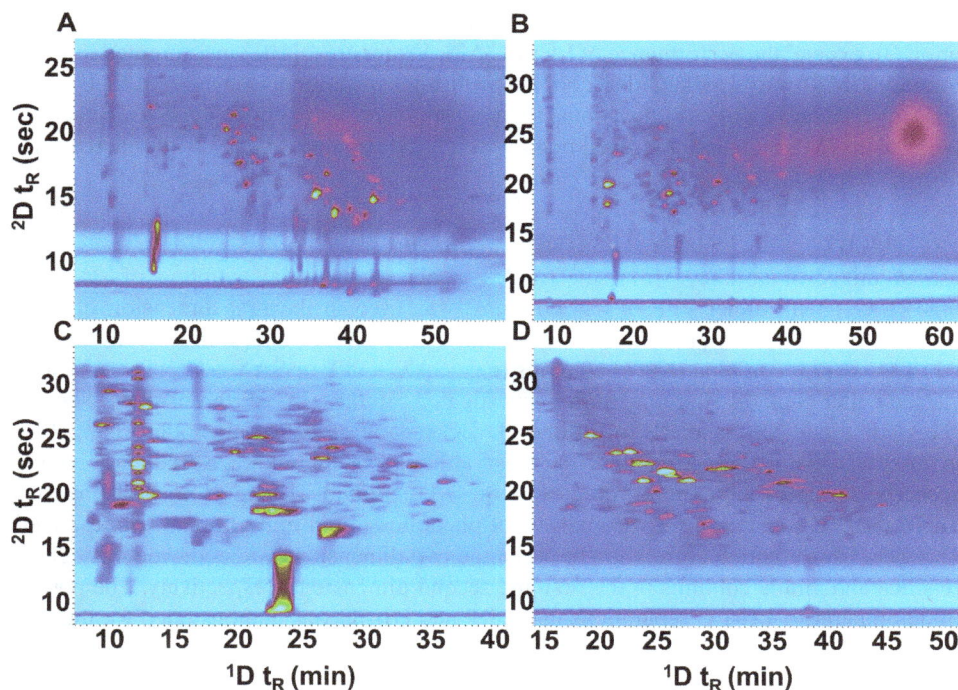

FIGURE 12.2 UV contour plots (280 nm) obtained for the online HILIC×RP separation of phenolic compounds in (A) chestnut; (B) grape seed; (C) wine and (D) rooibos tea. Chromatographic conditions: ¹D – Amide column (150 mm×1.0 mm i.d., 1.7 μm); Modulation: Fixed solvent modulation; Loop volume, 80 μL; ²D – C18 column (50 mm×3 mm i.d., 1.8 μm). For sample-specific conditions, refer to [101]. Reprinted with permission from P. Venter, M. Muller, J. Vestner, M. Stander, A. Tredoux, H. Pasch, A. de Villiers, Comprehensive three-dimensional LC×LC×ion-mobility spectrometry separation combined with high-resolution MS for the analysis of complex samples, Analytical Chemistry 90 (2018), 11643–11650. Copyright 2018 American Chemical Society.

of HILIC×RP [94], both offline and online methods were coupled with an online radical scavenging assay to evaluate the antioxidant capacity of cocoa, grape seed, and green tea phenolics.

The challenges associated with the online coupling of HILIC and RP have been overcome in several ways. For passive modulation approaches, it is common to use narrow-bore ¹D columns (e.g., 1.0 mm i.d.) operated at low flows (e.g., 10 μL/min.) to minimize fraction volumes [105, 114]. In some cases, groups have split the ¹D flow between the ¹D column outlet and the modulation interface to further reduce fraction volumes [113, 115]. Alternative modulation approaches have also been applied in recent years. Fixed solvent modulation (FSM), where the ¹D eluent is diluted with weak ²D mobile phase, was used in combination with a predictive kinetic optimization protocol to discover optimal conditions for the analysis of procyanidins [112]. This approach was subsequently extended to the online HILIC×RP-MS analysis of various phenolic classes, providing practical peak capacities of up to 3,000 [103]. Stationary phase assisted modulation (SPAM) has also been used, for example in a promising application for the separation and tentative identification of 121 polyphenols in an apple extract [88]. Two C18 trapping columns were used with an aqueous make-up flow added before the interface to increase analyte retention on the traps. The effects of different trapping columns and make-up flow rates on the effectiveness of SPAM has been investigated [108]. Finally, the advantages of SPAM were also demonstrated for the analysis of cocoa procyanidins [111].

For highly complex phenolic mixtures, MS detection is essential in combination with LC×LC separation. To expand the performance of LC×LC-MS, Venter *et al.* [101] incorporated ion-mobility spectrometry (IMS) into a HILIC×RP coupled with high resolution mass spectrometry (HRMS) methodology to achieve a three-dimensional separation system for phenolic analysis. Benefits of the incorporation of IMS included improved mass-spectral data quality, and a 13-fold increase in peak capacity.

12.3.1.3 Other

The use of mixed-mode-like (MM) selectivities, achieved by serial coupling of strong anion-exchange (SAX) with either PFP or CN columns, was used in the first dimension of an online LC×LC system by Li *et al.* [116]. When used with a C18 in the second dimension, their studies showed that different combinations of columns (e.g., SAX + PFP or SAX + CN) in the first dimension offered better peak distributions for the investigated samples. In a different study, Scoparo and coworkers combined SEC and RP for offline 2D-LC separation of tea phenolics [117]. The SEC elution order was mainly based on molecular mass, although as-yet unknown secondary interactions also appeared to affect the elution volumes of these compounds. Liang *et al.* [104] applied off-line HILIC×HILIC separation for the analysis of phenolics found in traditional Chinese medicine (TCM), where low retention correlation between the two dimensions was achieved through the use of bare silica and amide columns in the first and second dimensions, respectively. The potential of the offline preparative LC-LC separations using RP and HILIC was demonstrated for the isolation of eight flavonoid glycosides from *Sphaerophysa salsula* [118].

12.3.2 SAPONINS

Saponins are a diverse group of structurally related compounds comprised of steroid (C27) or triterpenoid (C30) aglycones linked to one or more saccharide moieties. They contribute to the health benefits of several herbal medicines [119], which has led to the extensive study of saponins, especially in TCMs [120]. Variations in both the aglycone and carbohydrate moieties as well as glycosylation position result in considerable structural diversity of saponins, making their analysis challenging.

Saponins are generally analyzed by RP [121], although the hydrophilic character of the glycosides also makes them amenable to HILIC separation. Indeed, online HILIC×HILIC has been applied for saponin analysis [122]. Relatively good usage of the separation space was achieved by utilizing amide and poly(2-hydroxyethyl aspartamide) columns, although the low peak capacity (about 100) reported can partially be ascribed to the inferior kinetic performance of HILIC compared to RP.

Jeong *et al.* [123] compared the performance of RP×RP and RP×HILIC systems, and concluded that the correlation of selectivities observed in the case of RP×RP was too high for the second dimension to provide a meaningful contribution to the separation of Platycodi Radix saponins. However, judicious selection of mobile phases [73, 124] and use of SGs [74] can be used to improve the orthogonality of RP×RP separations. For example, a RP×RP-MS method using MeOH/H_2O in the first dimension (C18 column) and ACN/H_2O in the second dimension (phenyl-hexyl column) proved efficient for both untargeted metabolic profiling [73] and characterization of saponins [124]. This is a surprising result, since it has been shown previously that using ACN-containing mobile phases suppresses some of the selectivity of aromatic phases for analytes rich with pi electrons and strongly dipolar functional groups [125].

Considering the popularity of HILIC×RP for the separation of phenolic glycosides (Section 12.3.2), it is not surprising that this combination has also been exploited for the analysis of saponins. *Panax ginseng*, an important component of TCM, has been analyzed by offline [126] and stop-flow [163] HILIC×RP separations using amide and C18 columns in the first and second dimensions, respectively. Online HILIC×RP using passive [107] and SPAM [108] modulation have been used for the separation of phenolics and saponins in licorice. More recently, Zhang *et al.* [128] reported the hyphenation of a HILIC×RP separation to IMS-MS detection for the analysis of ginseng saponins. The addition of IMS enabled the gas-phase separation of 10 isomeric pairs that were unresolved by 2D-LC-MS, and facilitated the detection of 201 saponins.

Schmitz and coworkers [129] developed a novel "at-column dilution" modulation approach to overcome the solvent incompatibility issues associated with coupling RP and HILIC. The system uses a transfer pump to flush collected fractions from the loops of a conventional 8-port / 2-position valve, which is connected with a tee-piece to the ^{2}D pump flow before the ^{2}D column. Dilution of fractions delivered to the ^{2}D column is regulated by the ratio of the flow rates of the transfer and ^{2}D pumps (see Section 4.4.5.4 for more discussion of this approach). This approach was successfully used in the RP×HILIC-MS analysis of saponins in ginseng [129] and phenolics in *Buddleja davidii* [130].

12.3.3 CAROTENOIDS

Carotenoids are natural pigments produced by plants, algae, and some microorganisms. They are responsible for the red, orange, and yellow colors of many fruits and vegetables and are used as natural dyes in the food industry [131]. Carotenoids consist of C40-tetraterpenoid structures with extensive conjugation, and can be broadly classified into carotenes, which are strictly hydrocarbons, and xanthophylls, which contain oxygen. More than 1100 naturally occurring carotenoids have been identified [132].

RP coupled with UV and/or MS detection is the primary method used for carotenoid analysis [133]. Carotenoids have distinctive UV-Vis spectra, and therefore diode-array detection provides both quantitative and qualitative information, but MS is required for detailed characterization. Both C18 and C30 stationary phases are used, with C30 providing improved separation of cis/trans isomers [134]. However, complete separation of all structural and conformational isomers using RP is rarely achieved for natural products, and samples are often saponified to release esterified carotenoids and simplify their analysis, a process that may affect the native carotenoid composition [135].

The first LC×LC analysis of carotenoids was performed in 2006 by Dugo *et al.* [136] on saponified citrus products. A bare silica NP column was used in the first dimension to separate different classes of carotenoids based on polarity, while in the second dimension a monolithic C18 column separated compounds within each polarity group based on hydrophobicity. This provided clear group-type separation of different carotenoid classes. The notoriously difficult coupling of NP and RP due to the mismatch in relative mobile phase elution strengths and solvent immiscibility was made possible in this study by: 1) coupling a 1.0 mm i.d. ^{1}D column operated at 10 µL/min with a 4.6 mm i.d. ^{2}D column operated at 4.7 mL/min; and 2) the use of a strong initial ^{2}D mobile phase composition (~76% ACN, 5% isopropanol) in the ^{2}D, which still provides reasonable retention of the highly lipophilic carotenoids by the RP column.

For the analysis of intact carotenoids [137–139], the method described above was adapted by replacing the bare silica ^{1}D column with a cyanopropyl column and a hexane/butyl acetate/acetone mobile phase, and supplementing UV detection with atmospheric pressure chemical ionization (APCI) MS. Further improvements to the method involved the use of SPP UHPLC columns in the second dimension [140, 141] (Figure 12.3) and reducing the sampling time [140] to achieve a practical peak capacity of 639. In an interesting application [142], the method was used to determine the stability of carotenoids in overripe fruit in order to evaluate the nutritional value of waste generated by food markets. More recently, Gallego *et al.* [143] reported an online NP×RP method for carotenoid analysis in less than 25 minutes using an amino and SPP C30 columns in the first and second dimensions, respectively.

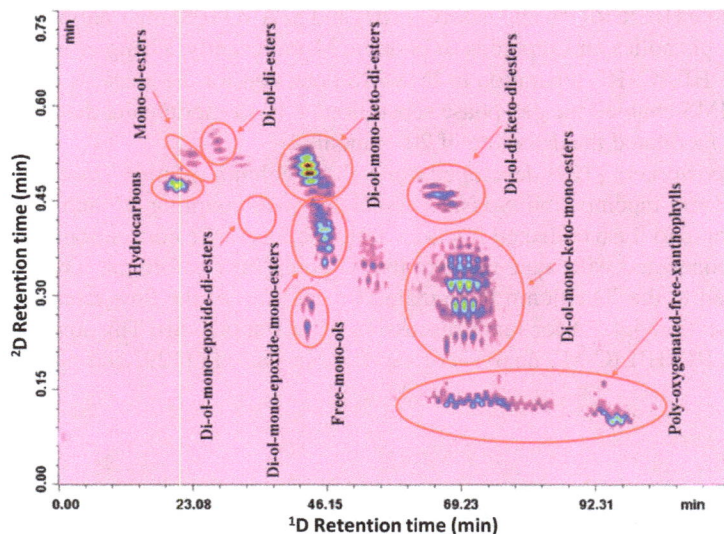

FIGURE 12.3 UV contour plot (450 nm) for the online NP×RP analysis of chili carotenoids. Chromatographic conditions: ^{1}D – Cyano column (250 mm×1.0 mm i.d., 5 μm); Gradient time, 65 min.; Flow rate, 10 μL/min.; Modulation, passive; Sampling time, 0.75 min.; ^{2}D – C18 column (30 mm×4.6 mm i.d., 2.7 μm); Gradient time, 0.71 min.; Flow rate, 4 mL/min. Reprinted from *Journal of Chromatography, A*, 1255, F. Cacciola, P. Donato, D. Giuffreda, G. Torre, P. Dugo, L. Mondello, Ultra high pressure in the second dimension of a comprehensive two-dimensional liquid chromatographic system for carotenoid separation in red chili peppers, 244–251, Copyright (2012), with permission from Elsevier.

12.3.4 Miscellaneous

2D-LC has also found extensive application in the analysis of food and natural product constituents beyond those discussed above. An interesting application of selective comprehensive 2D-LC (sLC×LC) exploiting commercial MHC instrumentation was reported by Lämmerhofer and coworkers [144] for the simultaneous analysis of fat- and water-soluble vitamins (see Figure 12.4). Polar vitamins were separated using a MM stationary phase in HILIC mode in the first dimension. The fat-soluble vitamins, which co-eluted close to the ^{1}D void volume, were automatically collected in "high resolution sampling" mode using two 14-port MHC valves. Upon completion of the HILIC separation, each of the 10 collected fractions were consecutively transferred to a ^{2}D C8 column using ASM. Because of the short sampling period, the 2D separation of apolar vitamins meets the criteria to be considered comprehensive, while the use of MHC loops decoupled the ^{2}D analysis time from the sampling period, allowing longer ^{2}D separations.

Medicinal plants used in traditional herbal remedies contain complex mixtures of many diverse analytes, and have therefore benefited from analysis by 2D-LC [145]. For example, Qiao *et al.* [146] reported a novel approach combining online MHC and comprehensive RP×RP to analyse the minor constituents in herbal medicines. By selectively removing five major components by heartcutting, up to 271 minor peaks could be detected in a *Pueraria lobata* root extract in 35 minutes (Figure 12.5). The combination of MHC and LC×LC has been extended to further TCMs [147]. Sheng *et al.* [148] used an online MHC RP-chiral separation for resolution of isomeric compounds found in a TCM that could not be resolved by RP×RP. Furthermore, offline HILIC×RP-MS has been used for the comprehensive profiling of polyphenols and ginkgolides in *Ginkgo biloba* [100] and for the characterization of multiple compound classes in the TCM Erzhi Pill [149].

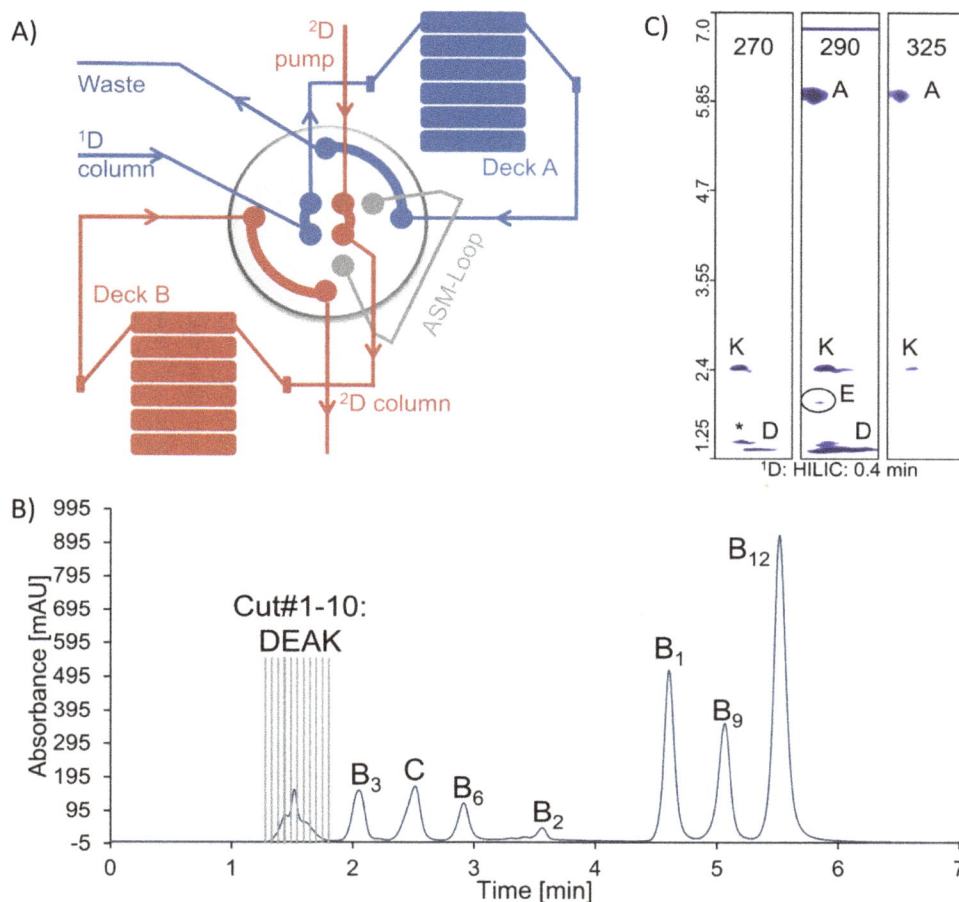

FIGURE 12.4 (A) Instrumental configuration used for the selective LC×LC analysis of water- and fat-soluble vitamins using ASM and MHC valves. (B) ¹D HILIC chromatogram for the target analytes. (C) Shows the contour plots for the high-resolution sampling analysis of apolar vitamins. Chromatographic conditions: ¹D – 2-pyridylurea column (150 mm×4.6 mm i.d., 5 µm); Gradient time, 10 min.; Flow rate, 1 mL/min.; Modulation, ASM (dilution factor 5); Sampling time, 0.04 min.; Loop volume, 40 µL; ²D – SPP C18 column (50 mm×2.1 mm i.d., 2.7 µm); Gradient time, 6.7 min.; Flow rate, 1.0 mL/min. Adapted by permission from Springer Nature: *Chromatographia*, Simultaneous Separation of Water- and Fat-Soluble Vitamins by Selective Comprehensive HILIC×RPLC (High-Resolution Sampling) and Active Solvent Modulation, S. Bäuer, W. Guo, S. Polnick, M. Lämmerhofer, Copyright 2019.

Extracts from *Stevia rebaudiana* plants are consumed as natural non-nutritional sweeteners. Detailed analysis of steviol glycosides, the compounds mainly responsible for the sweet taste, was performed by Fu *et al.* [150] using offline RP×HILIC, and subsequently scaled up for preparative isolation and characterization of some of the compounds by nuclear magnetic resonance (NMR) spectroscopy. Online LC×LC using C18 and primary amine columns in the first and second dimensions (the latter operated isocratically in mixed retention mode) has also been used for the separation of steviol glycosides [151], as has online HILIC×RP using secondary amine and C18 columns [105].

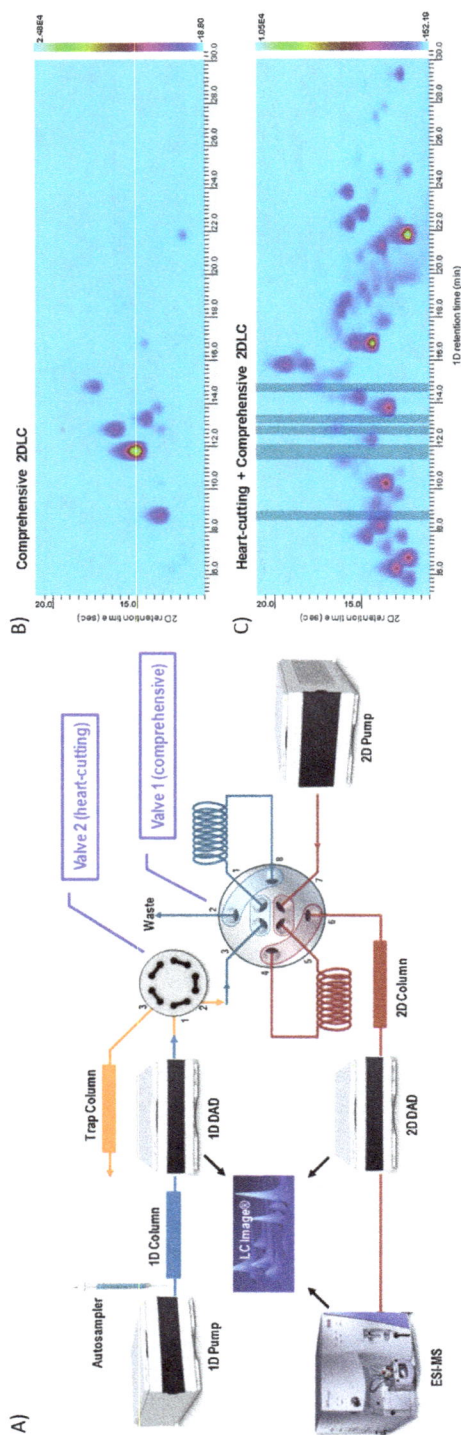

FIGURE 12.5 (A) Instrumental configuration used for the online MHC and RPxRP analysis of a *Pueraria lobata* root extract. (B) UV contour plot for RPxRP only, and (C) for the MHC and RPxRP analysis, where valve 2 was used to selectively remove major constituents. Chromatographic conditions: ^{1}D – CSH C18 column (100 mm×2.1 mm i.d., 1.7 μm); Gradient time, 31 min.; Flow rate, 0.1 mL/min; Modulation, passive; Sampling time, 0.5 min; Loop volumes, 80 μL; ^{2}D – Phenyl-hexyl column (50 mm×3.0 mm i.d., 2.7 μm); Gradient time, 0.4 min.; Flow rate, 2.5 mL/min. Adapted from *Journal of Chromatography, A*, 1362, X. Qiao, W. Song, S. Ji, Y. Li, Y. Wang, R. Li, R. An, D. Guo, M. Ye, Separation and detection of minor constituents in herbal medicines using a combination of heart-cutting and comprehensive two-dimensional liquid chromatography, 157–167, Copyright (2014), with permission from Elsevier.

12.4   FOOD ADDITIVES AND CONTAMINANTS

In addition to studying the composition of bioactive compounds, a second important goal of food and natural product analysis is to ensure the safety of these products. Increasingly 2D-LC has been applied in this area in recent years, primarily as a means to increase sensitivity and selectivity (especially to reduce matrix effects in MS detection) and reduce the need for sample clean-up steps.

From a food safety perspective, pesticide determination is among the most important analyses carried out by regulatory agencies. Kittlaus *et al.* [152] developed an online single heartcut HILIC-RP method coupled with tandem MS for the quantification of 300 pesticides in foods, where HILIC separation on a diol column replaces a liquid-liquid extraction that has been used traditionally to separate target analytes from the sample matrix. The method entails four steps: 1) apolar analytes eluting first from the HILIC column are transferred to a C8 trap; 2) the HILIC effluent is directed to the MS for the detection of polar pesticides; 3) analytes trapped in the C8 column are eluted into a ^{2}D RP column; and 4) apolar pesticides are detected by MS in the ^{2}D effluent without matrix interference (Figure 12.6). The efficiency of this method was compared to established multi-pesticide residue methods [153] and was recently used for the analysis of pesticides, mycotoxins and other contaminants in cereal samples [154, 155].

Online LC-LC has also been used for the quantification of the mycotoxins Ocratoxin A [156] and Aflatoxin B1 [157] in beverages and cereal products. Cheng *et al.* [158] investigated the hydrolysis products of aspartame sweetener in cola using a single heartcutting method combining RP and ligand exchange chromatography, with the latter hyphenated to online post-column derivatization to enable fluorescence detection of amino acid enantiomers.

Additives and preservatives are added to foods to improve organoleptic properties and extend shelf-life, but have to be regulated to ensure compliance with food safety standards. A stop-flow online heartcutting method using C4 and C18 columns in the first and second dimensions, along with a C18 trap column for modulation, was developed for the analysis of flavor enhancers and preservatives in yogurts [159] and milk powder [160]. This method allowed for the analysis of milk products without pre-treatment and effectively reduced interference from the endogenous proteins and lipids. Guo *et al.* [161] used a novel tolbutamide molecularly imprinted column in the first dimension of a LC-LC method to detect illegal sulfonylurea additives in Chinese herbal medicines. MHC has been used for the analysis of genotoxic pyrrolizidine alkaloids and their corresponding N-oxides in plants by RP-RP [162], where complementary selectivity was attained by using different stationary phases and acidic and basic mobile phases in the two dimensions.

12.5   CONCLUSIONS AND FUTURE OUTLOOK

Advances in food analysis techniques are largely driven by more stringent food safety requirements and growing regulatory demand for local and global trade. In addition, increased interest in natural health-promoting bioactive compounds has elicited demand for improved analytical methods to enable the detailed investigation of natural products. From the brief synopsis presented in this chapter, it is clear that 2D-LC has become a valuable tool in food and natural product analysis. The main benefits of 2D-LC exploited in this area are the increased resolving power (peak capacity) for complex mixtures, the ability to obtain chemical fingerprint profiles, and the potential to reduce laborious sample preparation steps.

For high throughput food safety and quality control analyses, the ability to avoid time-consuming sample clean-up methods (and their contribution to variability in analytical results) is arguably the most beneficial advantage of 2D-LC. Online heartcutting methods are especially suitable for such applications, and can be implemented easily with commercial instrumentation. Compared to LC×LC, method development is relatively straightforward, with the main challenge being identification of orthogonal separations with compatible mobile phases. SPAM can be used to improve sensitivity and minimize mobile phase mismatch problems for single heartcut methods, with the

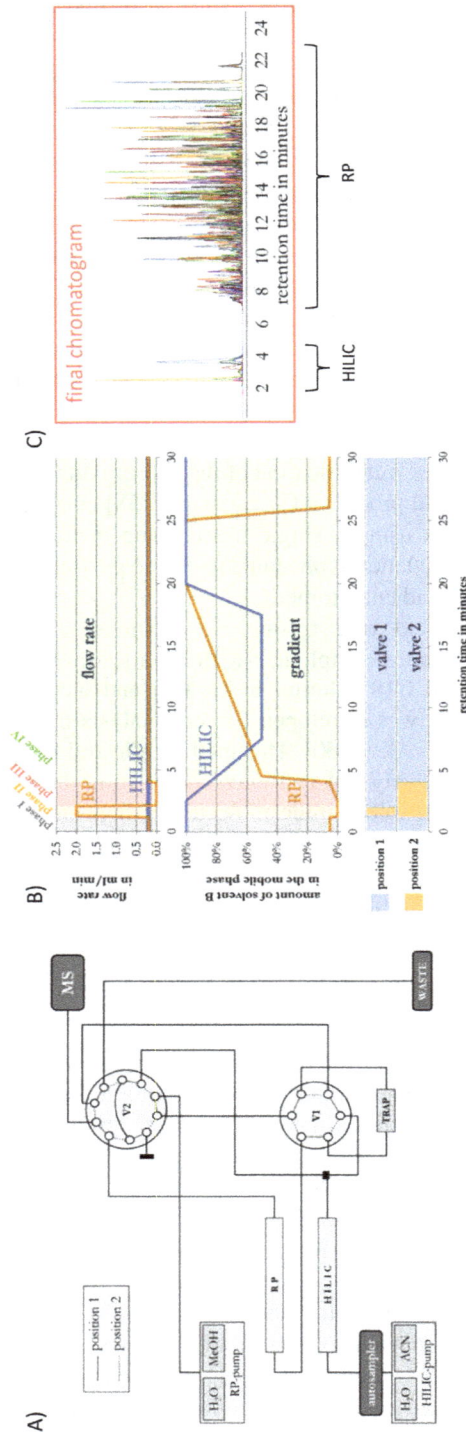

FIGURE 12.6 A) Schematic illustration of the instrument configuration used for the online HILIC-RP analysis of pesticides. B) Illustrates the method details. C) Shows an example MRM chromatogram. Apolar pesticides are trapped on a C8 trap and then separated by the ²D C18 column; polar pesticides are separated by the ¹D HILIC column. Chromatographic conditions: ¹D – Diol column (100 mm×2.1 mm i.d., 5 μm); Modulation, SPAM; Sampling time, 0.8 min.; Trapping column, C18 (12.5 mm×4.6 mm i.d.); ²D – C18 column (100 mm×2.1 mm i.d., 2.7 μm). Adapted from *Journal of Chromatography, A*, 1283, S. Kittlaus, J. Schimanke, G. Kempe, K. Speer, Development and validation of an efficient automated method for the analysis of 300 pesticides in foods using two-dimensional liquid chromatography–tandem mass spectrometry, 98–109, Copyright (2013), with permission from Elsevier.

trade-off being increased method complexity and potentially lower robustness. Since the introduction of commercial MHC 2D-LC instrumentation, this technique has also found growing application for somewhat more challenging food safety and quality control analyses.

Offline 2D-LC methods are less suited, and less used, for routine analyses. Offline 2D-LC is rather used where maximum peak capacity is sought, where typically time constraints are less important, such as for the detailed characterization of medicinal plant constituents. Offline heartcutting methods have also been exploited for the isolation of compounds from complex mixtures.

Online LC×LC offers the advantages of full automation and higher resolving power than 1D-LC, even at similar total analysis times. It is the most popular mode of LC×LC used for food and natural product analysis today, and its potential has been demonstrated for all of the major compound classes of interest in this area. When coupled with MS detection, LC×LC offers an exceptionally powerful methodology for the characterization of complex low volatility natural products. LC×LC contour plots facilitate fingerprint profiling of samples, which can be used to visually monitor degradation, determine authenticity, and to rapidly detect differences between samples.

Despite the success of online LC×LC, more widespread implementation of the technique is still hampered by several constraints. Method development in online LC×LC is much more complex than in 1D-LC (see Chapter 7 for more details); published methods are generally sample specific, such that their application to other samples containing the same compound classes may not be appropriate without significant adjustments. However, perhaps the main challenge in online LC×LC is mobile phase mismatch between the two dimensions. SPAM is arguably the most effective of the modulation methods to enable effective analyte focussing, although implementation of SPAM is fairly complex and the reliability of trapping columns limits its viability for routine analysis. Currently, solvent modulation techniques, such as active- or fixed solvent modulation, are simpler and more reliable alternatives.

A noteworthy trend in recent years has been the exploitation of the different forms of 2D-LC – single/multiple heartcutting, LC×LC, sLC×LC – and also combinations of these, to solve particular challenges encountered in food and natural product analysis. This versatility of 2D-LC means that the technique can be adapted to meet the requirements of a wide range of analyses, and bodes well for its further application in the field.

In summary, it is clear that 2D-LC has already greatly contributed towards the analysis of complex food and natural product samples. Future improvements to hardware (columns, modulators and detectors)[3] and software (method development and data-processing)[4] are expected to further improve the performance of 2D-LC and support its more widespread implementation. The full potential of 2D-LC in the field of food and natural product research is therefore yet to be realized.

NOTES

1   For further information on the different modes and hyphenation strategies of 2D-LC, readers are referred to Chapter 1.
2   All papers cited in this chapter can be found in the database of 2D-LC applications at www.multidlc.org/literature/2DLC-Applications.
3   Refer to Chapter 4 for further discussion on 2D-LC instrumentation.
4   Refer to Chapters 7 and 8 for further details.

REFERENCES

[1]   F. Cacciola, P. Donato, M. Beccaria, P. Dugo, L. Mondello, Advances in LC-MS for food analysis, LC-GC Eur. 25 (2012) 15–24.
[2]   M. Castro-Puyana, R. Pérez-Míguez, L. Montero, M. Herrero, Application of mass spectrometry-based metabolomics approaches for food safety, quality and traceability, Trends Analyt. Chem. 93 (2017) 102–118. https://doi.org/10.1016/j.trac.2017.05.004.

[3] G. Alvarez-Rivera, D. Ballesteros-Vivas, F. Parada-Alfonso, E. Ibañez, A. Cifuentes, Recent applications of high resolution mass spectrometry for the characterization of plant natural products, Trends Analyt. Chem. 112 (2019) 87–101. https://doi.org/10.1016/j.trac.2019.01.002.

[4] C. Aydoğan, Recent advances and applications in LC-HRMS for food and plant natural products: A critical review, Anal. Bioanal. Chem. 412 (2020) 1973–1991. https://doi.org/10.1007/s00 216-019-02328-6.

[5] T. Martinović, M. Šrajer Gajdošik, D. Josić, Sample preparation in foodomic analyses, Electrophoresis. 39 (2018) 1527–1542. https://doi.org/10.1002/elps.201800029.

[6] R. Gallego, M. Bueno, M. Herrero, Sub- and supercritical fluid extraction of bioactive compounds from plants, food-by-products, seaweeds and microalgae – An update, Trends Analyt. Chem. 116 (2019) 198–213. https://doi.org/10.1016/j.trac.2019.04.030.

[7] E.V.S. Maciel, A.L. de Toffoli, E.S. Neto, C.E.D. Nazario, F.M. Lanças, New materials in sample preparation: Recent advances and future trends, Trends Analyt. Chem. 119 (2019). https://doi.org/ 10.1016/j.trac.2019.115633.

[8] B.W.J. Pirok, D. Stoll R., P.J. Schoenmakers, Recent developments in two-dimensional liquid chromatography – Fundamental improvements for practical applications, Anal. Chem. 91 (2019) 240–263. https://doi.org/10.1021/acs.analchem.8b04841.

[9] P.Q. Tranchida, P. Donato, F. Cacciola, M. Beccaria, P. Dugo, L. Mondello, Potential of comprehensive chromatography in food analysis, Trends Analyt. Chem. 52 (2013) 186–205. https://doi.org/ 10.1016/j.trac.2013.07.008.

[10] F. Cacciola, P. Dugo, L. Mondello, Multidimensional liquid chromatography in food analysis, Trends Analyt. Chem. 96 (2017) 116–123. https://doi.org/10.1016/j.trac.2017.06.009.

[11] M.S. Franco, R.N. Padovan, B.H. Fumes, F.M. Lanças, An overview of multidimensional liquid phase separations in food analysis: Liquid phase separations, Electrophoresis. 37 (2016) 1768–1783. https:// doi.org/10.1002/elps.201600028.

[12] P.F. Brandão, A.C. Duarte, R.M.B.O. Duarte, Comprehensive multidimensional liquid chromatography for advancing environmental and natural products research, Trends Analyt. Chem. 116 (2019) 186–197. https://doi.org/10.1016/j.trac.2019.05.016.

[13] F. Cacciola, P. Donato, D. Sciarrone, P. Dugo, L. Mondello, Comprehensive liquid chromatography and other liquid-based comprehensive techniques coupled to mass spectrometry in food analysis, Anal. Chem. 89 (2017) 414–429. https://doi.org/10.1021/acs.analchem.6b04370.

[14] L. Montero, M. Herrero, Two-dimensional liquid chromatography approaches in Foodomics – A review, Anal. Chim. Acta. 1083 (2019) 1–18. https://doi.org/10.1016/j.aca.2019.07.036.

[15] W. Lv, X. Shi, S. Wang, G. Xu, Multidimensional liquid chromatography-mass spectrometry for metabolomic and lipidomic analyses, Trends Analyt. Chem. 120 (2019). https://doi.org/10.1016/ j.trac.2018.11.001.

[16] X. Shi, S. Wang, Q. Yang, X. Lu, G. Xu, Comprehensive two-dimensional chromatography for analyzing complex samples: recent new advances, Anal. Methods. 6 (2014) 7112–7123. https://doi.org/ 10.1039/C4AY01055H.

[17] F. Cacciola, F. Rigano, P. Dugo, L. Mondello, Comprehensive two-dimensional liquid chromatography as a powerful tool for the analysis of food and food products, Trends Analyt. Chem. (2020). https://doi.org/10.1016/j.trac.2020.115894.

[18] G. Marrubini, P. Appelblad, M. Maietta, A. Papetti, Hydrophilic interaction chromatography in food matrices analysis: An updated review, Food Chem. 257 (2018) 53–66. https://doi.org/10.1016/j.foodc hem.2018.03.008.

[19] C. Corradini, A. Cavazza, C. Bignardi, High-performance anion-exchange chromatography coupled with pulsed electrochemical detection as a powerful tool to evaluate carbohydrates of food interest: Principles and applications, International Journal of Carbohydrate Chemistry. 2012 (2012) 1–13. https://doi.org/10.1155/2012/487564.

[20] G. Nagy, T. Peng, N.L.B. Pohl, Recent liquid chromatographic approaches and developments for the separation and purification of carbohydrates, Anal. Methods. 9 (2017) 3579–3593. https://doi.org/ 10.1039/C7AY01094J.

[21] K. Klavins, D.B. Chu, S. Hann, G. Koellensperger, Fully automated on-line two-dimensional liquid chromatography in combination with ESI MS/MS detection for quantification of sugar phosphates in yeast cell extracts, Analyst. 139 (2014) 1512. https://doi.org/10.1039/c3an01930f.

[22] J. Ma, X. Hou, B. Zhang, Y. Wang, L. He, The analysis of carbohydrates in milk powder by a new "heart-cutting" two-dimensional liquid chromatography method, J. Pharm. Biomed. Anal. 91 (2014) 24–31. https://doi.org/10.1016/j.jpba.2013.11.006.

[23] M. Suto, H. Kawashima, N. Suto, Heart-cutting two-dimensional liquid chromatography combined with isotope ratio mass spectrometry for the determination of stable carbon isotope ratios of gluconic acid in honey, J. Chromatogr. A. (2019). https://doi.org/10.1016/j.chroma.2019.460421.

[24] A. Martín-Ortiz, A.I. Ruiz-Matute, M.L. Sanz, F.J. Moreno, M. Herrero, Separation of di- and trisaccharide mixtures by comprehensive two-dimensional liquid chromatography: Application to prebiotic oligosaccharides, Anal. Chim. Acta. 1060 (2019) 125–132. https://doi.org/10.1016/j.aca.2019.01.040.

[25] J.M. Bosque-Sendra, L. Cuadros-Rodríguez, C. Ruiz-Samblás, A.P. de la Mata, Combining chromatography and chemometrics for the characterization and authentication of fats and oils from triacylglycerol compositional data – A review, Anal. Chim. Acta. 724 (2012) 1–11. https://doi.org/10.1016/j.aca.2012.02.041.

[26] T. Cajka, O. Fiehn, Comprehensive analysis of lipids in biological systems by liquid chromatographymass spectrometry, Trends Analyt. Chem. 61 (2014) 192–206. https://doi.org/10.1016/j.trac.2014.04.017.

[27] B. Nikolova-Damyanova, Retention of lipids in silver ion high-performance liquid chromatography: Facts and assumptions, J. Chromatogr. A. 1216 (2009) 1815–1824. https://doi.org/10.1016/j.chroma.2008.10.097.

[28] L. Mondello, P.Q. Tranchida, V. Stanek, P. Jandera, G. Dugo, P. Dugo, Silver-ion reversed-phase comprehensive two-dimensional liquid chromatography combined with mass spectrometric detection in lipidic food analysis, J. Chromatogr., A. 1086 (2005) 91–98. https://doi.org/10.1016/j.chroma.2005.06.017.

[29] P. Dugo, T. Kumm, M. Lo Presti, B. Chiofalo, E. Salimei, A. Fazio, A. Cotroneo, L. Mondello, Determination of triacylglycerols in donkey milk by using high performance liquid chromatography coupled with atmospheric pressure chemical ionization mass spectrometry, J. Sep. Sci. 28 (2005) 1023–1030. https://doi.org/10.1002/jssc.200500025.

[30] P. Dugo, T. Kumm, B. Chiofalo, A. Cotroneo, L. Mondello, Separation of triacylglycerols in a complex lipidic matrix by using comprehensive two-dimensional liquid chromatography coupled with atmospheric pressure chemical ionization mass spectrometric detection, J. Sep. Sci. 29 (2006) 1146–1154. https://doi.org/10.1002/jssc.200500476.

[31] P. Dugo, T. Kumm, M.L. Crupi, A. Cotroneo, L. Mondello, Comprehensive two-dimensional liquid chromatography combined with mass spectrometric detection in the analyses of triacylglycerols in natural lipidic matrixes, J. Chromatogr. A. 1112 (2006) 269–275. https://doi.org/10.1016/j.chroma.2005.10.070.

[32] P. Dugo, T. Kumm, A. Fazio, G. Dugo, L. Mondello, Determination of beef tallow in lard through a multidimensional off-line non-aqueous reversed phase–argentation LC method coupled to mass spectrometry, J. Sep. Sci. 29 (2006) 567–575. https://doi.org/10.1002/jssc.200500342.

[33] E.J.C. van der Klift, G. Vivó-Truyols, F.W. Claassen, F.L. van Holthoon, T.A. van Beek, Comprehensive two-dimensional liquid chromatography with ultraviolet, evaporative light scattering and mass spectrometric detection of triacylglycerols in corn oil, J. Chromatogr., A. 1178 (2008) 43–55. https://doi.org/10.1016/j.chroma.2007.11.039.

[34] M. Navarro-Reig, J. Jaumot, T.A. van Beek, G. Vivó-Truyols, R. Tauler, Chemometric analysis of comprehensive LC×LC-MS data: Resolution of triacylglycerol structural isomers in corn oil, Talanta. 160 (2016) 624–635. https://doi.org/10.1016/j.talanta.2016.08.005.

[35] M. Holcapek, H. Velãnskã¡, M. Lãsa, P. Cesla, Orthogonality of silver-ion and non-aqueous reversedphase HPLC/MS in the analysis of complex natural mixtures of triacylglycerols, J. Sep. Sci. 32 (2009) 3672–3680. https://doi.org/10.1002/jssc.200900401.

[36] L. Mondello, M. Beccaria, P. Donato, F. Cacciola, G. Dugo, P. Dugo, Comprehensive two-dimensional liquid chromatography with evaporative light-scattering detection for the analysis of triacylglycerols in *Borago officinalis*: Liquid chromatography, J. Sep. Sci. 34 (2011) 688–692. https://doi.org/10.1002/jssc.201000843.

[37] Q. Yang, X. Shi, Q. Gu, S. Zhao, Y. Shan, G. Xu, On-line two dimensional liquid chromatography/mass spectrometry for the analysis of triacylglycerides in peanut oil and mouse tissue, J. Chromatogr. B. 895–896 (2012) 48–55. https://doi.org/10.1016/j.jchromb.2012.03.013.

[38] J. Hu, F. Wei, X.-Y. Dong, X. Lv, M.-L. Jiang, G.-M. Li, H. Chen, Characterization and quantification of triacylglycerols in peanut oil by off-line comprehensive two-dimensional liquid chromatography coupled with atmospheric pressure chemical ionization mass spectrometry: Liquid chromatography, J. Sep. Sci. 36 (2013) 288–300. https://doi.org/10.1002/jssc.201200567.

[39] M. Beccaria, R. Costa, G. Sullini, E. Grasso, F. Cacciola, P. Dugo, L. Mondello, Determination of the triacylglycerol fraction in fish oil by comprehensive liquid chromatography techniques with the support of gas chromatography and mass spectrometry data, Anal. Bioanal. Chem. 407 (2015) 5211–5225. https://doi.org/10.1007/s00216-015-8718-y.

[40] W.C. Byrdwell, Comprehensive dual liquid chromatography with quadruple mass spectrometry (LC1MS2×LC1MS2 = LC2MS4) for Analysis of *Parinari Curatellifolia* and other seed oil triacylglycerols, Anal. Chem. 89 (2017) 10537–10546. https://doi.org/10.1021/acs.analchem.7b02753.

[41] Y. Hirata, T. Hashiguchi, E. Kawata, Development of comprehensive two-dimensional packed column supercritical fluid chromatography, J. Sep. Sci. 26 (2003) 531–535. https://doi.org/10.1002/jssc.200390072.

[42] I. François, A. dos S. Pereira, P. Sandra, Considerations on comprehensive and off-line supercritical fluid chromatography×reversed-phase liquid chromatography for the analysis of triacylglycerols in fish oil, J. Sep. Sci. 33 (2010) 1504–1512. https://doi.org/10.1002/jssc.201000044.

[43] Y. Hirata, I. Sogabe, Separation of fatty acid methyl esters by comprehensive two-dimensional supercritical fluid chromatography with packed columns and programming of sampling duration, Anal. Bioanal. Chem. 378 (2004) 1999–2003. https://doi.org/10.1007/s00216-003-2487-8.

[44] I. François, P. Sandra, Comprehensive supercritical fluid chromatography×reversed-phase liquid chromatography for the analysis of the fatty acids in fish oil, J. Chromatogr. A. 1216 (2009) 4005–4012. https://doi.org/10.1016/j.chroma.2009.02.078.

[45] I. François, A. dos Santos Pereira, F. Lynen, P. Sandra, Construction of a new interface for comprehensive supercritical fluid chromatography×reversed-phase liquid chromatography (SFC×RPLC), J. Sep. Sci. 31 (2008) 3473–3478. https://doi.org/10.1002/jssc.200800267.

[46] R. Takahashi, M. Nakaya, M. Kotaniguchi, A. Shojo, S. Kitamura, Analysis of phosphatidylethanolamine, phosphatidylcholine, and plasmalogen molecular species in food lipids using an improved 2D high-performance liquid chromatography system, J. Chromatogr. B. 1077–1078 (2018) 35–43. https://doi.org/10.1016/j.jchromb.2018.01.014.

[47] C. Sun, Y.-Y. Zhao, J.M. Curtis, Characterization of phospholipids by two-dimensional liquid chromatography coupled to in-line ozonolysis–mass spectrometry, J. Agric. Food Chem. 63 (2015) 1442–1451. https://doi.org/10.1021/jf5049595.

[48] P. Dugo, N. Fawzy, F. Cichello, F. Cacciola, P. Donato, L. Mondello, Stop-flow comprehensive two-dimensional liquid chromatography combined with mass spectrometric detection for phospholipid analysis, J. Chromatogr. A. 1278 (2013) 46–53. https://doi.org/10.1016/j.chroma.2012.12.042.

[49] M. Navarro-Reig, J. Jaumot, R. Tauler, An untargeted lipidomic strategy combining comprehensive two-dimensional liquid chromatography and chemometric analysis, J. Chromatogr. A. 1568 (2018) 80–90. https://doi.org/10.1016/j.chroma.2018.07.017.

[50] P. Donato, G. Micalizzi, M. Oteri, F. Rigano, D. Sciarrone, P. Dugo, L. Mondello, Comprehensive lipid profiling in the Mediterranean mussel (Mytilus galloprovincialis) using hyphenated and multidimensional chromatography techniques coupled to mass spectrometry detection, Anal. Bioanal. Chem. 410 (2018) 3297–3313. https://doi.org/10.1007/s00216-018-1045-3.

[51] X. Zhang, A. Fang, C.P. Riley, M. Wang, F.E. Regnier, C. Buck, Multi-dimensional liquid chromatography in proteomics – A review, Anal. Chim. Acta. 664 (2010) 101–113. https://doi.org/10.1016/j.aca.2010.02.001.

[52] S. Di Palma, M.L. Hennrich, A.J.R. Heck, S. Mohammed, Recent advances in peptide separation by multidimensional liquid chromatography for proteome analysis, J. Proteomics. 75 (2012) 3791–3813. https://doi.org/10.1016/j.jprot.2012.04.033.

[53] Q. Wu, H. Yuan, L. Zhang, Y. Zhang, Recent advances on multidimensional liquid chromatography–mass spectrometry for proteomics: From qualitative to quantitative analysis – A review, Anal. Chim. Acta. 731 (2012) 1–10. https://doi.org/10.1016/j.aca.2012.04.010.

[54] P. Donato, F. Cacciola, L. Mondello, P. Dugo, Comprehensive chromatographic separations in proteomics, J. Chromatogr. A. 1218 (2011) 8777–8790. https://doi.org/10.1016/j.chroma.2011.05.070.

[55] V.-A. Duong, J.-M. Park, H. Lee, Review of three-dimensional liquid chromatography platforms for bottom-up proteomics, Int. J. Mass Spec. 21 (2020) 1524. https://doi.org/10.3390/ijms21041524.

[56] V. Guillén-Casla, M.E. León-González, L.V. Pérez-Arribas, L.M. Polo-Díez, Direct chiral determination of free amino acid enantiomers by two-dimensional liquid chromatography: Application to control transformations in E-beam irradiated foodstuffs, Anal. Bioanal. Chem. 397 (2009) 63–75. https://doi.org/10.1007/s00216-009-3376-6.

[57] Y. Miyoshi, M. Nagano, S. Ishigo, Y. Ito, K. Hashiguchi, N. Hishida, M. Mita, W. Lindner, K. Hamase, Chiral amino acid analysis of Japanese traditional Kurozu and the developmental changes during earthenware jar fermentation processes, J. Chromatogr. B. 966 (2014) 187–192. https://doi.org/10.1016/j.jchromb.2014.01.034.

[58] X. Wang, H. Wu, R. Luo, D. Xia, Z. Jiang, H. Han, Separation and detection of free D- and L-amino acids in tea by off-line two-dimensional liquid chromatography, Anal. Methods. 9 (2017) 6131–6138. https://doi.org/10.1039/C7AY01569K.

[59] A. Leitner, F. Castro-Rubio, M.L. Marina, W. Lindner, Identification of marker proteins for the adulteration of meat products with soybean proteins by multidimensional liquid chromatography–tandem mass spectrometry †, J. Proteome Res. 5 (2006) 2424–2430. https://doi.org/10.1021/pr060145q.

[60] S. Julka, K. Kuppannan, A. Karnoup, D. Dielman, B. Schafer, S.A. Young, Quantification of Gly m 4 Protein, a major soybean allergen, by two-dimensional liquid chromatography with ultraviolet and mass spectrometry detection, Anal. Chem. 84 (2012) 10019–10030. https://doi.org/10.1021/ac3024685.

[61] T. Zhou, S. Han, Z. Li, P. He, Purification and quantification of Kunitz Trypsin inhibitor in soybean using two-dimensional liquid chromatography, Food Anal. Meth. 10 (2017) 3350–3360. https://doi.org/10.1007/s12161-017-0902-6.

[62] X. Liu, D. Jiang, D.G. Peterson, Identification of bitter peptides in whey protein hydrolysate, J. Agric. Food Chem. 62 (2014) 5719–5725. https://doi.org/10.1021/jf4019728.

[63] J. Dai, R.J. Mumper, Plant phenolics: Extraction, analysis and their antioxidant and anticancer properties, Molecules. 15 (2010) 7313–7352. https://doi.org/10.3390/molecules15107313.

[64] C. Manach, A. Scalbert, C. Morand, C. Rémésy, L. Jiménez, Polyphenols: food sources and bioavailability, Am. J. Clin. Nutr. 79 (2004) 727–747. https://doi.org/10.1093/ajcn/79.5.727.

[65] O.M. Andersen, K.R. Markham, eds., Flavonoids: Chemistry, biochemistry and applications, 0 ed., CRC Press, 2005. https://doi.org/10.1201/9781420039443.

[66] K. Pyrzynska, A. Sentkowska, Recent developments in the HPLC separation of phenolic food compounds, Crit. Rev. Anal. Chem. 45 (2015) 41–51. https://doi.org/10.1080/10408347.2013.870027.

[67] P. Lucci, J. Saurina, O. Núñez, Trends in LC-MS and LC-HRMS analysis and characterization of polyphenols in food, Trends Analyt. Chem. 88 (2017) 1–24. https://doi.org/10.1016/j.trac.2016.12.006.

[68] A. de Villiers, P. Venter, H. Pasch, Recent advances and trends in the liquid-chromatography–mass spectrometry analysis of flavonoids, J. Chromatogr. A. 1430 (2016) 16–78. https://doi.org/10.1016/j.chroma.2015.11.077.

[69] F. Cacciola, S. Farnetti, P. Dugo, P.J. Marriott, L. Mondello, Comprehensive two-dimensional liquid chromatography for polyphenol analysis in foodstuffs, J. Sep. Sci. 40 (2016) 7–24. https://doi.org/10.1002/jssc.201600704.

[70] P. Jandera, T. Hájek, P. Česla, Comparison of various second-dimension gradient types in comprehensive two-dimensional liquid chromatography, J. Sep. Sci. 33 (2010) 1382–1397. https://doi.org/10.1002/jssc.200900808.

[71] G.M. Leme, F. Cacciola, P. Donato, A.J. Cavalheiro, P. Dugo, L. Mondello, Continuous vs. segmented second-dimension system gradients for comprehensive two-dimensional liquid chromatography of sugarcane (Saccharum spp.), Anal. Bioanal. Chem. 406 (2014) 4315–4324. https://doi.org/10.1007/s00216-014-7786-8.

[72] P. Donato, F. Rigano, F. Cacciola, M. Schure, S. Farnetti, M. Russo, P. Dugo, L. Mondello, Comprehensive two-dimensional liquid chromatography–tandem mass spectrometry for the simultaneous determination of wine polyphenols and target contaminants, J. Chromatogr. A. 1458 (2016) 54–62. https://doi.org/10.1016/j.chroma.2016.06.042.

[73] X. Qiao, W. Song, S. Ji, Q. Wang, D. Guo, M. Ye, Separation and characterization of phenolic compounds and triterpenoid saponins in licorice (Glycyrrhiza uralensis) using mobile phase-dependent reversed-phase×reversed-phase comprehensive two-dimensional liquid chromatography

coupled with mass spectrometry, J. Chromatogr. A. 1402 (2015) 36–45. https://doi.org/10.1016/j.chroma.2015.05.006.

[74] Y.F. Wong, F. Cacciola, S. Fermas, S. Riga, D. James, V. Manzin, B. Bonnet, P.J. Marriott, P. Dugo, L. Mondello, Untargeted profiling of Glycyrrhiza glabra extract with comprehensive two-dimensional liquid chromatography-mass spectrometry using multi-segmented shift gradients in the second dimension: Expanding the metabolic coverage, Electrophoresis. 39 (2018) 1993–2000. https://doi.org/10.1002/elps.201700469.

[75] K. Arena, F. Cacciola, D. Mangraviti, M. Zoccali, F. Rigano, N. Marino, P. Dugo, L. Mondello, Determination of the polyphenolic fraction of Pistacia vera L. kernel extracts by comprehensive two-dimensional liquid chromatography coupled to mass spectrometry detection, Anal. Bioanal. Chem. 411 (2019) 4819–4829. https://doi.org/10.1007/s00216-019-01649-w.

[76] M. Russo, F. Cacciola, K. Arena, D. Mangraviti, L. de Gara, P. Dugo, L. Mondello, Characterization of the polyphenolic fraction of pomegranate samples by comprehensive two-dimensional liquid chromatography coupled to mass spectrometry detection, Nat. Prod. Res. (2019) 1–7. https://doi.org/10.1080/14786419.2018.1561690.

[77] K. Arena, F. Cacciola, L. Dugo, P. Dugo, L. Mondello, Determination of the metabolite content of brassica juncea cultivars using comprehensive two-dimensional liquid chromatography coupled with a photodiode array and mass spectrometry detection, Molecules. 25 (2020) 1235. https://doi.org/10.3390/molecules25051235.

[78] K. Arena, F. Cacciola, F. Rigano, P. Dugo, L. Mondello, Evaluation of matrix effect in one-dimensional and comprehensive two-dimensional liquid chromatography for the determination of the phenolic fraction in extra virgin olive oils, J. Sep. Sci. 43 (2020) 1781–1789. https://doi.org/10.1002/jssc.202000169.

[79] M. Vergara-Barberán, J.A. Navarro-Huerta, J.R. Torres-Lapasió, E.F. Simó-Alfonso, M.C. García-Alvarez-Coque, Classification of olive leaves and pulp extracts by comprehensive two-dimensional liquid chromatography of polyphenolic fingerprints, Food Chem. 320 (2020) 126630. https://doi.org/10.1016/j.foodchem.2020.126630.

[80] S. Grutzmann Arcari, K. Arena, J. Kolling, P. Rocha, P. Dugo, L. Mondello, F. Cacciola, Polyphenolic compounds with biological activity in guabiroba fruits (*Campomanesia xanthocarpa* Berg.) by comprehensive two-dimensional liquid chromatography, Electrophoresis. 41 (2020) 1784–1792. https://doi.org/10.1002/elps.202000170.

[81] X. Dong, J. Yang, Q.-Y. Wang, X.-T. Zhen, F.-M. Liu, H. Zheng, J. Cao, Microextraction assisted multiple heart-cutting and comprehensive two-dimensional liquid chromatography hyphenated to Q-TOF/MS for the determination of multiclass compounds from Dendrobium species, Microchem. J. 157 (2020) 105097. https://doi.org/10.1016/j.microc.2020.105097.

[82] D. Li, O.J. Schmitz, Comprehensive two-dimensional liquid chromatography tandem diode array detector (DAD) and accurate mass QTOF-MS for the analysis of flavonoids and iridoid glycosides in Hedyotis diffusa, Anal. Bioanal. Chem. 407 (2014) 231–240. https://doi.org/10.1007/s00216-014-8057-4.

[83] S. Stephan, C. Jakob, J. Hippler, O.J. Schmitz, A novel four-dimensional analytical approach for analysis of complex samples, Anal. Bioanal. Chem. 408 (2016) 3751–3759. https://doi.org/10.1007/s00216-016-9460-9.

[84] A. de Villiers, K.M. Kalili, Comprehensive two-dimensional hydrophilic interaction chromatography x reversed phase liquid chromatography (HILIC x RP): Theory, practice and applications, in: Advances in Chromatography, CRC Press, Boca Raton, FL, 2016: pp. 217–299.

[85] K.M. Kalili, A. de Villiers, Off-line comprehensive two-dimensional hydrophilic interaction×reversed phase liquid chromatographic analysis of green tea phenolics, J. Sep. Sci. 33 (2010) 853–863. https://doi.org/10.1002/jssc.200900673.

[86] S.-H. Hsu, T. Raglione, S.A. Tomellini, T.R. Floyd, N. Sagliano Jr., R.A. Hartwick, Zone compression effects in high-performance liquid chromatography, J. Chromatogr., A. 367 (1986) 293–300. https://doi.org/10.1016/S0021-9673(00)94850-7.

[87] T. Beelders, K.M. Kalili, E. Joubert, D. de Beer, A. de Villiers, Comprehensive two-dimensional liquid chromatographic analysis of rooibos (*Aspalathus linearis*) phenolics: Liquid chromatography, J. Sep. Sci. 35 (2012) 1808–1820. https://doi.org/10.1002/jssc.201200060.

[88] E. Sommella, O.H. Ismail, F. Pagano, G. Pepe, C. Ostacolo, G. Mazzoccanti, M. Russo, E. Novellino, F. Gasparrini, P. Campiglia, Development of an improved online comprehensive hydrophilic interaction chromatography×reversed-phase ultra-high-pressure liquid chromatography platform for complex multiclass polyphenolic sample analysis, J. Sep. Sci. 40 (2017) 2188–2197. https://doi.org/10.1002/jssc.201700134.

[89] L. Montero, V. Sáez, D. von Baer, A. Cifuentes, M. Herrero, Profiling of Vitis vinifera L. canes (poly) phenolic compounds using comprehensive two-dimensional liquid chromatography, J. Chromatogr. A. 1536 (2018) 205–215. https://doi.org/10.1016/j.chroma.2017.06.013.

[90] L. Montero, M. Herrero, E. Ibáñez, A. Cifuentes, Profiling of phenolic compounds from different apple varieties using comprehensive two-dimensional liquid chromatography, J. Chromatogr. A. 1313 (2013) 275–283. https://doi.org/10.1016/j.chroma.2013.06.015.

[91] K.M. Kalili, A. de Villiers, Systematic optimisation and evaluation of on-line, off-line and stop-flow comprehensive hydrophilic interaction chromatography×reversed phase liquid chromatographic analysis of procyanidins. Part II: Application to cocoa procyanidins, J. Chromatogr. A. 1289 (2013) 69–79. https://doi.org/10.1016/j.chroma.2013.03.009.

[92] K.M. Kalili, J. Vestner, M.A. Stander, A. de Villiers, Toward unraveling grape tannin composition: Application of online hydrophilic interaction chromatography×reversed-phase liquid chromatography–time-of-flight mass spectrometry for grape seed analysis, Anal. Chem. 85 (2013) 9107–9115. https://doi.org/10.1021/ac401896r.

[93] L. Montero, M. Herrero, M. Prodanov, E. Ibáñez, A. Cifuentes, Characterization of grape seed procyanidins by comprehensive two-dimensional hydrophilic interaction×reversed phase liquid chromatography coupled to diode array detection and tandem mass spectrometry, Anal. Bioanal. Chem. 405 (2013) 4627–4638. https://doi.org/10.1007/s00216-012-6567-5.

[94] K.M. Kalili, S. De Smet, T. van Hoeylandt, F. Lynen, A. de Villiers, Comprehensive two-dimensional liquid chromatography coupled to the ABTS radical scavenging assay: a powerful method for the analysis of phenolic antioxidants, Anal. Bioanal. Chem. 406 (2014) 4233–4242. https://doi.org/10.1007/s00216-014-7847-z.

[95] L. Montero, M. Herrero, E. Ibáñez, A. Cifuentes, Separation and characterization of phlorotannins from brown algae *Cystoseira abies-marina* by comprehensive two-dimensional liquid chromatography: Liquid phase separations, Electrophoresis. 35 (2014) 1644–1651. https://doi.org/10.1002/elps.201400133.

[96] A.P. Sánchez-Camargo, L. Montero, A. Cifuentes, M. Herrero, E. Ibáñez, Application of Hansen solubility approach for the subcritical and supercritical selective extraction of phlorotannins from Cystoseira abies-marina, RSC Adv. 6 (2016) 94884–94895. https://doi.org/10.1039/C6RA16862K.

[97] C.M. Willemse, M.A. Stander, J. Vestner, A.G.J. Tredoux, A. de Villiers, Comprehensive two-dimensional hydrophilic interaction chromatography (HILIC)×Reversed-phase liquid chromatography coupled to high-resolution mass spectrometry (RP-LC-UV-MS) Analysis of anthocyanins and derived pigments in red wine, Anal. Chem. 87 (2015) 12006–12015. https://doi.org/10.1021/acs.analchem.5b03615.

[98] C.M. Willemse, M.A. Stander, A.G.J. Tredoux, A. de Villiers, Comprehensive two-dimensional liquid chromatographic analysis of anthocyanins, J. Chromatogr. A. 1359 (2014) 189–201. https://doi.org/10.1016/j.chroma.2014.07.044.

[99] W. Sun, L. Tong, J. Miao, J. Huang, D. Li, Y. Li, H. Xiao, H. Sun, K. Bi, Separation and analysis of phenolic acids from salvia miltiorrhiza and its related preparations by off-line two-dimensional hydrophilic interaction chromatography×reversed-phase liquid chromatography coupled with ion trap time-of-flight mass spectrometry, J. Chromatogr. A. 1431 (2016) 79–88. https://doi.org/10.1016/j.chroma.2015.12.038.

[100] S. Ji, D. He, T. Wang, J. Han, Z. Li, Y. Du, J. Zou, M. Guo, D. Tang, Separation and characterization of chemical constituents in Ginkgo biloba extract by off-line hydrophilic interaction×reversed-phase two-dimensional liquid chromatography coupled with quadrupole-time of flight mass spectrometry, J. Pharm. Biomed. Anal. 146 (2017) 68–78. https://doi.org/10.1016/j.jpba.2017.07.057.

[101] P. Venter, M. Muller, J. Vestner, M.A. Stander, A.G.J. Tredoux, H. Pasch, A. de Villiers, Comprehensive three-dimensional LC×LC×ion mobility spectrometry separation combined with high-resolution ms for the analysis of complex samples, Anal. Chem. 90 (2018) 11643–11650. https://doi.org/10.1021/acs.analchem.8b03234.

[102]　M. Liu, X. Huang, Q. Liu, M. Chen, S. Liao, F. Zhu, S. Shi, H. Yang, X. Chen, Rapid screening and identification of antioxidants in the leaves of *Malus hupehensis* using off-line two-dimensional HPLC-UV-MS/MS coupled with a 1,1'-diphenyl-2-picrylhydrazyl assay, J. Sep. Sci. 41 (2018) 2536–2543. https://doi.org/10.1002/jssc.201800007.

[103]　M. Muller, A.G.J. Tredoux, A. de Villiers, Application of kinetically optimised online HILIC×RP-LC methods hyphenated to high resolution MS for the analysis of natural phenolics, Chromatographia. 82 (2019) 181–196. https://doi.org/10.1007/s10337-018-3662-6.

[104]　Z. Liang, K. Li, X. Wang, Y. Ke, Y. Jin, X. Liang, Combination of off-line two-dimensional hydrophilic interaction liquid chromatography for polar fraction and two-dimensional hydrophilic interaction liquid chromatography×reversed-phase liquid chromatography for medium-polar fraction in a traditional Chinese medicine, J. Chromatogr., A. 1224 (2012) 61–69. https://doi.org/10.1016/j.chroma.2011.12.046.

[105]　F. Cacciola, P. Delmonte, K. Jaworska, P. Dugo, L. Mondello, J.I. Rader, Employing ultra high pressure liquid chromatography as the second dimension in a comprehensive two-dimensional system for analysis of Stevia rebaudiana extracts, J. Chromatogr. A. 1218 (2011) 2012–2018. https://doi.org/10.1016/j.chroma.2010.08.081.

[106]　T. Brazdauskas, L. Montero, P.R. Venskutonis, E. Ibañez, M. Herrero, Downstream valorization and comprehensive two-dimensional liquid chromatography-based chemical characterization of bioactives from black chokeberries (Aronia melanocarpa) pomace, J. Chromatogr. A. 1468 (2016) 126–135. https://doi.org/10.1016/j.chroma.2016.09.033.

[107]　L. Montero, E. Ibáñez, M. Russo, R. di Sanzo, L. Rastrelli, A.L. Piccinelli, R. Celano, A. Cifuentes, M. Herrero, Metabolite profiling of licorice (Glycyrrhiza glabra) from different locations using comprehensive two-dimensional liquid chromatography coupled to diode array and tandem mass spectrometry detection, Anal. Chim. Acta. 913 (2016) 145–159. https://doi.org/10.1016/j.aca.2016.01.040.

[108]　L. Montero, E. Ibáñez, M. Russo, L. Rastrelli, A. Cifuentes, M. Herrero, Focusing and non-focusing modulation strategies for the improvement of on-line two-dimensional hydrophilic interaction chromatography×reversed phase profiling of complex food samples, Anal. Chim. Acta. 985 (2017) 202–212. https://doi.org/10.1016/j.aca.2017.07.013.

[109]　J. Wang, L. Chen, L. Qu, K. Li, Y. Zhao, Z. Wang, Y. Li, X. Zhang, Y. Jin, X. Liang, Isolation and bioactive evaluation of flavonoid glycosides from Lobelia chinensis Lour using two-dimensional liquid chromatography combined with label-free cell phenotypic assays, J. Chromatogr. A. 1601 (2019) 224–231. https://doi.org/10.1016/j.chroma.2019.04.073.

[110]　A. de Villiers, D. Cabooter, F. Lynen, G. Desmet, P. Sandra, High performance liquid chromatography analysis of wine anthocyanins revisited: Effect of particle size and temperature, J. Chromatogr. A. 1216 (2009) 3270–3279. https://doi.org/10.1016/j.chroma.2009.02.038.

[111]　S. Toro-Uribe, L. Montero, L. López-Giraldo, E. Ibáñez, M. Herrero, Characterization of secondary metabolites from green cocoa beans using focusing-modulated comprehensive two-dimensional liquid chromatography coupled to tandem mass spectrometry, Anal. Chim. Acta. 1036 (2018) 204–213. https://doi.org/10.1016/j.aca.2018.06.068.

[112]　M. Muller, A.G.J. Tredoux, A. de Villiers, Predictive kinetic optimisation of hydrophilic interaction chromatography×reversed phase liquid chromatography separations: Experimental verification and application to phenolic analysis, J. Chromatogr. A. 1571 (2018) 107–120. https://doi.org/10.1016/j.chroma.2018.08.004.

[113]　K.M. Kalili, A. de Villiers, Systematic optimisation and evaluation of on-line, off-line and stop-flow comprehensive hydrophilic interaction chromatography×reversed phase liquid chromatographic analysis of procyanidins, Part I: Theoretical considerations, J. Chromatogr. A. 1289 (2013) 58–68. https://doi.org/10.1016/j.chroma.2013.03.008.

[114]　P.J. Schoenmakers, G. Vivó-Truyols, W.M.C. Decrop, A protocol for designing comprehensive two-dimensional liquid chromatography separation systems, J. Chromatogr. A. 1120 (2006) 282–290. https://doi.org/10.1016/j.chroma.2005.11.039.

[115]　M.R. Filgueira, Y. Huang, K. Witt, C. Castells, P.W. Carr, Improving peak capacity in fast online comprehensive two-dimensional liquid chromatography with post-first-dimension flow splitting, Anal. Chem. 83 (2011) 9531–9539. https://doi.org/10.1021/ac202317m.

[116] D. Li, R. Dück, O.J. Schmitz, The advantage of mixed-mode separation in the first dimension of comprehensive two-dimensional liquid-chromatography, J. Chromatogr. A. 1358 (2014) 128–135. https://doi.org/10.1016/j.chroma.2014.06.086.

[117] C.T. Scoparo, L.M. de Souza, N. Dartora, G.L. Sassaki, P.A.J. Gorin, M. Iacomini, Analysis of Camellia sinensis green and black teas via ultra high performance liquid chromatography assisted by liquid–liquid partition and two-dimensional liquid chromatography (size exclusion×reversed phase), J. Chromatogr., A. 1222 (2012) 29–37. https://doi.org/10.1016/j.chroma.2011.11.038.

[118] L. Jiao, Y. Tao, W. Wang, Y. Shao, L. Mei, Q. Wang, J. Dang, Preparative isolation of flavonoid glycosides from Sphaerophysa salsula using hydrophilic interaction solid-phase extraction coupled with two-dimensional preparative liquid chromatography, J. Sep. Sci. 40 (2017) 3808–3816. https://doi.org/10.1002/jssc.201700675.

[119] S.G. Sparg, M.E. Light, J. van Staden, Biological activities and distribution of plant saponins, J. Ethnopharmacol. 94 (2004) 219–243. https://doi.org/10.1016/j.jep.2004.05.016.

[120] C. Xu, W. Wang, B. Wang, T. Zhang, X. Cui, Y. Pu, N. Li, Analytical methods and biological activities of Panax notoginseng saponins: Recent trends, J. Ethnopharmacol. 236 (2019) 443–465. https://doi.org/10.1016/j.jep.2019.02.035.

[121] J. Negi, G.J. Pant, P. Singh, M.S. Rawat, High-performance liquid chromatography analysis of plant saponins: An update 2005–2010, Phcog Rev. 5 (2011) 155. https://doi.org/10.4103/0973-7847.91109.

[122] Y. Wang, X. Lu, G. Xu, Development of a comprehensive two-dimensional hydrophilic interaction chromatography/quadrupole time-of-flight mass spectrometry system and its application in separation and identification of saponins from Quillaja saponaria, J. Chromatogr. A. 1181 (2008) 51–59. https://doi.org/10.1016/j.chroma.2007.12.034.

[123] E.-K. Jeong, H.-J. Cha, Y.W. Ha, Y.S. Kim, I.J. Ha, Y.-C. Na, Development and optimization of a method for the separation of platycosides in Platycodi Radix by comprehensive two-dimensional liquid chromatography with mass spectrometric detection, J. Chromatogr. A. 1217 (2010) 4375–4382. https://doi.org/10.1016/j.chroma.2010.04.053.

[124] S. Wang, Q. Wang, X. Qiao, W. Song, L. Zhong, D. Guo, M. Ye, Separation and characterization of triterpenoid saponins in gleditsia sinensis by comprehensive two-dimensional liquid chromatography coupled with mass spectrometry, Planta Med. 82 (2016) 1558–1567. https://doi.org/10.1055/s-0042-110206.

[125] K. Croes, A. Steffens, D.H. Marchand, L.R. Snyder, Relevance of π-π and dipole-dipole interactions for retention on cyano and phenyl columns in reversed-phase liquid chromatography, J. Chromatogr. A. 1098 (2005) 123–130. https://doi.org/10.1016/j.chroma.2005.08.090.

[126] Q. Xing, T. Liang, G. Shen, X. Wang, Y. Jin, X. Liang, Comprehensive HILIC×RPLC with mass spectrometry detection for the analysis of saponins in Panax notoginseng, Analyst. 137 (2012) 2239–2249. https://doi.org/10.1039/c2an16078a.

[127] S. Wang, L. Qiao, X. Shi, C. Hu, H. Kong, G. Xu, On-line stop-flow two-dimensional liquid chromatography–mass spectrometry method for the separation and identification of triterpenoid saponins from ginseng extract, Anal. Bioanal. Chem. 407 (2015) 331–341. https://doi.org/10.1007/s00216-014-8219-4

[128] H. Zhang, J.-M. Jiang, D. Zheng, M. Yuan, Z.-Y. Wang, H.-M. Zhang, C.-W. Zheng, L.-B. Xiao, H.-X. Xu, A multidimensional analytical approach based on time-decoupled online comprehensive two-dimensional liquid chromatography coupled with ion mobility quadrupole time-of-flight mass spectrometry for the analysis of ginsenosides from white and red ginsengs, J. Pharm. Biomed. Anal. 163 (2019) 24–33. https://doi.org/10.1016/j.jpba.2018.09.036.

[129] Y. Chen, J. Li, O.J. Schmitz, Development of an at-column dilution modulator for flexible and precise control of dilution factors to overcome mobile phase incompatibility in comprehensive two-dimensional liquid chromatography, Anal. Chem. 91 (2019) 10251–10257. https://doi.org/10.1021/acs.analchem.9b02391.

[130] Y. Chen, L. Montero, J. Luo, J. Li, O.J. Schmitz, Application of the new at-column dilution (ACD) modulator for the two-dimensional RP×HILIC analysis of Buddleja davidii, Anal. Bioanal. Chem. 412 (2020) 1483–1495. https://doi.org/10.1007/s00216-020-02392-3.

[131] R. Arimboor, R.B. Natarajan, K.R. Menon, L.P. Chandrasekhar, V. Moorkoth, Red pepper (Capsicum annuum) carotenoids as a source of natural food colors: analysis and stability – a review, J. Food. Sci. Technol. 52 (2015) 1258–1271. https://doi.org/10.1007/s13197-014-1260-7.

[132] J. Yabuzaki, Carotenoids Database: Structures, chemical fingerprints and distribution among organisms, Database. 2017 (2017). https://doi.org/10.1093/database/bax004.

[133] D. Giuffrida, P. Donato, P. Dugo, L. Mondello, Recent analytical techniques advances in the carotenoids and their derivatives determination in various matrixes, J. Agric. Food. Chem. 66 (2018) 3302–3307. https://doi.org/10.1021/acs.jafc.8b00309.

[134] K.T. Amorim-Carrilho, A. Cepeda, C. Fente, P. Regal, Review of methods for analysis of carotenoids, Trends Analyt. Chem. 56 (2014) 49–73. https://doi.org/10.1016/j.trac.2013.12.011.

[135] E. Biehler, F. Mayer, L. Hoffmann, E. Krause, T. Bohn, Comparison of 3 spectrophotometric methods for carotenoid determination in frequently consumed fruits and vegetables, J. Food Sci. 75 (2010) C55–C61. https://doi.org/10.1111/j.1750-3841.2009.01417.x.

[136] P. Dugo, V. Škeříková, T. Kumm, A. Trozzi, P. Jandera, L. Mondello, Elucidation of carotenoid patterns in citrus products by means of comprehensive normal-phase×reversed-phase liquid chromatography, Anal. Chem. 78 (2006) 7743–7750. https://doi.org/10.1021/ac061290q.

[137] P. Dugo, M. Herrero, T. Kumm, D. Giuffrida, G. Dugo, L. Mondello, Comprehensive normal-phase×reversed-phase liquid chromatography coupled to photodiode array and mass spectrometry detection for the analysis of free carotenoids and carotenoid esters from mandarin, J. Chromatogr. A. 1189 (2008) 196–206. https://doi.org/10.1016/j.chroma.2007.11.116.

[138] P. Dugo, M. Herrero, D. Giuffrida, T. Kumm, G. Dugo, L. Mondello, Application of comprehensive two-dimensional liquid chromatography to elucidate the native carotenoid composition in red orange essential oil, J. Agric. Food Chem. 56 (2008) 3478–3485. https://doi.org/10.1021/jf800144v.

[139] P. Dugo, D. Giuffrida, M. Herrero, P. Donato, L. Mondello, Epoxycarotenoids esters analysis in intact orange juices using two-dimensional comprehensive liquid chromatography, J. Sep. Sci. 32 (2009) 973–980. https://doi.org/10.1002/jssc.200800696.

[140] F. Cacciola, P. Donato, D. Giuffrida, G. Torre, P. Dugo, L. Mondello, Ultra high pressure in the second dimension of a comprehensive two-dimensional liquid chromatographic system for carotenoid separation in red chili peppers, J. Chromatogr. A. 1255 (2012) 244–251. https://doi.org/10.1016/j.chroma.2012.06.076.

[141] F. Cacciola, D. Giuffrida, M. Utczas, D. Mangraviti, P. Dugo, D. Menchaca, E. Murillo, L. Mondello, Application of comprehensive two-dimensional liquid chromatography for carotenoid analysis in red mamey (pouteria sapote) fruit, Food Anal. Methods. 9 (2016) 2335–2341. https://doi.org/10.1007/s12161-016-0416-7.

[142] F. Cacciola, D. Giuffrida, M. Utczas, D. Mangraviti, M. Beccaria, P. Donato, I. Bonaccorsi, P. Dugo, L. Mondello, Analysis of the carotenoid composition and stability in various overripe fruits by comprehensive two-dimensional liquid chromatography, LC-GC Eur. 29 (2016) 252–257.

[143] R. Gallego, K. Arena, P. Dugo, L. Mondello, E. Ibáñez, M. Herrero, Application of compressed fluid-based extraction and purification procedures to obtain astaxanthin-enriched extracts from Haematococcus pluvialis and characterization by comprehensive two-dimensional liquid chromatography coupled to mass spectrometry, Anal. Bioanal. Chem. 412 (2020) 589–599. https://doi.org/10.1007/s00216-019-02287-y.

[144] S. Bäurer, W. Guo, S. Polnick, M. Lämmerhofer, Simultaneous separation of water- and fat-soluble vitamins by selective comprehensive HILIC×RPLC (High-Resolution Sampling) and active solvent modulation, Chromatographia. 82 (2019) 167–180. https://doi.org/10.1007/s10337-018-3615-0.

[145] S. Ji, S. Wang, H. Xu, Z. Su, D. Tang, X. Qiao, M. Ye, The application of on-line two-dimensional liquid chromatography (2DLC) in the chemical analysis of herbal medicines, J. Pharm. Biomed. Anal. 160 (2018) 301–313. https://doi.org/10.1016/j.jpba.2018.08.014.

[146] X. Qiao, W. Song, S. Ji, Y. Li, Y. Wang, R. Li, R. An, D. Guo, M. Ye, Separation and detection of minor constituents in herbal medicines using a combination of heart-cutting and comprehensive two-dimensional liquid chromatography, J. Chromatogr. A. 1362 (2014) 157–167. https://doi.org/10.1016/j.chroma.2014.08.038.

[147] X. Qiao, Q. Wang, W. Song, Y. Qian, Y. Xiao, R. An, D. Guo, M. Ye, A chemical profiling solution for Chinese medicine formulas using comprehensive and loop-based multiple heart-cutting

two-dimensional liquid chromatography coupled with quadrupole time-of-flight mass spectrometry, J. Chromatogr. A. 1438 (2016) 198–204. https://doi.org/10.1016/j.chroma.2016.02.034.

[148] N. Sheng, H. Zheng, Y. Xiao, Z. Wang, M. Li, J. Zhang, Chiral separation and chemical profile of Dengzhan Shengmai by integrating comprehensive with multiple heart-cutting two-dimensional liquid chromatography coupled with quadrupole time-of-flight mass spectrometry, J. Chromatogr. A. 1517 (2017) 97–107. https://doi.org/10.1016/j.chroma.2017.08.037.

[149] L. Fu, H. Ding, L. Han, L. Jia, W. Yang, C. Zhang, Y. Hu, T. Zuo, X. Gao, D. Guo, Simultaneously targeted and untargeted multicomponent characterization of Erzhi Pill by offline two-dimensional liquid chromatography/quadrupole-Orbitrap mass spectrometry, J. Chromatogr. A. 1584 (2019) 87–96. https://doi.org/10.1016/j.chroma.2018.11.024.

[150] Q. Fu, Z. Guo, X. Zhang, Y. Liu, X. Liang, Comprehensive characterization of *Stevia Rebaudiana* using two-dimensional reversed-phase liquid chromatography/hydrophilic interaction liquid chromatography: Liquid chromatography, J. Sep. Sci. 35 (2012) 1821–1827. https://doi.org/10.1002/jssc.201101103.

[151] J. Pól, B. Hohnová, T. Hyötyläinen, Characterisation of Stevia Rebaudiana by comprehensive two-dimensional liquid chromatography time-of-flight mass spectrometry, J. Chromatogr. A. 1150 (2007) 85–92. https://doi.org/10.1016/j.chroma.2006.09.008.

[152] S. Kittlaus, J. Schimanke, G. Kempe, K. Speer, Development and validation of an efficient automated method for the analysis of 300 pesticides in foods using two-dimensional liquid chromatography – tandem mass spectrometry, J. Chromatogr. A. 1283 (2013) 98–109. https://doi.org/10.1016/j.chroma.2013.01.106.

[153] S. Kittlaus, G. Kempe, K. Speer, Evaluation of matrix effects in different multipesticide residue analysis methods using liquid chromatography-tandem mass spectrometry, including an automated two-dimensional cleanup approach: Sample Preparation, J. Sep. Sci. 36 (2013) 2185–2195. https://doi.org/10.1002/jssc.201300044.

[154] M. Urban, S. Hann, H. Rost, Simultaneous determination of pesticides, mycotoxins, tropane alkaloids, growth regulators, and pyrrolizidine alkaloids in oats and whole wheat grains after online clean-up via two-dimensional liquid chromatography tandem mass spectrometry, J. Environ. Sci. Health B. 54 (2019) 98–111. https://doi.org/10.1080/03601234.2018.1531662.

[155] M. Kresse, H. Drinda, A. Romanotto, K. Speer, Simultaneous determination of pesticides, mycotoxins, and metabolites as well as other contaminants in cereals by LC-LC-MS/MS, J. Chromatogr. B. 1117 (2019) 86–102. https://doi.org/10.1016/j.jchromb.2019.04.013.

[156] C. Armutcu, L. Uzun, A. Denizli, Determination of Ochratoxin A traces in foodstuffs: Comparison of an automated on-line two-dimensional high-performance liquid chromatography and off-line immunoaffinity-high-performance liquid chromatography system, J. Chromatogr. A. 1569 (2018) 139–148. https://doi.org/10.1016/j.chroma.2018.07.057.

[157] A. Breidbach, F. Ulberth, Two-dimensional heart-cut LC-LC improves accuracy of exact-matching double isotope dilution mass spectrometry measurements of aflatoxin B1 in cereal-based baby food, maize, and maize-based feed, Anal. Bioanal. Chem. 407 (2015) 3159–3167. https://doi.org/10.1007/s00216-014-8003-5.

[158] C. Cheng, S.-C. Wu, Simultaneous analysis of aspartame and its hydrolysis products of Coca-Cola Zero by on-line postcolumn derivation fluorescence detection and ultraviolet detection coupled two-dimensional high-performance liquid chromatography, J. Chromatogr. A. 1218 (2011) 2976–2983. https://doi.org/10.1016/j.chroma.2011.03.033.

[159] X. Hou, J. Ma, X. He, L. Chen, S. Wang, L. He, A stop-flow two-dimensional liquid chromatography method for determination of food additives in yogurt, Anal. Methods. 7 (2015) 2141–2148. https://doi.org/10.1039/C4AY02855D.

[160] J. Ma, B. Zhang, Y. Wang, X. Hou, L. He, Determination of flavor enhancers in milk powder by one-step sample preparation and two-dimensional liquid chromatography: Liquid chromatography, J. Sep. Sci. 37 (2014) 920–926. https://doi.org/10.1002/jssc.201301367.

[161] P. Guo, X. Xu, G. Chen, K. Bashir, H. Shu, Y. Ge, W. Jing, Z. Luo, C. Chang, Q. Fu, On-line two dimensional liquid chromatography based on skeleton type molecularly imprinted column for selective determination of sulfonylurea additive in Chinese patent medicines or functional foods, J. Pharm. Biomed. Anal. 146 (2017) 292–301. https://doi.org/10.1016/j.jpba.2017.09.008.

[162] M.G.M. van de Schans, M.H. Blokland, P.W. Zoontjes, P.P.J. Mulder, M.W.F. Nielen, Multiple heart-cutting two dimensional liquid chromatography quadrupole time-of-flight mass spectrometry of pyrrolizidine alkaloids, J. Chromatogr. A. 1503 (2017) 38–48. https://doi.org/10.1016/j.chroma.2017.04.059.

[163] S. Wang, L. Qiao, X. Shi, C. Hu, H. Kong, G. Xu, On-line stop-flow two-dimensional liquid chromatography–mass spectrometry method for the separation and identification of triterpenoid saponins from ginseng extract, Anal. Bioanal. Chem. 407 (2015) 331–341. https://doi.org/10.1007/s00216-014-8219-4.

13 Application of Two-Dimensional Liquid Chromatography to the Analysis of Chiral and Structurally Similar Molecules

Michael Lämmerhofer and Carlos Calderón

CONTENTS

13.1 INTRODUCTION

Enantioselective analysis is an integral part of drug discovery, quality control of chiral drug substances, studies of drug metabolism and pharmacokinetics, and is nowadays routinely applied in the pharmaceutical industry. Enantioselective analysis is also highly important in agrochemistry and food analysis for food authentication and control of adulteration. Moreover, recent work has shown that unexpected enantiomers in biomaterials (such as D-amino acids) are quite useful as biomarkers of disease and for assessment of the quality of fermented food, to mention just a few fields of application, where enantioselective 2D-LC can address complicated analytical problems and make a

DOI: 10.1201/9781003090557-13

significant contribution. The most common approach of direct enantiomer separation makes use of chiral stationary phases (CSPs). Their implementation in enantioselective 2D-LC is the focus of this chapter.

In fact, enantioselective analysis is one of the applications in which 2D-LC was routinely used in practice already in its early days – i.e. in the 1980s and early 1990s [1]. It was popular in particular for bioanalytical applications and the reason for that is simple. Triple quadrupole instruments which nowadays dominate the field of quantitative bioanalysis of drugs and metabolites were not widely available in analytical labs at that time. For this reason, 2D-LC, at that times often called column switching or column coupling, was the most obvious approach to tackle the problem of enantioselective analysis of drugs and metabolites in complex bio-matrices such as plasma, urine, and tissue to remove matrix interferences and resolve metabolite peaks from chiral drugs allowing simultaneous enantiomer separation of parent drugs and metabolites in a second dimension that was enantioselective [1]. In this chapter, the focus is on modern implementations. Nowadays, 2D-LC is also frequently used for enantioselective amino acid analysis, impurity profiling of pharmaceuticals with multiple chiral centers or complex structures, and can be a valuable tool for the direct enantiomeric excess determination of chiral compounds from complex process samples without prior purification. Additionally, chiral stationary phases can be also quite useful even for separations of achiral compounds or substances that should be separable on achiral separation systems. Readers may also be interested in review articles that have addressed the topic of enantioselective analysis of chiral compounds by multi-dimensional LC [2, 3], and some have also covered enantioselective 2D-GC, 2D-LC-SFC, and 2D-TLC [4].

The fundamental aspects (e.g., undersampling, effective peak capacities, sample dilution effects) of 2D-LC discussed earlier in this book apply to enantioselective 2D-LC as well. Some peculiarities are briefly mentioned here. In enantioselective 2D-LC the primary motivation for adding a second dimension of separation is rarely focused on adding peak capacity. Rather, it is the insufficient selectivity of a 1D-LC separation that forces users toward 2D-LC. Complementarity in the separations used in the first and second dimensions is highly valued in enantioselective 2D-LC, as with other types of 2D-LC. However, this almost always develops naturally in cases where an achiral column is coupled with a chiral column – the achiral column does not exhibit any chiral recognition capability and therefore does not show any enantioselectivity. On the other hand, the chiral stationary phase rarely has the same chemoselectivity as the achiral stationary phase due to the multiple distinct functional groups and interaction sites that results in their having mixed-mode retention characteristics [5]. Finally, the most popular CSPs (i.e., polysaccharides, macrocyclic antibiotics, etc.) can be used in a variety of separation modes, including reversed-phase (RP), hydrophilic interaction chromatography (HILIC), polar organic, normal-phase (NP), and supercritical fluid chromatography (SFC); this flexibility makes these CSPs attractive complements to other more traditional stationary phases, and provides an avenue to avoid mobile phase mismatch between the two dimensions. A drawback of these CSPs compared to achiral columns is that they usually exhibit lower chromatographic efficiency, even when modern particles (sub-2µm totally porous particles [TPP] or sub-3µm superficially porous particles [SPP]) are employed, because enantioselective chromatography frequently suffers from slow surface adsorption-desorption kinetics due to multi-point interactions between the analyte and the chiral selector.

Although offline enantioselective 2D-LC is always an option, and examples of this can be found in the literature [1, 6, 7, 8], the remainder of this chapter is focused on online enantioselective 2D-LC for reasons discussed in Section 1.5.1.2.

13.2 ENANTIOSELECTIVE LC-LC

Enantioselective LC-LC has been carried out in single or multiple heartcut mode, usually with one enantioselective separation, but in rare cases, enantioselective separations in both dimensions. In the most common heartcut setup, the first dimension is achiral (usually an RP separation) and allows the isolation of the target analyte from potential interferences. A clean fraction of the peak containing the target analyte can then be transferred to the chiral column for enantiomer separation, which avoids interferences and potential contamination of expensive chiral columns with sample matrices. The reversed order (i.e., chiral first, achiral second) is rarely used although it has been utilized – e.g. to refocus the broad peaks from chiral ^{1}D separation using protein-type CSPs into a narrow zone on a ^{2}D RP column [9]. Another question in the context of heartcut techniques is related to the transfer volume. Traditionally, the term "heartcut" has been used to mean that only the center of the ^{1}D peak is transferred to the second dimension. However, very often in practice the entire peak from the ^{1}D separation is transferred to the ^{2}D column. In general, there is no selectivity for the two enantiomers on the achiral ^{1}D column. Hence, the chromatographic zones and peak profiles, respectively, of two enantiomers are completely superimposed and no discrimination between them occurs when only a small fraction of the ^{1}D peak is transferred to the second dimension – i.e., the enantiomeric ratio is the same across the entire ^{1}D peak regardless of the position at which the cut is made. To maximize detection sensitivity in the second dimension, it makes sense, in particular for quantitative analysis of enantiomers present at trace levels, to transfer the entire peak to the ^{2}D column. The downside of transferring large volumes is the increased risk that the resolution of the ^{2}D separation will be compromised due to mobile phase mismatch effects as discussed in Section 4.4.4.

13.2.1 Single Heartcut Achiral-Chiral 2D-LC

For many applications a single heartcut *achiral-chiral LC-LC* method is fully adequate – e.g. if the enantiomeric composition of a chiral drug in a biological matrix has to be determined. Accurate quantification of enantiomers by 1D-LC-ESI-MS/MS requires stable isotope-labeled internal standards for both enantiomers to adequately compensate for matrix effects. Using UV detection provides a means of accurate quantitation that does not rely on stable isotope-labeled standards. However, this approach requires complete chromatographic separation of the target enantiomers from matrix interferences, and LC-LC is often very effective for this purpose. For such applications, achiral-chiral LC-LC can be realized in its simplest form by online SPE with an achiral sorbent (e.g., C18 material, Oasis HLB, internal surface RP) coupled to a ^{2}D chiral column for enantiomer separation. This allows direct injection of biological samples (e.g., plasma) into the ^{1}D column for enantioselective bioanalytical applications [9–14]. In a more advanced version, an achiral LC separation in the first dimension is coupled with a chiral LC separation in the second dimension. For single heartcuts, an electronically controlled 6-port/2-position valve for direct transfer or indirect transfer (see Section 4.4 for more detail on these two modes of LC-LC) is sufficient and often supported by modern HPLC systems [1, 9]. Due to the technical simplicity and great effectiveness of these approaches, there are numerous studies described in the literature that utilize simple heartcut techniques for analysis of chiral compounds in complex matrices. A summary of these can be found in the review by M. E. León-González *et al.* [3]. RP-chiral LC-LC is an elegant way to remove (matrix) interferences in enantioselective analysis with direct injection of samples and has been utilized for: 1) combined assay and enantiomeric excess (EE) analysis of phenylalanine by determining the total concentration (D+L) from the ^{1}D UV detector and the EE from the ^{2}D detector [15]; 2) enantiomer ratio analysis in herbal drug extract [16]; 3) for the determination of chiral coumarins in *Citrus* essential oils [17]; 4) for chiral pesticide analysis (allethrin) [18]; 5) enantioselective process control [19, 20]; and 6) determination of enantiomerization kinetics of stereolabile drugs [21], to name just a few applications.

13.2.2 Multiple Heartcut Achiral-Chiral 2D-LC for Metabolism Studies

Achiral-chiral multiple heartcut 2D-LC (mLC-LC) can also be realized with a simple 2-position/6-port valve equipped with one sampling loop if the ^{2}D separation is fast and if there is enough time between [1]D peaks selected for sampling of two adjacent fractions. However, sophisticated multi-loop interfaces improve flexibility and provide users with the ability to analyze more [1]D fractions using a complementary [2]D separation. mLC-LC is a flexible and powerful tool for the bioanalytical characterization of the metabolism of chiral xenobiotics that usually occurs with high stereoselectivity. Phase 1 metabolism reactions catalyzed by cytochrome P450 (CYP) isoenzymes may introduce hydroxyl groups at specific positions leading to several pairs of enantiomers of constitutional isomers. The resulting highly complex mixtures of structural isomers cannot be fully resolved by 1D-LC using common enantioselective columns, even with the benefit of tandem MS detection. However, achiral-chiral mLC-LC coupled with tandem MS is an elegant way to fully resolve such mixtures of isomeric metabolites, and has been used to investigate the stereoselective metabolism of phenprocoumon [22, 23], warfarin (Wf) [6, 24, 25], and propranolol [26, 27]. For instance, Wf, a structural analog of phenprocoumon, is a widely used anticoagulant. Significant interindividual variability in the response to treatment is observed, which is due to extensive metabolism and genetic polymorphism of CYP 450 enzymes. As a result of this variability, the Wf dose has to be adjusted for each patient, and careful control of Wf level in the blood by continuous monitoring is required. Wf is biotransformed to several chiral hydroxylated metabolites (4/6/7/8/10-OH-Wf) (Figure 13.1), with potentially distinct stereoselective disposition in different individuals. Tandem MS is not able to distinguish the phase I metabolites which are hydroxylated at the same aromatic ring but at distinct position, viz. 6/7/8-OH-Wf. While chiral columns – here polysaccharide CSPs – often have adequate enantioselectivity

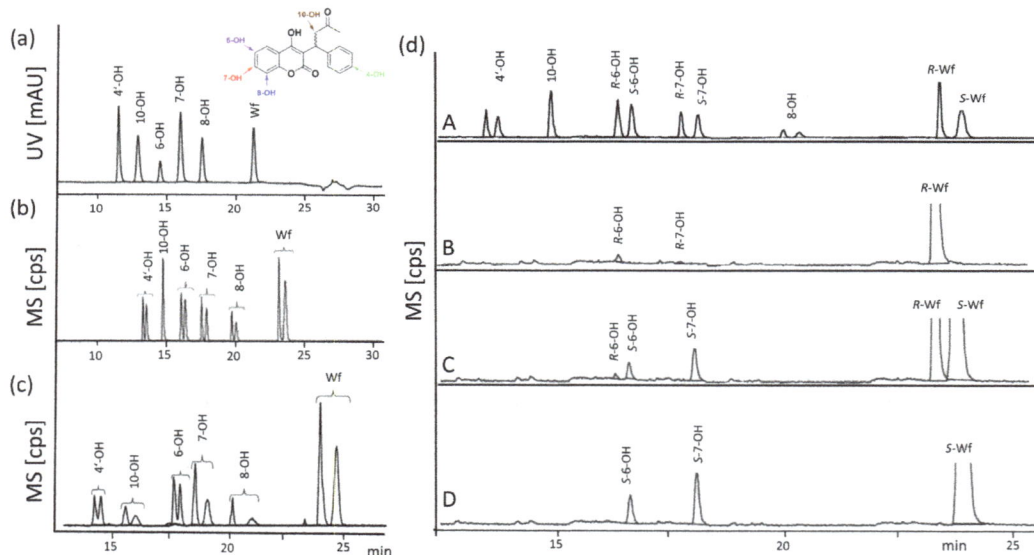

FIGURE 13.1 Achiral-chiral mLC-LC method for studying the stereoselective metabolism of warfarin. A) [1]D RPLC separation using ZORBAX Eclipse Plus Phenyl-Hexyl (150 mm × 2.1 mm i.d.; 1.8 μm); 25 °C; flow rate of 0.12 mL/min; B) [2]D chiral separation using Chiralcel OD-3R (3 μm, 50 mm × 4.6 mm i.d.), 30°C; flow rate of 0.8 ml/min; (c) [2]D chiral separation using tandem column of Chiralcel OD-3R and Chiralpak AD-3R (both 3 μm, 50 mm × 4.6 mm i.d.); 30 °C; flow rate of 0.8 ml/min; (d) Warfarin metabolite formation upon incubations with human liver microsomes over 120 min. (A) Standard mixture of warfarin and its metabolites at 100 ng/ml concentration. (B–D) Samples derived from incubations with human liver microsomes with (*R*)-, (*R/S*)-, and (*S*)-warfarin, respectively. Mobile phase (a, b, c, d): A: H_2O/formic acid (FA) (100:0.05, v/v); B: MeOH/FA (100:0.05, v/v) (both dimensions). Reprinted from S. Joseph, M. Subramanian, S. Khera, Simultaneous and stereospecific analysis of warfarin oxidative metabolism using 2D LC/Q-TOF, *Bioanalysis* 18(7) 2015, 2297–2309, with permission from Bioanalysis as agreed by Newlands Press Ltd.

to separate the enantiomers of the hydroxylated metabolites, the complex mixture of all isomers cannot be fully resolved by 1D-LC with a chiral column due to insufficient chemoselectivity and/ or peak capacity. For this reason, some of the peaks of the complex OH-Wf isomer mixture (e.g. the *R*-enantiomers of 6/7/8-OH-Wf) (partially) coelute. Moreover, they also cannot be distinguished by selective ion transitions in tandem MS, leading to a severe assay specificity problem. Such a problematic situation, or a similar one, occurs for many drugs (e.g. phenprocoumon and propranolol) and 2D-LC provides an adequate solution for them. Hence, a robust achiral-chiral mLC-LC method with a narrow-bore achiral ^{1}D phenylhexyl RP column, which exhibited exceptional selectivity for the constitutional HO-Wf isomers (Figure 13.1a), and a Chiralcel OD-3R column (Figure 13.1b) or even better a tandem column consisting of Chiralcel OD-3R and Chiralpak AD-3R (Figure 13.1d) in the second dimension provided full resolution of all isomers [24]. Fraction collection was time-based with a sampling time of 0.6 min using an 8-port/2-position valve that was equipped with an 80 μL loop (single loop setup). This mLC-LC method coupled with QTOF MS detection, revealed that the biotransformation of Wf in rat microsomal and plated hepatocyte incubations occurs stereoselectively with substrate-selectivity for the S-enantiomer (Figure 13.1c). Similarly, propranolol undergoes complex biotransformation reactions by CYP isoforms leading to 4/5/7-HO-propranolol, which cannot be distinguished by tandem MS alone without prior chromatographic separation (i.e., the selected-reaction monitoring [SRM] transitions for ring hydroxylated metabolites are all the same). Harps and coworkers developed a mLC-LC method for simultaneous quantitation of constitutional isomers and their enantiomers using a Poroshell 120 PhenylHexyl (100 mm × 2.1 mm i.d., 2.7 μm) column in first dimension and a Poroshell 120 Chiral-T (teicoplanin selector) (100 mm × 4.6 mm i.d., 2.7 μm) column in the second dimension [26]. UV detection was used following both the first and second dimensions, as well as triple quadrupole MS following the second dimension. In addition to fully resolving all of the isomers in a 10-min. analysis time, the mLC-LC approach also enabled a reduction in sample preparation time because samples could be injected directly following a simple protein precipitation step. Salts present in the matrix of a urine sample would be detrimental for the chiral separation if they were directly injected into a teicoplanin column, but in the LC-LC separation they elute near the dead time of the ^{1}D RP separation, and thus are not introduced into the teicoplanin column avoiding any negative impact on the ^{2}D chiral separation. This study also documented that such a 2D-LC method can be fully validated. It is well suited for setting a new standard in validated enantioselective bioanalysis assays.

13.2.3 MULTIPLE HEARTCUT ACHIRAL-CHIRAL 2D-LC FOR ENANTIOSELECTIVE AMINO ACID ANALYSIS

Enantioselective amino acid analysis is another field of application in which achiral-chiral mLC-LC has been used extensively [28–30]. Even with modern 1D-LC-MS/MS it is not a trivial task to quantify all of the enantiomers of all proteinogenic amino acids with adequate assay specificity using a single method. Matrix interferences, incomplete separation of isobaric amino acids (Ile/allo-Ile/Leu as well as Thr/allo-Thr), and matrix effects on detector response, are problems commonly encountered in comprehensive amino acid enantiomer analysis by 1D-LC separation coupled with detection by UV, FLD, and MS/MS, respectively. In a simple approach, achiral-chiral LC-LC with direct transfer using an RP and a chiral column (teicoplanin [31] and teicoplanin aglycon [32] CSPs) has been pursued for the direct analysis of single and multiple free amino acids (Tyr, Phe, Trp) in electron-beam irradiated foodstuffs [31] and homocysteine, methionine, and cysteine in human plasma [32]. In spite of the ability of the method to resolve them, no D-amino acids were ultimately found in the real biological samples. In a more advanced approach, Hamase and coworkers established a fully automated multi-loop mLC-LC setup with fluorescence detection in the second dimension for routine enantioselective amino acid analysis in biomarker research, food analysis, and studies on origin of life (Figure 13.2A) [33, 34]. The use of multiple loops for fraction storage gives

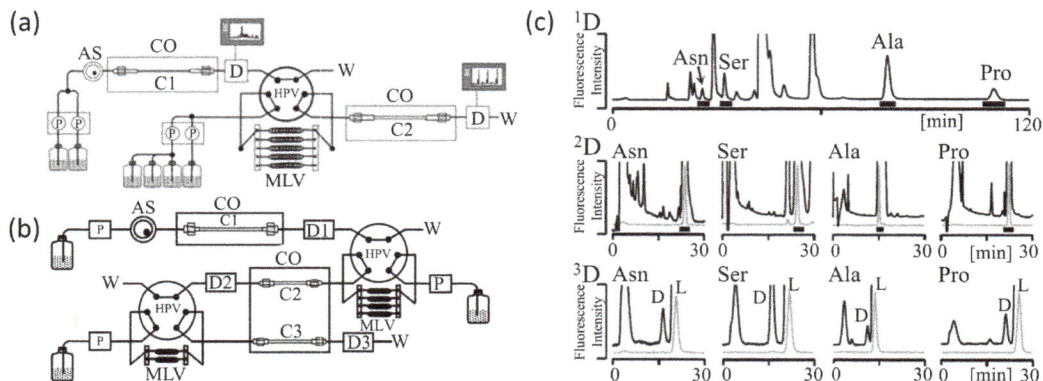

FIGURE 13.2 Achiral-chiral multi-loop mLC-LC method by Hamase and coworkers (a) 2D-LC setup, and (b) 3D-LC setup. (c) RP-AEX-chiral mLC-mLC-LC 3D-HPLC separations of NBD-Asn, Ser, Ala, and Pro in human plasma. In the second and third dimensions, solid lines represent 100× magnification of the gray lines. Caption: (a) P, pump; C1, reversed-phase column; C2, enantioselective column (KSAACSP-001S); CO, column oven; D, detector; HPV, high-pressure valve; W, waste. (b) Columns: C1, reversed-phase KSAARP (500 mm × 1.5 mm i.d.); C2, KSAAAX (150 mm × 1.5 mm i.d.) (anion-exchanger); C3, KSAACSP-001S (250 mm × 1.5 mm i.d.); multiloop valve between ^{1}D and ^{2}D with 3 loops of 500 µL and 1 loop of 800 µL; multiloop valve between ^{2}D and ^{3}D with 2 loops of 900 µL; flow rates: ^{1}D, 75 µL/min (15% MeCN with 0.05% TFA); ^{2}D, 200 µL/min (0.1% FA in a mixed solution of MeOH–MeCN (90/10, v/v)); ^{3}D, 200 µL/min (0.15% FA in a mixed solution of MeOH–MeCN (85/15, v/v)). Panel (a) is reprinted from ref. [34] with permission from K. Hamase. Panels (b) and (c) are reprinted with permission from F. Aogu, R. Koga, T. Akita, M. Mita, T. Kimura, K. Hamase, Three-dimensional high-performance liquid chromatographic determination of Asn, Ser, Ala and Pro enantiomers in the plasma of patients with chronic kidney disease, *Analytical Chemistry* 91 (2019) 11569–11575. Copyright 2019 American Chemical Society.

more time to develop the ^{2}D chromatogram and/or enables analysis of more fractions from the first dimension. In this approach the target amino acids analytes are derivatized prior to injecting them into the ^{1}D column using 4-fluoro-7-nitro-2,1,3-benzoxadiazole (NBD-F; reaction time is 2 min. at 60 °C). The NBD-amino acid derivatives can be detected by fluorescence with high sensitivity with excitation and emission wavelengths of 470 and 530 nm, respectively (LOQs typically in the range of 5 fmol on-column). In the first dimension, a long achiral microbore column (either 750 mm × 0.53 mm i.d. monolithic ODS [34], or 500 mm × 1.5 mm i.d. 3 µm particle ODS [35]) provides sufficient selectivity to remove most of the matrix interferences from the targeted NBD-amino acids. Fraction transfer from the first to the second dimension is accomplished using a multi-loop valve, usually equipped with four or five loops depending on the number of amino acids analyzed. Relatively large loop volumes are usually used to accommodate the entire ^{1}D peak volume, which is favorable for the detection of D-amino acids present at trace levels. For the ^{2}D chiral separation a variety of Pirkle type CSPs (e.g., Sumichiral OA-2500S (*N*-3,5-dinitrobenzoyl-1-naphthylglycineamide-bonded), and in-house prepared KSAACSP-001S (*N*-3,5-dinitrobenzoyl-*N*'-leucylamide-urea bonded)) or chiral anion-exchangers (Chiralpak QN-AX) have been evaluated [36, 37]. KSAACSP-001S provides baseline separation of all proteinogenic and metabolically relevant amino acids as NBD derivatives. It is possible to synthesize this phase in both enantiomeric forms, which was shown to be useful for validation of quantitative results by reversal of enantiomer elution orders through a switch in the configuration of the CSP (see for instance *N*-methyl-Asp and *N*-methyl-Glu on Sumichiral OA-2500S and OA-2500R in Fig. 4 of ref. [36]). Using the multi-loop mLC-LC system Hamase and coworkers analyzed a large variety of biological samples including rat urine [38], rat brain tissue and mantle of mussels [39], carbonaceous chondrite [37], food stuffs such as Japanese traditional black vinegars [40, 41] and other fermented food products including processed cheese and nam pla [41], brain of various mammalian species [42], human plasma of patients suffering from chronic kidney disease [43], urine of D-amino acid oxidase-deficient mice [44], and mammalian brain [45]; in each case a selected number of amino acids was targeted. The same 2D-LC system shown in Figure 13.2A, but equipped with an additional six-port/2-position valve (behind the two-loop interface) enabled switching between two different chiral columns (Chiralpak QD-AX for NBD-lactic acid and KSAACSP-001S for NBD-3-hydroxybutyric acid) for the ^{2}D separation. This approach has been utilized for the simultaneous detection of lactate and 3-hydroxybutyrate enantiomers after pre-column-derivatization with the fluorescent reagent 4-(*N*-chloroformylmethyl-*N*-methylamino)-7-nitro-2,1,3-benzoxadiazole (NBD-COCl) in human clinical samples [46].

13.2.4   MULTIPLE HEARTCUT 3D-LC FOR TRACE-LEVEL ENANTIOSELECTIVE AMINO ACID QUANTITATION IN BIOLOGICAL SAMPLES

Hamase and coworkers found that some interferences coeluted with target amino acids even after the ^{2}D separation. To resolve these interferences they extended the mLC-LC setup by adding a third separation dimension as shown in Figure 13.2B. In this work the ^{1}D, ^{2}D, and ^{3}D separations were RP (C18; 500 mm × 1.5 mm i.d., 3 μm), AEX (150 mm × 1.5 mm i.d.), and chiral (150 mm × 1.5 mm i.d.; Pirkle type KSAACSP-001S), respectively. This system was used for enantioselective analysis of amino acids [35, 47], dipeptides [48], and hydroxy carboxylic acids [34] in biological samples following derivatization with NBD-F. The mobile phases used in each dimension were finely tuned for optimal compatibility (^{1}D: 15% ACN with 0.05% TFA; ^{2}D: polar organic mode with MeOH/ACN, 90:10, v/v containing 0.1% formic acid; and ^{3}D: also polar organic with MeOH/ACN, 85:15, v/v containing 0.15% formic acid). As shown in Figure 13.2C, no unexpected peak broadening is observed in the second dimension, in spite of the strong TFA counterion present in the fractions injected into the second dimension. This is due to its low concentration and adequate refocusing of acidic analytes upon transfer from the aqueous ^{1}D separation to the nonaqueous conditions of the second dimension. Figure 13.2C shows that a number of interfering peaks were present in the ^{2}D chromatograms, but that the AEX separation effectively removed these, resulting in clean ^{3}D chromatograms with no compounds coeluting with the target amino acids. After validating the 3D-LC method it was applied to plasma samples of patients with different stages of chronic kidney disease, and the authors found correlations between the %D-amino acids (especially D-Asn and D-Ser) and the kidney function of the patients [35, 47].

13.2.5   SELECTIVE COMPREHENSIVE ACHIRAL-CHIRAL 2D-LC FOR ENANTIOSELECTIVE ANALYSIS OF ALL PROTEINOGENIC AMINO ACIDS

The mLC-LC methods described in the preceding sections have been used for enantioselective analysis of up to 13 amino acids in single separation [42, 37]. However, the full enantioselective analysis of all proteinogenic amino acids as required by pharmaceutical industry is still challenging. Recently, separation of proteinogenic amino acids as *N*-2,4-dinitrophenyl (DNP) amino acid derivatives has been demonstrated using mLC-LC and fast ^{2}D chiral separation on a short *tert*-butylcarbamoyl quinine-based core-shell column [49] (see Figure 1.8). The ^{1}D RP gradient mobile phase was found to work well with the nonaqueous polar organic ^{2}D eluent. Two different isocratic mobile phases were required to elute all enantiomer pairs in the second dimension within 120 s. A total of 29 fractions were transferred to the second dimension using a 12-loop interface and a software-optimized sampling program to make the most efficient use of the available ^{2}D time. The major challenge encountered in this analysis is the separation of the isobaric amino acids Leu/Ile/allo-Ile, which were not baseline resolved in the first dimension. Time-based sampling, which relies on high retention time precision (< 0.4% RSD in the first dimension), enabled positioning of sampling near the front and rear edges of the Leu/Ile/allo-Ile cluster to minimize coelution of isobaric amino acids in the second dimension. This method enables fully automated enantioselective quantitation of 25 amino acids (20 proteinogenic plus β-Ala, allo-Thr, homo-Ser, allo-Ile and Orn) using a single ^{2}D chiral column within total analysis time of 130 min. The method has been applied to determine the absolute configurations in non-ribosomal therapeutic peptides (bacitracin, gramicidin, etc., after hydrolysis of the peptides), and could be equally useful for determining the stereointegrity of synthetic therapeutic peptides.

13.2.6 HEARTCUT CHIRAL-ACHIRAL AND CHIRAL-CHIRAL 2D-LC

In the preceding sections we have discussed 2D-LC applications where the separations used in the first and second dimensions are achiral and chiral, respectively. This arrangement is by far the most frequently used one. The reverse arrangement (i.e., chiral first and achiral second) is more challenging to use because isomer ratios will be inaccurate unless the ^{1}D peaks are quantitatively transferred to the second dimension. When the chiral separation is used in the first dimension, isomers will be separated first before they are transferred to the second dimension. If a larger fraction of one of the isomers is transferred than the other, then the apparent isomer ratio observed at the ^{2}D detector will not be correct. The most robust way to prevent this is to choose conditions that provide nearly 100% transfer of each isomer, but this can be challenging in practice. Nevertheless, the feasibility and usefulness of the chiral-achiral arrangement has been demonstrated for the separation of Wf metabolites [6]. When used in the first dimension, a Chiralcel OD-3R column resolved seven out of the 12 components (Wf and its 4/6/7/8/10-HO-Wf metabolites) in 13 min. The remaining five components coeluted in two ^{1}D peaks, which were quantitatively transferred to an achiral Zorbax Eclipse Plus C18 column to finish the separation (with 16.2 min total analysis time). This analysis time was shorter than what could be achieved with achiral-chiral arrangement in which five fractions from the ^{1}D separation had to be analyzed in the second dimension; the combination of the larger number of fractions and a slow ^{2}D chiral separation led to a total analysis time of 204 min [6].

Also the great utility of chiral-chiral mLC-LC has been demonstrated in the pharmaceutical industry – e.g. for the simultaneous analysis of the various isomers of chiral drugs that contain multiple chiral centers such as Posaconazol [50]. For more details, interested readers are referred to Chapter 9 in this book.

13.3 ENANTIOSELECTIVE LC×LC

13.3.1 CHIRAL×ACHIRAL 2D-LC

Heartcut methods, especially those that rely on time-based sampling, are considered targeted assays. Preliminary information about the ^{1}D retention times of the target analytes is required before the method can be fully established. If a more comprehensive profile of the sample is required, enantioselective LC×LC is the preferred approach. However, so far only a few reports have addressed this topic.

Placement of the chiral separation in the second dimension of a LC×LC setup (i.e., achiral×chiral) requires that the ^{2}D separation be fast (< 1 min cycle time) to have a 2D separation with a reasonable analysis time and peak capacity. Although it has been documented that sub-minute and even sub-second 1D chiral separations [51–55] can be achieved on SPP CSPs, to the best of our knowledge an enantioselective LC×LC separation that leverages these technologies has not yet been realized in practice. On the other hand, Acquaviva and coworkers have reported the development of an enantioselective LC×LC separation with the opposite arrangement (chiral×achiral) for the enantioselective analysis of amino acids as DNP-derivatives in honey samples [56]. In this work, the ^{1}D column consisted of a stationary phase based on a quinine anion exchanger (100 mm × 2.1 mm i.d., 5 μm) and the ^{2}D separation used C18 or phenylbutyl columns (33 mm × 2.1 mm i.d.; 3 μm).

Although this column arrangement is the opposite of that preferred for enantioselective LC-LC separations, such a column arrangement has clear advantages in the case of LC×LC separations. The solute exchange (adsorption-desorption) kinetics are generally extremely slow in chiral separations on CSPs, which manifests as a steep C-term branch of the H/u-curves (i.e., van Deemter curves), and the optimal efficiency is usually found at relatively low flow rates [53, 57]. This makes the chiral separation the better choice for the first dimension where slow separations with low flow rates are generally preferred in LC×LC separations to minimize the effects of undersampling and volume overload on the ^{2}D column (see Sections 3.4 and 4.4 for more detail on these points). In sharp contrast, the adsorption-desorption kinetics in RP-type separations are generally fast, which enables extremely fast ^{2}D separations. In the work of Acquaviva the ^{2}D cycle time was just 15 s. Under these conditions many peaks were separated by the first dimension, and the second dimension effectively resolved coelution of different amino acids. This 2D-LC setup resulted in a characteristic peak pattern where both enantiomers of a particular amino acid eluted with the same ^{2}D retention time [56].

13.3.2 Chiral×Chiral 2D-LC

Chiral×chiral enantioselective LC×LC was found to be a useful modality in a number of applications. Three-micron TPP, core-shell, and sub-2 µm particle columns of short length (50 mm) can be operated at relatively high speed and may provide separations at timescales useful for ^{2}D chiral separations in LC×LC as first shown by Welch and coworkers [27]. New drug candidates have sometimes several chiral centers and then the analytical method development of assays which allow monitoring of all possible isomers is usually no longer trivial. To deal with this problem for a synthesis intermediate of a recently developed hepatitis C protease inhibitor that contained 3 stereogenic centers, thus 8 stereoisomers (4 pairs of enantiomers), a narrow-bore Chiralcel OJ-3R column (150 mm × 2.1 mm i.d.) operated at a low flow rate using a very slow gradient (130 min) as ^{1}D was combined with a short (50 mm × 4.6 mm i.d.) orthogonal Chiralcel OD-3R column at high flow rate in a comprehensive chiral×chiral LC×LC setup and enabled the complete separation of all stereoisomers (Figure 1.6.). The goal here is to overcome selectivity limitations, but not so much the gain of peak capacity. As shown by this example, orthogonal enantioselectivities can be achieved not only across distinct types of CSPs (e.g. when switching from polysaccharide to macrocyclic antibiotics CSPs) but also within the same CSP class by structural analogs such as polysaccharides differing in backbone (cellulose vs amylose), functional group (ester vs carbamate), and aromatic substituents (4-methyl, 3,5-dimethyl, 3,5-dichloro or mixed chloro/methyl), distinct macrocyclic antibiotics (vancomycin vs teicoplanin), or cinchonan CSPs with distinct carbamate residue (*tert*-butyl vs 2,6-diisopropylphenyl) [58], and so forth. However, when the latter two CSPs, *tert*-butyl carbamoylated quinine and corresponding 2,6-diisopropylphenylcarbamoylated quinine, were combined in a comprehensive chiral×chiral LC×LC approach, the peak pattern indicated significant correlation [58]. The combination of CSPs in which both the carbamate residue and the configurations of the selector backbone were altered was more effective [58].

One of the unique aspects of 2D-LC over 1D-LC is that 2D chromatograms sometimes exhibit structured peak patterns that facilitate identification of unknown compounds and interpretation of the chromatogram (see Section 1.6.4 for more detail on this feature of 2D separations). One of the most compelling examples of this is actually found in chiral LC×LC separations. Lämmerhofer and coworkers have shown that amino acid and peptide chirality can be readily determined from characteristic elution patterns in chiral×chiral separations in which the CSPs used in the two dimensions are chemically identical except for their stereochemistry (Figure 13.3). It follows that the CSPs are chemically equivalent but stereochemically orthogonal. This was shown for quinine and quinidine *tert*-butyl carbamate CSPs (tBuCQN-CSP and tBuCQD-CSP, respectively) using *N*-9-fluorenylmethoxycarbonyl (FMOC)-amino acids as analytes [59]. Elution orders for FMOC-amino

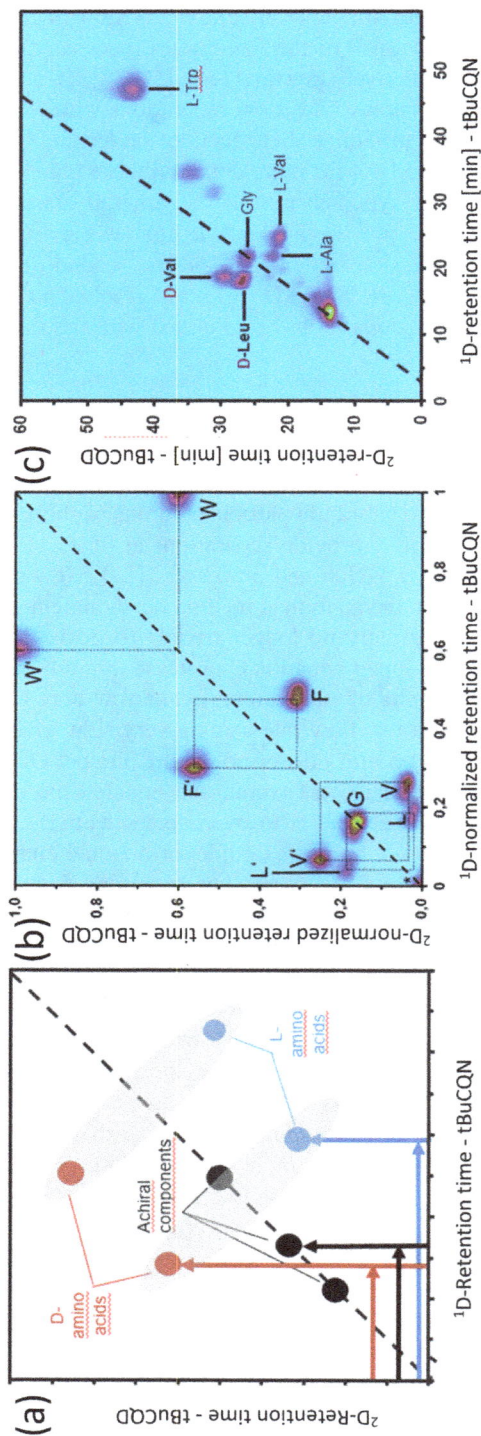

FIGURE 13.3 Chiralxchiral LCxLC approach with chemically equivalent columns differing just in the absolute configuration of the chiral selectors in [1]D and [2]D columns for "imaging" of amino acids and peptide stereochemistry. (a) General characteristic elution pattern with achiral analytes (reagent, reagent hydrolysate, achiral amino acids) eluting on a diagonal line (parity line) due to equal retention in [1]D and [2]D as well as cross peaks of opposite enantiomers due to reversed enantiomer elution order on [1]D and [2]D columns. (b) Practical implementation with racemic test mixture (one letter code for amino acids; dashed means D-configuration, L without dash). (c) Application to determination of absolute configurations of amino acids in non-ribosomal peptide gramicidin A1. Adapted with permission from U. Woiwode, R. Reischl, S. Buckenmaier, W. Lindner, M. Lämmerhofer, Imaging peptide and protein chirality via amino acid analysis by chiralxchiral two-dimensional correlation liquid chromatography, *Analytical Chemistry* 90 (2018), 7963–7971. Copyright 2018 American Chemical Society.

acid enantiomers (D- before L-enantiomers on tBuCQN-CSP and L- before D-enantiomers on tBuCQD-CSP) are consistently reversed on the two phases. When these phases are used in a LC×LC setup, peaks for achiral compounds (e.g., reagent peaks, FMOC-Gly) appear on a diagonal (parity) line in the 2D-contour plot because they have the same retention factor on both the ^{1}D and ^{2}D columns. In contrast, D- and L-amino acids are distributed either above or below the diagonal parity line due to their stronger retention on the tBuCQD-CSP or tBuCQN-CSP phases used in the second and first dimensions, respectively (Figure 13.3). The resulting chromatograms are easy to interpret for assignment of absolute configuration of amino acids.

13.4 ONLINE MULTI-COLUMN 2D-LC PLATFORM TECHNOLOGIES

Classical enantioselective 2D-LC approaches make use of a single ^{1}D and a single ^{2}D column. Since the scope of applicability for individual chiral columns is limited to certain analytes, it may be necessary to use more than one chiral column in the second dimension of a given 2D-LC method to completely resolve a complex mixture of chiral molecules. For this reason, flexible and versatile multi-column 2D-LC platforms have been established in the pharmaceutical industry and used to automate the discovery of optimal column combinations [25, 60].

Currently the pharmaceutical industry is more frequently developing chiral drugs with complex structures and multiple chiral centers. Requirements in impurity profiling demand that all stereo-isomeric impurities are controlled, in addition to achiral impurities, which presents a considerable analytical challenge. It is often impossible to separate all of these using one 1D-LC method and a single chiral column. To resolve this problem, an LC–mLC approach (here, the m stands for multiple ^{2}D columns) has been demonstrated that uses multiple chiral stationary phases in the second dimension [60]. A similar achiral-chiral LC-LC platform used a multi-column selection approach in the second (chiral) dimension with six different polysaccharide phases (Lux Amylose 1 and 2 as well as Cellulose 1, 2, 3 and 4; 150 mm × 4.6 mm i.d., 3 µm) for automated column screening [61]. A further extended multi-column 2D-LC platform allowed screening of multiple columns in both the first and second dimensions [25]. In this case, the system incorporated a mobile phase selection valve in the first dimension to enable automated mobile phase screening as well, and the second dimension was operated with volatile mobile phase components so that MS detection could be used for all ^{2}D separations. These multiple column 2D-LC approaches are emerging as powerful tools in the pharmaceutical industry that streamline automated column screening and high-throughput method development. Interested readers are referred to a more thorough and complementary discussion of these topics in Chapter 9 of this book.

13.5 ENANTIOSELECTIVE 2D CHROMATOGRAPHY BY LC-SFC

A fast ^{2}D separation is beneficial for multiple heartcut LC-LC and required for fully comprehensive LC×LC separations with practical analysis times. As discussed in Section 13.3, chiral LC separations are often too slow to be used regularly in the second dimension. Supercritical fluid chromatography (SFC) has been shown to be suitable for fast chiral separations [62] and is routinely used in pharmaceutical industry. Due to the lower viscosities of supercritical fluids compared to liquids (and thus higher analyte diffusivities in bulk mobile phase), the minima of H/u-curves are shifted to higher linear flow velocities and have smaller C-terms. This means that high mobile phase velocities can be used without significantly decreasing plate numbers [63]. These factors make SFC an attractive option for use in 2D separations with a chiral ^{2}D SFC separation, at least in principle.

Heinisch and coworkers evaluated the coupling of RPLC in the first dimension with SFC in the second dimension and passive (loop-based) modulation [64]. Contrary to LC×LC where partial loop filling is recommended to avoid analyte loss due to parabolic flow profile (see Section 4.4.13), in LC×SFC, full loop injection is mandatory to avoid problems with CO_2 depressurization after valve switching in case of partial loop filling. A selective comprehensive approach (sRPLC×chiral SFC) was used with gradient elution in the second dimension for the determination of the purity of a chiral API material. With this setup the ^{1}D RP separation resolved achiral impurities from the API, while ^{2}D SFC separation on a Chiralpak IC column enabled determination of the enantiomeric purities of the chiral API and residual chiral intermediate in the sample. The ^{2}D SFC also resolved another minor impurity that coeluted with the API peak in the ^{1}D RPLC separation; these capabilities clearly demonstrated the value of the sRPLC×SFC method.

Venkatramani *et al.* also developed a 2D separation approach with SFC in the second dimension for the determination of both overall chemical purity and enantiomeric purity of an API. However, rather than using open tubular loops in the interface, they used particle packed C18 trapping columns [65]. In a first study, two traps were mounted on an 8-port/2-position 2D-LC valve. Peaks of interest from the ^{1}D RP column, which provided overall chemical purity information, were effectively focused as sharp zones on the trapping columns from where they were eluted by ^{2}D SFC eluents onto the ^{2}D chiral column for enantiomeric excess determination. For applications in which several fractions had to be analyzed in the second dimension to determine their enantiomer ratios (e.g. drugs with several chiral centers), an array of trapping columns was connected to the 8-port/ 2-position valve (see Figure 13.4A) [65]. The system was further equipped with column selection valves that could accommodate six different chiral columns, which enabled efficient screening of different selectivities for the 2D separation. In the published example of an API with 3 stereogenic centers, four peaks corresponding to the stereoisomers that are diastereomeric to each other were successfully separated by the 1D achiral RPLC column (see Figure 13.4B). A heartcut of each stereoisomer peak was then transferred from the 1D RP column to one of four different trapping columns, which were then sequentially backflushed by the 2D SFC pump onto the chiral column (Chiralpak IC3) (Figure 13.4C). When coupled with MS detection, this system was applied in pre-clinical bioanalytical studies to determine the stereoselective disposition of a novel API after incubation with mouse hepatocytes [66]. In this case the ^{1}D RP separation resolved the API from matrix components as well as a metabolite of the API. Heartcuts of the API metabolite peaks were transferred first to the trap columns, and then to the ^{2}D chiral SFC separation. For this application it was found that different ^{2}D chiral columns were required for the API (Chiralpak IB-3) and the metabolite (Chiralpak AD-3).

13.6   ENANTIOSELECTIVE 2D CHROMATOGRAPHY BY MICROCHIP LC-LC

High-throughput enantioselective analysis is also a requirement in the process of developing chiral catalysts used for stereoselective synthesis. In this work, the chiral catalyst needs to be purified from educts, side products, and other process components and contaminants before determination of enantiomeric excess is possible. In the future, 2D-LC separations carried out on microchips could be a viable format for online reaction monitoring directly from process samples [67]. Lotter and coworkers proposed a chip layout for this purpose that contained two separation channels – one packed with C18 particles (^{1}D), and the other one packed with a chiral stationary phase (Chiralpak IB; ^{2}D). The end of the ^{1}D separation channel and the inlet to the ^{2}D separation channel were intersected by a channel allowing cross-flow. During the separation of the sample mixture by the ^{1}D column, the cross-flow was switched on such that ^{1}D effluent flowed to waste, and fresh mobile phase was delivered to the ^{2}D separation channel. When the peak of interest eluted from the ^{1}D separation channel the cross-flow was stopped, which resulted in the ^{1}D peak flowing into the ^{2}D separation channel. After transfer of a heartcut of the ^{1}D peak the cross-flow was switched on again, leading to separation of enantiomers by the ^{2}D column. The chip layout was then slightly modified to couple a flow reactor (in which warfarin was synthesized stereoselectively) with the 2D-chip LC and ESI-MS. A valve switch enabled a 5-nL transfer of reaction mixture from the flow reactor into the ^{1}D separation channel. Using this workflow, it was demonstrated that the chiral catalyst

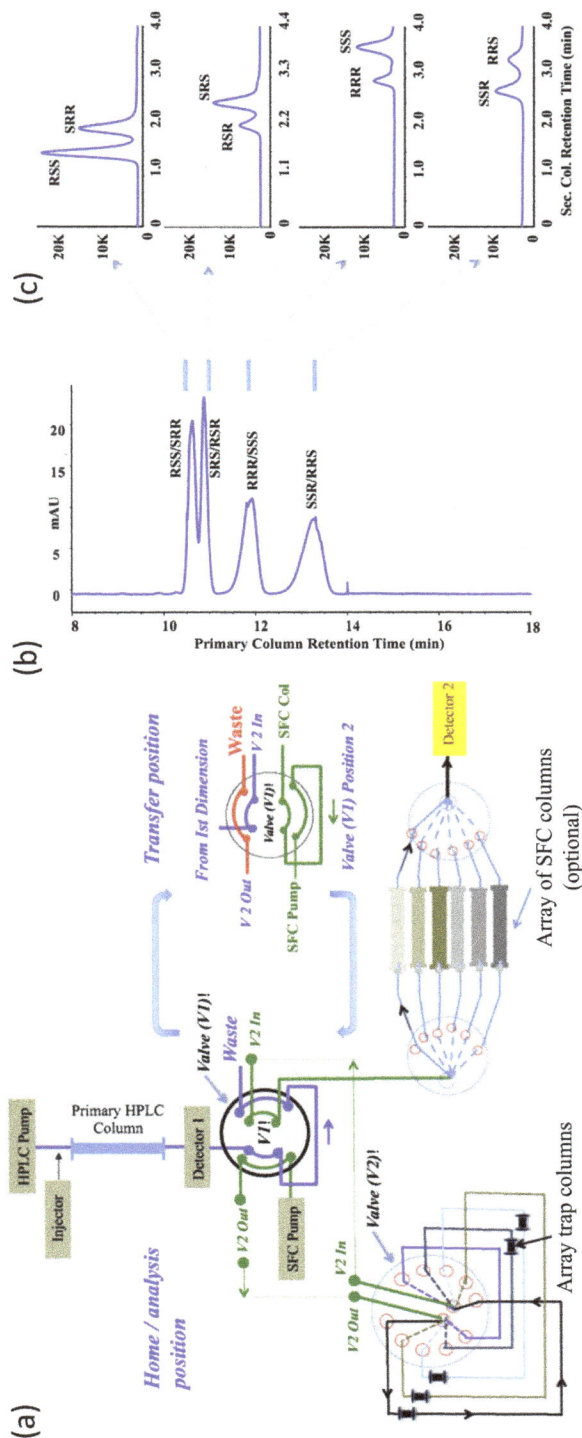

FIGURE 13.4 (a) Schematic diagram of achiral–chiral LC-SFC using an array of trapping columns and secondary SFC columns. The above setup enables transfer of multiple fractions to an array of secondary columns. (b) ¹D Achiral separation on an array of secondary columns. (b) ¹D Achiral separation on C18 column. (c) ²D Analysis of the backflushed heart cuts from ¹D by SFC on Chiralpak IC3. Reprinted from *Talanta*, 148, C. Venkatramani, M. Al-Sayah, G. Li, M. Goel, J. Girotti, L. Zang, L. Wigman, P. Yehl, N. Chetwyn, Simultaneous achiral-chiral analysis of pharmaceutical compounds using two-dimensional reversed phase liquid chromatography-supercritical fluid chromatography, 548–555, Copyright (2016), with permission from Elsevier.

(*1S,2S*)-diphenylethanediamine induced stereoselective formation of (*S*)-warfarin. This proof of principle study clearly indicated the high potential for 2D-chip LC to be used for cost-effective screening of enantioselective catalysts.

13.7 OTHER MULTI-DIMENSIONAL ISOMER SEPARATIONS WITH CHIRAL COLUMNS

In addition to their use in 2D-LC for separations of enantiomers, chiral stationary phases are also worth considering as complementary selectivity for separations of compounds that are difficult to resolve, but not necessarily enantiomers (e.g., other structural isomers, structural analogs). They have frequently shown improved selectivity over achiral phases for epimers, diastereomers, positional isomers, constitutional isomers, and even topoisomers, all of which are theoretically separable on achiral stationary phases.

Epimers are stereoisomers that differ only in one of several stereogenic centers. In general, achiral assays should be able to distinguish them. However, achiral RP is not always selective enough and chiral stationary phases may provide better epimer selectivity. For example, the epimer pair (*20S*)- and (*20R*)-ginsenoside Rh1 remained unresolved following LCxLC separation (C8 and C18 used in the first and second dimensions) and QTOF detection [68]. On the other hand, a C8-chiral (Chiralpak IC3) mLC-LC method successfully resolved this epimer pair and other isomer pairs (12 in total).

The potential of chiral LC methods to provide enhanced selectivity for diastereomers (stereoisomers which do not behave like mirror images) has been already documented for an intermediate of a synthetic anti-hepatitis C virus (HCV) therapeutic as shown in Figure 1.6 [27]. Compounds 13 and 14 in the 2D-contour plot are diastereomers. The Chiralcel OD-3R phase provides exceptional selectivity for this pair, which enables full resolution within 40 s analysis time in the second dimension. The final API has five stereogenic centers and impurities could arise from distinct configurations at these stereogenic centers (stereoisomerism) as well as methoxy substitution at four different aromatic positions (constitutional isomerism). Here again, resolution of two critical impurities required a single heartcut achiral–chiral LC-LC method with ¹D RP (Cortecs C18, 150 mmx2.1mm i.d., 1.6 μm) and 2D chiral (teicoplanin CSP, 50 mmx4.6 mm i.d., 1.9 μm) separations to completely resolve all nine closely related structural isomers in a single analysis [27].

The full separation of complex mixtures of constitutional isomers of small chemical building blocks such as halogenated aromatic compounds is another challenging task for achiral phases in 1D-LC [27]. High purity is required for such starting materials to avoid low overall yields and complicated reaction product mixtures at later stages. If achiral columns provide incomplete resolution, a heartcut or selective comprehensive 2D-LC method involving a chiral column may deliver a quick solution for this type of problem, as shown for example in Figure 13.5.

Peptide separations on common RP C18 columns also sometimes suffer from insufficient selectivity for isomeric species. Epimeric or diastereomeric impurities, Asp/iso-Asp isomers as well as deamidation impurities often present challenging problems. Some chiral columns such as CSPs based on macrocyclic antibiotics (vancomycin, teicoplanin) as well as cinchona alkaloid-derived anion exchanger and zwitterionic CSPs have shown great potential for providing selectivities complementary to typical C18 phases in such applications. On the other hand, chiral columns may show good enantio- and diastereoselectivity, but may fail to completely separate a complex mixture of isomeric peptides due to insufficient peak capacity. Thus, the combination of RP and chiral columns in a 2D-LC format may be a very powerful approach for challenging peptide separations, and this has been demonstrated for instance by Hamase and coworkers [48]. This group analyzed the sequence isomers Ser-Gly and Gly-Ser (as NBD derivatives) along with the corresponding enantiomers using a 3D-LC system (Figure 13.2B) in Japanese traditional amber rice vinegar. The sequence isomers were well resolved on the achiral RP column in the first dimension but showed similar retention on the chiral column. On the other hand, the achiral column could not separate the enantiomers. The

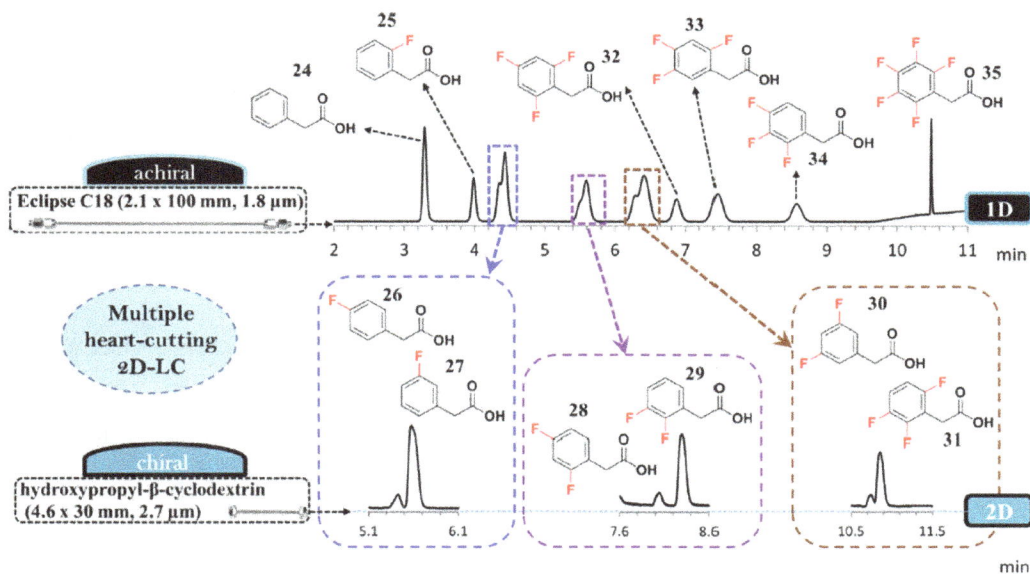

FIGURE 13.5 mLC-LC method for separation of complex mixture of fluorophenylacetic acid isomers. Conditions, first dimension (achiral): column, Zorbax RRHD Eclipse Plus C18 (100 mm × 2.1 mm i.d., 1.8 µm); temperature, 40 °C. Detection: UV 210 nm. Flow rate: 0.5 mL/ min. Eluent A, 0.1% H_3PO_4 in H_2O; eluent B, ACN. Step gradient: hold 20% B for 7 min; 7–9 min, 95% B. Conditions, second dimension (chiral): column, HPRSP (hydroxypropyl-beta-cyclodextrin) (DextroShell-RSP,4.6 mm×30 mm, 2.7 µm). Flow rate: 1 mL/min. Isocratic mobile phase: 95:5% H_3PO_4/ACN. Detection: UV 210 nm. Sampling frequency: 240 Hz. Reprinted with permission from C. Barhate, E. Regalado, N. Contrella, J. Lee, J. Jo, A. Makarov, D. Armstrong, C. Welch, Ultrafast chiral chromatography as the second dimension in two-dimensional liquid chromatography experiments, *Analytical Chemistry* 89 (2017), 3545–3553. Copyright 2017 American Chemical Society.

coupling of these columns in a 3D-LC setup gave adequate resolution of all isomers and clear separation from matrix interferences after an additional 2D anion-exchange separation. The Japanese vinegar contained relatively large amounts of the L-dipeptides and trace amounts of D-Ser-Gly.

2D-LC is also a powerful tool used in bioprocess control. In this context 2D-LC enables direct injection of bioprocess samples from the bioreactor without prior purification. For example, Mahut and coworkers demonstrated the separation of topological isomers (topoisomers) of a 3.5 kbp plasmid DNA (pDNA, cyclic dsDNA) using a 2D-LC with SEC and chiral columns in the first and second dimensions, respectively [69]. The pDNA eluted on the 1000 Å silica-based SEC column in the exclusion volume as the first peak while all other process constituents (host cell proteins, nuclease-digested RNA and DNA, and low molecular substances) at least partially permeated the pores. Thus, the process impurities were well resolved from the pDNA peak, which was transferred to a quinine carbamate column in the second dimension, commonly used for enantiomer separation of acidic chiral compounds, and has proven pDNA topoisomer selectivity. When the 2D quinine carbamate column was operated at elevated temperatures with NaCl buffers in anion-exchange mode, the natural pDNA topoisomers could be resolved from each other [69, 70]. When the quinine carbamate column in the 2D is run with a pH- instead of NaCl- gradient, it switches its selectivity from topoisomer selectivity to topological isoform selectivity allowing advanced separation of the impurities (linear, due to double strand break; open circular form, obtained by single strand break) from therapeutically relevant supercoiled (covalently closed circular) pDNA form [70]. This is just another example demonstrating the great utility of chiral columns in 2D-LC.

13.8 CONCLUDING REMARKS

Enantioselective achiral-chiral heartcut 2D-LC has a long tradition in enantioselective analysis and dates back to the 1980s and 1990s when it was frequently used for bioanalysis and tandem mass spectrometry (MS/MS) was not widely available. Nowadays, 2D-LC is often used for bioanalysis in combination with MS when multiple hydroxylated metabolite isomers cannot be distinguished by enantioselective 1D-LC-MS, for enantioselective amino acid analysis with fluorescence or MS detection, and for characterization and purity determination of complex drugs with multiple chiral centers or other structural isomers. The most widely used 2D-LC mode is achiral-chiral mLC-LC. Numerous examples of its use for enantioselective analysis of trace-level D-amino acids as biomarkers of disease have been reported.

Fast ^{2}D chiral separations on short columns packed with sub-3 μm SPP CSPs have enabled up to 30 fractions to be analyzed in the second dimension of 2D-LC separations of all proteinogenic amino acids, including isobaric ones. Early work on the use of SFC in the second dimension of enantioselective 2D-LC looks promising. Its use for chiral 1D-LC is widespread and well understood, and it can be used for fast chiral separations, which would be especially helpful in the second dimension.

In addition to their use in 2D-LC for separations of enantiomers, chiral stationary phases are also worth considering as complementary selectivity for separations of compounds that are difficult to resolve, but not necessarily enantiomers (e.g., structural isomers, structural analogs). They have been shown to provide improved selectivity over achiral phases for epimers, diastereomers, positional isomers, constitutional isomers, isomeric peptides, and even pDNA topoisomers, all of which are theoretically separable on achiral stationary phases.

Finally, flexible multi-column, multi-detector 2D-LC platforms that enable automated screening of multiple mobile phase and stationary phase combinations in both dimensions are becoming increasingly popular, and may foreshadow the future in the pharmaceutical industry for extended automated method development and improved quality of impurity profiling by 2D-LC.

The future of enantioselective 2D-LC could benefit from a further acceleration of 1D chiral separations that could be used in one or both dimensions of 2D-LC systems. Enantioselective SFC on short chiral SPP columns definitely holds great promise in this regard. Unfortunately, short chiral columns are typically limited to some standard formats such as 50 mm × 2.1 mm i.d. or 50 mm × 3.0 mm i.d. Chiral column suppliers are a bit conservative with respect to the column formats they offer, and 2D-LC is not in their focus or of interest as a significant market at this point. Customized chiral column dimensions would be of interest for 2D-LC users. Multi-column platforms with multiple chiral columns need to be more routinely implemented for enantioselective 2D-LC. Single chiral columns are limited in terms of the scope of their applicability, and more than one chiral column may be required in many instances to resolve all chiral constituents in one sample. Micro-LC pumps with tailored flow regimes for the ^{1}D separation are also not widely available or those available are not robust enough, e.g. for achieving high precision of < 0.5% with shallow gradients in multiple heartcut methods. Thus, enantioselective 2D-LC could also benefit from developments in this area. While multiple heartcut enantioselective LC methods can currently be applied on a routine basis with modern 2D-LC equipment in fields like enantioselective metabolism studies, enantioselective process control, and so forth, for more complex applications (especially full comprehensive 2D-LC) it is still necessary to develop, optimize, and publish more applications that reach further toward the ultimate goal to have generic methods or platform technologies that can serve as starting points for more efficient method development and optimization for specific analytical applications.

ACKNOWLEDGMENTS

We gratefully acknowledge support by Agilent Technologies through a research award. We are grateful to Dr. Stephan Buckenmaier as well as to Dr. Dwight R. Stoll for many fruitful discussions and collaborations.

REFERENCES

[1] K. Fried, I.W. Wainer, Column-switching techniques in the biomedical analysis of stereoisomeric drugs: Why, how and when, J. Chromatogr. B 689(1) (1997) 91–104. https://doi.org/10.1016/S0378-4347(96)00400-8.

[2] R.J. Soukup, D.W. Armstrong, Analysis of enantiomeric compounds using multidimensional liquid chromatography, in: S.A. Cohen, M.R. Schure (eds.), Multidimensional Liquid Chromatography2008, pp. 319–344. https://doi.org/10.1002/9780470276266.ch14.

[3] M.E. León-González, N. Rosales-Conrado, L.V. Pérez-Arribas, V. Guillén-Casla, Two-dimensional liquid chromatography for direct chiral separations: A review, Biomed. Chromatogr. 28(1) (2014) 59–83. https://doi.org/10.1002/bmc.3007.

[4] I. Ali, M. Suhail, H.Y. Aboul-Enein, Advances in chiral multidimensional liquid chromatography, TrAC, Trends Anal. Chem. 120 (2019). https://doi.org/ 10.1016/j.trac.2019.115634.

[5] S. Bäurer, M. Ferri, A. Carotti, S. Neubauer, R. Sardella, M. Lämmerhofer, Mixed-mode chromatography characteristics of chiralpak ZWIX(+) and ZWIX(–) and elucidation of their chromatographic orthogonality for LCxLC application, Anal. Chim. Acta 1093 (2020) 168–179. https://doi.org/10.1016/j.aca.2019.09.068.

[6] E.L. Regalado, J.A. Schariter, C.J. Welch, Investigation of two-dimensional high performance liquid chromatography approaches for reversed phase resolution of warfarin and hydroxywarfarin isomers, J. Chromatogr. A 1363 (2014) 200–206. https://doi.org/10.1016/j.chroma.2014.08.025.

[7] X. Wang, H. Wu, R. Luo, D. Xia, Z. Jiang, H. Han, Separation and detection of free d- and l-amino acids in tea by off-line two-dimensional liquid chromatography, Anal. Methods 9(43) (2017) 6131–6138. https://doi.org/10.1039/C7AY01569K.

[8] Ulrich Woiwode, Stefan Neubauer, Mike Kaupert, Wolfgang Lindner, M. Lämmerhofer, Trends in enantioselective high performance liquid chromatography, LCGC 30(6) (2017) 34–42.

[9] I.W. Wainer, The application of achiral/chiral coupled-column high-performance liquid chromatography to biomedical analysis, J. Pharm. Biomed. Anal. 7(9) (1989) 1033–1038. https://doi.org/10.1016/0731-7085(89)80042-1.

[10] R.N. Rao, K.N. Kumar, D.D. Shinde, Determination of rat plasma levels of sertraline enantiomers using direct injection with achiral–chiral column switching by LC–ESI/MS/MS, J. Pharm. Biomed. Anal. 52(3) (2010) 398–405. https://doi.org/10.1016/j.jpba.2009.09.020.

[11] R.F. Gomes, N.M. Cassiano, J. Pedrazzoli Jr., Q.B. Cass, Two-dimensional chromatography method applied to the enantiomeric determination of lansoprazole in human plasma by direct sample injection, Chirality 22(1) (2010) 35–41. https://doi.org/10.1002/chir.20701.

[12] Q.B. Cass, T. Ferreira Galatti, A method for determination of the plasma levels of modafinil enantiomers, (±)-modafinic acid and modafinil sulphone by direct human plasma injection and bidimensional achiral–chiral chromatography, J. Pharm. Biomed. Anal. 46(5) (2008) 937–944. https://doi.org/10.1016/j.jpba.2007.03.004.

[13] K.R.A. Belaz, Q.B. Cass, R.V. Oliveira, Determination of albendazole metabolites by direct injection of bovine plasma and multidimensional achiral–chiral high performance liquid chromatography, Talanta 76(1) (2008) 146–153. https://doi.org/ 10.1016/j.talanta.2008.02.013.

[14] Q.B. Cass, V.V. Lima, R.V. Oliveira, N.M. Cassiano, A.L.G. Degani, J. Pedrazzoli, Enantiomeric determination of the plasma levels of omeprazole by direct plasma injection using high-performance liquid chromatography with achiral–chiral column-switching, J. Chromatogr. B 798(2) (2003) 275–281. https://doi.org/10.1016/j.jchromb.2003.09.053.

[15] C.J. Venkatramani, L. Wigman, K. Mistry, N. Chetwyn, Simultaneous, sequential quantitative achiral–chiral analysis by two-dimensional liquid chromatography, J. Sep. Sci. 35(14) (2012) 1748–1754. https://doi.org/10.1002/jssc.201200005.

[16] Y.L. Song, W.H. Jing, G. Du, F.Q. Yang, R. Yan, Y.T. Wang, Qualitative analysis and enantiospecific determination of angular-type pyranocoumarins in Peucedani Radix using achiral and chiral liquid chromatography coupled with tandem mass spectrometry, J. Chromatogr. A 1338 (2014) 24–37. https://doi.org/10.1016/j.chroma.2014.01.078.

[17] P. Dugo, M. Russo, M. Sarò, C. Carnovale, I. Bonaccorsi, L. Mondello, Multidimensional liquid chromatography for the determination of chiral coumarins and furocoumarins in Citrus essential oils, J. Sep. Sci. 35(14) (2012) 1828–1836. https://doi.org/10.1002/jssc.201200078.

[18] F. Mancini, J. Fiori, C. Bertucci, V. Cavrini, M. Bragieri, M.C. Zanotti, A. Liverani, V. Borzatta, V. Andrisano, Stereoselective determination of allethrin by two-dimensional achiral/chiral liquid chromatography with ultraviolet/circular dichroism detection, J. Chromatogr. A 1046(1) (2004) 67–73. https://doi.org/10.1016/j.chroma.2004.06.075.

[19] S. Ma, N. Grinberg, N. Haddad, S. Rodriguez, C.A. Busacca, K. Fandrick, H. Lee, J.J. Song, N. Yee, D. Krishnamurthy, C.H. Senanayake, J. Wang, J. Trenck, S. Mendonsa, P.R. Claise, R.J. Gilman, T.H. Evers, Heart-cutting two-dimensional ultrahigh-pressure liquid chromatography for process development: Asymmetric reaction monitoring, Org. Process Res. Dev. 17(5) (2013) 806–810. https://doi.org/10.1021/op300266j.

[20] Q. Liu, X. Jiang, H. Zheng, W. Su, X. Chen, H. Yang, On-line two-dimensional LC: A rapid and efficient method for the determination of enantiomeric excess in reaction mixtures, J. Sep. Sci. 36(19) (2013) 3158–3164. https://doi.org/10.1002/jssc.201300412.

[21] G. Cannazza, U. Battisti, M.M. Carrozzo, L. Brasili, D. Braghiroli, C. Parenti, Evaluation of stereo and chemical stability of chiral compounds, Chirality 23(10) (2011) 851–859. https://doi.org/10.1002/chir.20941.

[22] B. Kammerer, R. Kahlich, M. Ufer, S. Laufer, C.H. Gleiter, Achiral–chiral LC/LC–MS/MS coupling for determination of chiral discrimination effects in phenprocoumon metabolism, Anal. Biochem. 339(2) (2005) 297–309. https://doi.org/10.1016/j.ab.2005.01.010.

[23] B. Kammerer, R. Kahlich, M. Ufer, A. Schenkel, S. Laufer, C.H. Gleiter, Stereospecific pharmacokinetic characterisation of phenprocoumon metabolites, and mass-spectrometric identification of two novel metabolites in human plasma and liver microsomes, Anal. Bioanal. Chem. 383(6) (2005) 909–917. https://doi.org/10.1007/s00216-005-0113-7.

[24] S. Joseph, M. Subramanian, S. Khera, Simultaneous and stereospecific analysis of warfarin oxidative metabolism using 2D LC/Q-TOF, Bioanalysis 7(18) (2015) 2297–2309. https://doi.org/10.4155/bio.15.119.

[25] H. Wang, H.R. Lhotka, R. Bennett, M. Potapenko, C.J. Pickens, B.F. Mann, I.A. Haidar Ahmad, E.L. Regalado, Introducing online multicolumn two-dimensional liquid chromatography screening for facile selection of stationary and mobile phase conditions in both dimensions, J. Chromatogr. A 1622 (2020). https://doi.org/10.1016/j.chroma.2020.460895.

[26] L.C. Harps, S. Schipperges, F. Bredendiek, B. Wuest, A. Borowiak, M.K. Parr, Two dimensional chromatography mass spectrometry: Quantitation of chiral shifts in metabolism of propranolol in bioanalysis, J. Chromatogr. A 1617 (2020). https://doi.org/10.1016/j.chroma.2019.460828.

[27] C.L. Barhate, E.L. Regalado, N.D. Contrella, J. Lee, J. Jo, A.A. Makarov, D.W. Armstrong, C.J. Welch, Ultrafast chiral chromatography as the second dimension in two-dimensional liquid chromatography experiments, Anal. Chem. 89(6) (2017) 3545–3553. https://doi.org/10.1021/acs.analchem.6b04834.

[28] A. Dossena, G. Galaverna, R. Corradini, R. Marchelli, Two-dimensional high-performance liquid chromatographic system for the determination of enantiomeric excess in complex amino acid mixtures: Single amino acid analysis, J. Chromatogr. A 653(2) (1993) 229–234. https://doi.org/10.1016/0021-9673(93)83178-U.

[29] T. Welsch, C. Schmidtkunz, B. Müller, F. Meier, M. Chlup, A. Köhne, M. Lämmerhofer, W. Lindner, A comprehensive chemoselective and enantioselective 2D-HPLC set-up for fast enantiomer analysis of a multicomponent mixture of derivatized amino acids, Anal. Bioanal. Chem. 388(8) (2007) 1717–1724. https://doi.org/10.1007/s00216-007-1399-4.

[30] F. Ianni, R. Sardella, A. Lisanti, A. Gioiello, B.T. Cenci Goga, W. Lindner, B. Natalini, Achiral–chiral two-dimensional chromatography of free amino acids in milk: A promising tool for detecting different levels of mastitis in cows, J. Pharm. Biomed. Anal. 116 (2015) 40–46. https://doi.org/10.1016/j.jpba.2014.12.041.

[31] V. Guillén-Casla, M.E. León-González, L.V. Pérez-Arribas, L.M. Polo-Díez, Direct chiral determination of free amino acid enantiomers by two-dimensional liquid chromatography: Application to control transformations in E-beam irradiated foodstuffs, Anal. Bioanal. Chem. 397(1) (2010) 63–75. https://doi.org/10.1007/s00216-009-3376-6.

[32] Z. Deáková, Z. Ďuračková, D.W. Armstrong, J. Lehotay, Two-dimensional high performance liquid chromatography for determination of homocysteine, methionine and cysteine enantiomers in human serum, J. Chromatogr. A 1408 (2015) 118–124. https://doi.org/ 10.1016/j.chroma.2015.07.009.

[33] K. Hamase, Sensitive two-dimensional determination of small amounts of D-amino acids in mammals and the study on their functions, Chem. Pharm. Bull. 55(4) (2007) 503–510. https://doi.org/10.1248/cpb.55.503.

[34] C. Ishii, A. Furusho, C.-L. Hsieh, K. Hamase, Multi-dimensional high-performance liquid chromatographic determination of chiral amino acids and related compounds in real world samples, Chromatography 41(1) (2020) 1–17. https://doi.org/10.15583/jpchrom.2020.004.

[35] A. Furusho, R. Koga, T. Akita, M. Mita, T. Kimura, K. Hamase, Three-dimensional high-performance liquid chromatographic determination of Asn, Ser, Ala, and Pro enantiomers in the plasma of patients with chronic kidney disease, Anal. Chem. 91(18) (2019) 11569–11575. https://doi.org/10.1021/acs.analchem.9b01615.

[36] R. Koga, H. Yoshida, H. Nohta, K. Hamase, Multi-dimensional HPLC analysis of metabolic related chiral amino acids – method development and biological/clinical applications, Chromatography 40(1) (2019) 1–8. https://doi.org/10.15583/jpchrom.2019.002.

[37] K. Hamase, Y. Nakauchi, Y. Miyoshi, R. Koga, N. Kusano, H. Onigahara, H. Naraoka, H. Mita, Y. Kadota, Y. Nishio, M. Mita, W. Lindner, Enantioselective determination of extraterrestrial amino acids using a two-dimensional chiral high-performance liquid chromatographic system, Chromatography 35(2) (2014) 103–110. https://doi.org/10.15583/jpchrom.2014.014.

[38] K. Hamase, Y. Miyoshi, K. Ueno, H. Han, J. Hirano, A. Morikawa, M. Mita, T. Kaneko, W. Lindner, K. Zaitsu, Simultaneous determination of hydrophilic amino acid enantiomers in mammalian tissues and physiological fluids applying a fully automated micro-two-dimensional high-performance liquid chromatographic concept, J. Chromatogr. A 1217(7) (2010) 1056–1062. https://doi.org/10.1016/j.chroma.2009.09.002.

[39] R. Koga, Y. Miyoshi, E. Negishi, T. Kaneko, M. Mita, W. Lindner, K. Hamase, Enantioselective two-dimensional high-performance liquid chromatographic determination of N-methyl-D-aspartic acid and its analogues in mammals and bivalves, J. Chromatogr. A 1269 (2012) 255–261. https://doi.org/10.1016/j.chroma.2012.08.075.

[40] Y. Miyoshi, M. Nagano, S. Ishigo, Y. Ito, K. Hashiguchi, N. Hishida, M. Mita, W. Lindner, K. Hamase, Chiral amino acid analysis of Japanese traditional Kurozu and the developmental changes during earthenware jar fermentation processes, J. Chromatogr. B. 966 (2014) 187–192. https://doi.org/10.1016/j.jchromb.2014.01.034.

[41] C. Ishii, T. Akita, M. Nagano, M. Mita, K. Hamase, Determination of chiral amino acids in various fermented products using a two-dimensional HPLC-MS/MS system, Chromatography 40(2) (2019) 83–87. https://doi.org/10.15583/jpchrom.2019.011.

[42] Y. Miyoshi, T. Oyama, Y. Itoh, K. Hamase, Enantioselective two-dimensional high-performance liquid chromatographic determination of amino acids; analysis and physiological significance of d-amino acids in mammals, Chromatography 35(1) (2014) 49–57. https://doi.org/10.15583/jpchrom.2014.005.

[43] T. Kimura, K. Hamase, Y. Miyoshi, R. Yamamoto, K. Yasuda, M. Mita, H. Rakugi, T. Hayashi, Y. Isaka, Chiral amino acid metabolomics for novel biomarker screening in the prognosis of chronic kidney disease, Scientific Reports 6(1) (2016). https://doi.org/10.1038/srep26137.

[44] R. Koga, Y. Miyoshi, Y. Sato, M. Mita, R. Konno, W. Lindner, K. Hamase, Enantioselective determination of citrulline and ornithine in the urine of D-amino acid oxidase deficient mice using a two-dimensional high-performance liquid chromatographic system, J. Chromatogr. A. 1467 (2016) 312–317. https://doi.org/10.1016/j.chroma.2016.07.053.

[45] A. Furusho, R. Koga, T. Akita, Y. Miyoshi, M. Mita, K. Hamase, Development of a highly-sensitive two-dimensional HPLC system with narrowbore reversed-phase and microbore enantioselective columns and application to the chiral amino acid analysis of the mammalian brain, Chromatography 39(2) (2018) 83–90. https://doi.org/10.15583/jpchrom.2018.007.

[46] S.-L. Liu, T. Oyama, Y. Miyoshi, S.-Y. Sheu, M. Mita, T. Ide, W. Lindner, K. Hamase, J.-A. Lee, Establishment of a two-dimensional chiral HPLC system for the simultaneous detection of lactate and 3-hydroxybutyrate enantiomers in human clinical samples, J. Pharm. Biomed. Anal. 116 (2015) 80–85. https://doi.org/10.1016/j.jpba.2015.05.036.

[47] A. Furusho, T. Akita, M. Mita, H. Naraoka, K. Hamase, Three-dimensional high-performance liquid chromatographic analysis of chiral amino acids in carbonaceous chondrites, J. Chromatogr. A. 1625 (2020). https://doi.org/10.1016/j.chroma.2020.461255.

[48] N. Sereekittikul, R. Koga, T. Akita, A. Furusho, R. Reischl, M. Mita, A. Fujii, K. Hashiguchi, M. Nagano, W. Lindner, K. Hamase, Multi-dimensional HPLC analysis of serine containing chiral dipeptides in Japanese traditional amber rice vinegar, Chromatography 39(2) (2018) 59–66. https://doi.org/10.15583/jpchrom.2018.002.

[49] U. Woiwode, S. Neubauer, W. Lindner, S. Buckenmaier, M. Lämmerhofer, Enantioselective multiple heartcut two-dimensional ultra-high-performance liquid chromatography method with a Coreshell chiral stationary phase in the second dimension for analysis of all proteinogenic amino acids in a single run, J. Chromatogr. A 1562 (2018) 69–77. https://doi.org/10.1016/j.chroma.2018.05.062.

[50] F. Xu, Y. Xu, G. Liu, M. Zhang, S. Qiang, J. Kang, Separation of twelve posaconazole related stereoisomers by multiple heart-cutting chiral–chiral two-dimensional liquid chromatography, J. Chromatogr. A 1618 (2020). https://doi.org/10.1016/j.chroma.2019.460845.

[51] M. Catani, O.H. Ismail, F. Gasparrini, M. Antonelli, L. Pasti, N. Marchetti, S. Felletti, A. Cavazzini, Recent advancements and future directions of superficially porous chiral stationary phases for ultrafast high-performance enantioseparations, Analyst 142(4) (2017) 555–566. https://doi.org/10.1039/C6AN02530G.

[52] D.C. Patel, Z.S. Breitbach, M.F. Wahab, C.L. Barhate, D.W. Armstrong, Gone in seconds: Praxis, performance, and peculiarities of ultrafast chiral liquid chromatography with superficially porous particles, Anal. Chem. 87(18) (2015) 9137–9148. https://doi.org/10.1021/acs.analchem.5b00715.

[53] K. Schmitt, U. Woiwode, M. Kohout, T. Zhang, W. Lindner, M. Lämmerhofer, Comparison of small size fully porous particles and superficially porous particles of chiral anion-exchange type stationary phases in ultra-high performance liquid chromatography: Effect of particle and pore size on chromatographic efficiency and kinetic performance, J. Chromatogr. A 1569 (2018) 149–159. https://doi.org/10.1016/j.chroma.2018.07.056.

[54] N. Khundadze, S. Pantsulaia, C. Fanali, T. Farkas, B. Chankvetadze, On our way to sub-second separations of enantiomers in high-performance liquid chromatography, J. Chromatogr. A 1572 (2018) 37–43. https://doi.org/10.1016/j.chroma.2018.08.027.

[55] C. Geibel, K. Dittrich, U. Woiwode, M. Kohout, T. Zhang, W. Lindner, M. Lämmerhofer, Evaluation of superficially porous particle based zwitterionic chiral ion exchangers against fully porous particle benchmarks for enantioselective ultra-high performance liquid chromatography, J. Chromatogr. A 1603 (2019) 130–140. https://doi.org/10.1016/j.chroma.2019.06.026.

[56] A. Acquaviva, G. Siano, P. Quintas, M.R. Filgueira, C.B. Castells, Chiral x achiral multidimensional liquid chromatography: Application to the enantioseparation of dintitrophenyl amino acids in honey samples and their fingerprint classification, J. Chromatogr. A. 1614 (2020). https://doi.org/10.1016/j.chroma.2019.460729.

[57] F. Gritti, G. Guiochon, Possible resolution gain in enantioseparations afforded by core–shell particle technology, J. Chromatogr. A. 1348 (2014) 87–96. https://doi.org/10.1016/j.chroma.2014.04.041.

[58] U. Woiwode, M. Ferri, N.M. Maier, W. Lindner, M. Lämmerhofer, Complementary enantioselectivity profiles of chiral cinchonan carbamate selectors with distinct carbamate residues and their implementation in enantioselective two-dimensional high-performance liquid chromatography of amino acids, J. Chromatogr. A. 1558 (2018) 29–36. https://doi.org/10.1016/j.chroma.2018.04.061.

[59] U. Woiwode, R.J. Reischl, S. Buckenmaier, W. Lindner, M. Lämmerhofer, Imaging peptide and protein chirality via amino acid analysis by chiralxchiral two-dimensional correlation liquid chromatography, Anal. Chem. 90(13) (2018) 7963–7971. https://doi.org/10.1021/acs.analchem.8b00676.

[60] J. Lin, C. Tsang, R. Lieu, K. Zhang, Fast chiral and achiral profiling of compounds with multiple chiral centers by a versatile two-dimensional multicolumn liquid chromatography (LC–mLC) approach, J. Chromatogr. A 1620 (2020). https://doi.org/10.1016/j.chroma.2020.460987.

[61] R.S. Hegade, K. Chen, J.-P. Boon, M. Hellings, F. Lynen, Development of an achiral-chiral 2-dimensional heart-cutting platform for enhanced pharmaceutical impurity analysis, J. Chromatogr. A(2020). https://doi.org/10.1016/j.chroma.2020.461425.

[62] O.H. Ismail, G.L. Losacco, G. Mazzoccanti, A. Ciogli, C. Villani, M. Catani, L. Pasti, S. Anderson, A. Cavazzini, F. Gasparrini, Unmatched kinetic performance in enantioselective supercritical fluid chromatography by combining latest generation whelk-O1 chiral stationary phases with a low-dispersion in-house modified equipment, Anal. Chem. 90(18) (2018) 10828–10836. https://doi.org/10.1021/acs.analchem.8b01907.

[63] K.W. Phinney, Enantioselective separations by packed column subcritical and supercritical fluid chromatography, Anal. Bioanal. Chem. 382(3) (2005) 639–645. https://doi.org/10.1007/s00 216-005-3074-y.

[64] M. Iguiniz, E. Corbel, N. Roques, S. Heinisch, On-line coupling of achiral reversed phase liquid chromatography and chiral supercritical fluid chromatography for the analysis of pharmaceutical compounds, J. Pharm. Biomed. Anal. 159 (2018) 237–244. https://doi.org/10.1016/j.jpba.2018.06.058.

[65] C.J. Venkatramani, M. Al-Sayah, G. Li, M. Goel, J. Girotti, L. Zang, L. Wigman, P. Yehl, N. Chetwyn, Simultaneous achiral-chiral analysis of pharmaceutical compounds using two-dimensional reversed phase liquid chromatography-supercritical fluid chromatography, Talanta 148 (2016) 548–555. https://doi.org/10.1016/j.talanta.2015.10.054.

[66] M. Goel, E. Larson, C.J. Venkatramani, M.A. Al-Sayah, Optimization of a two-dimensional liquid chromatography-supercritical fluid chromatography-mass spectrometry (2D-LC-SFC-MS) system to assess "in-vivo" inter-conversion of chiral drug molecules, J. Chromatogr. B 1084 (2018) 89–95. https://doi.org/10.1016/j.jchromb.2018.03.029.

[67] C. Lotter, E. Poehler, J.J. Heiland, L. Mauritz, D. Belder, Enantioselective reaction monitoring utilizing two-dimensional heart-cut liquid chromatography on an integrated microfluidic chip, Lab on a Chip 16(24) (2016) 4648–4652. https://doi.org/10.1039/C6LC01138A.

[68] N. Sheng, H. Zheng, Y. Xiao, Z. Wang, M. Li, J. Zhang, Chiral separation and chemical profile of Dengzhan Shengmai by integrating comprehensive with multiple heart-cutting two-dimensional liquid chromatography coupled with quadrupole time-of-flight mass spectrometry, J. Chromatogr. A 1517 (2017) 97–107. https://doi.org/10.1016/j.chroma.2017.08.037.

[69] M. Mahut, E. Haller, P. Ghazidezfuli, M. Leitner, A. Ebner, P. Hinterdorfer, W. Lindner, M. Lämmerhofer, Topology-selective chromatography reveals plasmid supercoiling shifts during fermentation and allows rapid and efficient preparation of topoisomers, Angew. Chem. Int. Ed. 51(1) (2012) 267–270. https://doi.org/10.1002/anie.201106495.

[70] M. Mahut, A. Gargano, H. Schuchnigg, W. Lindner, M. Lämmerhofer, Chemoaffinity material for plasmid DNA analysis by high-performance liquid chromatography with condition-dependent switching between isoform and topoisomer selectivity, Anal. Chem. 85(5) (2013) 2913–2920. https://doi.org/10.1021/ac3034823.

Glossary

A	A term of the van Deemter equation
A_{Desm}	A coefficient in Desmet's eddy dispersion equation
A_{Gidd}	A coefficient in Giddings' eddy dispersion equation
A_{Knox}	A coefficient in Knox's eddy dispersion equation
ADC	antibody-drug conjugate
ADCC	antibody-dependent cellular cytotoxicity
AEX	anion exchange
ALS	alternating least squares
API	active pharmaceutical ingredient
ASO	antisense oligonucleotide
AUC	analytical ultracentrifugation
b	dimensionless gradient slope
B	longitudinal diffusion coefficient of the van Deemter equation
$\mathbf{c}$	representation of chromatographic profile ($I \times 1$ or $J \times 1$ vector, for first or second dimension separation, respectively)
C	slow interphase mass transfer coefficient of the van Deemter equation
C_{Desm}	C coefficient used in Desmet's mass transfer equation
C_{Gidd}	C coefficient used in Giddings' mass transfer equation
CD	circular dichroism
CDC	compliment dependent cytotoxicity
CE	capillary electrophoresis
CEX	cation exchange
cIEF	capillary isoelectric focusing
c_i	ith element of chromatographic profile vector, $\mathbf{c}$
$c_{j,n}$	jth element of chromatographic profile vector for the nth component
COW	correlation optimized warping
C_p	heat capacity of the mobile phase
d	duty cycle
^2DF	second dimension dilution factor
DF_{2D}	overall dilution factor in two-dimensional separation
DNA	deoxyribonucleic acid
^1D	first dimension of separation
^2D	second dimension of separation
D	dimensionality of separation
DAD	diode array detection
DAR	drug-to-antibody ratio
d_{core}	diameter of particle core in superficially porous particles
D_{eff}	effective diffusion coefficient
DLS	dynamic light scattering
D_m	molecular diffusion coefficient in the mobile phase
DP	drug product
d_p	particle diameter
d_p^*	optimum particle diameter used in K-S-H optimization
d_{part}	particle diameter used by Desmet
D_{pz}	molecular diffusion coefficient in the porous zone

D_{rad} radial dispersion coefficient
D_s molecular diffusion coefficient in the stationary phase
DSC   differential scanning calorimetry
E Knox's separation impedance
ELISA   enzyme linked immunosorbent assay
Ellman's   free thiol quantitation
EMT   effective medium theory
f number of first dimension fractions, $f = {}^1t_g \big/ {}^2t_c$
f_{cov} fractional coverage
F flow rate
F^* optimized mobile phase flow rate
FACS   fluorescence assisted cell sorting
Fc   fragment crystallizable
FNP   fully non-porous particle
FTIR   Fourier transform infrared
γ_e obstruction/tortuosity factor in the interstitial space
γ_i obstruction/tortuosity factor in the intraparticle space
G(p)   gradient compression factor
GRAM   generalized rank annihilation
h reduced plate height
h_{min} minimum reduced plate height
h_{opt} optimized reduced plate height
$H_{viscous}$ contribution to the plate height due to viscous self-heating of the mobile phase
H   height equivalent to a theoretical plate
HDX   hydrogen-deuterium exchange
HETP   height equivalent to a theoretical plate
HOS   higher order structure
IdeS   immunoglobulin G-degrading enzyme of Streptococcus pyogenes
IEF   isoelectric focusing
$IKSFA$ iterative key set factor analysis
IP   ion pairing
IPR   ion pairing reagent
IPRP   ion pairing reversed-phase chromatography
k phase retention factor (previously denoted k')
K phase distribution equilibrium constant
K_c proportionality constant in Kozeny-Carman equation
k_e retention factor at the column exit during a gradient elution program
k_0 retention factor at initial percentage of organic modifier in a gradient elution program
k_g retention factor at t_g in gradient elution
KinExA   kinetic exclusion assay
k_w retention factor 100% aqueous mobile phase, in reversed-phase separations
k'' zone retention factor
${}^2k_{1e}$ retention factor of solute on the ^{2}D column as it enters that column while in the ^{1}D eluent
${}^2k_{2e}$ retention factor of solute on the ^{2}D column as it enters that column while in the ^{2}D eluent
KSH   Knox-Saleem-Halasz
L column length
L^* optimal column length

LC-mLC	heartcut 2D-LC with multiple columns used in the second dimension
LSST	linear solvent strength theory
LOQ	limit of quantitation
m	number of analytes in a mixture
mAb	monoclonal antibody
MALDI	matrix assisted laser desorption ionization
MALS	multi-angle light scattering
MAM	multi-attribute method
MFI	microflow imaging
MHC	multiple heartcut 2D-LC (also mLC-LC)
MM	mixed mode
MMT	monomethoxytrityl
M_R	modulation ratio, $M_R = \dfrac{4\,^1\sigma}{t_s}$
MCR	multivariate curve resolution
MSX	methionine sulfoximine
n_c	peak capacity
$n_{c,grad}$	gradient peak capacity
$n_{c,iso}$	isocratic peak capacity
$n_{c,1D}$	peak capacity of 1D separation
^1n_c	peak capacity of first dimension of 2D separation
^1n_c	corrected first dimension peak capacity
$^1n_{c,0.9}$	^1n_c producing 90% of limiting corrected 2D peak capacity
$^1n_c^\infty$	limit of ^1n_c for long first dimension gradient time
^2n_c	peak capacity of second dimension of 2D separation
n_{lim}	limiting peak capacity
$n_{c,2D}$	2D peak capacity
$n'_{c,2D}$	corrected 2D peak capacity
$n^*_{c,2D}$	effective 2D peak capacity
$n'^{,\infty}_{c,2D}$	limit of $n'_{c,2D}$ for long first dimension gradient time
N	number of theoretical plates
N^*	plate number achievable with Poppe optimal velocity and length (two-parameter optimization)
N^*_{lim}	maximum plate number in the limit of long analysis time (two-parameter optimization)
N^{o*}_{lim}	plate number in the limit of short analysis time (two-parameter optimization)
N_{max}	plate number achievable under K-S-H optimized conditions
N_{opt}	plate number at the van Deemter optimum velocity
NGS	next generation sequencing
NR	non-reduced
p	number of peaks observed in a chromatogram (can contain any number of analytes)
ΔP	pressure drop across column or length of connecting tubing
PEG	polyethylene glycol
P_{max}	maximum available system pressure drop
PARAFAC	parallel factor analysis
POE	poly(oxyethylene)
q	rate of heat production due to viscous flow
qPCR	quantitative polymerase chain reaction
r	radial position within the column relative to the central axis

R_{col}	radius of column
R_{out}	inside radius of packed column
Re	Reynold's number
R_s	resolution
R_s^*	average minimum resolution of SOT
s	number of singlet (i.e., pure) peaks observed in a chromatogram
$s_{l,n}$	lth element of spectrum vector for the nth component
S	solvent strength parameter – slope of a plot of ln k vs. ϕ in reversed-phase chromatography
$\mathbf{S}$	diagonal matrix of singular values in SVD
SDS-PAGE	sodium dodecylsulfate – polyacrylamide gel electrophoresis
SEC	size exclusion chromatography
Sh	Sherwood number – dimensionless number that relates the rate of transport by diffusive and convective processes
Sh_{WG}	Sherwood number based on the Wilson-Geankopolis correlation
Sh_{Desm}	Sherwood number based on Desmet's calculation
siRNA	small interfering RNA
SOT	statistical overlap theory
SPP	superficially porous particle
SPR	surface plasmon resonance
SVD	singular value decomposition
$\mathbf{s}_n$	representation of spectrum for the nth component ($L \times 1$ vector)
t	time
T	temperature
^2t_c	second dimension cycle time
t_g	gradient time
t_m or t_0	column dead time; also known as void time, or hold-up time
TPP	totally porous particle
t_r	retention time under gradient elution conditions
t_R	IUPAC symbol for isocratic retention time
$^2t_{re-eq}$	re-equilibration time of second dimension
TRLC	temperature responsive liquid chromatography
t_s	sampling interval
TSA	total sialic acid
T_w	column wall temperature
u	linear velocity
u_e	interstitial linear mobile phase velocity
u_e^*	optimal interstitial linear velocity
$u_{e,opt}$	van Deemter optimum velocity
u_m or u_0	measured linear velocity of the mobile phase
u_{Rw}	velocity of a retained analyte at the column wall
$\mathbf{U}$	matrix of left singular vectors in SVD
V	molar volume
$\mathbf{V}$	matrix of right singular vectors in SVD
V_{inj}	injection volume
V_m	volume of mobile phase in a column
V_s	volume of stationary phase in a column
WAX	weak anion exchange
w	peak width, equal to 4σ for Gaussian peaks
$x_{i,j}$	detector signal at ith first dimension time and jth second dimension time

$x_{i,j,k}$	detector signal at ith first dimension time, jth second dimension time and for the kth sample
$x_{i,j,k,l}$	detector signal at ith first dimension time, jth second dimension time, the kth sample and the lth wavelength
$\mathbf{X}$	detector signal of 2D chromatographic peak ($I \times J$ matrix)
$\underline{\mathbf{X}}$	detector signal of 2D chromatographic peak for multiple samples and multiple wavelengths ($I \times J \times K \times L$ data array)
$\otimes$	outer product
α	chromatographic selectivity; saturation factor in Statistical Overlap Theory; coefficient of thermal expansion of mobile phase
a_1	$= S\Delta\phi$
a_2	$= S\Delta\phi t_o$
α_{Knox}	exponent in used in equation used to calculate h_{eddy} based on Knox equation
α_{Desm}	exponent in used in equation used to calculate h_{eddy} based on Desmet equation
β	first dimension peak broadening factor
$<\beta>$	average first dimension peak broadening factor
δ^2_{det}	detector broadening factor
δ^2_{inj}	injection induced broadening factor taken as 12 for a perfect pulse
ε_e	interstitial porosity
ε_i	intraparticle porosity
ε_T	total porosity ($\varepsilon_T = \varepsilon_e + \varepsilon_i(1-\varepsilon_e)$)
η	mobile phase viscosity
κ	fitting coefficient in expression for undersampling factor $<\beta>$
λ	fraction of second dimension cycle devoted to running second dimension gradient; ratio of $\dfrac{\varepsilon_e}{\varepsilon_T}$; thermal conductivity; factor related to column packing quality in van Deemter eddy dispersion term
v	reduced mobile phase velocity
v_e	reduced interstitial mobile phase velocity
$v_{e,opt}$	van Deemter optimum reduced interstitial mobile phase velocity
v_e^*	two-parameter optimized reduced interstitial mobile phase velocity
$v_e^{\mp}$	K-S-H optimized reduced interstitial mobile phase velocity
σ	standard deviation of peak
σ_m	standard deviation of first dimension peak after sampling
$^1\sigma^*$	standard deviation of sampled ^{1}D peak
$<^1\sigma^*>$	average standard deviation of multiple sampled ^{1}D peaks
ϕ	sampling phase; volume fraction of organic modifier in reversed-phase chromatography (0-1 scale)
ϕ_0	mobile phase composition at the start of a gradient elution program
ϕ_f	mobile phase composition at the end of a gradient elution program
Φ	dimensionless flow resistance factor
χ	solvent self-association factor used in Wilke-Chang correlation

Index

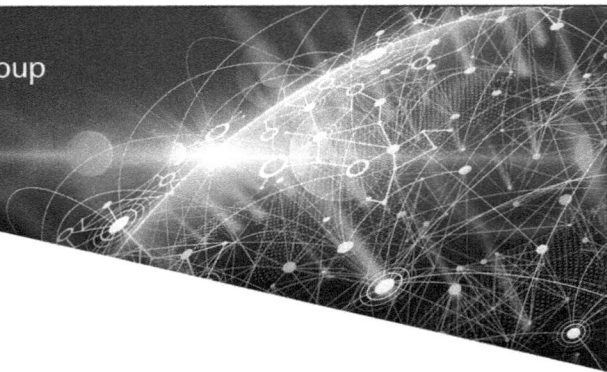

For Product Safety Concerns and Information please contact our EU
representative GPSR@taylorandfrancis.com
Taylor & Francis Verlag GmbH, Kaufingerstraße 24, 80331 München, Germany

www.ingramcontent.com/pod-product-compliance
Lightning Source LLC
Chambersburg PA
CBHW080700220326
41598CB00033B/5269

9 780367 547745